Liquid and Crystal Nanomaterials for Water Pollutants Remediation

Editors

Uma Shanker

Department of Chemistry
Dr B R Ambedkar National Institute of Technology Jalandhar
Jalandhar, Punjab, India

Manviri Rani

Department of Chemistry
Malaviya National Institute of Technology
JLN Road, Malaviya Nagar
Rajasthan, India

CRC Press is an imprint of the
Taylor & Francis Group, an **informa** business
A SCIENCE PUBLISHERS BOOK

Cover credit: Thanks to Dr Manviri Rani and Mrs Jyoti Yadav for designing the cover image of the book

First edition published 2022
by CRC Press
6000 Broken Sound Parkway NW, Suite 300, Boca Raton, FL 33487-2742

and by CRC Press
2 Park Square, Milton Park, Abingdon, Oxon, OX14 4RN

CRC Press is an imprint of Taylor & Francis Group, LLC

Library of Congress Cataloging-in-Publication Data

Names: Shanker, Uma, 1985- editor. | Rani, Manviri, 1983- editor.
Title: Liquid and crystal nanomaterials for water pollutants remediation / editors, Uma Shanker, Department of Chemistry, Dr B R Ambedkar National Institute of Technology Jalandhar, Jalandhar, Punjab, India, Manviri Rani, Department of Chemistry, Malaviya National Institute of Technology, JLN Road, Malaviya Nagar, Rajasthan, India.
Description: First edition. | Boca Raton : CRC Press/Taylor & Francis Group, [2022] | "A Science Publishers Book." | Includes bibliographical references and index.
Identifiers: LCCN 2021040438 | ISBN 9780367549879 (hbk)
Subjects: LCSH: Water--Purification--Materials. | Nanostructured materials--Environmental aspects. | Nanoparticles--Environmental aspects. | Liquid crystals--Environmental aspects.
Classification: LCC TD477 .L57 2022 | DDC 628.1/62--dc23/eng/20211027
LC record available at https://lccn.loc.gov/2021040438

ISBN: 978-0-367-54987-9 (hbk)
ISBN: 978-0-367-54990-9 (pbk)
ISBN: 978-1-003-09148-6 (ebk)

DOI: 10.1201/9781003091486

Typeset in Times Lt Std
by TVH Scan

Preface

Nanoscience technology is playing a vital role in multidisciplinary research due to unique characteristics developed at the nanoscale compared to the bulk materials. In view of such excellent properties, like high surface area, semiconducting nature and non-toxicity, nanotechnology has emerged as the most promising science that can curb the pollution caused by various contaminants. One of the most popular fields widely adopted is photocatalysis of nanomaterials that involve the photo-conduction worth in the efficient removal/degradation of noxious pollutants. Liquid and crystal nanomaterials aim for products and processes that are ecofriendly, economically sustainable, safe, energy-efficient, decrease waste and diminish greenhouse gas emissions. Such products and processes are based on renewable materials and/or have a low net impact on the environment. Recently, liquid and crystal nanomaterials of various systems are the new trend in the remediation of pollutants from the environment. These photoactive nanomaterials under UV/sunlight irradiation yield reactive radical species that quickly eradicate the pollutants by a redox mechanism. In view of environmental concern, involvement of such efficient technology in the remediation of water pollutants is highly imperative. Such methodologies are able to minimize environmental hazards, unintentional threats to human health and next-generation improved implications. This book is proposed for a large and broad readership, including researchers, scientists, academicians and readers from diverse backgrounds across various fields, such as nanotechnology, chemistry, agriculture, environmental science, water engineering, waste management and energy. It could also serve as a reference book for graduates and post-graduate students, faculties, environmentalists and industrial personnel who are working in the area of green technologies. This book will also assist to develop the policy and implementation to lower the risk of exposure to liquid and crystal nanomaterials. These aspects need to be pursued in the future for investigating and assessing health risks due to exposure to liquid and crystal nanomaterials. Overall, the aim of this handbook is to focus on recent developments in the area of fabrication of liquid and crystal nanomaterials and their further utilization in the removal of water pollutants.

This book is on liquid and crystal nanomaterials and aim at products and processes that are ecofriendly, economically sustainable, energy-efficient and decrease waste. In this book, comprehensive information on the introduction, generation and application of such nanomaterials for the removal of water-pollutant has been provided. In this book, Environment, Health and Safety issues of nanomaterials has also been offered. These aspects need to be pursued in the future for investigating and assessing health risks due to exposure to these materials.

Preface

[illegible]

Contents

Liquid-Crystal Nanomaterials: Introduction, Design and Properties

Anas Saifi[1,2], Charu Negi[1], Atul Pratap Singh[3] and Kamlesh Kumar[1,2,*]

1. INTRODUCTION

The state of the matter that exhibits intermediate properties of amorphous liquids and those of the solid crystals are termed as Liquid Crystals (LCs). They are also called 'mesogen'. It has been observed in certain materials that a solid to a liquid phase is not a single transition. Rather, it is a cascade of phase transition in between the two as shown in Figure 1. These intermediate phases exhibit the physical properties between an amorphous liquid and a solid crystal. Due to the crystalline nature, the medium shows optical birefringence as seen in other solids. At the same time, the liquidity shows the flow property but does not follow the shearing property like any other liquids. The most important characteristic of LCs is their anisotropic nature where the orientation of all the molecules is not within the same direction but may change with time, and this direction is referred to as a director. It is an average molecular orientation represented by ($\vec{n}$), a unit vector that aligns with the primary direction of axes of the main molecules in LC compounds. This leads to the assumption of the most probable direction of molecular orientation in LC nanomaterials. They exhibit low positional or orientational order of the molecular center of mass or their long molecular axis respectively, which in turn show anisotropic properties, such as dielectric anisotropy $\Delta\varepsilon$, birefringence Δn, viscosity η, threshold voltage V_{th}, refractive index n or the response time τ while retaining the flowability (Dierking 2018, Li 2014b, Kelker and Hatz 1981).

Previously, only three states were known, i.e., solid, liquid and gas. In 1989, F. Reinitzer investigated the nature of cholesterol, which led to the discovery of liquid crystals (Reinitzer 1989). He contemplated the 'double melting' behavior of cholesteryl benzoate. At 145.5°C, material crystals melted into a cloudy fluid and turn into a clear solution when further heated to 178.5°C. This cloudy fluid was first proposed by F. Renitzer as a new phase of matter for which he received the credit of LC phase discovery. In the 1960s, a French theoretical physicist, Pierre-Gilles de Gennes, working in the field of superconductivity and magnetism, had shown interest in exploring LCs and observed compelling analogies between superconductors and LCs as well as magnetic materials (Lavrentovich 1994). For his research contribution in LCs, he was awarded the Nobel Prize in 1991. The work of Pierre-Gilles de Gennes has profoundly influenced and persuaded the

[1] CSIR-Central Scientific Instruments Organisation, Sector-30, Chandigarh 160030, India.
[2] Academy of Scientific and Innovative Research (AcSIR), Ghaziabad, 201002, India.
[3] Department of Chemistry, Chandigarh University, Gharuan, Mohali, Punjab, India.
[*] Corresponding author: kamlesh.kumar@csio.res.in, kksaini@gmail.com.

modern development of LCs. For the first time in 2007, Liquid Crystal Display (LCD) televisions surpassed conventional Cathode Ray Tube (CRT) units in sales across the world. The development in the field of LCs has shown their application in numerous areas, such as laser (Coles and Morris 2010, Wallace et al. 2021), sensors (Qi et al. 2020), light modulators (Li et al. 2020), biomedical applications (Lin 2020, Mei et al. 2018, Yang et al. 2021), organic transistors (Han et al. 2021), switchable windows (Murray et al. 2017), etc.

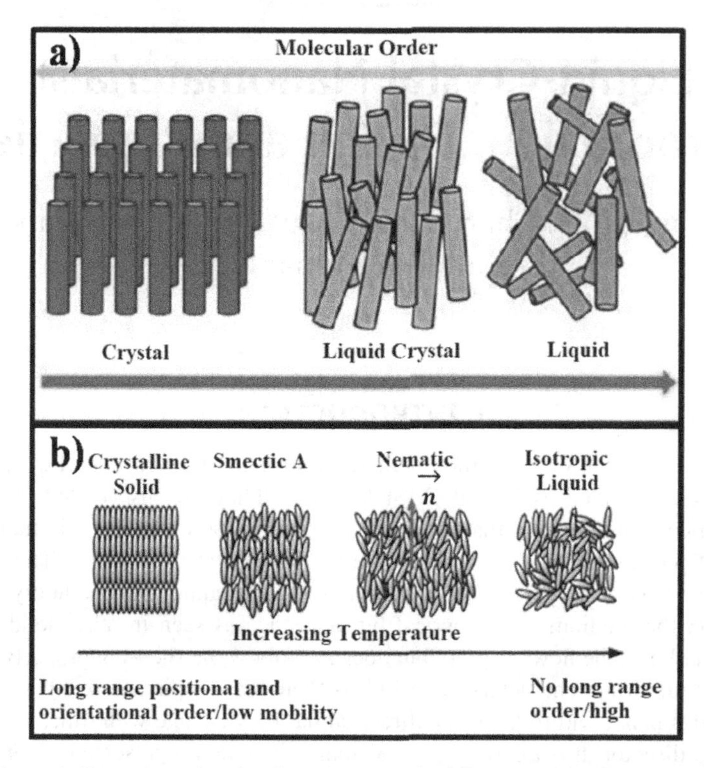

FIGURE 1 (a) Schematic illustration of molecular orientation in different states of a material composed of an elongated anisotropic molecule (Bisoyi and Li 2016), (b) temperature dependent phases of materials (Bai and Abbott 2011). (Reprinted with permission from a) (Hari Krishna Bisoyi and Quan Li, Light-Driven Liquid Crystalline Materials: From Photo-Induced Phase Transitions and Property Modulations to Applications (Chemical Reviews: American Chemical Society, 2016), 116, 24, 15089-15166.) b) (Yiqun Bai and Nicholas L. Abbott, Recent Advances in Colloidal and Interfacial Phenomena Involving Liquid Crystals (Langmuir: American Chemical Society, 2011), 27, 10, 5719-5738).

Despite their tremendous success, the research work in LCs has not reached its zenith. It has revived itself by paving the way into other research fields because of its self-organization and optoelectronic properties. When applied with an external stimulus, e.g., magnetic field, electrical field, light, etc., they exhibit response with preferred orientations (Dierking 2018). With LCDs becoming ubiquitous in our daily lives, the R&D of LCs is progressing rapidly into the forefront of nanoscience. Nanoscience implies the study of matter with at least one dimension ranging from 1-100 nm. It has revolutionized scientific research and expanded significantly in areas ranging from energy conversion to biomedicine fields (Lagerwall and Scalia 2014, Blanc et al. 2013, Stamatoiu et al. 2011). Nanomaterials have distinct properties relative to their bulk counterparts due to their tunable physical, chemical and biological properties, such as higher specific surface area, size distribution and higher optical absorbance and emission. With the advancement of nanoscience, LC materials experimental research has also changed. Various nanoparticles (NPs) (Garbovskiy and Glushchenko 2010, Mertelj and Lisjak 2017), carbon nanotubes (Yadav and Singh 2016) and quantum dots (Mirzaei et al. 2012) are gaining popularity because of their potential applications. They have been used as a doping agent to improve the physical properties of LC materials. The doping of LCs with nanomaterials to enhance the physical, electrical and optical properties as per the requirement proved to be a novel method. This method allows the tuning of LC parameters without synthesizing the new LC material. Therefore, the combination of nanomaterial with unique properties of LCs is challenging as well as intriguing from both technological and fundamental points of view. These composite materials can endow multifunctional systems with potential end-use applications (Bisoyi and Kumar 2011, Kumar 2014, Nealon et al. 2012).

Conventional black and white displays had a small viewing angle and low contrast ratio. Two modes (a) the birefringence and (b) the guest-host have been proposed to resolve this issue. In birefringence mode, the electric field control phase delay of the light polarized between the two components is controlled by the electric field, but this control is confined to a very narrow portion of visible light; therefore, this mode has drastic drawbacks, such as lack of natural colors, viewing angle-dependent display colors, thickness-dependent color patches, etc. In 1968, L. A. Zanoni and G. H. Heilmeier introduced the guest-host mode to resolve the shortcomings of the birefringence mode (Heilmeier and Zanoni 1968). The concept of guest-host mode is described as the dispersion (doping) of any guest entity, such as NPs, polymer, dye, carbon nanotubes, quantum dots, etc., into the host LC material. Over the last decade, the guest-host system has been regarded as the best prospect for tuning the physical properties of LC materials through molecular structure engineering to prevent the complex synthesis of new material. Figure 2 shows gold NPs doped-LC mixture based on the guest-host model considered as an ideal model for the incorporation of NPs into the LC matrix where 4-Cyano-4'-pentylbiphenyl (5CB) is a host material.

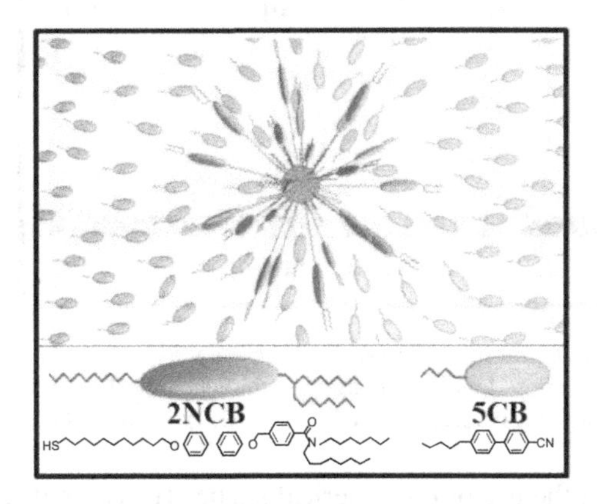

FIGURE 2 Schematic illustration of gold NPs doped-LC matrix based on guest-host interaction (Lesiak et al. 2019). (Reprinted with permission from Lesiak et al., Self-Organized, One-Dimensional Periodic Structures in a Gold Nanoparticle-Doped Nematic Liquid Crystal Composite (ACS Nano: American Chemical Society, 2019), 13, 9, 10154-10160).

Since then, much effort has been put into the study of dispersion of NPs in LCs, i.e. guest-host system. These systems are primarily studied for different aspects of the development of materials, such as (i) tuning of LC properties with the addition of nanomaterials, (ii) introduction of novel functionalities to the LC and (iii) the self-organization of the LC state. These systems help in forming the ordered structures or to transfer orders into dispersed nanomaterials. Generally, inorganic or mineral nanomaterials are used in dispersions such as ferroelectric and dielectric particles, like TiO_2, graphene oxide, carbon-based nanotubes and gold NPs (Dierking and Al-Zangana 2017).

As insignificant or irrelevant it may sound, a homogeneous substance containing LCs and nanomaterials is by no means easy to synthesize (Saliba et al. 2013). The interaction between the LCs and nanomaterials is one of the main parameters. A weak interaction can often lead to a dissolution of the LC phase or segregation of nanomaterials in the defects of the LC phase. Conversely, a strong interaction may lead to the perfect dispersion of nanomaterials in the LC phase as shown in Figure 3. Some methods for instance ligand exchange processes can be employed for the coating of nanomaterial with LC ligands, which can provide more complex LC-NP hybrids.

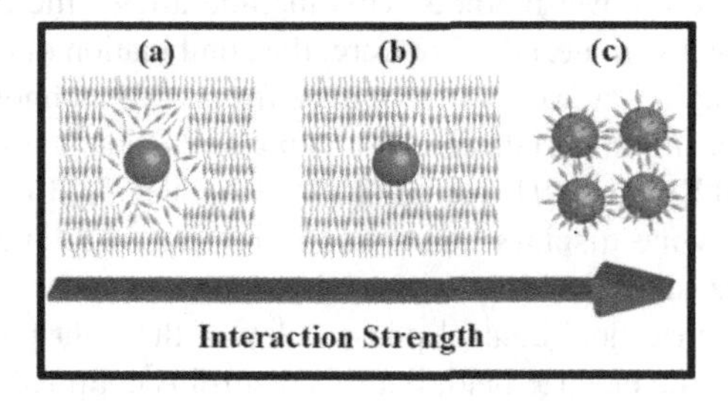

FIGURE 3 Schematic illustration of interaction strength dependent assembly of NPs (spherical) and LCs (rod-like) (Saliba et al. 2013). (Reprinted with permission from Saliba et al., Liquid crystalline thermotropic and lyotropic nanohybrids (Nanoscale: Royal Society of Chemistry, 2013), 5, 6641-6661).

Sometimes, a direct growth method, an *in situ* approach, is also feasible in which a complete dispersion of NP precursor should be attained initially. If the interactions of the precursors are weak, it may lead to dispersion within the LC defects. Conversely, if the precursor interactions are strong, then it may lead to homogeneous dispersions as shown in Figure 4. In addition, if the precursor is linked covalently with the LC moiety, it may exhibit the properties of that LC.

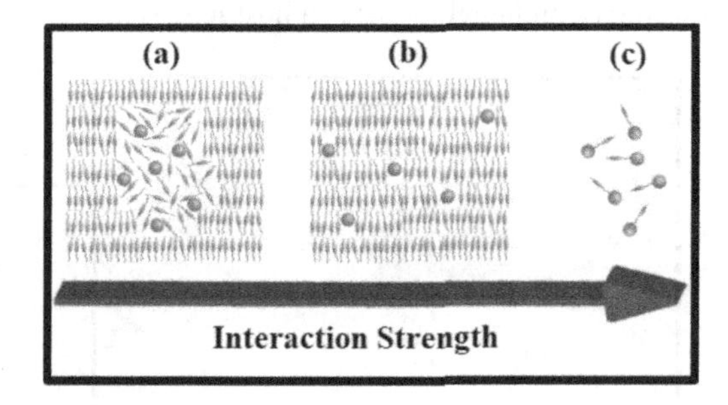

FIGURE 4 Schematic illustration of interaction strength dependent assembly of precursor of NPs (spherical) and (rod-like) (Saliba et al. 2013). (Reprinted with permission from Saliba et al., Liquid crystalline thermotropic and lyotropic nanohybrids (Nanoscale: Royal Society of Chemistry, 2013), 5, 6641-6661).

It has been observed that the use of the spherical shape NPs can reduce the order parameter of LC molecules, whereas the same can be enhanced using the rod shape NPs because of their dipole moment coupling and long-range dimensional correlation. When compared with the pristine LCs, the LC-NP hybrid has displayed its importance in different ways. Various intriguing and fascinating applications of LC-NPs hybrid system were shown by the research community in the field of LC across the globe, such as self-assembled LC matrix via QDs (Rodarte et al. 2013), LC composite fibers (Kye et al. 2015), optical waveguides and switches (Fan et al. 2012), photovoltaic solar cell (Sun et al. 2015), impurity-free non-volatile memory devices (Chandran et al. 2013), modern optoelectronic and photonic devices (Ganguly et al. 2012), FLC diffraction grating and Fresnel lenses (Fan et al. 2012), etc.

1.1 Classification of Liquid-Crystal Nanomaterials

Liquid crystals are categorized into two general classes, i.e. thermotropic and lyotropic materials. Thermotropic materials are those which show LC state purely on the variation of temperature, while lyotropic materials mainly depend on the variation of solvent.

(a) Thermotropic Liquid Crystals

When the temperature drives the phase transition of the LC material, these types of LCs are called 'Thermotropic LC' (Lavrentovich 1994, Dierking 2003). When the temperature of the LC material increased from its crystal phase, a cascade of different LC phases is achieved before its actual transition to an isotropic phase and vice versa. The temperature where the phase transition occurs is known as 'Phase transition temperature'. LC phases are stable during the heating process above the crystal-mesophase transition temperature and known as 'enantiotropic'. In contrast, the LC phases are metastable during the cooling process below the isotropic-mesophase transition temperature and termed 'monotropic'. The structure of all the thermotropic LCs consists of a flexible periphery, which is usually aliphatic groups, and a central rigid core, which is often aromatic groups.

(b) Lyotropic Liquid Crystals

When the solvent drives the phase transitions of LC materials, these types of LC are termed 'Lyotropic LCs' (Petrov 1999, Neto and Salinas 2005, Hara 2019). Lyotropic LCs consist of amphiphilic molecules, where the opposite ends of an individual molecule are 'hydrophilic' and 'hydrophobic' as shown in Figure 5. Owing to the interaction of the LC compound with an appropriate solvent at a specific concentration, the molecules order themselves into various shapes, like rods, discs and spheres, and are called 'micelles', e.g. detergent and soap form micelle in a certain concentration of water.

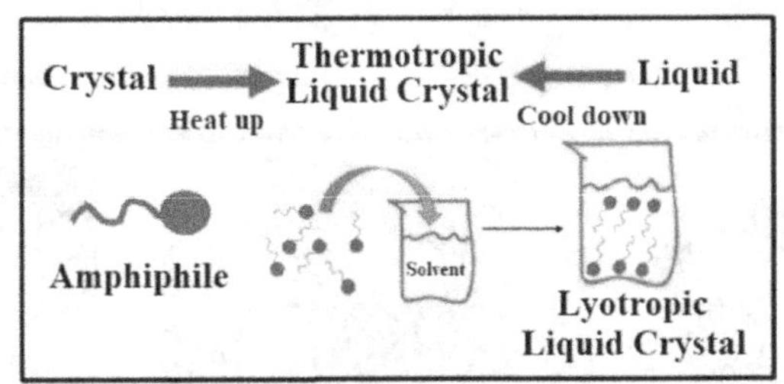

FIGURE 5 Schematic representation of the formation of thermotropic and lyotropic LC (Mo, Milleret and Nagaraj 2017). (Reprinted with permission from Mo et al., Liquid crystal nanoparticles for commercial drug delivery (Liquid Crystals Reviews: Taylor & Francis, 2017), 5, 2, 69-85).

At present, thermotropic LCs are primarily used for technological applications like modern LCDs, whereas lyotropic LCs are essential for biological systems such as membranes. Shape anisotropy is one of the major properties for showing the LC phase. Based on the shape, thermotropic LCs are further divided into three categories, i.e. (i) calamitic (rod shape), (ii) discotic (disc shape) and (iii) bent-core LCs (Stephen and Straley 1974). Different phases and subphases are shown in Figure 6,

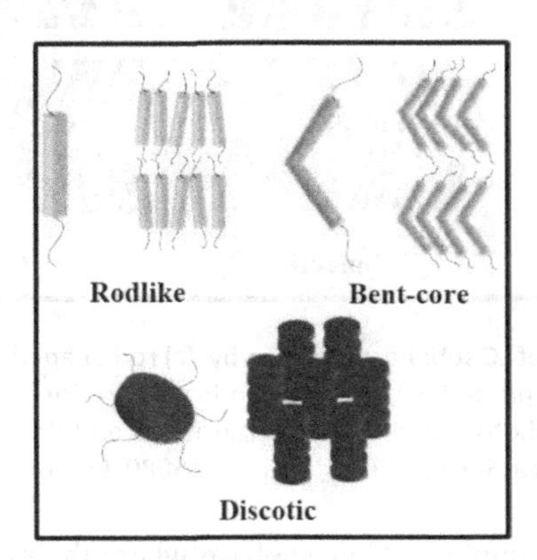

FIGURE 6 Structural representation of thermotropic LCs i.e. rod-like, bent-core and discotic mesogens (Nealon et al. 2012). (Reprinted with permission from Nealon et al., Liquid-crystalline nanoparticles: Hybrid design and mesophase structures (Beilstein Journal of Organic Chemistry: Beilstein Institute for the Advancement of Chemical Sciences, 2012), 8, 349-370).

and the potential application of these LCs are in the field of displays (Kawamoto 2012), robotics (McCracken et al. 2021), sensors (Popov et al. 2018), optical couplers (Brake and Abbott 2007), etc.

Calamitic LCs depict a rod-like structure that has a high length-to-breadth ratio. It consists of an aromatic central rigid core and an aliphatic flexible periphery. The position ordering or orientation is provided by the central core, whereas the required fluidity is achieved through the aliphatic chains. For instance, cyanobiphenyls are ubiquitous in displays having a rigid polarizable core which make them crystalline, and a flexible aliphatic tail reduces the degree of order due to which they are able to flow as well. Based on the molecular arrangement, the calamitic molecules consist of various subphases. These sub-phases are nematic, chiral nematic or cholesteric and smectic phase as shown in Figure 7.

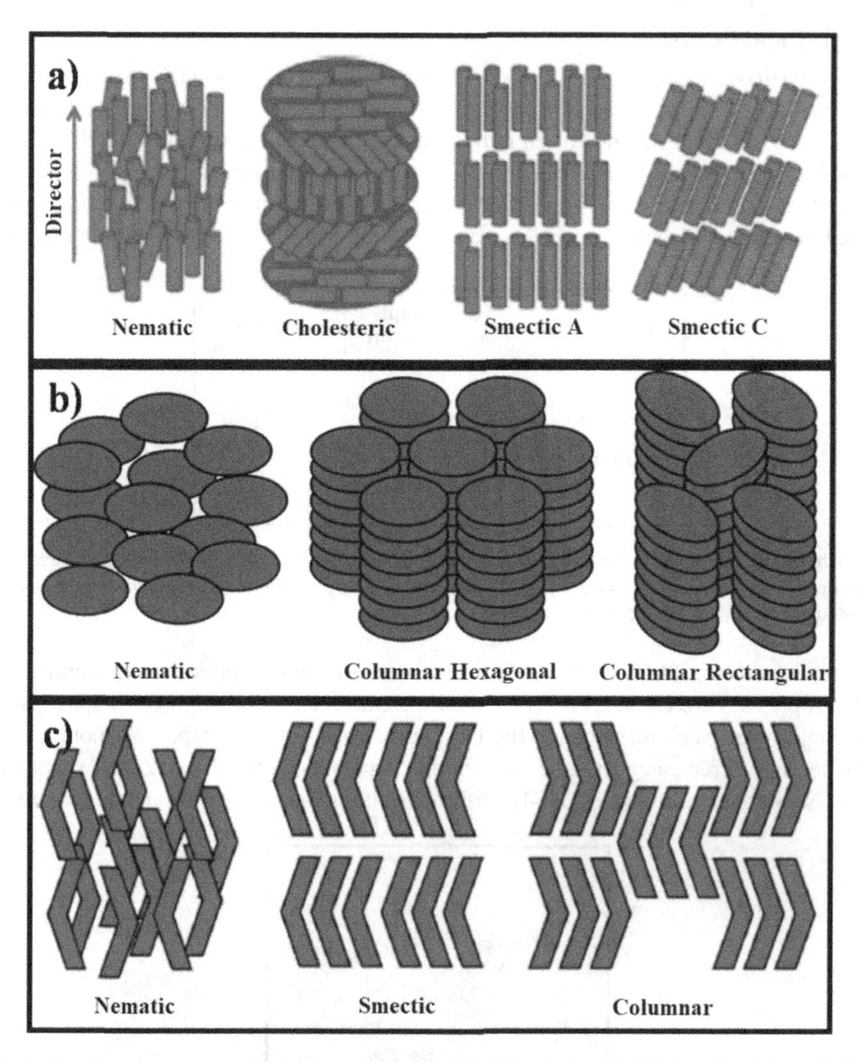

FIGURE 7 Molecular arrangement of LC subphases formed by, (a) rod-shaped, (b) disc-shaped, and (c) bent-core compounds (Bisoyi and Li 2016). (Reprinted with permission from Hari Krishna Bisoyi and Quan Li, Light-Driven Liquid Crystalline Materials: From Photo-Induced Phase Transitions and Property Modulations to Applications (Chemical Reviews: American Chemical Society, 2016), 116, 24, 15089-15166).

(i) Nematic Phase: It is the simplest LC mesophase where the molecule has only orientational order. The self-organizing characteristic of the molecule in the nematic phase has preferred direction which is often called as director ($\vec{n}$). Among all the LC phases, the nematic phase is highly symmetric; however, it is less ordered. The director becomes a unit vector since it has no average electrical polarity and is proportionate in the opposite direction, i.e. ($n = -n$).

(ii) Chiral Nematic Phase: The chiral nematic also known as the cholesteric phase (N^*) and self-organizes in a helical manner where the nematic phase transformation to a chiral phase occurs due to chirality. This chirality is achieved either through the same molecule or through some external doping agent eliciting the helical arrangement. The chirality in the molecule leads to the variation in the direction of the director from one layer to the next, creating a helical arrangement where the helical axis and the molecular axis are perpendicular to each other.

(iii) Smectic Phase: Unlike the nematic phase, the smectic phase mesogens exhibit orientational order as well as 1-D positional order. Therefore, the LC molecules are capable of aligning in a layer by layer order, and these layers are termed smectic layers. The smectic phases are further categorized in different types based on the sub-letters, i.e. SmA, SmB and SmC. These subphases can be examined by the relative direction of the director w.r.t. the layer normal.

SmA phase = when the director n is parallel to the smectic layer normal,
SmC phase = when the director n makes an angle θ with the smectic layer normal.
SmB phase = when the hexagonal symmetry is introduced into the SmA, thus slightly increasing the order.

Besides, if these LC phases consist of chiral molecules, they are denoted as SmA* and SmC*.

To enhance the existing properties of LC materials (nematic and smectic) various doping agents have been used. In addition to wet-lab experimental studies, extensive research has been carried out by experts in simulation and theoretical work to apprehend the self-assembly and immersion properties of nanoparticles in the LC matrix. Some of the experimental results exhibit enhanced properties, while some other reported work does not exhibit any change or show even worsen properties of these LC materials. For instance, the doping of 4-Cyano-4'-pentylbiphenyl (5CB) into ferroelectric NPs barium titanate ($BaTiO_3$) was reported contrastingly by various research groups. A decrease in Fréedericksz threshold V_{th} was reported by while Glushchenko et al. reported no change and contrarily an increase in V_{th} was reported by Klein et al. It shows that there is a scope of systematic experiments to ascertain the effect of nanomaterials on the LC properties.

2. DESIGN OF LIQUID-CRYSTAL NANOMATERIALS

It is fundamental to coat the nanoparticles (NPs) with the ideal material in order to display a thermotropic liquid-crystalline phase. The material should at least form a liquid phase at ambient temperature, preferably less than 250°C. Considering the two usually encountered morphologies of NPs, i.e. pseudospherical polyhedral and anisotropic rod/needle-like as well as plate-like shapes, it may be strategically possible to induce mesophase behavior in these hybrid NPs. The intrinsic shape of rod/needle-like NPs with high aspect ratios leads to the formation of liquid crystalline phases, including smetics and nematics, which are controlled by inter-particle interactions. The self-assembly of rod-like and disc-like NPs has been extensively studied, and this can be achieved by dilution with a solvent. The reduction in inter-particle interactions and enhancement of fluidity in the system can be facilitated by coating the NPs with a suitable organic sheath. The lack of naturally preferred orientation in the case of pseudospherical polyhedral NPs makes it difficult to arrange them into the liquid crystalline phases, even though the regular two-dimensional into hexagonal lattice has been observed for simple monolayer protected NPs on surfaces. Simple gold NPs-hybrids coated with simple linear thiols have not yet exhibited liquid crystalline phases, even though theoretical studies suggest that they should exhibit spontaneous symmetry in solution or at low temperatures. Thus, it is important for the ligand to impart adequate anisotropy into the hybrid system, forcing the pseudo-spherical particles to configure themselves into self-assembled structures while preserving their liquid state. Consequently, their mesogenic nature, i.e. the mobility and the orientational flexibility of the ligands on the surface of the NPs, hold equal importance as their chemical structure (Nealon et al. 2012, Mischler et al. 2012).

During the study of LCs, it is natural to question whether all molecules can form a liquid crystalline phase. Indeed, all the molecules cannot display liquid crystalline nature. To exhibit a liquid crystalline phase, the molecules must possess the following features:

1. The molecules should possess an anisotropic shape and should be relatively thin.
2. The length of the molecule should be at least 1.3 nm, following the presence of a long alkyl group on various room temperature LCs.
3. Except for the bowlics, the structure of the molecule should not be branched. To avoid metastable monotropic liquid crystalline phases, the molecule should possess a preferably low melting point. Technologically, low-temperature mesomorphic behavior proves to be more beneficial, which is promoted by alkyl terminal groups.
4. For a molecule to exhibit liquid crystalline behavior, it is necessary to possess a structurally rigid core and flexible end groups.

The controlled self-organization of metallic NPs of definite shape and size in organic matrixes has gained considerable attention from researchers due to their synergistic property in combination with the self-organized state. Cseh et al. studied the design of liquid crystalline gold NP system which exhibited a thermotropic nematic phase in bulk. The principal challenge in the design of the system was to accommodate the spherical structure of the NPs with the structure of the nematic phase. This was described by the absence of positional order and the presence of long-range order only in one direction. Figure 8 shows the structure of the mesogen which was chosen to achieve the nematic phase behavior. The rod-like structure is comprised of three aromatic rings connecting flexible alkyl chains of eleven and eight methylene groups at the terminals of the mesogen. This mesogen could be attached to the gold NPs by the terminal thiol group at the moiety consisting of eleven methylene groups. These materials would further help in the promotion of the nematic phase in organic-inorganic hybrids, dendrimers and polymers (Cseh and Mehl 2006).

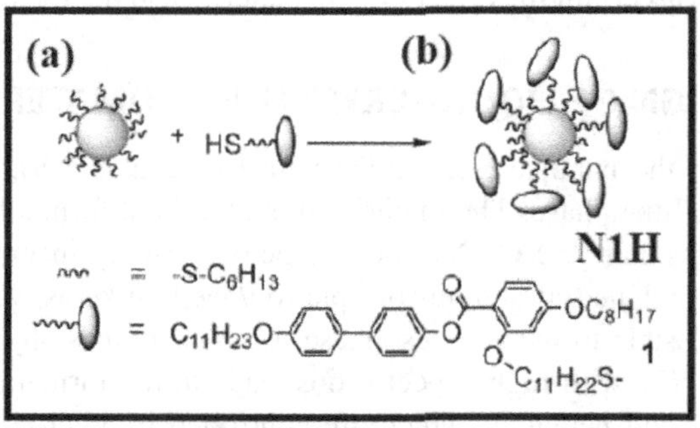

(a) (b)

+ HS

N1H

= -S-C_6H_{13}

= $C_{11}H_{23}O$ ... OC_8H_{17}

1

$C_{11}H_{22}S$-

FIGURE 8 Different segments of the molecular structure to design the liquid crystalline molecules, (a) NPs, and (b) exchange reaction yielding N1H (Cseh and Mehl 2006). (Reprinted with permission from Liliana Cseh and Georg H. Mehl, The Design and Investigation of Room Temperature Thermotropic Nematic Gold Nanoparticles (Journal of the American Chemical Society: American Chemical Society, 2006), 128, 41, 13376-13377).

It can be predicted that the interaction between the ligand nature, shape and size of the NP understudy plays a major role in determining the mesogenic properties of these hybrid systems. It can also be observed that the structure of the NP plays a vital role in discovering the mesomorphic properties of the systems reported so far. The phase behavior can also be explored based on the influence of NP composition. Evidently, the research area on LC NPs provides researchers an opportunity to synthesize new hybrid materials. The ease of modification in LC NPs enables them to tackle compelling challenges in the future, such as energy conservation, energy storage, robotics and biotechnology. The ordered structures can be achieved by modifying the self-assembly of NPs and integrating it with various processing steps and defect tolerance of liquid crystalline state having variation in the structure and functionality of the organic coating.

3. PROPERTIES OF LIQUID-CRYSTAL NANOMATERIALS

Nanomaterials have proved to be quite appealing in scientific research due to their characteristic properties, which are otherwise not observable in bulk materials. These mainly include electrical, mechanical, optical and magnetic properties besides others (Hong 2019). Each of these properties results in unique nanomaterials which find potential applications in many fields, such as photocatalysis, energy conversion, bioengineering, cosmetics, food industries and treatment of industrial effluents (Cheng 2014, Zhang et al. 2019, Fernando et al. 2015, Sudha et al. 2018). Engineered nanomaterials are specially synthesized to utilize the maximum advantageousness of their particle size and distinct properties, which are not usually observed in bulk materials. The synthesis techniques and conditions can influence the particle size of the nanomaterials, which in turn can affect their physiochemical properties (Khan et al. 2019, Sudha et al. 2018). Based on the precursors used to synthesize nanomaterials, properties like color, absorption, transparency, solubility, melting point, conductivity and catalytic behavior can be modified by altering the particle size (Yokoyama et al. 2008). The change in fundamental properties with respect to particle size is known as the 'quantum size effect'. This is mainly due to (a) high surface energy, (b) a large fraction of surface atoms, i.e. large surface area, (c) reduced imperfections and (d) spatial confinement that is absent in corresponding bulk materials. With decreasing particle size, there is an observable change in the physical behavior of the nanomaterials showing transition from classical physics to quantum physics (Sudha et al. 2018). Some of these fundamental properties are discussed in the section below.

3.1 Optical and Electronic Properties

The optical properties of nanomaterials are of key interest owing to their technical applications. Apart from their economic importance, the scientific background of these properties is essential to comprehend the behavior of nanomaterials (Vollath 2013). The optical properties of nanomaterials are dependent on their electronic structures; therefore, the analysis of the optical properties can lead to a better understanding of their structure (Zhang 2018). Also, the optoelectronic properties of nanomaterials are interrelated to a great extent. The electronic properties of nanomaterials mainly involve the movement of charged carriers. The change in the dimensions of the material down to nanometre range results in quantum size effect and quantum confinement effect as illustrated in Figure 8 (Khan et al. 2019, Zhang 2018). For instance, the optical properties of quantum dots partially occur due to quantum confinements (Hong 2019).

Recently, studies based on semiconductor NPs have broadened owing to their size-dependent nature which is the consequence of the quantum confinement effect. The interaction between matter and light on the nanoscale basis can be possible due to (a) the interaction of light with nanoscale material, (b) confinement of light to the dimensions that are smaller than the wavelength of light and (c) nanoscale confinement of a photo-process when photochemistry is induced or a light-induced phase change occurs (Sudha et al. 2018).

The quantum confinement effect in semiconductor nanomaterials is size-dependent, where quantum confinement increases the bandgap. There is an inter-dependence between the optical absorption and the size of the NPs. Semiconductor nanocrystals display size and shape-dependent properties with simple surface modifications (Figure 9), thereby finding potential applications in solar cells, a light-emitting device, optoelectronics, thermoelectrics, sensing, luminescent solar concentrators, drug-delivery and catalysis (Berends and Donega 2017).

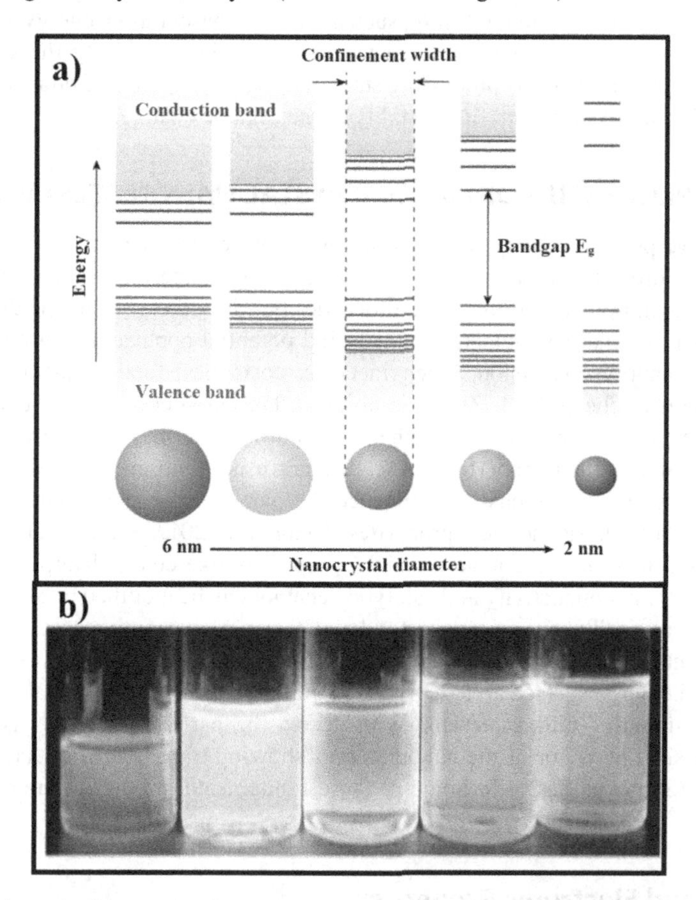

FIGURE 9 (a) Schematic illustration of quantum confinement effect depicting increasing band gap (CB: conduction band and VB: valence band) of semiconductor nanocrystal with decreasing particle size, and (b) colloidal dispersions of CdSe quantum dots with decreasing particle size (from 6 to 2 nm) under UV illumination (Donega 2011). (Reprinted with permission from Celso de Mello Donegá, Synthesis and properties of colloidal heteronanocrystals (Chemical Society Reviews: American Chemical Society, 2011), 40, 1512-1546).

In a crystalline nanomaterial, the crystal symmetry, the lattice constant and the atomic orbitals of the constituent atoms determine its electronic structure. The electrons in the conduction band are delocalized in infinite periodic potentials. In the case of nanostructures, one or more dimensions have limited sizes, creating confinement for electrons all along with the directions of those dimensions (Figure 10). In two-dimensional (2D) nanostructures, such as nanowells, the electrons are confined in one dimension and free in the remaining two dimensions. In the case of one-dimensional (1D) nanostructures, the electrons are confined in two dimensions and free in one dimension, as in the case of nanowires. In zero-dimensional (0D) nanostructures, the electrons are confined in all three dimensions. Further confinement of electrons leads to smaller dimensions, thereby entering the realms of molecules and atoms (Cheng 2014).

Optical processes such as absorption and emission of light in materials are directly related to their electronic state. Owing to the unique electronic structures and density of states of nanomaterials, their optical properties considerably differ from that of the bulk materials, thereby enabling a range of novel and improved devices for optoelectronic applications (Zhang 2018, Cheng 2014). In a recent study, Ahmed et al. investigated the effect of particle size on the optical properties of zinc oxide (ZnO) quantum dots. The ultrasmall ZnO quantum dots showed optical confinement for particles below 9 nm in diameter. This reduction in particle size led to an enhanced photocatalytic behavior of the quantum dots as a result of increased surface area and quantum confinement effect as well (Ahmed and Edvinsson 2020).

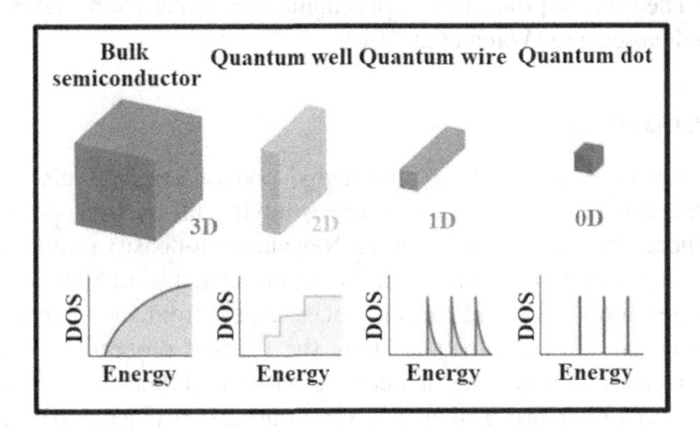

FIGURE 10 Schematic representation of energy level structure with reducing dimensionality, from 3D (bulk semiconductor) → 2D (quantum well) → 1D (quantum wire) → 0D (quantum dot); DOS representing the density of electronic states (Rabouw and de Mello Donega 2016). (Reprinted with permission from Freddy T. Rabouw and Celso de Mello Donega, Excited-State Dynamics in Colloidal Semiconductor Nanocrystals. (Topics in Current Chemistry: Springer Nature, 2016), 374, 58).

The optical properties of noble metal-based nanomaterials are size-dependent, and they exhibit a strong UV-visible excitation band. This excitation band is the result of constant incident photon frequency with collective excitation of the electrons in the conduction band and is known as the localized surface plasmon resonance (LSPR) as illustrated in Figure 11 (Khan et al. 2019).

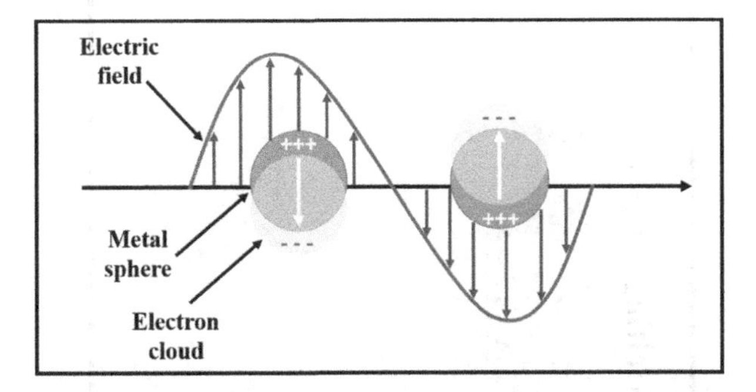

FIGURE 11 Schematic illustration of gold nanospheres exhibiting localized surface plasmon resonance (Wang et al. 2019). (Reprinted with permission from Wang et al., Modulating Catalytic Performance of Metal-Organic Framework Composites by Localized Surface Plasmon Resonance (ACS Catalysis: American Chemical Society, 2019), 9, 12, 11502-11514).

The two most noteworthy features of surface plasmon modes are (a) distinctively large field enhancement and (b) extremely compact spatial confinement of light energy. Plasmonic resonance in metallic nanomaterials at visible and near-IR wavelengths results in exceptional enhancement of field owing to their minute structure. The amplified interaction between light and matter proves to be beneficial in applications where strong light intensity is necessary, as in chemical and bio-detection of trace level chemicals, solar cells and photodetectors. The resonant wavelength of the localized surface plasmons greatly depends on the size, shape and dielectric functions of the metal as well as the dielectric surroundings (Cheng 2014).

The plasmonic nanostructured metals, such as Ag and Au, can be excited upon visible light irradiation to produce hot carriers, which can further be utilized by semiconductors, like metal-organic frameworks (MOFs), thereby promoting efficient electron-hole separation. The catalytic performance of MOFs can be further improved by the incorporation of plasmonic NPs into them (Wang et al. 2019). The tunable properties of plasmonic NPs make them highly applicable in the fields of catalysis and biosensing (Wang et al. 2019).

3.2 Thermal Properties

To explore nature and develop structural knowledge about nanomaterials, it is important to comprehend their thermal properties. For instance, studying the melting points can help us in understanding the underlying interactions between NPs and high-density grain boundaries. Studies on thermal conductivity help in revealing the information related to the heat resistance reflection of phonons from grain boundaries and the interfaces. Apart from theoretical and fundamental significance, it is necessary to measure and study the thermal properties of nanomaterials for applications. Nanomaterials with low thermal conductivities find their use as insulators. Also, the thermal properties govern their use in thermoelectric conversion (Zhang 2018). The modulation in surface energy and inter-atomic spacing as a function of size affects the material properties of nanomaterials. In the case of bismuth NPs, the melting point, which is a bulk thermodynamic property, has been found to decrease abruptly for particle sizes lesser than 10 nm, as illustrated in Figure 12 (Roumanille et al. 2017). It has also been shown that the melting point of metallic nanocrystals which are embedded into a continuous matrix have higher melting points for smaller particle size (Putnam et al. 2006, Thangadurai et al. 2020).

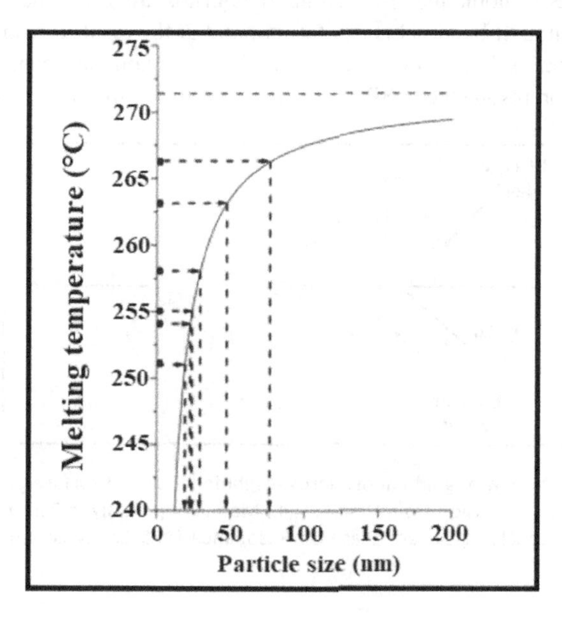

FIGURE 12 Schematic representation of the calculation of particle size of bismuth NPs from theoretical model and experimental measurements (Roumanille et al. 2017). (Reprinted with permission from Roumanille et al., $Bi_2(C_2O_4)_3{\cdot}7H_2O$ and $Bi(C_2O_4)OH$ Oxalates Thermal Decomposition Revisited. Formation of Nanoparticles with a Lower Melting Point than Bulk Bismuth (Inorganic Chemistry: American Chemical Society, 2017), 56, 16, 9486-9496).

The heat of sublimation or the cohesive energy is an important thermal property that is responsible for the strength of metallic bonds since it is the energy required to break down the metallic crystal into individual atoms (Sudha et al. 2018). Li et al. theoretically calculated the modeling of shape and size-dependent cohesive energy of NPs and thus concluded that the cohesive energy of the free NPs decreases with a decrease in the particle size (Li 2014a). The cohesive energy is used to predict the specific heat, melting enthalpy, melting temperature and melting entropy of the nanomaterials. The depression in melting point confirms the reduction in particle size during the preparation of nanomaterials. Owing to a larger surface-to-volume ratio than the bulk materials alters the thermal and thermodynamic properties of nanomaterials, thereby exhibiting a more significant change in their melting points (Sudha et al. 2018). The thermal conductivity of nanomaterials depends on factors such as grain size, grain boundary, surface interaction, doping, structural defects and temperature (Qiu et al. 2020). Metal NPs have higher thermal conductivities than fluids in solid form. Metal oxides like alumina (Al_2O_3) have greater thermal conductivity than water. Fluids containing dispersed solid particles are supposed to exhibit considerably enhanced thermal conductivities in comparison to conventional heat transfer fluids. Nanofluids are prepared by dispersing nanomaterials into desired solvents such as water, ethylene glycol or oils. Owing to the fact that heat transfer occurs at the surface of the NPs, it is, therefore, preferable to use particles with the large total surface area which also enhances the stability of the suspension (Khan et al. 2019).

3.3 Mechanical Properties

One of the most important factors to determine the potential applications of structural material is its mechanical property. A lot of investigation has already been done on the mechanical properties of metals, non-metals and composites. Theories of deformation, dislocation, fracture, etc., have already been developed, which can explain the experimental results very well. They can also be used to design specific materials of desired properties. The decrease in crystallite size leads to an increase in the hardness of the crystalline material and a considerable increase in the mechanical strength in moving from bulk material to nanometer scale. At a high temperature of about 50% over the melting point, nanoscale ceramic materials exhibit a unique superplastic phenomenon, thus providing the possibility of forming and processing ceramics (Yokoyama 2008). Based on previous studies it has been shown that (a) the elastic moduli of nanocrystalline materials are 30-50% lower than those of conventional materials; (b) the hardness of these nanoscale materials decreases with the grain size; (c) the strength and hardness of pure metals with the grain size of 10 nm are 2-7 times greater than that of metals with grain sizes greater than 1 mm and; (d) nanocrystalline materials exhibit superior toughness due to super-plasticity (Zhang 2018).

The distinctive mechanical properties of nanomaterials allow their potential applications in fields such as nanofabrication, tribology, nanomanufacturing and surface engineering. Various mechanical parameters, such as hardness, elastic modulus, adhesion, friction, stress and strain, can be studied to comprehend the mechanical nature of nanomaterials as illustrated in Figure 13 (Guo et al. 2013). Besides these, lubrication, coagulation and surface coating also adds to the mechanical properties of nanomaterials. In a greased or lubricated contact, the difference in the stiffness between the NPs and the contacting external surface controls whether the particles are indented into the plane surface or deformed when significantly immense pressure is present at the contact. Controlling the mechanical properties of nanomaterials helps in enhancing their interaction with any kind of surface, thereby improving the surface quality (Khan et al. 2019, Guo et al. 2013).

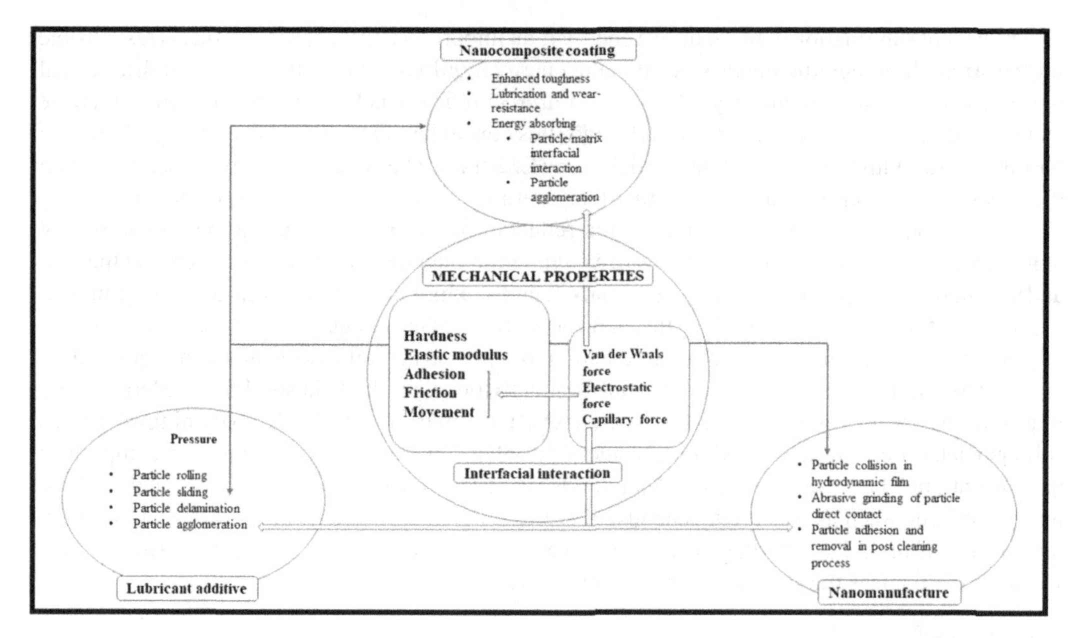

FIGURE 13 Schematic illustration of the mechanical properties of nanomaterials and their applications (Guo, Xie, and Luo 2013). (Reprinted with permission from Guo et al., Mechanical properties of nanoparticles: basics and applications (Journal of Physics D: Applied Physics: IOP Sciences, 2014), 47, 013001).

3.4 Magnetic Properties

Magnetic NPs are used in numerous applications including homogeneous and heterogeneous catalysis, ferrofluids, bio-processing, data storage, magnetic resonance imaging (MRI) as well as wastewater remediation. The large surface-to-volume ratio renders a substantial proportion of atoms having distinct magnetic coupling with nearby atoms resulting in distinctive magnetic properties. Figure 14 depicts the gradual increase in the saturation magnetization of magnetite NPs as a function of their particle size (Mohapatra et al. 2018).

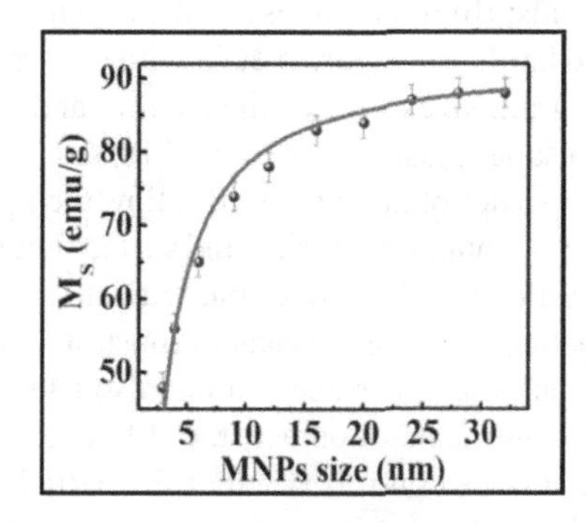

FIGURE 14 Graphical representation of the variation of the saturation magnetization of magnetite NPs with particle size (Mohapatra et al. 2018). (Reprinted with permission from Mohapatra et al., Size-dependent magnetic and inductive heating properties of Fe_3O_4 nanoparticles: scaling laws across the superparamagnetic size (Physical Chemistry Chemical Physics: Royal Society of Chemistry, 2018), 20, 12879-12887).

While the bulk ferromagnetic materials usually form several magnetic domains, magnetic NPs often comprise a single domain and exhibit superparamagnetic behavior. The overall magnetic coercivity is lowered in this case, and the magnetizations of different particles are distributed randomly due to thermal fluctuations, aligning only in the presence of an applied magnetic field (Thangadurai et al. 2020). Uneven electronic distribution results in magnetic property in nanomaterials. These properties are highly dependent on the synthetic route, and the precursors used (Khan et al. 2019, Zhang et al. 2019). With the effects of small particle size and large surface area, magnetic NPs exhibit interesting size-dependent magnetic properties (Figure 15) (Beygi and Sajjadi 2018).

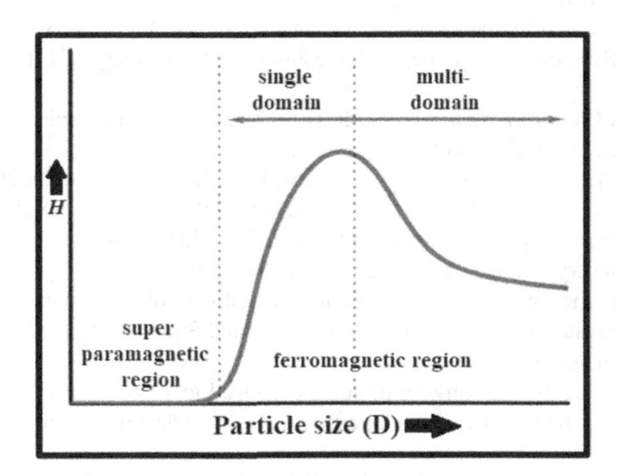

FIGURE 15 Plot of Magnetic coercivity versus particle size (Zhang, Yang, and Guan 2019). (Reprinted with permission from Zhang et al., Applications of Magnetic Nanomaterials in Heterogeneous Catalysis (ACS Applied Nano Materials: American Chemical Society, 2019), 2, 8, 4681-4697).

Acknowledgments

The authors thank SERB-DST (SRG/2019/001628) and CSIR (MLP-2006) for financial support.

References

Ahmed, T. and Edvinsson, T. 2020. Optical quantum confinement in ultrasmall ZnO and the effect of size on their photocatalytic activity. J. Phys. Chem. C. 124(11): 6395-6404.

Bai, Y. and Abbott, N.L. 2011. Recent advances in colloidal and interfacial phenomena involving liquid crystals. Langmuir 27(10): 5719-5738.

Berends, A.C. and Donega, C.D.M. 2017. Ultrathin one- and two-dimensional colloidal semiconductor nanocrystals: Pushing quantum confinement to the limit. J. Phys. Chem. Lett. 8(17): 4077-4090.

Beygi, H. and Sajjadi, S.A. 2018. Magnetic properties of crystalline nickel and low phosphorus amorphous Ni1-xPx nanoparticles. Mater. Chem. Phys. 204: 403-409.

Bisoyi, H.K. and Kumar, S. 2011. Liquid-crystal nanoscience: An emerging avenue of soft self-assembly. Chem. Soc. Rev. 40: 306-319.

Bisoyi, H.K. and Li, Q. 2016. Light-driven liquid crystalline materials: From photo-induced phase transitions and property modulations to applications. Chem. Rev. 116(24): 15089-15166.

Blanc C., Coursault, D. and Lacaze, E. 2013. Ordering nano- and microparticles assemblies with liquid crystals. Liq. Cryst. Rev. 1(2): 83-109.

Brake, J.M. and Abbott, N.L. 2007. Coupling of the orientations of thermotropic liquid crystals to protein binding events at lipid-decorated interfaces. Langmuir 23(16): 8497-8507.

Chandran, A., Prakash, J., Ganguly, P. and Biradar, A.M. 2013. Zirconia nanoparticles/ferroelectric liquid crystal composites for ionic impurity-free memory applications. RSC Adv. 3(38): 17166-17173.

Cheng, X. 2014. Nanostructures: Fabrication and applications. pp. 348-375. *In*: Feldman, M. [ed.]. Nanolithography: The Art of Fabricating Nanoelectronic and Nanophotonic Devices and Systems. Woodhead Publishing Limited, Amsterdam, Netherlands.

Coles, H. and Morris, S. 2010. Liquid-crystal lasers. Nat. Photonics 4(10): 676-685.

Cseh, L. and Mehl, G.H. 2006. The design and investigation of room temperature thermotropic nematic gold nanoparticles. J. Am. Chem. Soc. 128(41): 13376-13377.

Dierking, I. 2003. The blue phases. pp. 43-50. *In*: Dierking, I. [ed.]. Textures of Liquid Crystals. John Wiley & Sons, New Jersey, USA.

Dierking, I. and Al-Zangana, S. 2017. Lyotropic liquid crystal phases from anisotropic nanomaterials. Nanomaterials (Basel) 7(10): 305.

Dierking, I. 2018. Nanomaterials in liquid crystals. Nanomaterials (Basel) 8(7): 453.

Donega, C.D.M. 2011. Synthesis and properties of colloidal heteronanocrystals. Chem. Soc. Rev. 40(3): 1512-1546.

Fan, F., Srivastava, A.K., Chigrinov, V.G. and Kwok, H.S. 2012. Switchable liquid crystal grating with sub millisecond response. Appl. Phys. Lett. 100(11): 111105.

Fernando, K.A.S., Sahu, S., Liu, Y., Lewis, W.K., Guliants, E.A., Jafariyan, A., et al. 2015. Carbon quantum dots and applications in photocatalytic energy conversion. ACS Appl. Mater. Interfaces 7(16): 8363-8376.

Ganguly, P., Joshi, T., Singh, S., Haranath, D. and Biradar, A.M. 2012. Electrically modulated photoluminescence in ferroelectric liquid crystal. Appl. Phys. Lett. 101(26): 262902.

Garbovskiy, Y.A. and Glushchenko, A.V. 2010. Liquid crystalline colloids of nanoparticles: Preparation, properties, and applications. pp. 1-74. *In*: Camley, R.E. and Stramps, R.L. [eds.]. Solid State Physics. Elsevier, Amsterdam, Netherlands.

Gennes, P.G. de and Prost, J. 1994. The physics of liquid crystals. Liq. Cryst. Today 4(3): 7.

Guo, D., Xie, G. and Luo, J. 2013. Mechanical properties of nanoparticles: Basics and applications. J. Phys. D. Appl. Phys. 47(1): 013001.

Han, M.J., Lee, D.W., Lee, E.K., Kim, J.K., Jung, Y.J., Kang, H., et al. 2021. Molecular orientation control of liquid crystal organic semiconductor for high-performance organic field-effect transistors. ACS Appl. Mater. Interfaces 13(9): 11125-11133.

Hara, M. 2019. Mesostructure and orientation control of lyotropic liquid crystals in a polysiloxane matrix. Poly. J. 51(10): 989-996.

Heilmeier, G.H. and Zanoni, L.A. 1968. Guest-host interactions in nematic liquid crystals: A new electro-optic effect. Appl. Phys. Lett. 13: 91-92.

Hong, N.H. 2019. Introduction to nanomaterials: Basic properties, synthesis, and characterization. pp. 1-19. *In*: Hong, N.H. [ed.]. Nano-Sized Multifunctional Materials. Elsevier, Amsterdam, Netherlands.

Kawamoto, H. 2012. The history of liquid-crystal display and its industry. Third IEEE History of ELectro-technology Conference (HISTELCON), 1-6.

Kelker, H. and Hatz, R. 1981. Handbook of Liquid Crystals. John Wiley & Sons, New Jersey, USA.

Khan, I., Saeed, K. and Khan, I. 2019. Nanoparticles: Properties, applications and toxicities. Arab. J. Chem. 12(7): 908-931.

Kumar, S. 2014. Discotic liquid crystal-nanoparticle hybrid systems. NPG Asia Mater. 6: e82.

Kye, Y., Kim, C. and Lagerwall, J. 2015. Multifunctional responsive fibers produced by dual liquid crystal core electrospinning. J. Mater. Chem. C Mater. 3(34): 8979-8985.

Lagerwall, J.P.F. and Scalia, G. 2014. Liquid Crystals with Nano and Microparticles, Series in Soft Condensed Matter. World Scientific Publishing, Singapore.

Lavrentovich, O.D. 1994. The Physics of Liquid Crystals, Second Edition, by PG de Gennes and J Prost. Liq. Cryst. Today. 4(3): 7.

Lesiak, P., Bednarska, K., Lewandowski, W., Wójcik, M., Polakiewicz, S., Bagiński, M., et al. 2019. Self-organized, one-dimensional periodic structures in a gold nanoparticle-doped nematic liquid crystal composite. ACS Nano. 13(9): 10154-10160.

Li, Q. 2014a. Nanoscience with Liquid Crystals: From Self-Organized Nanostructures to Applications. Springer International Publishing, Switzerland.

Li, X. 2014b. Modeling the size- and shape-dependent cohesive energy of nanomaterials and its applications in heterogeneous systems. Nanotechnology 25(18): 185702.

Li, Y., Zhang, J., Li, Y., Wang, Z., Yin, C., et al. 2020. Generalized phase calibration method of liquid crystal spatial light modulator with absolute reference system of obnoxious background light. Opt. Lasers. Eng. 132: 106132.

Lin, S.Y. 2020. Thermoresponsive gating membranes embedded with liquid crystal(s) for pulsatile transdermal drug delivery: An overview and perspectives. J. Control Release 319: 450-474.

McCracken, J.M., Donovan, B.R., Lynch, K.M. and White, T.J. 2021. Molecular engineering of mesogenic constituents within liquid crystalline elastomers to sharpen thermotropic actuation. Adv. Funct. Mater. 31(16): 2100564.

Mei, L., Xie, Y., Huang, Y., Wang, B., Chen, J., Quan, G., et al. 2018. Injectable *in situ* forming gel based on lyotropic liquid crystal for persistent postoperative analgesia. Acta Biomater. 67: 99-110.

Mertelj, A. and Lisjak, D. 2017. Ferromagnetic nematic liquid crystals. Liq. Cryst. Rev. 5(1): 1-33.

Mirzaei, J., Reznikov, M. and Hegmann, T. 2012. Quantum dots as liquid crystal dopants. J. Mater. Chem. 22: 22350-22365.

Mischler, S., Guerra, S. and Deschenaux, R. 2012. Design of liquid-crystalline gold nanoparticles by click chemistry. Chem. Commun. 48(16): 2183-2185.

Mo, J., Milleret, G. and Nagaraj, M. 2017. Liquid crystal nanoparticles for commercial drug delivery. Liq. Cryst. Rev. 5(2): 69-85.

Mohapatra, J., Zeng, F., Elkins, K., Xing, M., Ghimire, M., Yoon, S., et al. 2018. Size-dependent magnetic and inductive heating properties of Fe_3O_4 nanoparticles: Scaling laws across the superparamagnetic size. Phys. Chem. Chem. Phys. 20(18): 12879-12887.

Murray, J., Ma, D. and Munday, J.N. 2017. Electrically controllable light trapping for self-powered switchable solar windows. ACS Photonics 4(1): 1-7.

Nealon, G.L., Greget, R., Dominguez, C., Nagy, Z.T., Guillon, D., Gallani, J.-L., et al. 2012. Liquid-crystalline nanoparticles: Hybrid design and mesophase structures. Beilstein J. Org. Chem. 8: 349-370.

Neto, A.M.F. and Salinas, S.R.A. 2005. Phase transitions between periodically organized lyotropic phases. pp. 138-162. *In*: Neto, A.M.F. and Salinas, S.R.A. [eds.]. The Physics of Lyotropic Liquid Crystals: Phase Transitions and Structural Properties. Oxford University Press, Oxford, England.

Petrov, A.G. 1999. Lyotropic mesogens and lyotropic polymorphism. pp. 11-81. *In*: Petrov, A.G. [ed.]. The Lyotropic States of Matter: Molecular Physics and Living Matter Physics. CRC Press, London.

Popov, N., Honaker, L.W., Popova, M., Usoltseva, N., Mann, E.K., Jakli, A., et al. 2018. Thermotropic liquid crystal-assisted chemical and biological sensors. Materials 11(1): 20.

Putnam, S.A., Cahill, D.G., Braun, P.V., Ge, Z. and Shimmin, R.G. 2006. Thermal conductivity of nanoparticle suspensions. J. Appl. Phys. 99(8): 084308.

Qi, L., Liu, S., Jiang, Y., Lin, J.-M., Yu, L. and Hu, Q. 2020. Simultaneous detection of multiple tumor markers in blood by functional liquid crystal sensors assisted with target-induced dissociation of aptamer. Anal. Chem. 92(5): 3867-3873.

Qiu, L., Zhu, N., Feng, Y., Michaelides, E.E., Żyła, G., Jing, D., et al. 2020. A review of recent advances in thermophysical properties at the nanoscale: From solid state to colloids. Phys. Rep. 843: 1-81.

Rabouw, F.T. and Donega, C.D.M. 2016. Excited-state dynamics in colloidal semiconductor nanocrystals. Top. Curr. Chem. 374(5): 58.

Reinitzer, F. 1989. Contributions to the knowledge of cholesterol. Liq. Cryst. 5(1): 7-18.

Rodarte, A.L., Pandolfi, R.J., Ghosh, S. and Hirst, L.S. 2013. Quantum dot/liquid crystal composite materials: Self-assembly driven by liquid crystal phase transition templating. J. Mater. Chem. C Mater. 1(35): 5527-5532.

Roumanille, P., Baco-Carles, V., Bonningue, C., Gougeon, M., Duployer, B., Monfraix, P., et al. 2017. $Bi_2(C_2O_4)_3 \cdot 7H_2O$ and $Bi(C_2O_4)OH$ oxalates thermal decomposition revisited: Formation of nanoparticles with a lower melting point than bulk Bismuth. Inorg. Chem. 56(16): 9486-9496.

Saliba, S., Mingotaud, C., Kahn, M.L. and Marty, J.D. 2013. Liquid crystalline thermotropic and lyotropic nanohybrids. Nanoscale 5(15): 6641-6661.

Stamatoiu, O., Mirzaei, J., Feng, X. and Hegmann, T. 2011. Nanoparticles in liquid crystals and liquid crystalline nanoparticles. pp. 331-393. *In*: Tschierske, C. [ed.]. Liquid Crystals. Springer, Berlin, Heidelberg.

Stephen, M.J. and Straley, J.P. 1974. Physics of liquid crystals. Rev. Mod. Phys. 46(4): 617-704.

Sudha, P.N., Sangeetha, K., Vijayalakshmi, K. and Barhoum, A. 2018. Nanomaterials history, classification, unique properties, production and market. pp. 341-384. *In*: Barhoum, A. and Makhlouf, A.S.H. [eds.]. Emerging Applications of Nanoparticles and Architecture Nanostructures. Elsevier, Amsterdam, Netherlands.

Sun, K., Xiao, Z., Lu, S., Zajaczkowski, W., Pisula, W., Hanssen, E., et al. 2015. A molecular nematic liquid crystalline material for high-performance organic photovoltaics. Nat. Commun. 6(1): 6013.

Thangadurai, T.D., Manjubaashini, N., Thomas, S. and Maria, H.J. 2020. Nanostructured Materials. Springer, Cham, Switzerland.

Vollath, D. 2013. Nanomaterials: An Introduction to Synthesis, Properties and Applications (Second Edition). John Wiley & Sons, New Jersey, USA.

Wallace, J.U., Marshall, K.L., Batesky, D.J., Kosc, T.Z., Hoffman, B.N., Papernov, S., et al. 2021. Highly saturated glassy liquid crystal films having nano- and microscale thicknesses for high-power laser applications. ACS Appl. Nano Mater. 4(1): 13-17.

Wang, M., Tang, Y. and Jin, Y. 2019. Modulating catalytic performance of metal–organic framework composites by localized surface plasmon resonance. ACS Catal. 9(12): 11502-11514.

Yadav, S.P. and Singh, S. 2016. Carbon nanotube dispersion in nematic liquid crystals: An overview. Prog. Mater. Sci. 80: 38-76.

Yang, X., Zhong, X., Zhang, J. and Gu, J. 2021. Intrinsic high thermal conductive liquid crystal epoxy film simultaneously combining with excellent intrinsic self-healing performance. J. Mater. Sci. Technol. 68: 209-215.

Yokoyama, T., Masuda, H., Suzuki, M., Ehara, K., Nogi, K., Fuji, M., et al. 2008. Basic properties and measuring methods of nanoparticles. pp. 3-47. *In*: Naito, M., Yokoyama, T., Hosokawa, K. and Nogi, K. [eds.]. Nanoparticle Technology Handbook (Third Edition). Elsevier, Amsterdam, Netherlands.

Zhang, B. 2018a. Electrical properties of nanometer materials. pp. 337-385. *In*: Zhang, B. [ed.]. Physical Fundamentals of Nanomaterials. Elsevier, Amsterdam, Netherlands.

Zhang, B. 2018b. Mechanical properties of nanomaterials. pp. 211-250. *In*: Zhang, B. [ed.]. Physical Fundamentals of Nanomaterials. Elsevier, Amsterdam, Netherlands.

Zhang, B. 2018c. Optical properties of nanomaterials. pp. 291-335. *In*: Zhang, B. [ed.]. Physical Fundamentals of Nanomaterials. Elsevier, Amsterdam, Netherlands.

Zhang, B. 2018d. Thermal properties of nanomaterials. pp. 251-289. *In*: Zhang, B. [ed.]. Physical Fundamentals of Nanomaterials. Elsevier, Amsterdam, Netherlands.

Zhang, Q., Yang, X. and Guan, J. 2019. Applications of magnetic nanomaterials in heterogeneous catalysis. ACS Appl. Nano Mater. 2(8): 4681-4697.

Recent Development in Nanomaterials: Industrial Scale Fabrication and Applications

Ankita Dhillon[1], Meena Nemiwal[2] and Dinesh Kumar[3,*]

1. INTRODUCTION

The development of nanotechnology has influenced the facet of every human being. It is an integrated part of science that includes multidirectional phases of nanoparticles. Nanoparticles have numerous advanced and novel physical properties that differ largely from their bulk counterpart. Nanomaterials' fascinating and wonderful properties, such as surface-enhanced Rayleigh scattering, surface-enhanced Raman scattering (SERS), surface plasmon light scattering, and surface plasmon resonance (SPR), are due to the reach of the size of particles in the nano regime. The fraction of atoms on the material's surface reaches the maximum, resulting in high surface activity (Akhtar et al. 2013). The numerous benefits of nanoparticles over bulk materials include high surface activity, lower defectiveness, and surface plasmon resonance properties (Jain et al. 2007). These properties have been extensively used in various remedial processes, water purification, sensing methods, and biological applications (Dhillon and Kumar 2015a, Dhillon and Kumar 2015b, Dhillon et al. 2016a, Dhillon et al. 2016b, Dhillon and Kumar 2018a, Dhillon and Kumar 2018b, Ghadimi et al. 2020, El-Berry et al. 2021).

The creation of nanoparticles revolves around two mainstream approaches: bottom-up and top-down (Figure 1) (Khan and Khan 2020). Several physical or chemical methods are used to reduce bulk material into nanosize to fabricate the metallic nanoparticles through the top-down approach (Khan and Khan 2020). Molding materials into designed shape-order, cutting, and milling is done using externally controlled tools in a top-down approach. Many methods of physical synthesis belonging to this category are lithography (Bayat and Tajalli 2020), pyrolysis (Abe and Laine 2020), thermolysis (He et al. 2021), and radiation-induced methods (El-Berry et al. 2021).

Initially, nanoparticles are built to fabricate the atoms, molecules, or clusters in the bottom-up approach (Charitidis et al. 2014). This approach is also considered a self-assembly approach. Nanoparticles originally manufactured are assembled into the product with the help of different synthetic methods. The bottom-up method delivers an improved fortuity in obtaining metal-based nanoparticles with lowered defects, more homogenous confirmation, and an extensively reasonable economical price. Various fabrication processes, such as chemical, electrochemical, microwave, and sonochemical synthesis, use the bottom-up approach (Figure 1) (Siddique and Numan 2021).

[1] Department of Chemistry, Banasthali Vidyapith, Rajasthan 304022, India.
[2] Department of Chemistry, Malaviya National Institute of Technology, Jaipur 302017, India.
[3] School of Chemical Sciences, Central University of Gujarat, Gandhinagar 382030, India.
[*] Corresponding author: dinesh.kumar@cug.ac.in.

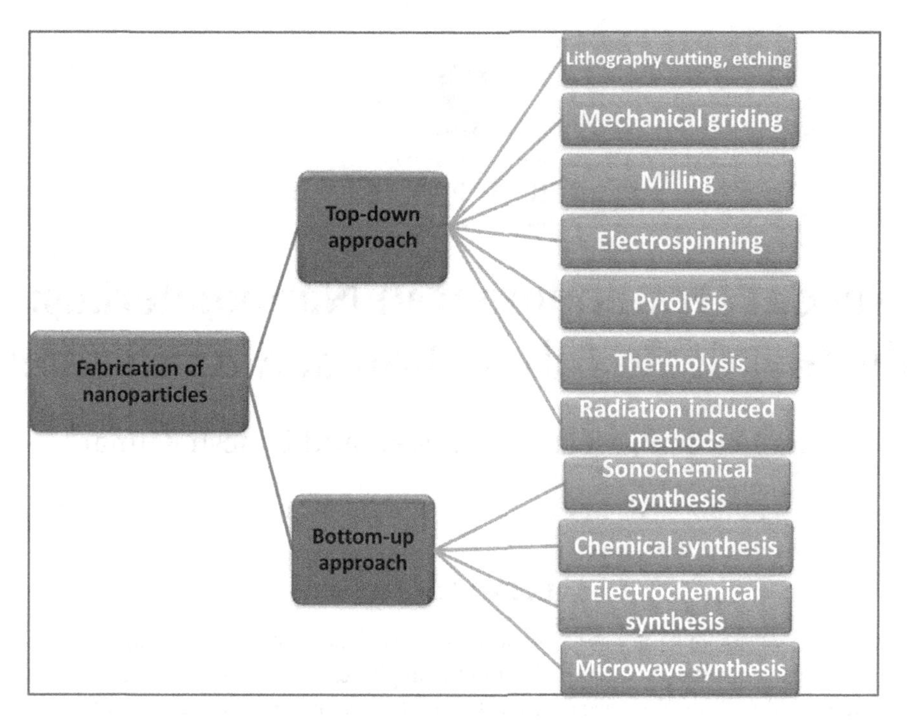

FIGURE 1 General methods for the nanoparticle's fabrication.

Nanomaterials synthesized using these techniques have been employed to remediate aquatic pollutants (Ghadimi et al. 2020, Dhillon et al. 2017a, Chen et al. 2020a, Valluri et al. 2020, Hastings et al. 2020, Dhillon et al. 2017b, Dhillon et al. 2018, Dhillon and Kumar 2017). High surface activity, low defectiveness, selectivity, and good stability of nanomaterials are widely used in various processes, such as water purification, sensing methods, and the remediation of various toxic environmental pollutants.

This chapter explores various industrially important nanostructures, such as carbon-based, inorganic, and metal-based. In addition, the properties of nanomaterials are observed. Thereafter, nanostructured materials recently manufactured with precise forms and dimensions are also discussed in various fields, mainly in the remediation of water pollutants. Next, the continued improvement of modern nanomaterials will address challenges and future approaches in the form of high throughput materials.

2. INDUSTRIAL SCALE FABRICATION OF NANOMATERIALS

2.1 Synthesis Involving Top-Down Routes

2.1.1 Solid-Phase Methods

2.1.1.1 Mechanical Attrition

For the development of alloys and phase mixtures, mechanical attrition has been industrialized since 1970. The mechanical milling process also has a significant advantage of working at low temperatures, making the slow growth of newly formed grains (Sahai et al. 2021). This method can prepare interface-boundary designs and unconventional nanomaterials with a precise design of grain. A glass structure hardly differentiates a nanocrystalline structure in many situations, so studies are directed to develop boundaries among nanophase domains (Chen et al. 2020b).

Mechanical attrition leads to severe plastic deformation, so on nanometer scales, continuous refinement of the powder particles' internal structure is performed (Sahai et al. 2021, Chen et al. 2020b). A rise in the temperature between 100°C and 200°C can be seen during the process (Mancillas-Salas et al. 2020). Simultaneously, mechanical milling can chemically react to the environment and the milled powders due to its sensitive nature to contamination and atmospheric control (Chen et al. 2020b). To produce various nanopowders, mechanochemical processing (MCP) has evolved into a fresh and economical approach, including a second mechanical apathy method. A typical ball mill can be utilized as a chemical reactor with an ambient temperature in MCP. Its utilization during the chemical reaction enhances the kinetics of the reaction because of blending and refining the grain structure at the nano level. Therefore, a suitable precursor and a reasonably reactive gas, like O_2, N_2, and other atmospheric gases, are chosen (Sahai et al. 2021). The capability of reactive milling exhibit by numerous metals, such as Ti, Zr, W, Fe, V, Hf, Mo, and Ta (Chen et al. 2020b, Valluri et al. 2020, Hastings et al. 2020); these can be converted into a variety of metal oxides of nanocrystalline nature. An organic fluid can be used in wet milling. By this method, a nanocrystalline metal-ceramic composite can be established from metal powder (Chen et al. 2020a).

Using various prototypes, like chlorides, fluorides, hydroxides, carbonates, and sulfates, we can manufacture products and lower reaction costs (Sahai et al. 2021). Numerous ball mills are utilized in mechanical attrition, such as shaker mills, planetary mills, tumbler mills, attrition, and vibratory mills (Wainwright et al. 2020). The powdered substance is kept in a closed vessel containing hard metal-covered balls. Both the mass and the balls' velocity govern their kinetic energy (Chen et al. 2020a).

2.2 Bottom-Up Routes

2.2.1 Vapour Phase Methods

2.2.1.1 Aerosol-Based Methods

Aerosol-based methods widely achieve the industrial fabrication of nanoparticles. Here, particles possibly present as solid or liquid systems and being held in the air or another gas environment, these can be termed aerosol. When it comes to size, these can even range up to 100 nm from particles to molecules. Although various applications of aerosols were known years ago, their functioning and basic science come to light recently. For instance, for boosting manufacturing paints and plastics' strength, pigments such as particles of carbon black and titania were utilized. Optical fibers are also produced by an identical technique (Nadal et al. 2021).

The employment of the spraying process is both to dry up wet ingredients and to coat surfaces. Pyrolysis of the reactant begins after the starting materials are drizzled on a hot surface. As a result, the alternation of particles is witnessed. For the sake of accomplishing the growth of CNTs, the spray gun method was similarly used as a way for decomposing the catalyst precursors like the iron chlorides (III). Methods involving thin film over various substrates due to catalyst decomposition are modest, suitable, and economical.

2.2.1.2 Atomic or Molecular Condensation

Since first reported in 1930, gas-phase condensation is inevitably the oldest for producing metal nanoparticles (Kurapov et al. 2021). A gas condensation procedure's elementary component includes a vacuum compartment incorporated with a heating component, a vaporized metal apparatus, apparatus for the collection of powder, and a driving system. The process involves a large amount of material that is to be heated up to an adequately high temperature within a vacuum compartment to generate atomized and vaporized reaction products. After that, the atomized products are made to enter a chamber containing a gaseous atmosphere. Under these conditions' nucleation that is followed by nanoparticle formation is achieved after instant cooling down of metal atoms. Metal nanoparticles are produced in the system after introducing oxygen as a reactive gas. Intense care should be given during rapid oxidation that might cause the particle's compacting (Nadal et al. 2021, Kurapov et al. 2021).

2.2.1.3 Arc Discharge Generation

An electric arc can be used as a source of energy to vaporize metals. The principle of this technique is charging two electrodes in an inert gas made up of the metal that must be vaporized. Until the breakdown voltage is not reached, a large amount of current is applied. The arc which is established across the two electrodes vaporizes a little bit of vaporization of one electrode to another. Being characterized by its comparative reproducibility, a tiny quantity of metal nanoparticles is produced in this method (Özkan et al. 2020). Though arc-discharge is a quite popular technique, the control of arcs' elevated temperatures results in high evaporation amount, which is of utmost significance at an industrial scale (Lin et al. 2021). The applied electric current and the composition of carrier gas are the two important factors regulating the primary particles' size. Nitrogen comes out to be the most reasonably affordable transporter gas, putting the electric current as of the lone trail behind as a parameter of this process. The higher the electric current, the bigger the particle size production that ultimately results in an increased production rate.

2.2.1.4 Laser Ablation Method

The important components making up the laser ablation technique are a powerful laser beam, an optical focusing system, and a metal-target feeding device. The beam of the laser is sharply focused on the target. A superior jet of evaporated material (plume) is expelled at a right angle to the target surface, extending above the target in the gas sphere. The formed particles are then transported to the product collector along with the carrier gas. The boon in this method is the usage as precursors of metals and their oxides and the formation of highly crystalline materials.

Ambient air, water, and argon in the ablation chamber are the experimental conditions upon which the concentration of developed particles and their size depend. The laser's operating parameters, such as energy, wavelength, pulse duration, speed of scanning beam, and repetition time and target material, are also influential factors. Experimental studies depicted that the developed nanoparticles in the ablation chamber air had a mass hundred times greater than their water mass. On the other hand, their mass in argon gas had a 100 times greater magnitude than that of the ambient air. For instance, nickel nanoparticle formation was evaluated using a nano-second laser, and it came out to be lower, in contrast, to be in an argon environment. Nevertheless, agglomerates can get formed due to evaporated material in elevated concentrations in the plume. Due to its high operation cost and low outcome, this technique is usually not utilized, particularly at industrial scale production (Lin et al. 2021, Ahmed et al. 2020).

2.2.1.5 Plasma Process

The plasma process is usually broadly categorized into plasma spray synthesis and microwave plasma process. Particles hold up electric charges and emerge in the plasma zone itself in the case of the microwave plasma process. Subsequently, due to charged particles' advantage, there is a reduction in agglomeration and coagulation (Graves et al. 2020, Tsyganov et al. 2020). As compared to the chemical vapor deposition, lower reaction temperatures can be accomplished as the reactants hold a charge and are dissociated where the particles remain electrically charged. The high production rate, the low size distribution of particles, and the compatibility for un-agglomerated particles' production show the usefulness of this method (Tsyganov et al. 2020). For producing nanoparticles, the plasma spray technique is usually utilized straight in an open atmosphere. As the surge and velocity of nanoparticles are exceedingly high, the demand for collecting the produced nanoparticles deliberately arises. The soberness, affordability, and mass production capability are advantageous for this method (Graves et al. 2020).

2.2.1.6 Chemical Vapor Deposition

Initiating chemical exchanges among a gaseous reactant and surface of a substrate for the deposition of a solid thin film on a substrate's surface is referred to as chemical vapor deposition (CVD). Here, the activation is attained either at high temperatures (thermal CVD) or using plasma-enhanced chemical vapor deposition (PECVD). In PECVD, a lower temperature is required in contrast to the thermal CVD process utilizing high temperatures. This material processing technology is commonly utilized for its low-cost, high yield, and it is easy to scale up. Chemical vapor deposition is one of its kind production process in several industrial areas, like semiconductor and ceramic industries (Hong et al. 2020). Generally, a CVD system on an industrial scale constitutes a loading and unloading system for the treatment of a gas transport system, a reaction compartment, substrate transport, vacuum development, energy production, process automation as well as exhaust gas (for example, before emptying the by-products into the atmosphere, NaOH and liquid N_2 are trapped). The reaction chamber is functional at a high-temperature range (500-1,200°C) into which the gases are delivered. Nitrogen and argon (inert gases) are generally used as carrier gases. The gases react to form a deposition layer on the substrate's surface as they get into contact with a heated-up surface while passing through the reactor. The two most crucial factors are pressure and temperature upon which this method operates (Mwafy 2020).

CVD is a popularly used technique for manufacturing CNTs and is a solely effective method for extending the production amount to arc discharge and laser-assisted approaches (Mwafy 2020, Park et al. 2020). As hinted earlier, CVD is a robust method with multiple benefits extending from uniformity in the thickness of coatings, number of choices utilizing a chemical precursor without strictly regulating the surface structure, and product orientation by applying an ultra-high vacuum environment. However, the CVD process is not a hundred percent safe technique and can result in various hazardous side-products (Park et al. 2020).

2.2.2 Liquid Phase Methods

2.2.2.1 Sol-Gel Method

To produce colloidal nanoparticles from liquid media, this method is a well-known large-scale fabrication method. It has been worked upon and has progressed to produce advanced nanomaterials and coatings (Parashar et al. 2020b). The sol-gel process solely relies upon hydrolysis or condensation reactions, as illustrated in Figure 2. Nanosized particles can be made to precipitate by using the appropriate amounts of reactants. This process has multiple benefits, like no elevation of temperature during processing, versatility, and is handy in shaping and embedding. Alkoxides are the most common precursors to make oxides because of their easy availability and highly reactive M-OR bond that permits the alternation. There is a minimal or negligible risk of releasing nanoparticles on drying the solution thereupon (Parashar et al. 2020b).

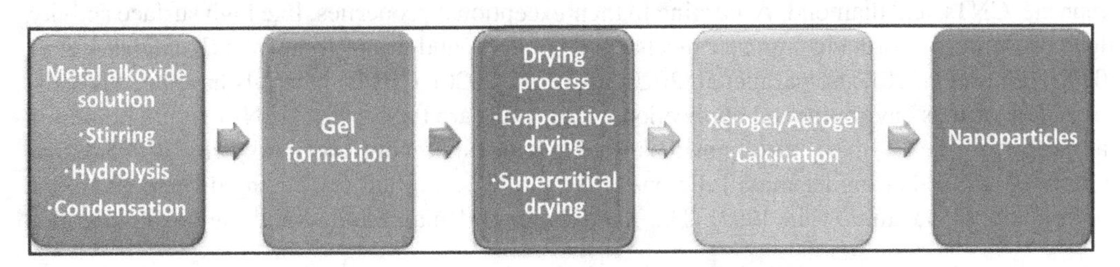

FIGURE 2 Key steps of the sol-gel method.

2.2.2.2 *Solvothermal Method*

For making both materials like crystalline oxide and non-oxides, a solvothermal method is the most relevant method. This method can produce various crystalline solids consisting of materials with silicates and high porosity, such as zeolites (Van Tran et al. 2020), oxides, and non-oxide nanoparticles (Karthikeyan et al. 2020). Semiconductors are examples of non-oxide nanoparticles with ample uses and are formed by a solvothermal method (Lv et al. 2020). Other examples are carbon nanotubes, diamonds, and carbides (Yoshimura et al. 2017). The major plus point of this technique is the elimination issues regarding the solvent, i.e. its hazardous nature and its incapability to dissolve the salts properly. The problem of the high cost of solvent also has been overcome in this method. Solvents are put in an enclosed vessel fairly above their boiling points in the solvothermal method. In this way, the congenital pressure of the attached vessel is supported. Organic solvents can be used to make metastable phases more stable and disperse non-oxide nanocrystallites. The method is referred to as hydrothermal in case the solvent is water. The method often includes auxiliary processes like hydrolysis, oxidation, and thermolysis (Agarwal et al. 2019). For making fine powders and oxides, hydrothermal synthesis is highly suitable. Usually, the pressure here surpasses the ambient pressure, and operating temperatures are generally lower than 100°C.

Hydrothermal synthesis is ready to establish itself even on an industrial platform as the recent discoveries or breakthroughs regarding its reactor's design (Lv et al. 2020). This suggests that it can be employed for mass-production at a large scale, even at 100 tons per annum. A continuous hydrothermal synthesis mixes supercritical water along with a metal salt solution to produce nanoparticle material. Instead of slowly and continuously heating the vessel's whole content, two fluids are differently heated and mixed; the ionic product of water increases when the water is heated in the direction of its critical point.

2.2.2.3 *Sonochemical Method*

Sonochemistry is defined as the chemistry branch where a chemical reaction occurs between molecules because of the applied ultrasound radiations (Jameel et al. 2020). The sonochemical method's chief motive is the formation, expansion, and destruction of the bubbles due to ultrasonic radiations in a liquid. This phenomenon is known as acoustic cavitation. Although this phenomenon is thought to be evaded during the reactors' building, it is the main reason behind sonochemical processing. It can regulate the effect of the reaction and not on the reactor (Jameel et al. 2020), as unique conditions of the process such as elevated temperatures (1,500 K), pressure conditions (>20 MPa), and exceptional cooling rates can make the development of nanoparticles of different sizes than other common materials (Jameel et al. 2020). This is the reason why this method is used mainly for manufacturing unusual nanomaterials. Its most striking feature is its highly affordable economic value (Jameel et al. 2020).

2.3 Synthesis of Carbon-Based Nanomaterials

The carbon-based nanomaterials (CBNs) involve various carbon allotropes, like fullerene, graphene, CNTs, and diamond. According to their exceptional properties, like high surface activity, high adsorption performance, greater mechanical strength, and minor footprints (Khandaker et al. 2020, Thines et al. 2017, Selvaraj et al. 2020, Roy et al. 2020), CBMs materials have been broadly utilized in various environmental applications. Fullerenes are the type of CBNs that are present in various forms like C_{60} mass (also known as Buckminsterfullerene) (Andrievsky et al. 2005, Fagan et al. 1992) and other higher mass fullerenes, like C_{70}, C_{76}, C_{78} and C_{80} having diverse geometric conformations (Bowmar et al. 1993). The Krätschmer-Huffman method and the gas combustion method (Lamb and Huffman 1993) are the frequently used research methodologies for the synthesis of fullerenes. The former method involves graphite electrodes with helium, resulting in 70% of C_{60} and 15% of C_{70}. The latter involves the burning of steady-flowing hydrocarbon fuel in the presence of oxygen at low pressure. This method is advantageous in developing large clusters of fullerenes. Their variation in size can be effortlessly organized by adjusting combustion parameters (Bogdanov et al. 2000). The gas combustion method is a popular method to fabricate fullerene and produce tons of fullerenes per year (Takehara et al. 2005).

Based on fabrication procedures, CNTs and carbon nanofibres (CNFs) have variable lengths and can be up to several hundred micrometers in diameter (Teo et al. 2003). Single-walled carbon nanotubes (SWCNTs) are sole cylindrical graphene sheets, while multi-walled carbon nanotubes (MWCNTs) are numerous concentric and coaxial rolled-up graphene sheets. The CNFs are developed through graphene nanocones or "cups" (Kharisov and Kharissova 2019). They have diameters from 70 to 200 nm (Li et al. 2021). Both CNTs and CNFs have numerous environmental applications because of their exceptional physicochemical, electrical, and mechanical properties. The most dominant method for CNTs and CNFs large-scale fabrication is chemical vapor deposition (CVD), which involves the gaseous hydrocarbon's breakdown over transition metal-catalyst (Hitchman et al. 1979, Ravindra et al. 2014). Although the CVD method has the advantage of easy scaling up, the resultant product's purity has decreased with increasing yield. The fabrication of both CNTs and CNFs at an industrial scale can be more difficult as they are synthesized in large reaction vessels resulting in product impurity and broad diametric distribution (Charitidis et al. 2014). During the past decade, commercial activity related to CNT has been reported markedly; as a result, there is increased growth in production capacity (Mwafy 2020).

2.4 Synthesis of Inorganic Nanomaterials

Much research attention has been gained by metal oxides-based nanoparticles due to their numerous exclusive properties (Nikam et al. 2018), which results in numerous environmental applications. They offer many special properties, like high surface energies and activities and plasmon excitation (Gupta et al. 2020, Salem and Fouda 2020). The most common synthesis method involves metal complexes reduction (Brock 2004) in reducing agents in aqueous solutions. However, both the reducing agent and solvent can be the same in nonaqueous systems (Brock 2004). Various precursors like $HAuCl_4$, $RhCl_3$, etc., reduction reagents like sodium citrate, hydroxylamine hydrochloride, etc., and stabilizers like polyvinyl alcohol are utilized in the fabrication of MNPs (Brock 2004). Among a range of fabrication paths, the liquid-phase precipitation method involving low-cost raw materials has been extensively utilized for silver and gold NPs synthesis.

2.4.1 Gold Nanoparticles

After AuNPs discovery, the research has been directed toward finding the relations between various development parameters, like absorption maximum, size, the concentration of reactants, and production strategies (Lee et al. 2020, Tiyyagura et al. 2020, Yu et al. 2020a). In this case, for the AuNPs synthesis, several strategies are being recounted, such as reduction, seed growth, and photochemical reduction (Akintelu et al. 2020, Ohara et al. 2020). In conventional approaches, two methods, including Turkevich-Frens (Figure 3a) and Brust-Schiffrin (Figure 3b) method, are commonly utilized for AuNPs synthesis. Additionally, seed-mediated growth process (Figure 3c) and green synthesis (Figure 3d) are some new methods that have also been developed for the fabrication of AuNPs. Researchers have also mostly focused on the AuNPs production with customizable properties leading to various applications (Schmarsow et al. 2020).

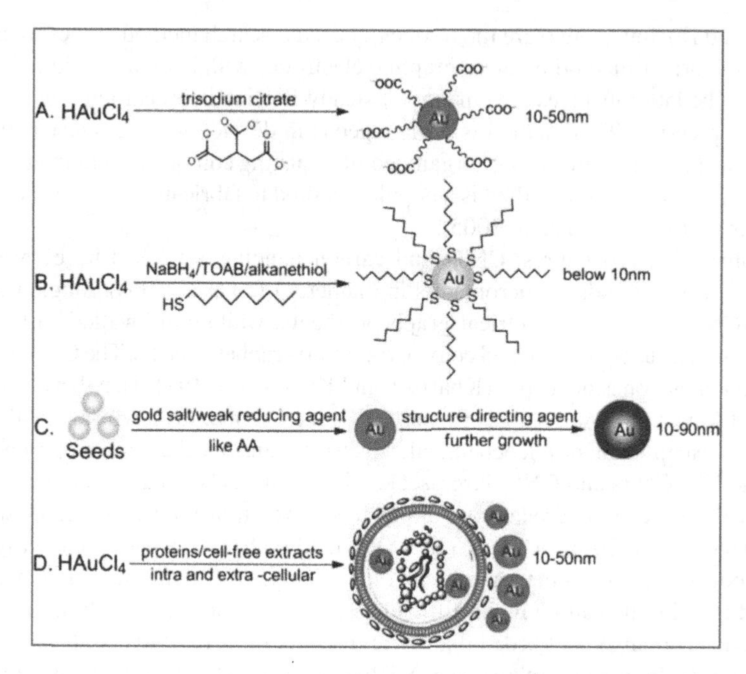

FIGURE 3 (a) Turkevich-Frens method of AuNPs fabrication, (b) Brust-Schiffrin method of AuNPs fabrication, (c) seed-mediated growth method of AuNPs fabrication, (d) fabrication of AuNPs by green approach (Qin et al. 2018).

2.4.1.1 Sodium Citrate Reduction Method

The sodium citrate reduction method is also known as the classical-citrate method (Méndez et al. 2020). This method involves sodium citrate being added into the systematically stirred boiling solution of chloroauric acid, which shows serial color changes from light yellow to ruby red. This is due to the utilization of reducing and capping agents, such as citrate to the chloroauric acid solution that results in reduced zerovalent gold atoms (Sangwan and Seth 2021). Forming AuNPs involves firstly nucleation, then growth, and finally probable agglomeration (Grys et al. 2020). However, due to the narrow range of synthesis size, this method was confined and several studies were devoted to addressing this issue.

Later, various developments were reported regarding the creation of different types of AuNPs. These developments were achieved mainly in three ways: (1) by monitoring reducing agent or the whole mixed reagents pH and change in the size and stability were achieved; (2) preparation were simplified by altering the reducing agent or using an extra stabilizing agent; and (3) some other novel approaches were applied (Trzciński et al. 2020).

In a study, stabilized AuNPs (around 5 nm size) were obtained by enclosing in citrate and polythiophene sulfonate poly [2-(3-thienyl)ethyloxy-4-butylsulfonate] sodium salt (Sanfelice et al. 2016). Also, in this method, by changing the $HAuCl_4$ and sodium citrate molar ratio, the size can be controlled. Furthermore, the citrate ions stabilizer can be effortlessly swapped by other ligands and concents for widespread sensing applications were taken. Though, there are limitations that need to be controlled: (1) the aggregation AuNPs are beyond the size range of 10 and 50 nm; (2) the stability for AuNPs is a function of the stabilizer and involves environmental tempted aggregation, which restricts them as ideal structures for further applications; and (3) the AuNPs toxicity is associated to the stabilizer. Still, this technique is widely utilized to fabricate AuNPs, even after it has appeared unsatisfactory.

2.4.1.2 Seed-Mediated Growth Method

The synthesis of AuNPs with controlled sizes has been examined for the era. The above-discussed method has the limitation of rapid growth to some extent or relatively narrow size distribution. Seed mediated growth method, also known as the step-by-step seed-mediated growth approach, is an alternative method to fabricate the AuNPs of customized size ranges. In 1989, Wiesner and Wokaun (1989) involved the addition of $HAuCl_4$ solution to the gold seeds formed by $HAuCl_4$ reduction using phosphorus (Priecel et al. 2016). In this method, the AuNPs with small sizes are fabricated using powerful reducing agents such as $NaBH_4$ or citrate salt, reducing the gold salts. Lastly, a structure-directing agent is supplemented to avoid additional nucleation and growth of AuNPs (Rodríguez-Fernández et al. 2006, Brown et al. 2000). In this approach, the improved physical properties of AuNPs were achieved than via other methods. Later, Jana and coworkers enhanced the approach by utilizing an alternative reducing regent, namely ascorbic acid, and a stabilizer, namely hexadecyl trimethyl ammonium bromide (CTAB), to produce AuNPs of the unvarying size of up to 40 nm (Jana et al. 2001). Subsequently, the Kwon group utilized the identical procedure in developing controlled size AuNPs (Kwon et al. 2007). It controlled many seeds and growth conditions that inhibited any secondary nucleation compared with other methods.

2.4.2 Silver Nanoparticles

The synthesis of silver nanoparticles (AgNPs) with controlled sizes has been examined for the era. Due to their major ratio of surface to bulk silver atoms, mostly silver nanoparticles contain a high percentage of silver oxide. The physicochemical properties of AgNPs result in various environmental applications. Due to the large effective scattering cross-section of individual silver and the surface plasmon resonance, molecular labeling can be achieved (Ibrahim et al. 2021, Elmehbad and Mohamed 2020). A typical synthesis of AgNPs involves reducing silver salt using a reducing agent; for example, sodium borohydride, polyvinyl alcohol, citrate, poly (vinylpyrrolidone), cellulose, and stabilization are commonly achieved using bovine serum albumin.

Starch can also be used as the stabilizer and β-d-glucose to reduce sugar to develop AgNPs (Olad et al. 2018). AgNPs of uniform size distribution were developed as both size and shape have been revealed to influence its efficiency.

2.4.3 Platinum Nanoparticles

Sputter deposition, laser ablation, vapor deposition, arc discharge, flame pyrolysis, and ball milling are the physical methods of synthesizing PtNPs (Jeyaraj et al. 2019). Among these, laser ablation is an efficient but costly technique. Here, instead of using electric heating, a laser beam is used. The energy production rate is high in the case of laser ablation, and the energy efficiency is good. The absorbed laser energy is utilized by laser flux resources (Prasetya and Khumaeni 2018, Mamonova et al. 2021, Lin et al. 2020). The reduction of particles' size and mixing of particles into new phases are involved in the milling process. Generally, the commonly used chemical reducing agent is $C_6H_5O_7Na_3$, N_2H_4, and $NaBH_4$. The reactants' solubility can be increased, and a low temperature is required during the solvothermal processes. As a result, it is a low-temperature technique involving polar solvents under pressure. In inert gas condensation (IGC) approaches, vacuum chambers having inert gas (at 100 Pa pressure) for the evaporation of metals. For the high-quality silver and PtNPs fabrication, IGC is an exceedingly effective method (Zheng and Branicio 2020). This method involves a chamber where an interatomic collision occurs between gas atoms. For the condensation as crystals, the evaporated metal atoms lose their kinetic energy, thereby accumulating liquid nitrogen. Chemical preparation involves various chemical reactions and chemical compositions to synthesize metal NPs (Siddique and Numan 2021). For instance, the metal ions are chemically reduced inside reversed micelles involving a nonpolar solvent. It is the most frequently utilized approach for metal nanoparticles (MNPs) synthesis (Jamkhande et al. 2019). Both the reversed micelles volume and water ratio determine the final particle size.

Suitable shape, size, temperature, appropriate solvent, and reducing/capping agent need to be well-thought-out for the synthesis of PtNPs. Aggregation can be avoided by this method. The inexpensive nature of the process, ease of functionality, adaptable surface chemistry, thermal stability, high production, low dispersion, and size-controlled synthesis are various chemical methods. The employment of harmful chemicals and organic solvents and reduced purity are downsides of chemical methods. Several studies have reported the chemical synthesis of PtNPs (Yuan et al. 2020, Wu et al. 2021, Quinson et al. 2021, Rondelli et al. 2017) that involved three critical modules for the fabrication are metal salt solutions, capping or stabilizing reagents, and reducing agents. Colloidal solution of PtNPs is produced by chemically reducing the metallic ion using a suitable reducing agent (Figure 4) (Jeyaraj et al. 2019). Sodium borohydride, elemental hydrogen, potassium bitartrate, trisodium citrate, and ascorbate are commonly used reducing agents. The reaction temperature (Shameli et al. 2010), a reducing agent (Salvador et al. 2021), and the concentration of the precursor (Figure 5) determines the morphology of the synthesized PtNPs (Jeyaraj et al. 2019).

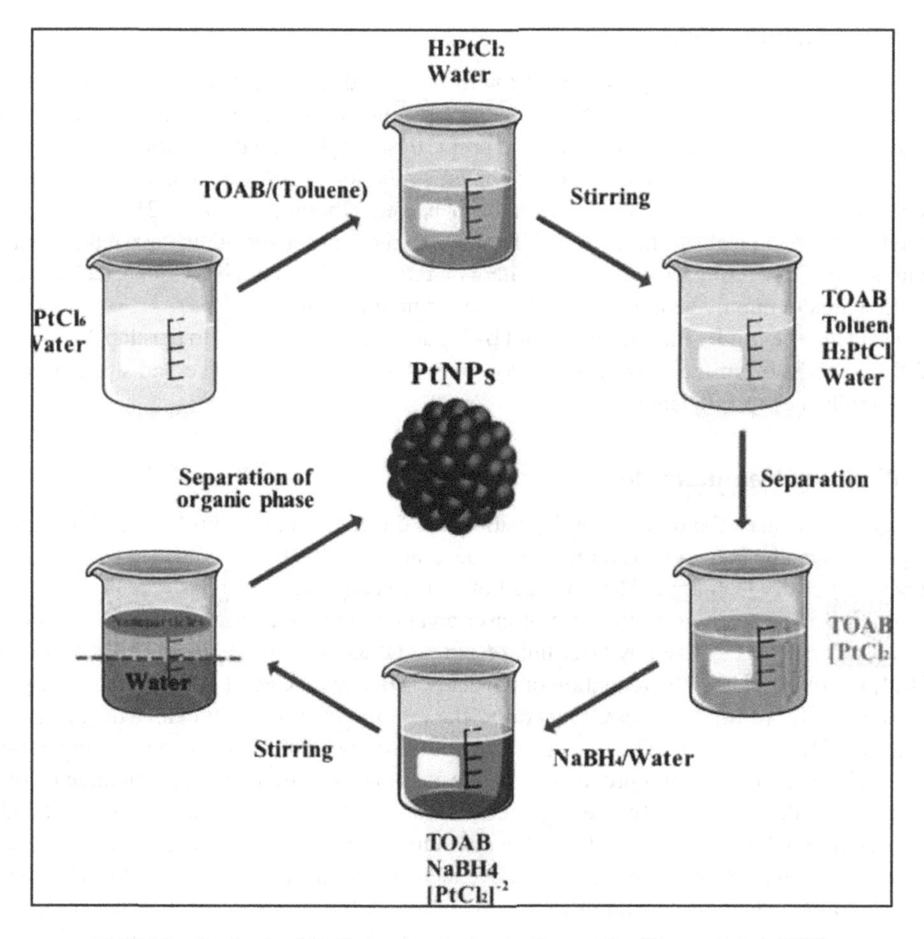

FIGURE 4 Synthesis of PtNPs by chemical reduction method (Jeyaraj et al. 2019).

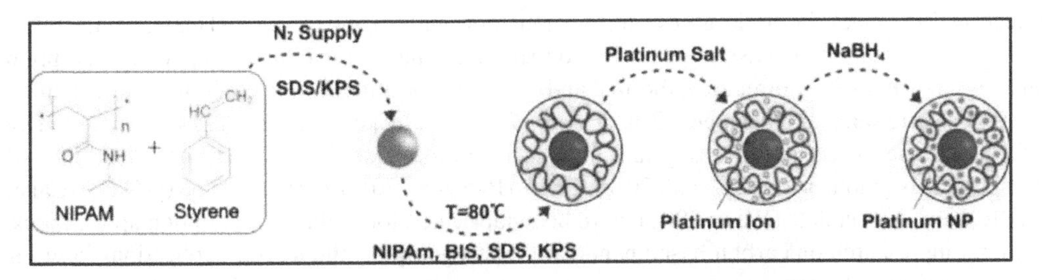

FIGURE 5 Chemical synthesis of doped PtNPs (Jeyaraj et al. 2019).

3. PROPERTIES OF NANOMATERIALS

The fundamental properties of nanoparticles closely determine their development, fate, performance, and environmental toxicity. Many nanoparticles' aspects can simply be neglected because of their deceptive talent to act more like molecules than bigger colloidal suspensions (Wu et al. 2020, Mazari et al. 2021, Zeng et al. 2021). A straightforward explanation of particle behavior in suspension is discussed here.

Several factors like buoyancy, gravitational forces, and Brownian motion control the process of diffusion (Phan et al. 2020, Medhi et al. 2020, Khan et al. 2019). Together they can be taken into description employing Einstein's law of diffusion:

$$Df = kT$$

where D = diffusion coefficient, f = the frictional coefficient, k = Boltzmann constant and T = temperature.

Where nanoparticle's frictional coefficient is calculated using Stokes law:

$$f = 6\pi\eta a$$

where η = medium viscosity and a = radius of the particle.

It has been found that the nanoparticles' small size is related to the formation of stable dispersions because of the substantial amount of Brownian motion. However, a slight collision between them will lead to accumulation and precipitation. Therefore, stabilization of dispersed NPs is essential, which can be achieved by rendering their close approach either by charge or by steric stabilization. The act of stabilization by these two factors is different. For instance, the charge stabilized colloids of lower concentrations are more dramatically influenced by ionic concentration variation than sterically stabilized colloids. This can be explained as increasing ionic strength reduces the efficacy of the Coulomb repulsion. The exchange of polyvalent ions may occur with the monovalent ions on the colloids' surface (Loza et al. 2020). As a result, rapid precipitation and aggregation of particles can occur, which is much less susceptible in the case of sterically stabilized colloids (Loza et al. 2020). Due to high surface activity, a gold nanoparticle with 5 nm size will have 31% of surface atoms. However, this drops to 3.4% at 50 nm and 0.2% at 1 micron (Sau et al. 2001).

4. APPLICATIONS OF NANOMATERIALS

4.1 Carbonaceous Nanomaterials

Carbon-based nanomaterials have fascinated scientists around the world due to their unprecedented number of applications such as water treatment (Madima et al. 2020, Baragau et al. 2021, decolorizing (Madima et al. 2020), solvent recovery (Baragau et al. 2021), and military usage to protect against attack by toxic gases, nuclear reactors, and air purification (Huang et al. 2021). They are the popular nano adsorbents owing to their abundant accessibility, exceptional adsorption capacities, inexpensive nature, great thermal and chemical constancies, and environmentally benign regarding wastewater management (Zhang et al. 2017). Activated carbon (AC) has been widely utilized because of its good porosity and high surface area. Additionally, various other kinds of carbon, such as graphene (Kyzas et al. 2018), CNTs (Ren et al. 2011), carbon nitride (g-C_3N_4) (Cai et al. 2016), and carbon dots (Wang 2005), have been fabricated for pollutant adsorption applications. Proliferating research on carbon-based nanomaterials is going on about their applications in water contaminants remediation.

Graphene is 2-dimensional (2D) (with carbon atoms sp^2 hybridized) single atom thick nanosheets and possesses a honeycomb-like arrangement. Graphene has gained researchers' attention worldwide because of its extraordinary properties comprising exceptional mechanical strength, great thermal and chemical constancy, high current density, etc. They have a wide range of applications as sensors, energy storage, lubrication, membrane separation techniques, etc. (Song et al. 2018). Their negatively charged oxygen functionalities can positively charge water contaminants (Minitha et al. 2017). This can be explained as the crossing of electric double-layer by heavy metal ions and anchoring to GO surface, disturbing their surface potential. Lately, ecofriendly rGO has fascinated researchers in water treatment (Arias et al. 2020). In this manner, methylene blue (MB) adsorption onto ecofriendly rGO has been investigated. The MB adsorption increased with pH up to 9, while decreased on further increase in pH. The rGO presented a good adsorption performance of 121.95 mg g^{-1} within 30 minutes. The adsorption was exothermic, where both physisorptions–chemisorption of MB involving H-bonding, electrostatic, and π-π interactions occurred (Arias et al. 2020). In a comparative study, polymeric nanocomposites polyaniline/graphene oxide (PANI/GO) and polyaniline/reduced graphene oxide (PANI/rGO) were studied for MB sorption (El-Sharkaway et al. 2019). The PANI/rGO presented good dye adsorption performance (19.2 mg g^{-1}) compared to PANI/GO (14.2 mg g^{-1}) as determined by the Langmuir model (El-Sharkaway et al. 2019). Wang et al. fabricated GO nanosheets to evaluate Cu(II)'s adsorptive activity from wastewater (Wang et al. 2018). The presence of an ample amount of oxygen-containing species resulted in enhanced Cu(II) adsorption. Furthermore, increasing the solution pH presented increased Cu(II) adsorption, and Cu(II) precipitation occurred under alkaline conditions. The presence of formate inhibited Cu(II) adsorption, whereas it was promoted by benzoate (Wang et al. 2018).

Although graphene has excellent adsorbent properties, it is yet inclined toward agglomeration because of stacking and strong interplanar interactions. In this regard, the GO surface was modified with a cationic surfactant, cetyltrimethylammonium bromide (CTAB), via non-covalent interactions. The developed rGO/CTAB had enhanced direct red 80 (213 mg g^{-1} and direct red 23 (79 mg g^{-1}) anionic dye adsorption performance using quaternary ammonium functional groups of surfactants (Mahmoodi et al. 2017). Different forms of graphene-like pristine graphene, rGO, and functionalized graphene nanoribbons were investigated to absorb sodium diclofenac drug (s-DCF) (Jauris et al. 2016). The sorption experiments using rGO presented good adsorption performance (59.67 mg g^{-1}) toward s-DCF drug. Additionally, the simulation experiments provided a physisorption type of s-DCF adsorption on selected graphene surfaces. Binding energies regarding the s-DCF adsorption onto selected graphene surfaces are most likely to increase with increased functional group moieties (Jauris et al. 2016). In the process of adsorbent recycling, recovery of GO from wastewater is a challenging task. The issue can be resolved by integrating GO or rGO nanosheets with magnetic (Fe_3O_4) nanoparticles that allow their easy retrieval with an external magnet. Boruah et al. efficiently utilized Fe_3O_4/reduced graphene oxide (rGO) nanocomposite for ametryn, prometryn, simazine, simeton, and atrazine pesticides recovery from wastewater (Boruah et al. 2017). Adsorption isotherm thermodynamics studies have shown physisorption processes for pesticide uptake with good adsorption performance (63.7 mg g^{-1}) toward ametryn (Boruah et al. 2017). Natural organic matter, such as fulvic acid (FA) adsorption, was achieved by iron-functionalized reduced graphene oxide (frGO)-coated sand (Ray et al. 2017). Likewise, Abdelsalam et al. applied the DFT treatment for heavy metals adsorption on graphene quantum dots (GQDs (Abdelsalam et al. 2019). Adsorption occurred via both physical and chemical interactions. The chemically modified flakes involving chemically adsorbed metals showed the highest adsorption energy (Abdelsalam et al. 2019).

One of the allotropic forms of elemental carbon is CNTs. The structure of the CNTs can be considered elongated fullerenes present in the tubular arrangement (Zadeh et al. 2021). They can be visualized as cylindrical tubes of graphene nanosheets and are classified as SWCNTs and MWCNTs (Peng et al. 2014, Alshehri et al. 2016). A comparative study between MWCNT and powder activated carbon (PAC) was carried out for Direct Blue 53 dye (DB-53) from wastewater. The higher sorption performance by MWCNT (409.4 mg g^{-1}) toward dye was noticed compared to PAC (135.2 mg). The possible mechanism of DB-53 dye adsorption on MWCNT involves the protonation of MWCNT by immersing them in an acidic solution. The positively charged surface of MWCNT electrostatically attracts the negatively charged groups under acidic conditions (Prola et al. 2013). Similar behavior was noted in dioxin adsorption on MWCNTs and ACs using the temperature-programmed desorption technique. The strong interactions of dioxin's benzene rings and the surface of the CNTs demonstrated that higher dioxin adsorption using CNTs compared to ACs (Long and Yang 2001). Various carbon nanomaterials, such as CNTs, CNF, and graphene nanoplates (GNPS), were used to develop hybrid aerogels using a simple freeze-dring process to remove organic dyes. The experimental studies showed the hybrid aerogel of CNF to GnP having a mass ratio of 3:1 showed both methylene blue (MB) and Congo red (CR) dye effective adsorption (Yu et al. 2020b). Tang et al. prepared chitosan/carboxylated carbon nanotubes (CS-CCN) composite aerogels for the effective U(VI) sorption from radioactive wastewater. During fabrication, the carboxylation process introduced carboxyl groups onto CNTs generating a high affinity for U(VI) ions and advanced aerogels' strength (Tang et al. 2020).

It has been found that even if MWCNT-based adsorbents have effective adsorption properties they are difficult to separate, thereby limiting their practical water treatment applications. The drawback can be overcome by using magnetic adsorbents having easy solid-liquid separation. Therefore, Fayazi developed magnetic multi-walled carbon nanotubes (MMWCNTs) followed by coating with a thin sulfur layer, presenting S-MMWCNTs adsorbent for excellent Hg(II) adsorption performance. Further, the thermodynamic studies presented the endothermic and spontaneous nature of Hg(II) adsorption. The regeneration and reusability experiments presented repeated Hg(II) removal due to the easy separability of magnetic adsorbent (Fayazi et al. 2016).

4.2 Metal/Metal Oxides Nanomaterials

In a recent study, three types of Fe-Mn binary oxides having a combination of ferrihydrite, goethite, or hematite were developed, and their comparative studies with corresponding precursor oxide forms were studied for As(III) removal. The experimental setup showed both synergistic and antagonistic effects.

Both acid orange 7, reactive black 5, and direct blue 71 dyes and heavy metal ions remediation from wastewaters were achieved using mesoporous silica-alumina oxide sorbent. The adsorbent presented the highest capacity toward Co(II) and Ni(II) ions and Reactive Black 5 anions (Wawrzkiewicz et al. 2017). Experimental studies have revealed that even though high magnetization can improve nanoadsorbents separation, it does not affect adsorption capacity. Increasing magnetic properties could result in fast nanoadsorbents agglomeration because of lower surface area, thereby lesser contaminant removal performance. Consequently, a striking balance of ferrites' magnetic and adsorption properties vital to enhance their removal performance. In this regard, core-shell bimagnetic nanoparticles ($CoFe_2O_4@\gamma\text{-}Fe_2O_3$) with two altered sizes were utilized for Cr(VI) remediation. The isotherm studies supported the Freundlich model demonstrating the formation of multilayers on nanoadsorbents surface. The adsorbent presented the highest removal performance at pH = 2.5 (Campos et al. 2019). Lately, several reports are available on the plant-based fabrication of iron oxide NPs. Though very few studies are presented using the hematite phase having non-uniform structure and large size deviation of particles, none of them explained the antibacterial activity. For the first time, Pallela et al. utilized *Sida cordifolia* plant extract to manufacture iron oxide (α-Fe_2O_3) NPs for antibacterial performance. The presence of phytochemicals in extract functioned as a potential reducing and stabilizing agent in Sida cordifolia mediated iron oxide development, which can also incorporate synergetic antibacterial effectiveness because of metabolites' enhanced medicinal properties. Out of all tested species like *B. subtilis, S. aureus, E. coli*, and *K. pneumonia*, the developed NPs presented the highest inhibition toward the growth of *B. subtilis* (Pallela et al. 2019). Lately, Parashar et al. developed magnetic iron oxide nanoparticles for the possible adsorption of oxcarbazepine (10,11-dihydro-10-oxo-5Hdibenzazepine-5-carboxamide), an antiepileptic drug from aqueous solution. The highest removal efficiency toward oxcarbazepine was observed under acidic conditions (pH 2.5) with adsorption equilibrium achieved within 10 minutes. The spent nanomaterial is annealed at elevated temperature; therefore, the cost-effective recovery, management, and disposal of spent adsorbent is an indicator of resourceful utilization of nanotechnology in wastewater treatment (Parashar et al. 2020a).

4.3 Nanocomposites

Earlier discussion of nanocomposites' (NCs) properties demonstrated their potential and several benefits toward water remediation applications (Zhou et al. 2020). To develop a cleaner eco-friendly adsorbent, Song et al. loaded Fe_3O_4 on the surface of halloysite nanotubes (HNTs) using one pot co-precipitation procedure to formulate Fe_3O_4/HNTs NCs (Fe_3O_4/HNTs) for both As(III) and As(V) remediation. Among different morphologies, the cactus-like Fe_3O_4/HNTs exhibited good reusability of the used adsorbent. The developed composite showed adsorption performance of 408.71 mg g^{-1} and 427.72 mg g^{-1} for As(III) and As(V), respectively (Song et al. 2019).

Arancibia-Miranda et al. synthesized imogolite and magnetic imogolite-Fe oxide NCs of various ratios for aqueous Cu, Cd, and As pollutants uptake. It was found that the Fe oxide, mainly magnetite, preferentially adsorbed the contaminants than the pristine imogolite (Arancibia-Miranda et al. 2020). Even though several mixed metal oxides NCs have been witnessed for bacterial disinfectants and fluoride remediation, very few examples are present for the simultaneous efficient fluoride adsorption and bacteria remediation from drinking water. Thus, our research group has carried out both bacterial and fluoride remediation using Ca–Ce nanoadsorbent. The 50% bacterial inhibition (IC_{50}) at 31.5 and 27.0 µg/mL was achieved using Ca–Ce NC for *E. coli* and *S. aureus* cells, respectively (Dhillon et al. 2015). The green approach was used to create a photocatalyst with good performance, and static reusability has prompted bacterial cellulose (BC)-based nanocomposite membranes to degrade photocatalytic dyes. UV radiation, coated with a functional layer of polydopamine (PDA) together with titanium dioxide (TiO_2) nanoparticles, was used. The developed BC/PDA/TiO_2 nanocomposite good adsorption performance toward methyl orange, rhodamine B, methylene blue, and photocatalytic dye degradation using UV radiation. Also, the as-prepared nanocomposite offered exceptional consistency and reusability even after five cyclic of utilization. Further, the authors believe that the photocatalytic reactivity of developed can be efficiently utilized toward the degradation of phenolic compounds by optimizing and modifying the existing method (Yang et al. 2020). Volatile organic compounds (VOCs) like Toluene are serious environmental and human health threats. Therefore, the ultrasonication-supported sol-gel process has been utilized by immobilizing cerium dioxide (CeO_2)/TiO_2 on activated carbon fiber (ACF) for toluene degradation. Characterization studies of composite have shown that the surface of TiO_2/ACF is covered by dispersed CeO_2, and Ce^{4+} ions, which does not replace Ti^{4+} ions in the lattice. Compared to TiO_2/ACF composites, CeO_2/TiO_2/ACF nanocomposites presented considerably high adsorption and photocatalytic activity with ultraviolet irradiation for toluene remediation. The composite demonstrated maximum removal efficiency of 90% by irradiation of UV for 6 hours under other optimization parameters. The composites further showed good recyclability even after three adsorptions and photodegradation cycles (Li et al. 2020).

The material surface wettability is an important criterion for oil removal, which depends on the material's chemical conformation and its surface geometrical microstructure. A recent report presented the hydrothermal fabrication of polydimethylsiloxane@zinc oxide tetrapod@iron oxide nanohybrid as an efficient, bio-inspired, and magnetic nanosorbent for oil removal. The as-prepared nanohybrid having Fe_2O_3 nanorods coated with ZnO tetrapods showed an exceptional adsorption performance (1,135 mg g^{-1}) and significant magnetic separation. Therefore, the sorbent presented good removal efficiency and enhanced magnetic properties for oily wastewater management (Sharma et al. 2019). Although nanocomposites are efficient candidates for scientific communities for treating several types of chemical contaminants, some limitations and challenges need to be overcome. This is where there is a need for researchers to advance nanosized materials to overcome the existing problems.

4.4 Noble Metal Nanomaterials

Noble metal nanomaterials present a wide variety of treatment applications in different fields because of their exclusive properties. The noble metal nanomaterials have an extremely diverse field of applications, especially in the area of disinfection. In this regard, Prabhakar et al. addressed some of the problems of existing disinfection materials like low antimicrobial efficiency, high cost, etc., by developing a microfluidic device having numerous physiochemical effects for efficient and rapid disinfection. The device was firstly embedded with multiple germicidal UV-LEDs via an innovative soft-lithography procedure. After that, the fabricated silver nanoparticles were restrained inside its internal microchannel surface. It was found that silver NPs as well as germicidal UV light had a synergistic effect on the disruption of bacterial cells.

Moreover, the system ensured both restrained silver leakage and unnecessary UV light human exposure (Prabhakar et al. 2020). In a recent study, the modified hummers method was utilized for GO synthesis. The heat treatment method for rGO development and silver different concentrations was integrated into GO nanosheets using the hydrothermal method. The finally developed Ag decorated rGO photocatalyst (Ag/rGO) supported enhanced photocatalytic activity with increasing doping ratio (Ikram et al. 2020).

Wang et al. utilized a porous covalent organic framework (COF) as a support material for *in situ* Ag NPs growth using a single-step solution infiltration process to address the instability and aggregation problems in acidic solution. The developed Ag NPs@COF composite showed good removal performance, high selectivity, and constancy toward Hg(II) removal (Wang et al. 2019). However, gold is one of the most stable elements. Its ions are toxic for almost all organisms because of its high oxidizability. Su et al. (2020) synthesized lipoic acid (LA) functionalized elongated tetrahexahedral (ETHH) Au NPs where LA functioned as a natural antioxidant having a terminal carboxylic acid and a dithiolane ring for the fabrication of ETHH-LA Au NPs. The result demonstrated higher antimicrobial activity of Au nanoparticles against *B. subtilis* in comparison to *E. coli*. The hybrid material had considerably enhanced antimicrobial activity against both bacteria compared to the precursor. Further, cell membrane fatty acids oxidation and lipid peroxides production was brought out by the ETHH Au nanoparticles suggesting their potential as antimicrobial agents (Su et al. 2020).

For the first time, Liu et al. carried out both sensing and uptake of Hg(II) and Ag(I) ions by using the oligonucleotide functionalized magnetic silica sphere (MSS)@Au NPs. The developed system presented excellent selective sensing and capture of Hg(II) and Ag(I) ions. Furthermore, using an external magnetic field, the Hg(II) and Ag(I) ions can be effectually remediated from treated solutions. The spent MSS@Au NPs were easily recycled (more than 80%) using cysteine, thereby supporting its large-scale applications (Liu et al. 2014).

A platinum complex, namely cisplatin, has been a famous drug for cancer for over 40 years. Yet, its high toxicities and developed resistance have mostly bound its scientific enactment. These limitations were addressed by developing more inert octahedral Pt(IV) prodrugs having various ligands modification for targeting numerous kinds of malignant tumor cells by changing cytostatic pathways and so on. There are various reviews and articles reported regarding the recent development in the design and fabrication of Pt(IV) prodrugs. This reveals that much work to be done in the systemic toxicity assessment of hypotoxic Pt(IV) complexes, their absorption, delivery, breakdown, and elimination properties in the human body (Zhang et al. 2019, Lozada et al. 2020, Rébé et al. 2020).

In a recent study by Narasaiah and Mandal, a novel and ecofriendly route were employed to fabricate Pd NPs by empty cotton boll peels as agricultural waste. The cotton peels extract having phytochemicals primarily acts as a reducing agent by reducing Pd NPs and a stabilizing agent via surface capping of Pd NPs. The developed Pd NPs catalytically reduced all azo-dyes using $NaBH_4$ in an aqueous medium (Narasaiah and Mandal 2020).

5. CHALLENGES AND FUTURE PERSPECTIVES

For developments in lab-based research and for examining proportional competence of new generation nanomaterials, the following points need to be considered:

1. A greener approach is required for the large-scale fabrication of nanomaterials, such as the utilization of existing industrial and natural wastes are good precursor materials in this path.
2. From an industrial point of view, it is advantageous to have a collective experimental study including chemical engineering, nanotechnology, bioprocess engineering, and plant physiology for the complete range of examination of synthesized nanoparticles.
3. More research utilizing hydraulic residence time (HRT) of contaminants is required considering diverse environmental variables for the nanoadsorbents' *in situ* applications.
4. Designing appropriate columns for the removal kinetics' optimization must be considered to study the influence of NP porosity, filling, and particle density.
5. Proper toxic sludge management generated in the form of spent nanomaterials by both lab-scale practices and industrial scale fabrication and application must be considered.
6. Compared to conventional adsorbents, the production cost of NPs is considerably huge.
7. A compulsory examination of nanoadsorbents removal performance under various experimental conditions, like competing ions, aerobic, and anaerobic conditions, is required for the factors' evaluation concerning the NPs recontamination, oxidation, or phase transformation.

Conclusions

Compared to conventional technologies, nanomaterials have appeared promising candidates to combat water pollution because of their exceptional physical and chemical properties. In recent years, nanotechnology is an evolving scientific area of having growing applications of excellent nanomaterials to the industrial ground. The present chapter covered various aspects of recent developments on various industrial synthesis methods to fabricate nanomaterials, their properties, and specific applications in various areas, chiefly in the remediation of water pollutants. Nanomaterials' development about their superior reusability, lower treatment costs, high stability, and easier separation requires more comprehensive research.

References

Abdelsalam, H., Teleb, N.H., Yahia, I.S., Zahran, H.Y., Elhaes, H. and Ibrahim, M.A. 2019. First principles study of the adsorption of hydrated heavy metals on graphene quantum dots. J. Phys. Chem. Solids 130: 32-40. doi.org/10.1016/j.jpcs.2019.02.014.

Abe, Y. and Laine, R.M. 2020. Photocatalytic $La_4Ti_3O_{12}$ nanoparticles fabricated by liquid-feed flame spray pyrolysis. Ceram. Int. 46: 18656-18660.

Agarwal, S., Rai, P., Gatell, E.N., Llobet, E., Güell, F., Kumar, M., et al. 2019. Gas sensing properties of ZnO nanostructures (flowers/rods) synthesized by hydrothermal method. Sens. Actuators B Chem. 292: 24-31.

Ahmed, M.K., El-Naggar, M.E., Aldalbahi, A., El-Newehy, M.H. and Menazea, A.A. 2020. Methylene blue degradation under visible light of metallic nanoparticles scattered into graphene oxide using laser ablation technique in aqueous solutions. J. Mol. Liq. 315: 113794.

Akhtar, M.S., Panwar, J. and Yun, Y.-S. 2013. Biogenic synthesis of metallic nanoparticles by plant extracts. ACS Sustain. Chem. Eng. 1: 591-602. doi.org/10.1021/sc300118u.

Akintelu, S.A., Olugbeko, S.C. and Folorunso, A.S. 2020. A review on synthesis, optimization, characterization and antibacterial application of gold nanoparticles synthesized from plants. Int. Nano Lett. 10: 1-12.

Alshehri, R., Ilyas, A.M., Hasan, A., Arnaout, A., Ahmed, F. and Memic, A. 2016. Carbon nanotubes in biomedical applications: Factors, mechanisms, and remedies of toxicity: Miniperspective. J. Med. Chem. 59: 8149-8167. doi.org/10.1021/acs.jmedchem.5b01770.

Andrievsky, G., Klochkov, V. and Derevyanchenko, L. 2005. Fullerenes, nanotubes. Carbon Nanostruct. 13: 363.

Arancibia-Miranda, N., Manquián-Cerda, K., Pizarro, C., Maldonado, T., Suazo-Hernández, J., Escudey, M., et al. 2020. Mechanistic insights into simultaneous removal of copper, cadmium and arsenic from water by iron oxide-functionalized magnetic imogolite nanocomposites. J. Hazard. Mater. 2020: 122940. doi.org/10.1016/j.jhazmat.2020.122940.

Arias, F., Guevara, M., Tene, T., Angamarca, P., Molina, R., Valarezo, A., et al. 2020. The adsorption of methylene blue on eco-friendly reduced graphene oxide. Nanomater. 10: 681. 10.3390/nano10040681.

Baragau, I.A., Lu, Z., Power, N.P., Morgan, D.J., Bowen, J., Diaz, P., et al. 2021. Continuous hydrothermal flow synthesis of S-functionalised carbon quantum dots for enhanced oil recovery. Chem. Eng. J. 405: 126631.

Bayat, F. and Tajalli, H. 2020. Nanosphere lithography: The effect of chemical etching and annealing sequence on the shape and spectrum of nano-metal arrays. Heliyon 6: 03382.

Bogdanov, A.A., Deininger, D. and Dyuzhev, G. A. 2000. Development prospects of the commercial production of fullerenes. Tech. Phys. 45: 521-527. doi.org/10.1134/1.1259670.

Boruah, P.K., Sharma, B., Hussain, N. and Das, M.R. 2017. Magnetically recoverable Fe_3O_4/graphene nanocomposite towards efficient removal of triazine pesticides from aqueous solution: Investigation of the adsorption phenomenon and specific ion effect. Chemosphere 168: 1058-1067. doi.org/10.1016/j.chemosphere.2016.10.103.

Bowmar, P., Kurmoo, M., Green, M.A., Pratt, F.L., Hayes, W., Day, P., et al. 1993. Raman and infrared studies of single-crystal C_{60} and derivatives. J. Phys. Condens. Matter. 5: 2739. doi.org/10.1088/0953-8984/5/17/008.

Brock, S.L. 2004. Nanostructures and Nanomaterials: Synthesis, Properties and Applications by Guozhang Cao (University of Washington). Imperial College Press (distributed by World Scientific): London. 2004. xiv+ 434 pp. $78.00. ISBN 1-86094-415-9: 14679-14679. doi.org/10.1021/ja0409457.

Brown, K.R., Walter, D.G. and Natan, M.J. 2000. Seeding of colloidal Au nanoparticle solutions. 2. Improved control of particle size and shape. Chem. Mater. 12: 306-313.

Cai, X., He, J., Chen, L., Chen, K., Li, Y., Zhang, K., et al. 2016. A 2D-g-C_3N_4 nanosheet as an eco-friendly adsorbent for various environmental pollutants in water. Chemosphere 171: 192-201. doi.org/10.1016/j.chemosphere.2016.12.073.

Campos, A.F.C., de Oliveira, H.A.L., da Silva, F.N., da Silva, F.G., Coppola, P., Aquino, R., et al. 2019. Core-shell bimagnetic nanoadsorbents for hexavalent chromium removal from aqueous solutions. J. Hazard. Mater. 362: 82-91. doi.org/10.1016/j.jhazmat.2018.09.008.

Charitidis, C.A., Georgiou, P., Koklioti, M.A., Trompeta, A.F. and Markakis, V. 2014. Manufacturing nanomaterials: From research to industry. Manuf. 1: 11.

Chen, C., Tang, W., Li, X., Wang, W. and Xu, C. 2020a. Structure and cutting performance of Ti-DLC films prepared by reactive magnetron sputtering. Diam. Relat. Mater. 104: 107735.

Chen, N., Li, H.N., Wu, J., Li, Z., Li, L., Liu, G., et al. 2020b. Advances in micro milling: From tool fabrication to process outcomes. Int. J. Mach. Tools Manuf. 160: 103670.

Dhillon, A. and Kumar, D. 2015a. Development of a nanoporous adsorbent for the removal of health–hazardous fluoride ions from aqueous systems. J. Mater. Chem. A 3: 4215-4228.

Dhillon, A. and Kumar, D. 2015b. Nanocomposite for the detoxification of drinking water: Effective and efficient removal of fluoride and bactericidal activity. New J. Chem. 39: 9143-9154.

Dhillon, A., Nair, M., Bhargava, S.K. and Kumar, D. 2015. Excellent fluoride decontamination and antibacterial efficacy of Fe–Ca–Zr hybrid metal oxide nanomaterial. J. Colloid Interface Sci. 457: 289-297. doi.org/10.1016/j.jcis.2015.06.045.

Dhillon, A., Nair, M. and Kumar, D. 2016a. Analytical methods for determination and sensing of fluoride in biotic and abiotic sources: A review. Anal. Methods 8: 5338-5352. https://doi.org/10.1039/C6AY01534D.

Dhillon, A., Sharma, T.K., Soni, S.K. and Kumar, D. 2016b. Fluoride adsorption on a cubical ceria nanoadsorbent: Function of surface properties. RSC adv. 6: 89198-89209. https://doi.org/10.1039/C6RA16962G.

Dhillon, A. and Kumar, D. 2017. Dual adsorption behaviour of fluoride from drinking water on Ca-$Zn(OH)_2CO_3$ adsorbent. Surf. Interfaces 6: 154-161.

Dhillon, A., Prasad, S. and Kumar, D. 2017a. Recent advances and spectroscopic perspectives in fluoride removal. Appl. Spectrosc. Rev. 52: 175-230.

Dhillon, A., Soni, S.K. and Kumar, D. 2017b. Enhanced fluoride removal performance by Ce–Zn binary metal oxide: Adsorption characteristics and mechanism. J. Fluor. Chem. 199: 67-76.

Dhillon, A. and Kumar, D. 2018a. Chitosan-based natural biosorbents: Novel search for water and wastewater desalination and heavy metal detoxification. pp. 123-143. *In*: Mishra, Shiwani Bhardwaj and Mishra, Ajay Kumar [eds.]. InBio-and Nanosorbents from Natural Resources. Springer, Cham.

Dhillon, A. and Kumar, D. 2018b. Recent advances and perspectives in polymer-based nanomaterials for Cr(VI) removal. pp. 29-46. *In*: Hussain, Chaudhery Mustansar and Misha, Ajay Kumar [eds.]. New Polymer Nanocomposites for Environmental Remediation. Elsevier.

Dhillon, A., Choudhary, B.L., Kumar, D. and Prasad, S. 2018. Excellent disinfection and fluoride removal using bifunctional nanocomposite. Chem. Eng. J. 1: 193-200.

El-Berry, M.F., Sadeek, S.A., Abdalla, A.M. and Nassar, M.Y. 2021. Microwave-assisted fabrication of copper nanoparticles utilizing different counter ions: An efficient photocatalyst for photocatalytic degradation of safranin dye from aqueous media. Mater. Res. Bull. 133: 111048.

El-Sharkaway, E.A., Kamel, R.M., El-Sherbiny, I.M. and Gharib, S.S. 2019. Removal of methylene blue from aqueous solutions using polyaniline/graphene oxide or polyaniline/reduced graphene oxide composites. Environ. Technol. 2019: 1-9. doi.org/10.1080/09593330.2019.1585481.

Elmehbad, N.Y. and Mohamed, N.A. 2020. Designing, preparation and evaluation of the antimicrobial activity of biomaterials based on chitosan modified with silver nanoparticles. Int. J. Biol. Macromol. 151: 92-103.

Fagan, P.J., Calabrese, J.C. and Malone, B. 1992. Metal complexes of buckminsterfullerene (C60). Acc. Chem. Res. 25: 134-142.

Fayazi, M., Taher, M.A., Afzali, D., Mostafavi, A. and Ghanei-Motlagh, M. 2016. Synthesis and application of novel ion-imprinted polymer coated magnetic multi-walled carbon nanotubes for selective solid phase extraction of lead (II) ions. Mater. Sci. Eng. C 60: 365-373. doi.org/10.1016/j.msec.2015.11.060.

Ghadimi, M., Zangenehtabar, S. and Homaeigohar, S. 2020. An overview of the water remediation potential of nanomaterials and their ecotoxicological impacts. Water 12: 1150.

Graves, B., Engelke, S., Jo, C., Baldovi, H.G., de La Verpilliere, J., De Volder, M., et al. 2020. Plasma production of nanomaterials for energy storage: Continuous gas-phase synthesis of metal oxide CNT materials via a microwave plasma. Nanoscale 12: 5196-5208.

Grys, D.B., de Nijs, B., Salmon, A.R., Huang, J., Wang, W., Chen, W.H., et al. 2020. Citrate coordination and bridging of gold nanoparticles: The role of gold adatoms in AuNP aging. ACS Nano. 14: 8689-8696.

Gupta, A., Tandon, M. and Kaur, A. 2020. Role of metallic nanoparticles in water remediation with special emphasis on sustainable synthesis: A review. Nanotechnol. Environ. Technol. 5: 1-13.

Hastings, D., Schoenitz, M. and Dreizin, E.L. 2020. Zirconium-boron reactive composite powders prepared by arrested reactive milling. J. Energ. Mater. 38: 142-161.

He, H., Li, L., Liu, Y., Kassymova, M., Li, D., Zhang, L., et al. 2021. Rapid room-temperature synthesis of a porphyrinic MOF for encapsulating metal nanoparticles. Nano Res. 14: 444-449.

Hitchman M.L., Kane, J. and Widmer, A.E. 1979. Polysilicon growth kinetics in a low pressure chemical vapour deposition reactor. Thin Solid Films 59: 231-247. https://doi.org/10.1016/0040-6090(79)90296-7.

Hong, Y.L., Liu, Z., Wang, L., Zhou, T., Ma, W., Xu, C., et al. 2020. Chemical vapor deposition of layered two-dimensional $MoSi_2N_4$ materials. Science 369: 670-674.

Huang, X., Gu, X., Zhang, H., Shen, G., Gong, S., Yang, B., et al. 2021. Decavanadate-based clusters as bifunctional catalysts for efficient treatment of carbon dioxide and simulant sulfur mustard. J. CO_2 Util. 45: 101419.

Ibrahim, S., Ahmad, Z., Manzoor, M.Z., Mujahid, M., Faheem, Z. and Adnan, A. 2021. Optimization for biogenic microbial synthesis of silver nanoparticles through response surface methodology, characterization, their antimicrobial, antioxidant, and catalytic potential. Sci. Rep. 11: 1-18.

Ikram, M., Raza, A., Imran, M., Ul-Hamid, A., Shahbaz, A. and Ali, S. 2020. Hydrothermal synthesis of silver decorated reduced graphene oxide (rGO) nanoflakes with effective photocatalytic activity for wastewater treatment. Nanoscale Res. Lett. 15: 1-11. doi.org/10.1186/s11671-020-03323-y.

Jain, P.K., Huang, X., El-Sayed, I.H. and El-Sayed, M.A. 2007. Review of some interesting surface plasmon resonance-enhanced properties of noble metal nanoparticles and their applications to biosystems. Plasmonics 2: 107-118. doi.org/10.1007/s11468-007-9031-1.

Jameel, M.S., Aziz, A.A. and Dheyab, M.A. 2020. Comparative analysis of platinum nanoparticles synthesized using sonochemical-assisted and conventional green methods. Nano-Struct. Nano-Objects 23: 100484.

Jamkhande, P.G., Ghule, N.W., Bamer, A.H. and Kalaskar, M.G. 2019. Metal nanoparticles synthesis: An overview on methods of preparation, advantages and disadvantages, and applications. J. Drug Deliv. Sci. Technol. 53: 101174.

Jana, N.R., Gearheart, L. and Murphy, C.J. 2001. Seeding growth for size control of 5-40 nm diameter gold nanoparticles. Langmuir 17: 6782-6786.

Jauris, I.M., Matos, C.F., Saucier, C., Lima, E.C., Zarbin, A.J.G., Fagan, S.B., et al. 2016. Adsorption of sodium diclofenac on graphene: A combined experimental and theoretical study. Phys. Chem. Chem. Phys. 18: 1526-1536. doi.org/10.1039/C5CP05940B.

Jeyaraj, M., Gurunathan, S., Qasim, M., Kang, M.H. and Kim, J.H. 2019. A comprehensive review on the synthesis, characterization, and biomedical application of platinum nanoparticles. Nanomaterials 9: 1719.

Karthikeyan, C., Arunachalam, P., Ramachandran, K., Al-Mayouf, A.M. and Karuppuchamy, S. 2020. Recent advances in semiconductor metal oxides with enhanced methods for solar photocatalytic applications. J. Alloys Compd. 828: 154281.

Khan, I., Saeed, K. and Khan, I. 2019. Nanoparticles: Properties, applications, and toxicities. Arab. J. Chem. 12: 908-931.

Khan, N.T. and Khan, M.J. 2020. Metallic nanoparticles fabrication methods: A brief overview. SunKrist Nanotechnol. Nanosci. J. 2: 1002.

Khandaker, T., Hossain, M.S., Dhar, P.K., Rahman, M., Hossain, M. and Ahmed, M.B. 2020. Efficacies of carbon-based absorbents for carbon dioxide capture. Processes 8: 654.

Kharisov, B.I. and Kharissova, O.V. 2019. Less-common carbon nanostructures. pp. 111-302. *In*: Carbon Allotropes: Metal-Complex Chemistry, Properties, and Applications. Springer, Cham.

Kurapov, Y.A., Litvin, S.E., Belyavina, N.N., Oranskaya, E.I., Romanenko, S.M. and Stelmakh, Y.A. 2021. Synthesis of pure (ligandless) titanium nanoparticles by EB-PVD method. J. Nanopart. Res. 23: 1-13.

Kwon, K., Lee, K.Y., Lee, Y.W., Kim, M., Heo, J., Ahn, S.J., et al. 2007. Controlled synthesis of icosahedral gold nanoparticles and their surface-enhanced Raman scattering property. J. Phys. Chem. C 111: 1161-1165.

Kyzas, G.Z., Deliyanni, E.A., Bikiaris, D.N. and Mitropoulos, A.C. 2018. Graphene composites as dye adsorbents. Chem. Eng. Res. Des. 129: 75-88. doi.org/10.1016/j.cherd.2017.11.006.

Lamb, L.D. and Huffman, D.R. 1993. Fullerene production. J. Phys. Chem. Solids 54: 1635-1643. doi.org/10.1016/0022-3697(93)90277-X.

Lee, K.X., Shameli, K., Yew, Y.P., Teow, S.Y., Jahangirian, H., Rafiee-Moghaddam, R., et al. 2020. Recent developments in the facile bio-synthesis of gold nanoparticles (AuNPs) and their biomedical applications. Int. J. Nanomedicine 15: 275.

Li, Y., Li, W., Liu, F., Li, M., Qi, X., Xue, M., et al. 2020. Construction of CeO_2/TiO_2 heterojunctions immobilized on activated carbon fiber and its synergetic effect between adsorption and photodegradation for toluene removal. J. Nanopart. Res. 22: 122. doi.org/10.1007/s11051-020-04860-4.

Li, Y., Xiao, Z., Wu, H., Zhong, H., Liu, Y., Zhao, G., et al. 2021. Controlled synthesis of uniform cup-stacked carbon nanotubes for energy applications. J. Alloys Comps. 865: 158912.

Lin, F., Liu, Y., Li, X. and Bai, C. 2021. The numerical and experimental investigation of particle size distribution produced by an electrical discharge process. Mater. 14: 287.
Lin, Z., Yue, J., Liang, L., Tang, B., Liu, B., Ren, L., et al. 2020. Rapid synthesis of metallic and alloy micro/nanoparticles by laser ablation towards water. Appl. Surf. Sci. 504: 144461.
Liu, M., Wang, Z., Zong, S., Chen, H., Zhu, D., Wu, L., et al. 2014. SERS detection and removal of mercury (II)/silver (I) using oligonucleotide-functionalized core/shell magnetic silica sphere@Au nanoparticles. ACS Appl. Mater. Interfaces 6: 7371-7379. doi.org/10.1021/am5006282.
Long, R.Q. and Yang, R.T. 2001. Carbon nanotubes as superior sorbent for dioxin removal. J. Am. Chem. Soc. 123: 2058-2059. doi.org/10.1021/ja003830l.
Loza, K., Heggen, M. and Epple, M. 2020. Synthesis, structure, properties, and applications of bimetallic nanoparticles of noble metals. Adv. Funct. Mater. 30: 1909260.
Lozada, I.B., Huang, B., Stilgenbauer, M., Beach, T., Qiu, Z., Zheng, Y., et al. 2020. Monofunctional platinum(II) anticancer complexes based on multidentate phenanthridine-containing ligand frameworks. Dalton T. 49: 6557-6560. doi.org/10.1039/D0DT01275K.
Lv, Y., Duan, S. and Wang, R. 2020. Structure design, controllable synthesis, and application of metal-semiconductor heterostructure nanoparticles. Prog. Nat. Sci. 30: 1-12.
Madima, N., Mishra, S.B., Inamuddin, I. and Mishra, A.K. 2020. Carbon-based nanomaterials for remediation of organic and inorganic pollutants from wastewater: A review. Environ. Chem. Lett. 18: 1169-1191.
Mahmoodi, N.M., Maroofi, S.M., Mazarji, M. and Nabi-Bidhendi, G. 2017. Preparation of modified reduced graphene oxide nanosheet with cationic surfactant and its dye adsorption ability from colored wastewater. J. Surfactants Deterg. 20: 1085-1093. doi.org/10.1007/s11743-017-1985-1.
Mamonova, D.V., Vasileva, A.A., Petrov, Y.V., Danilov, D.V., Kolesnikov, I.E., Kalinichev, A.A., et al. 2021. Laser-induced deposition of plasmonic Ag and Pt nanoparticles, and periodic arrays. Materials 14: 10.
Mancillas-Salas, S., Hernández-Rodrígueza, P., Reynosa-Martínez, A.C. and López-Honorato, E. 2020. Production of aluminum nanoparticles by wet mechanical milling. MRS Adv. 5: 3133-3140.
Mauter, M.S. and Elimelech, M. 2008. Environmental applications of carbon-based nanomaterials. Environ. Sci. Technol. 42: 5843-5859. doi.org/10.1021/es8006904.
Mazari, S.A., Ali, E., Abro, R., Khan, F.S.A., Ahmed, I., Ahmed, M., et al. 2021. Nanomaterials: Applications, waste-handling, environmental toxicities, and future challenges: A review. J. Environ. Chem. Eng. 9(2): 105028. https://doi.org/10.1016/j.jece.2021.105028.
Medhi, R., Marquez, M.D. and Lee, T.R. 2020. Visible-light-active doped metal oxide nanoparticles: Review of their synthesis, properties, and applications. ACS Appl. Nano Mater. 3: 6156-6185.
Minitha, C.R., Lalitha, M., Jeyachandran, Y.L., Senthilkumar, L. and Rajendra Kumar R.T. 2017. Adsorption behaviour of reduced graphene oxide towards cationic and anionic dyes: Co-action of electrostatic and π-π interactions. Mater. Chem. Phys. 194: 243-252. doi.org/10.1016/j.matchemphys.2017.03.048.
Mwafy, E.A. 2020. Eco-friendly approach for the synthesis of MWCNTs from waste tires via chemical vapor deposition. Environ. Nanotechnol. Monit. Manag. 14: 100342.
Méndez, E., Fagúndez, P., Sosa, P., Gutiérrez, M.V. and Botasini, S. 2020. Experimental evidences support the existence of an aggregation/disaggregation step in the Turkevich synthesis of gold nanoparticles. Nanotechnology 32: 045603.
Nadal, E., Milaniak, N., Glenat, H., Laroche, G. and Massines, F. 2021. A new approach for synthesizing plasmonic polymer nanocomposite thin films by combining a gold salt aerosol and an atmospheric pressure low-temperature plasma. Nanotechnology 32(17). https://doi.org/10.1088/1361-6528/abdd60.
Narasaiah, B.P. and Mandal, B.K. 2020. Remediation of azo-dyes based toxicity by agro-waste cotton boll peels mediated palladium nanoparticles. J. Saudi Chem. Soc. 24: 267-281. doi.org/10.1016/j.jscs.2019.11.003.
Nikam, A.V., Prasad, B.L.V. and Kulkarni, A.A. 2018. Wet chemical synthesis of metal oxide nanoparticles: A review. Cryst. Eng. Comm. 20: 5091-5107.
Ohara, Y., Akazawa, K., Shibata, K., Hirota, T., Kodama, Y., Amemiya, T., et al. 2020. Seed-mediated gold nanoparticle synthesis via photochemical reaction of benzoquinone. Colloids Surf. A Physicochem. Eng. Asp. 586: 124209.
Olad, A., Ghazjahaniyan, F. and Nosrati, R. 2018. A facile and green synthesis route for the production of silver nanoparticles in large scale. Int. J. Nanosci. Nanotechnol. 14: 289-296.
Özkan, E., Cop, P., Benfer, F., Hofmann, A., Votsmeier, M., Guerra, J.M., et al. 2020. Rational synthesis concept for cerium oxide nanoparticles: On the impact of particle size on the oxygen storage capacity. J. Phys. Chem. C 124: 8736-8748.

Pallela, P.N.V.K., Ummey, S., Ruddaraju, L.K., Gadi, S., Cherukuri, C.S.L., Barla, S., et al. 2019. Antibacterial efficacy of green synthesized α-Fe_2O_3 nanoparticles using Sida cordifolia plant extract. Heliyon 5: e02765. doi.org/10.1016/j.heliyon.2019.e02765.

Parashar, A., Sikarwar, S. and Jain, R. 2020a. Removal of pharmaceuticals from wastewater using magnetic iron oxide nanoparticles (IOPs). Int. J. Environan. Ch. 2020: 1-17. doi.org/10.1080/03067319.2020.17 16977.

Parashar, M., Shukla, V.K. and Singh, R. 2020b. Metal oxides nanoparticles via sol-gel method: A review on synthesis, characterization and applications. J. Mater. Sci.: Mater. 31: 3729-3749.

Park, J.H., Park, J., Lee, S.H. and Kim, S.M. 2020. Continuous synthesis of high-crystalline carbon nanotubes by controlling the configuration of the injection part in the floating catalyst chemical vapor deposition process. Carbon Lett. 30: 613-619.

Peng, L.-M., Zhang, Z. and Wang, S. 2014. Carbon nanotube electronics: Recent advances. Mater. Today 17: 433-442. doi.org/10.1016/j.mattod.2014.07.008.

Phan, T.T.V., Huynh, T.C., Manivasagan, P., Mondal, S. and Oh, J. 2020. An up-to-date review on biomedical applications of palladium nanoparticles. Nanomaterials 10: 66.

Prabhakar, A., Agrawal, M., Mishra, N., Roy, N., Jaiswar, A., Dhwaj, A., et al. 2020. Cost-effective smart microfluidic device with immobilized silver nanoparticles and embedded UV-light sources for synergistic water disinfection effects. RSC Adv. 10: 17479-17485. 10.1039/D0RA00076K.

Prasetya, O.D. and Khumaeni, A. 2018. Synthesis of colloidal platinum nanoparticles using pulse laser ablation method. In AIP Conference Proceedings 2014: 020050.

Priecel, P., Salami, H.A., Padilla, R.H., Zhong, Z. and Lopez-Sanchez, J.A. 2016. Anisotropic gold nanoparticles: Preparation and applications in catalysis. Chin. J. Catal. 37: 1619-1650.

Prola, L.D.T, Machado, F.M., Bergmann, C.P., de Souza, F.E., Gally, C.R., Lima, E.C., et al. 2013. Adsorption of Direct Blue 53 dye from aqueous solutions by multi-walled carbon nanotubes and activated carbon. J. Environ. Manage. 130: 166-175. doi.org/10.1016/j.jenvman.2013.09.003.

Qin, L., Zeng, G., Lai, C., Huang, D., Xu, P., Zhang, C., et al. 2018. "Gold rush" in modern science: Fabrication strategies and typical advanced applications of gold nanoparticles in sensing. Coord. Chem. Rev. 359: 1-31.

Quinson, J., Simonsen, S.B., Theil Kuhn, L. and Arenz, M. 2021. Commercial spirits for surfactant-free syntheses of electro-active platinum nanoparticles. Sustain. Chem. 2: 1-7.

Ravindra, C., Costas, A., Georgiou, P., Koklioti, M.A., Trompeta, A.-F. and Markakis, V. 2014. Manufacturing nanomaterials: From research to industry. Manuf. Rev. 1: 19. doi.org/10.1051/mfreview/2014009.

Ray, S.K., Majumder, C. and Saha, P. 2017. Functionalized reduced graphene oxide (fRGO) for removal of fulvic acid contaminant. RSC Adv. 7: 21768-21779. 10.1039/C7RA01069A.

Ren, X., Chen, C., Nagatsu, M. and Wang, X. 2011. Carbon nanotubes as adsorbents in environmental pollution management: A review. Chem. Eng. J. 170: 395-410. doi.org/10.1016/j.cej.2010.08.045.

Rodríguez-Fernández, J., Pérez-Juste, J., García de Abajo, F.J. and Liz-Marzán, L.M. 2006. Seeded growth of submicron Au colloids with quadrupole plasmon resonance modes. Langmuir 22: 7007-7010.

Rondelli, M., Zwaschka, G., Krause, M., Rötzer, M.D., Hedhili, M.N., Högerl, M.P., et al. 2017. Exploring the potential of different-sized supported subnanometer Pt clusters as catalysts for wet chemical applications. ACS Catal. 7: 4152-4162.

Roy, K., Mukherjee, A., Maddela, N.R., Chakraborty, S., Shen, B., Li, M., et al. 2020. Outlook on the bottleneck of carbon nanotube in desalination and membrane-based water treatment: A review. J. Environ. Chem. Eng. 8: 103572.

Rébé, C., Demontoux, L., Pilot, T. and Ghiringhelli, F. 2020. Platinum derivatives effects on anticancer immune response. Biomolecules 10: 13. doi.org/10.3390/biom10010013.

Sahai, K., Narayan, A. and Yadava, V. 2021. Micro-milling processes: A review. pp. 403-411. *In*: Ranganath, M., Mathiyazhagan, Singarikaliyan and Kumar, Harish [eds.]. Advances in Manufacturing and Industrial Engineering. https://doi.org/10.1007/978-981-15-8542-5_35.

Salem, S.S. and Fouda, A. 2020. Green synthesis of metallic nanoparticles and their prospective biotechnological applications: An overview. Biol. Trace Elem. Res. 199: 344-370.

Salvador, M., Gutiérrez, G., Noriega, S., Moyano, A., Blanco-López, M.C. and Matos, M. 2021. Microemulsion synthesis of superparamagnetic nanoparticles for bioapplications. Int. J. Mol. Sci. 22: 427.

Sanfelice, R.C., Pavinatto, A., Goncalves, V.C., Correa, D.S., Mattoso, L.H. and Balogh, D.T. 2016. Synthesis of a nanocomposite containing a water-soluble polythiophene derivative and gold nanoparticles. J. Polym. Sci., Part B: Polym. Phys. 54: 1245-1254.

Sangwan, S. and Seth, R. 2021. Synthesis, characterization and stability of gold nanoparticles (AuNPs) in different buffer systems. J. Clust. Sci. 1-16. https://doi.org/10.1007/s10876-020-01956-8.

Sau, T.K., Pal, A. and Pal, T. 2001. Size regime dependent catalysis by gold nanoparticles for the reduction of eosin. J. Phys. Chem. B 105: 9266-9272. doi.org/10.1021/jp011420t.

Schmarsow, R.N., dell'Erba, I.E., Villaola, M.S., Hoppe, C.E., Zucchi, I.A. and Schroeder, W.F. 2020. Effect of light intensity on the aggregation behavior of primary particles during *in situ* photochemical synthesis of gold/polymer nanocomposites. Langmuir 36: 13759-13768.

Selvaraj, M., Hai, A., Banat, F. and Haija, M.A. 2020. Application and prospects of carbon nanostructured materials in water treatment: A review. J. Water Process. Eng. 33: 100996.

Shameli, K., Ahmad, M.B., Yunus, W.Z.W., Ibrahim, N.A. and Darroudi, M. 2010. Synthesis and characterization of silver/talc nanocomposites using the wet chemical reduction method. Int. J. Nanomedicine 5: 743.

Sharma, M., Joshi, M., Nigam, S., Avasthi, D.K., Adelung, R., Srivastava, S.K., et al. 2019. Efficient oil removal from wastewater based on polymer coated superhydrophobic tetrapodal magnetic nanocomposite adsorbent. Appl. Mater. Today 17: 130-141. doi.org/10.1016/j.apmt.2019.07.007.

Siddique, J.A. and Numan, A. 2021. Perspective future development of nanomaterials. pp. 319-343. *In*: Mubarak, Nabisab Mujawar, Khalid, Mohammad, Walvekar, Rashmi and Numan, Arshid. Contemporary Nanomaterials in Material Engineering Applications. Springer, Cham.

Song, N., Gao, X., Ma, Z., Wang, X., Wei, Y. and Gao, C. 2018. A review of graphene-based separation membrane: Materials, characteristics, preparation and applications. Desalin. 437: 59-72. doi.org/10.1016/j.desal.2018.02.024.

Song, X., Zhou, L., Zhang, Y., Chen, P. and Yang, Z. 2019. A novel cactus-like Fe_3O_4/Halloysite nanocomposite for arsenite and arsenate removal from water. J. Clean. Prod. 224: 573-582. doi.org/10.1016/j.jclepro.2019.03.230.

Su, C., Huang, K., Li, H.H., Lu, Y.G. and Zheng, D.L. 2020. Antibacterial properties of functionalized gold nanoparticles and their application in oral biology. J. Nanomater. 2020: 1-13.

Takehara, H., Fujiwara, M., Arikawa, M., Diener, M.D. and Alford, J.M. 1993. Experimental study of industrial scale fullerene production by combustion synthesis. Carbon 43: 311-319. doi.org/10.1016/j.carbon.2004.09.017.

Tang, X., Zhou, L., Le, Z., Wang, Y., Liu, Z., Huang, G., et al. 2020. Preparation of porous chitosan/carboxylated carbon nanotube composite aerogels for the efficient removal of uranium (VI) from aqueous solution. Int. J. Biol. Macromol. doi.org/10.1016/j.ijbiomac.2020.05.179.

Teo, K.B.K, Singh, C., Chhowalla, M. and Milne, W.I. 2003. Catalytic synthesis of carbon nanotubes and nanofibers. ENN 10: 1-22.

Thines, R.K., Mubarak, N.M., Nizamuddin, S., Sahu, J.N., Abdullah, E.C. and Ganesan, P. 2017. Application potential of carbon nanomaterials in water and wastewater treatment: A review. J. Taiwan Inst. Chem. Eng. 72: 116-133. doi.org/10.1016/j.jtice.2017.01.018.

Tiyyagura, H.R., Majerič, P., Anžel, I. and Rudolf, R. 2020. Low-cost synthesis of AuNPs through ultrasonic spray pyrolysis. Mater. Res. Express 7: 055017.

Trzciński, J.W., Panariello, L., Besenhard, M.O., Yang, Y., Gavriilidis, A. and Guldin, S. 2020. Synthetic guidelines for the precision engineering of gold nanoparticles. Curr. Opin. Chem. 29: 59-66.

Tsyganov, D., Bundaleska, N., Dias, A., Henriques, J., Felizardo, E., Abrashev, M., et al. 2020. Microwave plasma-based direct synthesis of free-standing N-graphene. Phys. Chem. Chem. Phys. 22: 4772-4787.

Valluri, S.K., Schoenitz, M. and Dreizin, E. 2020. Preparation and characterization of silicon-metal fluoride reactive composites. Nanomaterials 10: 2367.

Van Tran, T., Nguyen, H., Le, P.H.A., Nguyen, D.T.C., Nguyen, T.T., Van Nguyen, C., et al. 2020. Microwave-assisted solvothermal fabrication of hybrid zeolitic–imidazolate framework (ZIF-8) for optimizing dyes adsorption efficiency using response surface methodology. J. Environ. Chem. Eng. 8: 104189.

Wainwright, E.R. and Weihs, T.P. 2020. Microstructure and ignition mechanisms of reactive aluminum–zirconium ball milled composite metal powders as a function of particle size. J. Mater. Sci. 55: 14243-14263.

Wang, H., Hu, X., Guo, Y., Qiu, C., Long, S., Hao, D., et al. 2018. Removal of copper ions by few-layered graphene oxide nanosheets from aqueous solutions: External influences and adsorption mechanisms. J. Chem. Technol. Biotechnol. 93: 2447-2455. doi.org/10.1002/jctb.5601.

Wang, J. 2005. Carbon-nanotube based electrochemical biosensors: A review. Electronalysis (N.Y.N.Y.) 17: 7-14. doi.org/10.1002/elan.200403113.

Wang, L., Cheng, C., Tapas, S., Lei, J., Matsuoka, M., Zhang, J., et al. 2015. Carbon dots modified mesoporous organosilica as an adsorbent for the removal of 2, 4-dichlorophenol and heavy metal ions. J. Mater. Chem. A 3: 13357-13364. doi.org/10.1039/C5TA01652E.

Wang, L., Xu, H., Qiu, Y., Liu, X., Huang, W., Yan, N., et al. 2019. Utilization of Ag nanoparticles anchored in covalent organic frameworks for mercury removal from acidic waste water. J. Hazard. Mater. 2019: 121824. doi.org/10.1016/j.jhazmat.2019.121824.

Wawrzkiewicz,M.,Wiśniewska,M.,Wołowicz,A.,Gun'ko,V.M.andZarko,V.I.2017.Mixedsilica-aluminaoxideas sorbentfordyesandmetalionsremovalfromaqueoussolutionsandwastewaters.Micropor.Mesopor.Mat.250: 128-147. doi.org/10.1016/j.micromeso.2017.05.016.

Wiesner, J. and Wokaun, A. 1989. Anisometric gold colloids. Preparation, characterization, and optical properties. Chem. Phys. Lett. 157: 569-575.

Wu, C., Zheng, D., Wang, X., Zhao, D., Wang, X., Pei, W., et al. 2021. Effects of high magnetic field on the growth and magnetic property of L10-FePtCu nanoparticles. J. Magn. Magn. 526: 67731.

Wu, Q., Miao, W.S., Zhang, Y.D., Gao, H.J. and Hui, D. 2020. Mechanical properties of nanomaterials: A review. Nanotechnol. Rev. 9: 259-273.

Yamamoto, M. and Nakamoto, M. 2003. Novel preparation of monodispersed silver nanoparticles via amine adducts derived from insoluble silver myristate in tertiary alkylamine. J. Mater. Chem. 13: 2064-2065. doi.org/10.1039/B307092A.

Yang, L., Chen, C., Hu, Y., Wei, F., Cui, J., Zhao, Y., et al. 2020. Three-dimensional bacterial cellulose/polydopamine/TiO_2 nanocomposite membrane with enhanced adsorption and photocatalytic degradation for dyes under ultraviolet-visible irradiation. J. Colloid Interface Sci. 562: 21-28. doi.org/10.1016/j.jcis.2019.12.013.

Yu, Y., Yang, T. and Sun, T. 2020a. New insights into the synthesis, toxicity and applications of gold nanoparticles in CT imaging and treatment of cancer. Nanomedicine 15: 1127-1145.

Yu, Z., Hu, C., Dichiara, A.B., Jiang, W. and Gu, J. 2020b. Cellulose Nanofibril/Carbon Nanomaterial hybrid aerogels for adsorption removal of cationic and anionic organic dyes. Nanomater. 10: 169. doi.org/10.3390/nano10010169.

Yuan, M.M., Zou, J., Huang, Z.N., Peng, D.M. and Yu, J.G. 2020. PtNPs-GNPs-MWCNTs-β-CD nanocomposite modified glassy carbon electrode for sensitive electrochemical detection of folic acid. Anal. Bioanal. Chem. 412: 2551-2564.

Zadeh, Z.E., Solouk, A., Shafieian, M. and Nazarpak, M.H. 2021. Electrospun polyurethane/carbon nanotube composites with different amounts of carbon nanotubes and almost the same fiber diameter for biomedical applications. Mater. Sci. Eng. C 118: 111403.

Zeng, M., Chen, M., Huang, D., Lei, S., Zhang, X., Wang, L., et al. 2021. Engineered two-dimensional nanomaterials: An emerging paradigm for water purification and monitoring. Mater. Horiz. https://doi.org/10.1039/D0MH01358G.

Zhang, S., Wang, X. and Guo, Z. 2019. Rational design of anticancer platinum (IV) prodrugs. Adv. Inorg. Chem. 75: 149-182. Academic Press. doi.org/10.1016/bs.adioch.2019.10.009.

Zhang, X., Gao, B., Creamer, A.E., Cao, C. and Li, Y. 2017. Adsorption of VOCs onto engineered carbon materials: A review. J. Hazard. Mater. 338: 102-123. doi.org/10.1016/j.jhazmat.2017.05.013.

Zheng, K. and Branicio, P.S. 2020. Synthesis of metallic glass nanoparticles by inert gas condensation. Phys. Rev. Mater. 4: 076001.

Zhou, J., Zhou, X., Yang, K., Cao, Z., Wang, Z., Zhou, C., et al. 2020. Adsorption behavior and mechanism of arsenic on mesoporous silica modified by iron-manganese binary oxide (FeMnOx/SBA-15) from aqueous systems. J. Hazard. Mater. 384: 121229. doi.org/10.1016/j.jhazmat.2019.121229.

Alumina-Based One-Dimensional (1D) Nanomaterials: Synthesis and Environmental Remediation

Abhipsa Mahapatra, Manamohan Tripathy and Garudadhwaj Hota*

1. INTRODUCTION

Currently, nanomaterials are at the cutting edge of material science research and have several applications in our daily lives, which include energy, environmental, and biomedical applications (J. Xu et al. 2018). Over the last few decades, one-dimensional (1D) materials in nano-dimension, like nanorods, nanowhiskers, nanowires, nanotubes, and nanofibers, have been used for several applications. Carbon nanotubes are a common example of these high aspect ratios of one-dimensional nanomaterials. However, it is observed that 1D metal oxide nanostructures are efficient electron transport and also show optical excitation properties. These 1D nanostructures show outstanding structural flexibility due to which it inherits the fascinating properties from their bulk counterpart. Various synthetic techniques have been used to prepare 1D nanostructures, such as hydrothermal or solvothermal, self-assembly, sol-gel, chemical vapor deposition (CVD), template-directive method, electrospinning, and a few more. Among all the techniques hydrothermal, electrospinning, and sol-gel techniques are more common. So we mostly emphasize giving comprehensive information about these methods. The hydrothermal technique is more convenient to control particle size and morphological characteristics of the nanostructure via controlling the experimental conditions (Ángel et al. 2017). Also, the sol-gel method is a simple and economic technique used for synthesizing 1D nanostructure. Apart from this, electrospinning is found to be a simple, versatile, most reliable, cost-effective, and industry-viable technique for the fabrication of a 1D nanofiber matrix. Continuous nanofibers with uniform surface morphology in a specified longitudinal direction can be fabricated by the electrospinning method. It is a non-equilibrium electro-hydrodynamic technique that can be used for the fabrication of inorganic and ceramic nanofiber materials (Liu et al. 2019). Due to greater porosity and diameter in nano-range, nanofibers possess a greater specific surface area compared to other 1D nanostructures. On account of its specific surface area and porosity, it helps for the promotion of an increase in adsorption capacity and also shows good catalytic activity. Figure 1 displays the SEM images of PVP-aluminium acetate composite fibers and calcined fibers at 1,000°C

Department of Chemistry, NIT Rourkela, Odisha, India 769008.

* Corresponding author: garud@nitrkl.ac.in, garud31@yahoo.com.

(Mahapatra et al. 2011). After calcination, these material shows crystalline properties as confirmed from the XRD analysis, which enhances their stability under normal environmental circumstances. Another prominent property of 1D nanomaterials is their high aspect ratio and porosity, which enhances the potential of adsorbents for the removal of pollutants, such as heavy metals, organic pollutants (Awad et al. 2020, Saxena et al. 2020, Vidal and Moraes 2019), inorganic anions (Bagheri et al. 2020), and other organisms.

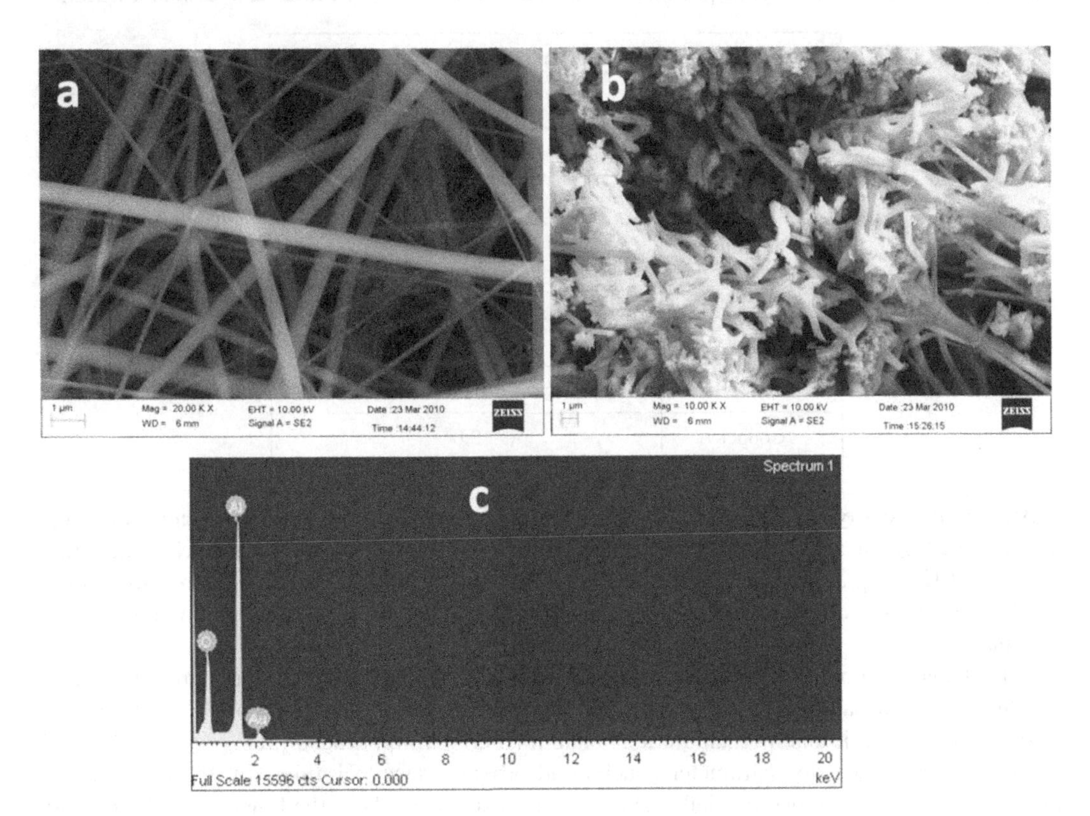

FIGURE 1 SEM images of (a) PVP-aluminium acetate composite, (b) calcined at 1,000°C, and (c) EDAX of calcined alumina (Mahapatra et al. 2011).

Rapidly increasing water contamination has turned into a major threat globally. Pollutants in wastewater must be removed before water discharge or recycling. Adsorption is considered to be superior to other conventional techniques because of its low cost, flexibility, simple design, ease of operation, and insensitivity to toxic pollutants (Peng et al. 2015). So, it allows high removal percentage with minimum secondary sludge generation. Nanomaterials possess greater reusable characteristics, so they can be effectively used for industrial wastewater and potable water decontamination (Wadhawan et al. 2020). It is useful for physical, chemical, and biological processes, and the inversion process of adsorption is desorption that represents the transfer of adsorbate ions from the adsorbent surface to the solution. For the past few years, 1D nanomaterials are being explored in water treatment applications owing to their beneficial properties. According to World Health Organization (WHO), the maximum tolerable concentration of different cations and anions lies way below than to what is present in our regular drinking water (WHO/UNICEF 2017). Table 1 demonstrates the permissible limit of inorganic pollutants present in the natural aqueous system as per the WHO guideline. To meet the growing demands for clean water, it is essential to search for an effective material and process for wastewater treatment (Yu et al. 2017).

TABLE 1 WHO recommended permissible limit of few inorganic pollutants in drinking water.

Pollutants	Permissible Limit (mgL^{-1})
Arsenic	0.01
Fluoride	1.5
Lead	0.01
Chromium	0.05
Cadmium	0.003
Mercury	0.006
Nickel	0.07
Copper	2.0

Among different types of 1D nanoadsorbents alumina-based 1D materials in nano-dimension have attracted significant research attention due to their exceptional physicochemical characteristics (Li et al. 2012, Peng et al. 2015). Therefore, here we have focused to gives a comprehensive idea about the synthesis, characterization, and adsorptive applications of various alumina-based 1D nanomaterials.

This chapter summarizes the basic synthetic strategy required for the fabrication of various 1D alumina-based nano-architecture and their effectiveness toward decontamination of various persistent toxic organic and inorganic water contaminants by a simple adsorption process. Also, the effect of various adsorption parameters, such as adsorbent dosage, pH, time, initial concentration, and temperature, are discussed briefly. The comprehensive idea about the kinetics, isotherm, and thermodynamics investigation of the adsorption processes are also elucidated. Finally, the conclusion and prospects for the development of advanced 1D alumina-based nanomaterials are also presented. The structure of some 1D alumina-based NMs can be placed in the introduction.

2. SYNTHESIS STRATEGIES OF 1D ALUMINA-BASED NANOADSORBENTS

Alumina is a ceramic material with several crystalline phases. Among them α, β, and γ is the most important and are stable phases. Figure 2 shows the comparison between three alumina fiber samples (γ-Al_2O_3) using different polymer precursors calcined at 900°C. Usually, alumina shows sharp peaks of crystallinity after sintering at around 1,200°C or more. Alumina has many unique properties that show varied types of applications in different areas. It is used for the synthesis of cutting tools, support materials for catalysis, reinforcing components, electronic devices, and many more (Roque-Ruiz et al. 2019a). The 1D alumina-based nanoadsorbents can be synthesized by various synthetic techniques. The key techniques most widely used include sol-gel, hydrothermal/solvothermal, and electrospinning. Here, we discuss the details of these techniques.

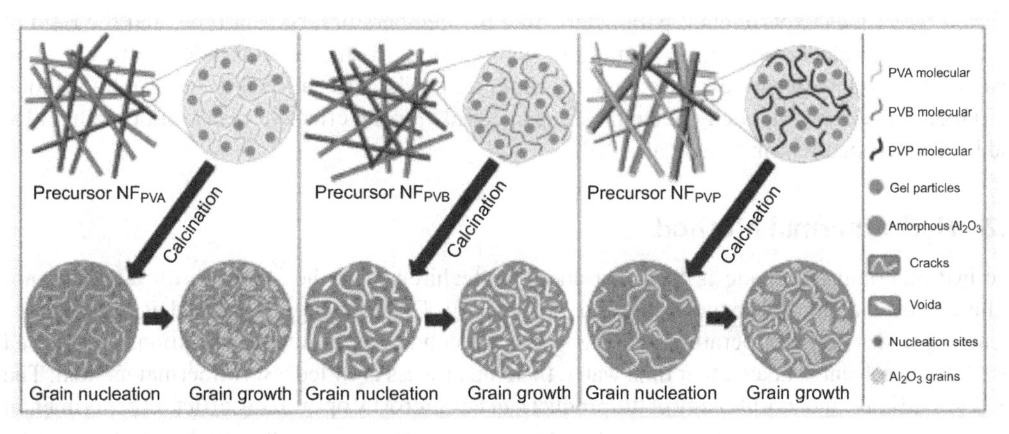

FIGURE 2 The morphology, microstructure, and mechanical strength of γ-Al_2O_3 nanofibers prepared using three different polymers after calcination at 900°C (Song et al. 2020).

2.1 Sol-Gel Method

For the fabrication of various types of nano-architecture in simple and cost-effective ways, the sol-gel technique is the most appropriate and convenient colloidal technique. This process involves the chemical transformation of a liquid (i.e., sol) into a gel state and with subsequent post-treatment and transformation into solid oxide materials. In a typical synthetic procedure, a colloidal suspension (i.e., sol) is obtained from the precursors due to its hydrolysis and poly-condensation process. The precursor materials are mostly inorganic metal salts or metal oxides. Figure 3 shows the flow diagram of sol-gel mediated synthesis. The key advantages of this process are to obtain high purity and uniform nanostructure at low-temperature conditions.

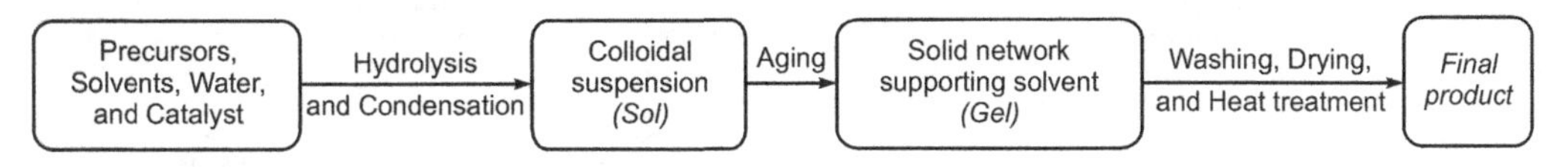

FIGURE 3 Representation of 1D Nanomaterials synthesis using the Sol-Gel method.

Ruiz et al. fabricated dense alumina fibers by sol-gel and electrospinning method using aluminium nitrate. In this study, the use of two polymers, i.e. PVP and Pluronic 127 is done. Formation of stable fibers is achieved after sintering at 800°C for γ & δ-Al_2O_3 and 1,000°C for α-Al_2O_3 (Roque-Ruiz et al. 2019b).

Different types of 1D alumina-based nanocomposite were synthesized by Leon et al. using the sol-gel process. Here, they used 'Pluronics' [(EO)n(PO)m(EO)n], a polymer material, as a structural controlling group. Three forms of Pluronics were used having the average molecular weight of 2,900, 5,800, and 8,350 $gmol^{-1}$ with code of L64, P123, and F68, respectively. In a brief synthetic procedure: aluminum-tri-sec butoxide (10 mL) and Pluronics (1 g) were dissolved with 2-propanol (166 mL) under vigorous stirring at 273 K for 2 hours. After that, the solution obtained by mixing H_2O (11.5 mL) and HNO_3 (0.57 mL, 70%) was added drop by drop while stirring. The obtained sol was kept undisturbed (for 24 hours at 273 K) to form the gel. Then, to remove the solvent the obtained gel was integrated into a crystallizer for 48 hours at 343 K. Afterward, the obtained solids product was calcined (for 4 hours at 773 K) to obtain the final product (De León et al. 2014).

Teoh et al. also synthesized alumina nanofibers with an enhanced aspect ratio (>300 m^2g^{-1}) in bulk quantities by sol-gel techniques. The uniformity and formation of the nanofibers were mostly affected by various experimental parameters, like pH, temperature, solvent type, and the ratio of the binary water-alcohol system to aluminium isopropoxide (Teoh et al. 2007). Recently, Qin et al. developed sol-gel mediated reverse micelles techniques for the fabrication of γ-Al_2O_3 ultrafiltration (UF) membrane. They applied the synthesized materials for the effective rejection of methyl blue and bovine serum albumin (Qin et al. 2019).

2.2 Hydrothermal Method

The hydrothermal technique is the most adequate technique for the synthesis of various nano-architectures with control shape, sizes, and morphology. This technique is defined as the reactions occurring under high temperature and high-pressure conditions in aqueous solutions in a closed vessel. If the solvent is taken other than water, then the process is called a solvothermal method. The size, crystallinity, and morphology of the product can be control by optimizing different experimental parameters, like experimental time, reaction temperature, pressure, precursor composition, nature of solvents, etc.

Recently, Hu et al. have synthesized the alumina nanorods by the hydrothermal method. At first, the metal precursors (i.e., $Al(NO_3)_3 \cdot 9H_2O$), urea, and polyethylene glycol (PEG) were mixed in a 1: 1: 2.5 ratio with continuous stirring (60 minutes). Then the hydrothermal reaction was conducted in a stainless steel autoclave. The obtained product was washed properly with ethanol and water and subsequently dried at 80°C for 24 hours. Then the obtained powders were ground properly and the subsequent high-temperature treatment results in the alumina nanorods. The detailed synthetic procedure is presented in Figure 4. To expand the ionic conductivity of the obtained materials it was applied over the poly (propylene carbonate) based solid electrolyte (Hu et al. 2020).

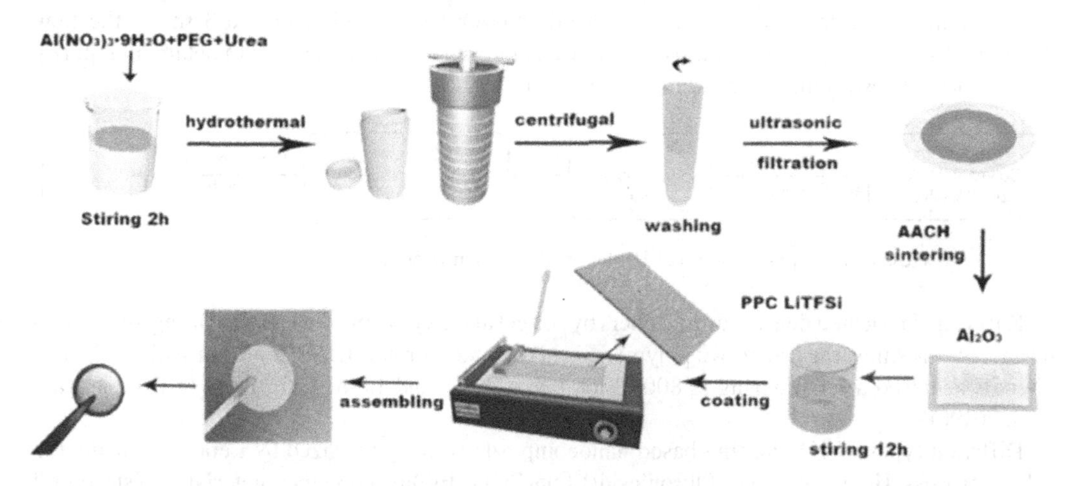

FIGURE 4 Representation of the hydrothermal synthesis of Al_2O_3 nanorods and the composite solid electrolyte membrane (Hu et al. 2020).

Yuan et al. also synthesized γ-Al_2O_3 nanorods by hydrothermal method and subsequent calcination. The TEM images of the γ-Al_2O_3 nanorods are presented in Figure 5 (Yuan et al. 2015). Selim et al. also adopted a solvothermal approach for the synthesis of single-crystal γ-Al_2O_3 nanorods with 10 nm diameter by using $AlCl_3 \cdot 6H_2O$ as the metal precursors (Selim et al. 2020). To obtain the desired morphology, they have investigated the effect of experimental parameters, like temperature, pH, reaction time, and the amount of surfactant. After hydrothermal reaction of precursors materials at 200°C for 24 hours, γ-AlOOH nanorods were obtained which on calcination at 500°C for 3 hours gives γ-Al_2O_3 nanorods.

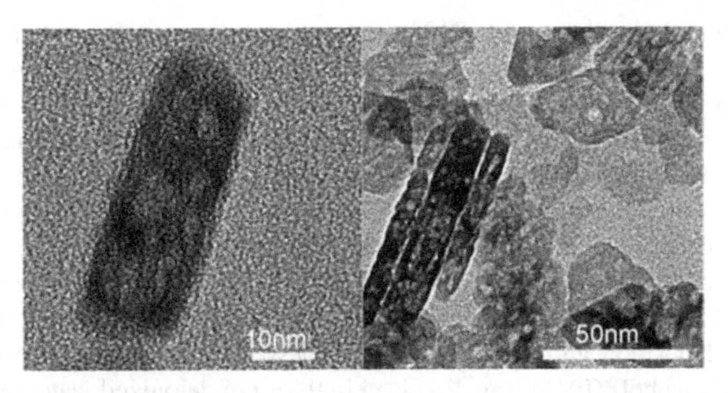

FIGURE 5 TEM images of the γ-Al_2O_3 nanorods prepared by hydrothermal method (Yuan et al. 2015).

Li et al. synthesized different morphology of boehmite (AlOOH) using the hydrothermal process (Li et al. 2019). The morphology of the obtained materials mostly affected by the pH of the precursors' solution and changes to 1D rods or 2D sheets of different sizes. The obtained materials were applied for the adsorption of Cr(VI) and nitrates from the water system. The details about the adsorption process are discussed later.

2.3 Electrospinning Method

Electrospinning is a promising synthetic approach for the synthesis of a 1D nanofibers matrix on a commercial scale. The basic instrumental components required for this method are a high voltage DC power supply, a syringe, a syringe pump capable to supply microliters of a solution, a metallic needle, and a metal collector. The general principle of this technique includes the production of the nonwoven mesh of micro or nanofibers by applying a high voltage electric power supply between the metallic needle and ground metallic collector (Pereao et al. 2017, Wang et al. 2020). The detailed instrumental setup for this technique is presented in Figure 6.

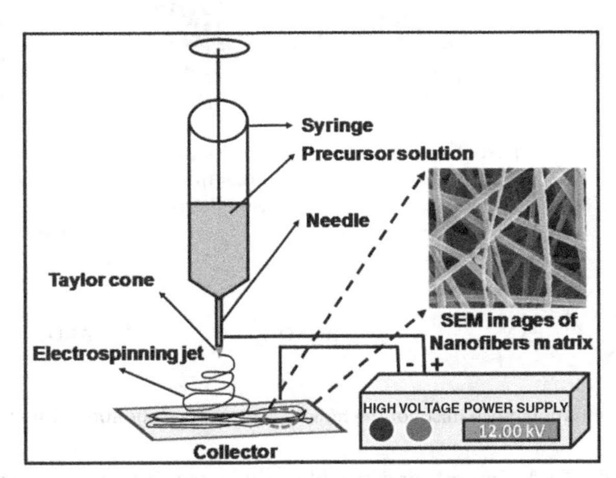

FIGURE 6 Electrospinning setup for the synthesis of 1D nanofibers mats.

Recently Ruiz et al. have synthesized alumina nanofibers by sol-gel and subsequent electrospinning techniques by using aluminum nitrate nonahydrate as precursors. They also used polyvinyl pyrrolidone (PVP) as a spinning polymer, Pluronic 127 as an additive, and distilled water and ethanol as solvents (Roque-Ruiz et al. 2019c). Peng et al. have also synthesized γ-Al_2O_3 nanofibers by electrospinning process. Here, they also used aluminum nitrate nonahydrate as precursor material and Polyacrylonitrile (PAN) as the polymer source (Peng et al. 2015). Mahapatra et al. have synthesized alumina nanofibers by electrospinning the PVP/aluminum acetate composite solution followed by heat treatment at 1,000°C. The diameters of the obtained fibers were 100 to 500 nm. They have also reported on the preparation of Fe_2O_3-Al_2O_3 composite nanofibers using the electrospinning method. They have first prepared AlOOH (Boehmite) nanoparticles using the sol-gel method and then impregnated them with PVP-iron acetylacetonate composite solution followed by electrospinning. Then, Fe_2O_3-Al_2O_3 composite nanofibers were obtained by sintering the as spun boehmite-PVP-iron acetylacetonate nanofibers matrix at 1,000°C (Mahapatra et al. 2013a, 2013b).

Similarly, Li et al. have synthesized core-shell nanofibers of α-Fe_2O_3-γ-Al_2O_3 by electrospinning followed by vapor deposition and heat treatment process. In a brief synthetic procedure at first 0.65 g of Poly (vinyl alcohol) (PVA) was dissolved in 9.35 g of deionized water and heated at 80°C for 2 hours under stirring. After cooling to room temperature about 0.385 g of ammonium ferric citrate was added and stirred for a further 3 hours. Afterward, in a glass syringe, the prepared solution was taken and the electrospinning process was carried out by applying an 18 kV DC power supply in between the metallic needle and grounded metallic collector. The grounded collector and the metallic needle were fixed at a distance of 15 cm. Then to the obtain ammonium ferric citrate/PVA composite, nanofibers aluminum vapor was deposited. After that, the ammonium ferric citrate/PVP-Al composite nanofibers were calcined at 600 and 800°C for 2 hours to obtain the core-shell α-Fe_2O_3-γ-Al_2O_3 composite (Li et al. 2014).

Also, Liu et al. fabricated hollow and porous alumina nanofibers by electrospinning techniques using a single-spinneret system (Liu et al. 2013). They used polyacrylonitrile (PAN) as the polymer precursors and aluminum nitrate ($Al(NO_3)_3$) as the metal precursors. After the formation of the homogeneous precursors' solution, the electrospinning process was conducted and then the spinning fibers were dried in a hot air oven for few hours at 60°C. Then the fibers were calcined at 500°C, 1,000°C, and 1,300°C with a heating rate of 5°C/minutes. The formation mechanism of the obtained hollow fibers was based on the 'Kirkendall effect' as depicted in Figure 7.

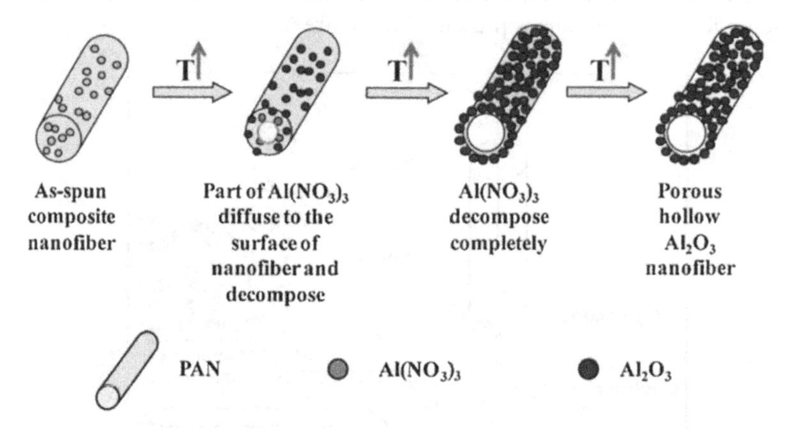

FIGURE 7 Formation mechanism of the hollow and porous nanofibers (Liu et al. 2013).

Figure 8 indicates the SEM images and the fiber diameter distribution curve of the obtained nanofibers calcined at different temperatures. It is evident that the obtained fibers are continuous and randomly oriented. With the increase in calcination temperature the fiber diameter decreases, which is mostly due to the decomposition of the precursor materials at the higher temperature. From the overall study, it is clear that the calcination temperature plays the crucial role to control the crystallinity, microstructure, and the diameter of the alumina nanofibers.

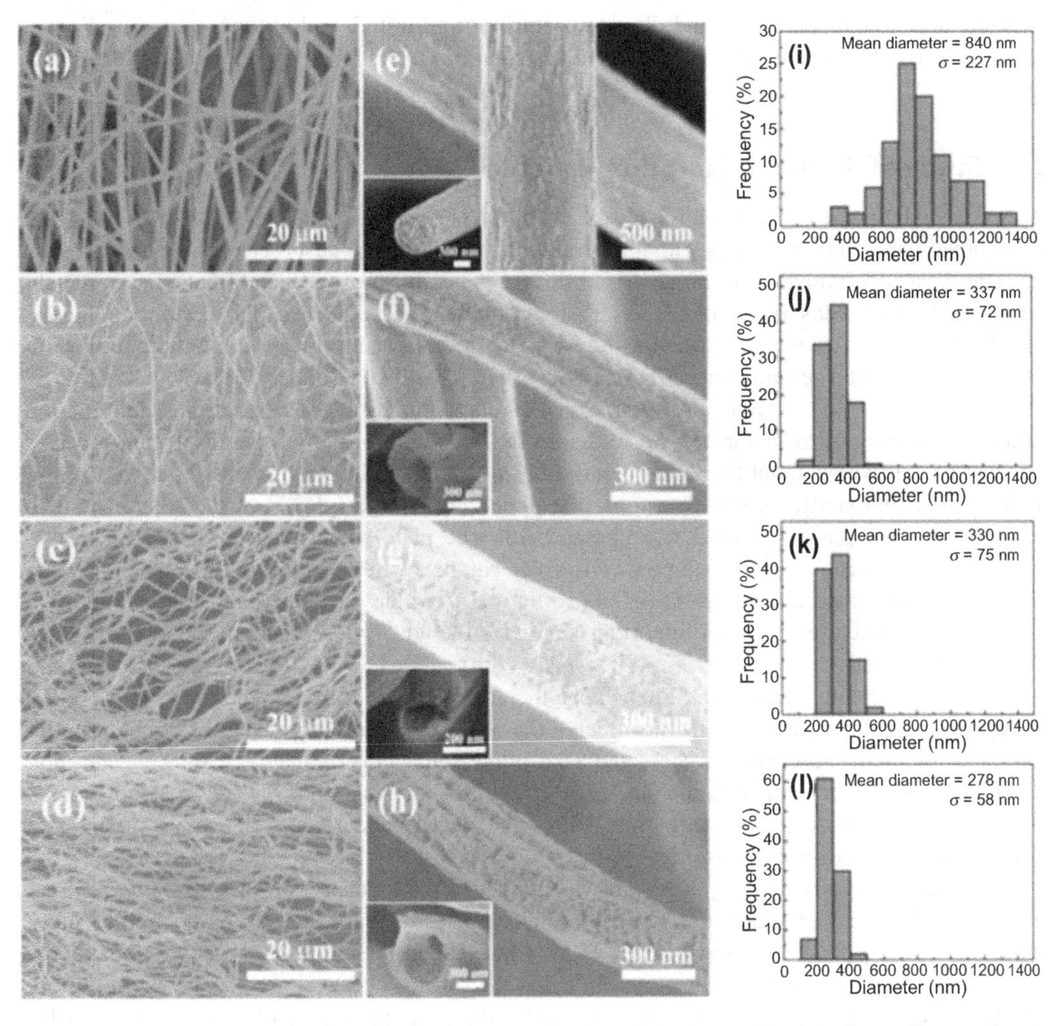

FIGURE 8 SEM micrograph of (a and e) $Al(NO_3)_3$/PAN composite; SEM micrograph calcined nanofibers at (b and f) 500°C, (c and g) 1,000°C, and (d and h) 1,300°C; diameter distribution curve of (i) $Al(NO_3)_3$/PAN composite; diameter distribution curve calcined nanofibers at (j) 500°C, (k) 1,000°C, and (l) 1,300°C (Liu et al. 2013).

3. ADSORPTION OF INORGANIC AND ORGANIC CONTAMINANTS BY 1D ALUMINA-BASED NANOMATERIALS

Mostly the batch adsorption experimental studies for removal of various toxic inorganic and organic pollutants were done for the evaluation of adsorption efficiency of different alumina-based nanomaterials. The overall effect of every parameter, like adsorbent amount, pH, temperature, initial concentration, and contact time, etc., is discussed here briefly. The percentage of removal (R%) of adsorbate and equilibrium adsorption capability (q_e, in mg g^{-1}) can be evaluated by the following two equations (1 and 2) (Tripathy et al. 2020, Tripathy and Hota 2020).

$$R\% = \left(\frac{C_o - C_t}{C_o}\right)100 \tag{1}$$

$$q_e = \frac{(C_o - C_e)V}{m} \tag{2}$$

In the above equation C_e, C_o and C_t are the equilibrium, initial, and at time t concentration of adsorbate (mgL^{-1}), V is the volume of the aqueous solution (L), and m is the mass of the adsorbents (g).

3.1 Study of Several Adsorption Variables

From the study of the adsorbent dosage experiment of various reported literature, it is found that the removal percentage increases with an increase in the amount of adsorbent dosage and remains constant after a certain amount of adsorbent dosage. The above experimental outcome is mostly due to the complete adsorption of the adsorbate on the adsorbent surface. The certain amount of adsorbent dosage after which the removal percentage remains constant is called the equilibrium adsorbent dosage at a certain concentration of adsorbate (Vidal and Moraes 2019).

The effect of solution pH is an essential criterion for the adsorption of cationic and anionic pollutants on the adsorbent surface. So to determine the effect of pH on the adsorption process, we most have to know about the Zeta-potential of the adsorbent by Zeta-potential analysis. From the Zeta-potential study, we can determine the point zero charge (pH_{PZC}) of the adsorbent. The adsorbent surface possesses a positive charge below the pH_{PZC} value and possesses a negative charge above the pH_{PZC} value. So, anionic adsorbate can be adsorbed at pH below the pH_{PZC} value and cationic adsorbate can adsorb at pH above the pH_{PZC} value by the strong electrostatic force of attraction. Therefore, pH is an important factor for the adsorption process (Sahoo et al. 2019, Tripathy et al. 2020).

Also, the minimum time required to achieve the adsorption equilibrium and the kinetics of the adsorption process is determined by the effect of the time experiment. However, the effect of initial concentration on the adsorption process and adsorption isotherm is evaluated by the concentration effect experiment. The temperature effect study gives prior information about the thermodynamics of the adsorption process (Sahoo et al. 2019). The details about the kinetics, isotherm, and thermodynamics study are discussed later.

The boehmite synthesized by Li et al. was applied for the adsorption of Cr(VI) and nitrate (Li et al. 2019). They proposed the adsorption mechanism by characterized the boehmite before and after the adsorption process by SEM, TEM, EDX, IR, Raman, and XPS analytical methods. From the characterization result, they conclude that the surface hydroxyl groups play an important for the adsorption process as presented in Figure 9. Also, the appearance of Cr signal in the XPS spectra and its fitting data confirm the adsorption and reduction of Cr(VI). The Cr $2p_{1/2}$ spectra having two peaks at 586.9 and 588.4 eV ascribed to the Cr(III) and Cr(VI) species. Also, the peaks at 577.2 and 579.2 eV for Cr $2p_{3/2}$ specifies to Cr(III) and Cr(VI). Finally, from their overall study, it is evident that the morphology of the adsorbent plays a significant role in the efficient adsorption of contaminants from the polluted water.

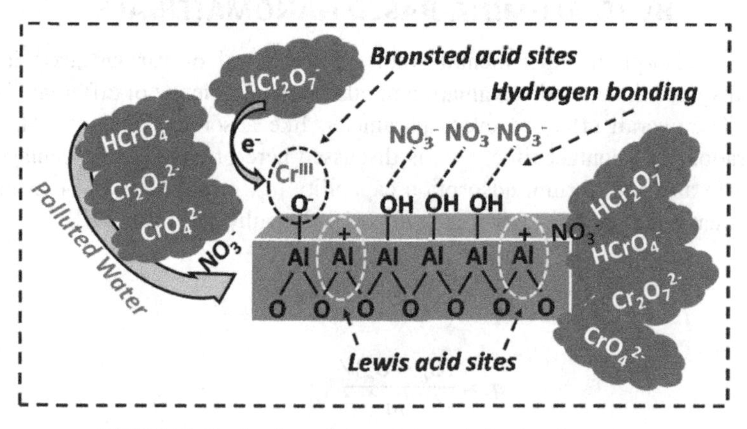

FIGURE 9 Mechanism of chromium and nitrate adsorption.

Recently, Prathna et al. have worked on iron oxide-alumina nanocomposites possessing sorption capacity toward both arsenic and fluoride. The sorption capacity of iron oxide nanoparticles was 1.47 mg g^{-1} whereas nanocomposite was 4.82 mg g^{-1} (TC et al. 2018). Compared to traditional organic pollutants, heavy metal ions are difficult to degrade into cleaning products, which can be accumulated in the living organism. Zou et al. have worked on a review regarding this environmental remediation and application of nanoscale iron oxide and its composite for heavy metal ions (Zou et al. 2016). The alumina nanofibers prepared by the electrospinning method were applied for Cr(VI) and F^- remediation. Figure 10 represents the SEM images and EDAX spectra of the alumina nanofibers before and after the adsorption studies. It was observed that alumina nanofibers retain their fiber morphology after adsorption and exhibit a maximum uptake of 6.8 mg g^{-1} of Cr(VI) & 1.2 mg g^{-1} for F^- ion (Figure 11) (Mahapatra et al. 2013a).

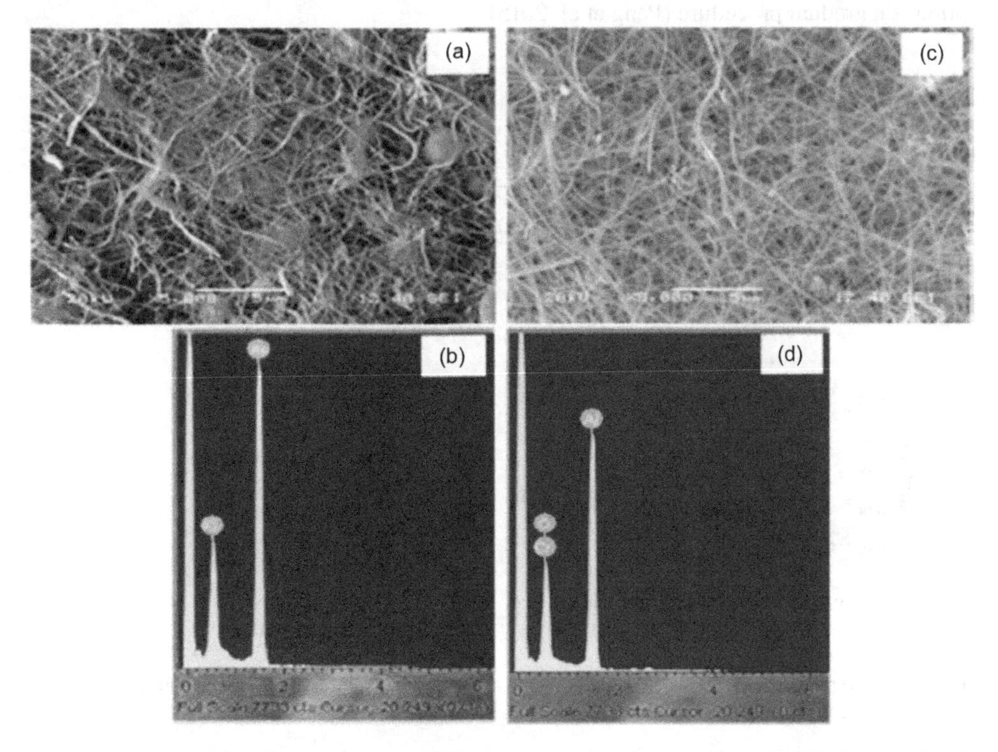

FIGURE 10 Representation of SEM images and EDAX spectra of alumina nanofibers (a, b) before and (c, d) after adsorption of Cr(VI) (Mahapatra et al. 2013a).

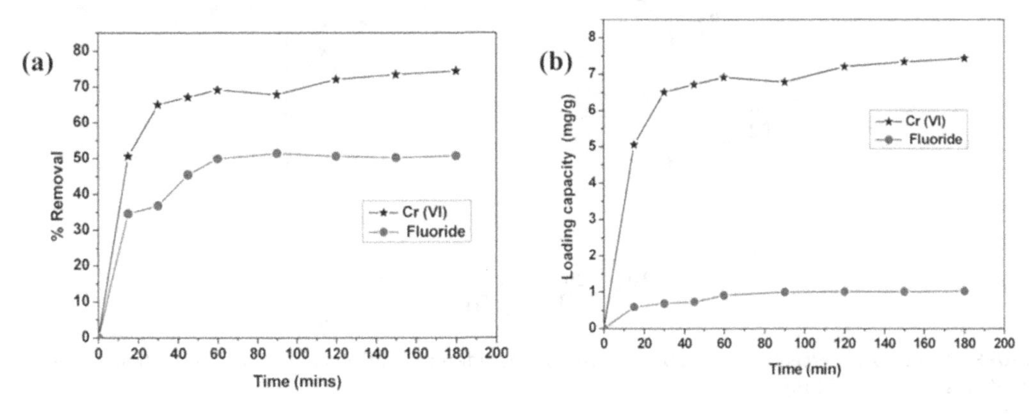

FIGURE 11 Percentage removal (a) and loading capacity (b) of Cr(VI) and F^- ions as a function of time (Mahapatra et al. 2013a).

Again we can take the example of hollow and porous γ-Al_2O_3 nanofibers synthesized by Peng et al. to understand the adsorption process. Here, they applied the γ-Al_2O_3 nanofibers for the adsorption of three different types of dyes such as congo red (CR), methyl blue (MB), and acid fuchsine (AF). UV-vis absorption spectroscopy at a definite time interval was used to evaluate the removal ratio of the adsorbate. The noticeable decrease in absorbance specifies the effective adsorption of dye on the adsorbent surface (Figure 12). The absorbance of CR, MB, and AF is taken as 500, 600, and 544 nm respectively to quantify the adsorption performance. Figure 13 represents the experimental results of the effect of the contact time study. During the CR and MB adsorption, the rate of adsorption is very fast (up to 5 minutes), and then it slows down from 5 minutes to 60 minutes. At 60 minutes of adsorption time, about 96.52% and 92.38% of CR and MB were adsorbed respectively. About 91.70% of AF was removed within 60 minutes, indicating that the AF adsorption is a gradual procedure (Peng et al. 2015).

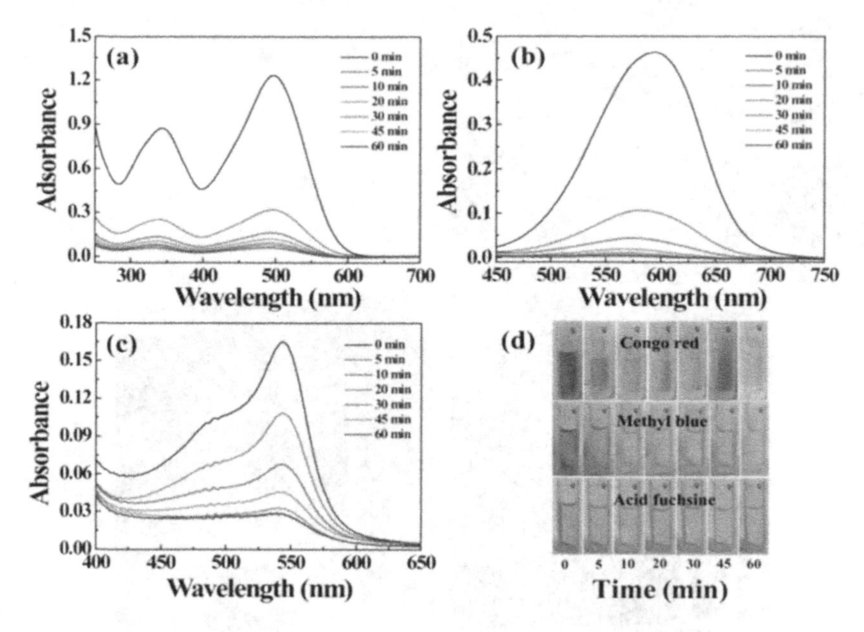

FIGURE 12 UV-vis spectra of CR (a), MB (b), and AF (c) at a different time interval. (d) Digital photography during adsorption at different time intervals (Peng et al. 2015).

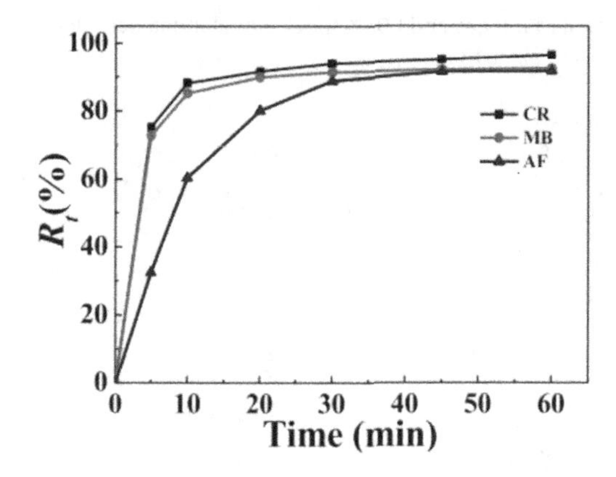

FIGURE 13 Removal percentage of CR, MB, and AF dyes at different time intervals by the porous hollow γ-Al_2O_3 nanofibers (Peng et al. 2015).

Several adsorbents have been studied for their capability to sorb numerous contaminants from wastewater. Different experimental outcomes for the synthesis and adsorption of various pollutants are elucidated in Table 2 where different adsorbents, their synthesis procedure, adsorbate, kinetic study, and also the sorption efficiency of their best-fitted model is represented.

TABLE 2 Adsorption of different pollutants by alumina-based nanomaterials, their kinetics, isotherm, and maximum adsorption capacity.

Sl. No.	Adsorbents	Synthetic Procedure	Adsorbate	Kinetics	Isotherm	Maximum Adsorption Capacity	References
1	α-Fe_2O_3-γ-Al_2O_3 core-shell nanofibers	Electrospinning	Cr(VI)	pseudo-2^{nd}order	Freundlich		(Li et al. 2014)
2	Iron Oxide/ Alumina Nanocomposites	Precipitation	As(III) As(V) F		Freundlich	1,000 μg g^{-1} 2,500 μg g^{-1} 4.82 μg g^{-1}	(Dehghani et al. 2020)
3	MWCNTs γ-alumina	Commercially available	Co(II)	pseudo-2^{nd}order	DR	78.94 mg g^{-1} 75.78 mg g^{-1}	(Dehghani et al. 2020)
4	HIAGO	Precipitation	F	pseudo-2^{nd}order	Langmuir	22.03 mg g^{-1}	(Kanrar et al. 2016)
5	Hollow γ-Al_2O_3 nanofibers	Electrospinning	CR MB AF	pseudo-2nd order pseudo-1^{st}order			(Peng et al. 2015)
6	Nano-γ-Al_2O_3		Cr^{6+} Ni^{2+} Pb^{2+} Cd^{2+}	pseudo-2nd order pseudo-1^{st}order	Langmuir Freundlich	13.3 mg g^{-1} 6 mg g^{-1} 1.1 mg g^{-1} 0.33 mg g^{-1}	(Shokati Poursani et al. 2015)
7	Rod-like Alumina	Hydrothermal	Cr(VI)			52.1 mg g^{-1}	(X. Xu et al. 2018)
8	Al/Fe dispersed in porous granular ceramics	Impregnation followed by precipitation	F	pseudo-2nd order	Langmuir	1.79 mg g^{-1}	(Chen et al. 2011)

3.2 Kinetics, Isotherms and Thermodynamics Study

Investigation of adsorption kinetics is a significant criterion to examine the time required for the adsorption process and the required mechanism. The commonly used kinetics model to study the adsorption rate are pseudo-first-order, pseudo-second-order, intra-particle diffusion, and double exponential model. The details about these models are discussed here. The mathematical expression for these models is given below (Largitte and Pasquier 2016, Tripathy et al. 2020, Yousefi et al. 2018).

Pseudo-first-order model

$$\log(q_e - q_t) = \log q_e - \frac{k_1 t}{2.303} \quad (3)$$

Pseudo-second-order model

$$\frac{t}{q_t} = \frac{1}{k_2 q_e^2} + \frac{t}{q_e} \quad (4)$$

In the above equations, q_e(mg g^{-1}) is the equilibrium adsorption capacity, q_t(mg g^{-1}) is the adsorption capacity at time t (min), k_1 (min^{-1}), and k_2 (g mg^{-1} min^{-1}) is the pseudo-first and pseudo-second-order rate constants, respectively. For the pseudo-first-order-kinetic model the slope and intercept obtained from the plot of $\log(q_e - q_t)$ versus t give the value of k_1 and q_e, respectively. For the pseudo-second-order kinetic model, the slope and intercept obtained from the t/q_t versus t plot give the value of q_e and k_2, respectively. The pseudo-first-order model indicates the physisorption mechanism, while the pseudo-second-order model indicates the chemisorption process.

Intra-particle diffusion model

$$q_t = k_{\text{id}} t^{0.5} + C \quad (5)$$

where $t^{0.5}$ (in s) is the half lifetime, intra-particle diffusion constant is k_{id} (mg g^{-1} min$^{0.5}$), and C is the boundary width. The slope and intercept of the linear plot of q_t versus $t^{0.5}$ give the value of k_{id} and C.

Double exponential model

$$q_t = q_e - D_2 m_{\text{ads}} e^{(-k_{D_2} t)} \quad (6)$$

$$\text{In}(q_e - q_t - D^2 m_{\text{ads}} e^{(-k_{D_2} t)}) = \text{In} D_1 m_{\text{ads}} - k_{D_1} t \quad (7)$$

where k_{D_1} and k_{D_2} are diffusion parameters (min^{-1}), which control the overall kinetics. Sorption rate parameters (m mol^{-1}) for rapid and slow steps are D_1 and D_2, respectively, and the mass of adsorbent is m_{ads}. Equation (6) can be simplified for the determination of D_2 and k_{D_2}, where the exponential term is assumed to be negligible. Further, the parameters D_1 and k_{D_1} can be obtained from the plot of linearized Equation (7) concerning the known values of D_2 and k_{D_2}. This model is helpful to outline the adsorption of adsorbate through a two-step mechanism. The rapid uptake of adsorbate via both external and internal diffusion is the first step. In the second step, the uptake of adsorbate slows down and finally reaches equilibrium. That is why the mechanism is called a 'double exponential model'.

At constant temperature, the adsorption isotherm study correlates the adsorption process with their concentrations. To elucidate the mechanism of adsorption, the commonly used isotherm models are Langmuir, Freundlich, Temkin, and Dubinin-Radushkevich (DR). The mathematical expression of all the above models and their detailed significance are discussed here (Tripathy et al. 2020, Yousefi et al. 2018).

Langmuir isotherm model

$$\frac{C_e}{q_e} = \frac{1}{q^0 b} + \frac{C_e}{q^0} \quad (8)$$

C_e(mgL^{-1}) represents the concentration of adsorbate at equilibrium. At equilibrium, q_e (mg g^{-1}) is adsorbate adsorbed at equilibrium; q^0 (mg g^{-1}) is adsorbate adsorbed at the maximum adsorption capacity. The b (L mg^{-1}) is the Langmuir constant. This model indicates the monolayer adsorption of adsorbate species with a fixed number of adsorption sites. The feasibility of the adsorption process will have estimated by determining the separation factor (R_L) value, which is the essential feature of the Langmuir isotherm.

Separation factor (R_L)

$$R_L = \frac{1}{1 + bC_0} \quad (9)$$

where C_0 (mgL^{-1}) is the initial adsorbate concentration and b is Langmuir constant. The values of R_L are classified into different types, such as $R_L = 0$, $R_L = 1$, $0 < R_L < 1$, and $R_L > 1$, indicates that the process of adsorption is irreversible, linear, favorable, and unfavorable, respectively. However, the Langmuir model does not give prior information about the roughness and heterogeneous adsorption process.

Freundlich isotherm model

$$\log q_e = \log k_f + \frac{\log C_e}{n} \quad (10)$$

In this equation, the Freundlich constant is k_f (mgg^{-1}) and n is the Freundlich exponent. The smaller value of 1/n indicates the stronger adsorption process. This model describes the roughness and multilayer adsorption process.

Temkin isotherm model

$$q_e = B_1 \text{In} K_T + B_1 \text{In} C_e \quad (11)$$

where B_1: Temkin constant and K_T (L mg^{-1}): equilibrium binding constant. If the K_T increases with an increase in temperature, the electrostatic force of attraction is the key binding factor between the adsorbate and adsorbent.

D-R isotherm model

$$\text{In} q_e = \text{In} q_m - K\in^2 \quad (12)$$

$$\in = RT \text{In}\left(1 + \frac{1}{C_e}\right) \quad (13)$$

$$E = (-2K)^{-0.5} \quad (14)$$

where q_m is theoretical adsorption capacity (mg g^{-1}), $\in$ is D-R isotherm constant, R universal gas constant (8.314 JK^{-1} mol^{-1}), T is temperature (K), K is constant related to adsorption energy, and E is free energy of adsorption (kJ mol^{-1}). If the value of E is between 8-16 kJ mol^{-1}, then it specifies the chemisorption process; but if it is less than 8 kJ mol^{-1}, then it specifies the physisorption process.

The effect of solution temperature or thermodynamics study is also an important criterion to illustrate the adsorption mechanism. Initially, the increase in temperature decreases the solution viscosity, which results in an increase in the diffusion rate of adsorbate on the adsorbent surface, which leads to a rise in removal percentage. However, further temperature change can influence the adsorption process, which depends on the absorption or release of heat. For an exothermic process, the adsorption capacity decreases with an increase in temperature, while for an endothermic process it increases. Apart from this, the temperature study of an adsorption process gives major information about different thermodynamics parameters such as the change in standard Gibbs free energy (ΔG), change in enthalpy (ΔH), and change in entropy (ΔS). The negative value of ΔG and the positive value of ΔH shows the spontaneous and endothermic nature of the adsorption. However, the positive value of ΔS indicates the increase in disorder during adsorption at the solid-liquid interface (Panda et al. 2019).

After the kinetics, isotherm, and thermodynamics investigations, the adsorption mechanism can be demonstrated. To further support the adsorption process, researchers characterize the adsorbent system after the adsorption process by various analytical techniques. The bonding configuration, oxidation state, and functionality of the adsorbent can be analyzed by XPS and FTIR techniques. The morphological characteristics of the adsorbent can be investigated by FE-SEM and TEM techniques (Pham et al. 2019, Sahoo et al. 2019, Zhao et al. 2020).

Conclusions and Prospects

This chapter gives comprehensive information on the synthesis and adsorptive applications of the 1D alumina-based nano-architecture and their applicability toward the decontamination of various inorganic and organic pollutants. The effect of different adsorption variables, like adsorbent dosage, pH, time, and concentration are also interpreted. Also, the details about the kinetics, isotherm, and thermodynamics study are represented properly. Generally, this chapter helps the riders to synthesize 1D alumina-based nanoadsorbents for water treatment applications. Even if several 1D alumina-based nanoadsorbents have been developed earlier, still a lot of research is needed to explore its applicability. No such reports have available for the development of functionalized 1D alumina-based materials. Details mechanistic investigation is needed after the adsorption process. Also, the applicability of the adsorbents toward the real water body needs to be carryout. Therefore, the researcher can proceed with their work to fulfill the above literature gap in the future.

Acknowledgments

The author would like to acknowledge NIT Rourkela, Odisha for providing the infrastructure and research facility to carry out this work.

References

Ángel, M., Zavala, L., Alejandro, S. and Morales, L. 2017. Synthesis of stable TiO_2 nanotubes: Effect of hydrothermal treatment, acid washing and annealing temperature. Heliyon 3(11): e00456. https://doi.org/10.1016/j.heliyon.2017.e00456.

Awad, A.M., Jalab, R., Benamor, A., Nasser, M.S., Ba-Abbad, M.M., El-Naas, M., et al. 2020. Adsorption of organic pollutants by nanomaterial-based adsorbents: An overview. J. Mol. Liq. 301: 112335. https://doi.org/10.1016/j.molliq.2019.112335.

Bagheri, H., Fakhri, H., Ghahremani, R., Karimi, M., Madrakian, T. and Afkhami, A. 2020. Nanomaterial-based adsorbents for wastewater treatment. pp. 467-485. *In*: Phuong Nguyen Tri, Trong-On Do and Tuan Anh Nguyen. [eds.]. Smart Nanocontainers. Elsevier Inc. https://doi.org/10.1016/b978-0-12-816770-0.00028-9.

Chen, N., Zhang, Z., Feng, C., Zhu, D., Yang, Y. and Sugiura, N. 2011. Preparation and characterization of porous granular ceramic containing dispersed aluminum and iron oxides as adsorbents for fluoride removal from aqueous solution. J. Hazard. Mater. 186: 863-868. https://doi.org/10.1016/j.jhazmat.2010.11.083.

Dehghani, M.H., Yetilmezsoy, K., Salari, M., Heidarinejad, Z., Yousefi, M. and Sillanpää, M. 2020. Adsorptive removal of cobalt(II) from aqueous solutions using multi-walled carbon nanotubes and γ-alumina as novel adsorbents: Modelling and optimization based on response surface methodology and artificial neural network. J. Mol. Liq. 299: 112154. https://doi.org/10.1016/j.molliq.2019.112154.

De León, J.N.D., Petranovskii, V., De Los Reyes, J.A., Alonso-Nuñez, G., Zepeda, T.A., Fuentes, S., et al. 2014. One dimensional (1D) γ-alumina nanorod linked networks: Synthesis, characterization and application. Appl. Catal. A Gen. 472: 1-10. https://doi.org/10.1016/j.apcata.2013.12.005.

Hu, X. yu, Liu, Q. yao, Jing, M. xiang, Chen, F., Ju, B. wei, Tu, F. yue, et al. 2020. Controllable preparation of alumina nanorods with improved solid electrolyte electrochemical performance. Ceram. Int. 46: 16224-16234. https://doi.org/10.1016/j.ceramint.2020.03.178.

Li, J., Li, M., Yang, X., Zhang, Y., Liu, X., Liu, F., et al. 2019. Morphology-controlled synthesis of boehmite with enhanced efficiency for the removal of aqueous Cr(VI) and nitrates. Nanotechnology 30: 195702.

Kanrar, S., Debnath, S., De, P., Parashar, K., Pillay, K., Sasikumar, P., et al. 2016. Preparation, characterization and evaluation of fluoride adsorption efficiency from water of iron-aluminium oxide-graphene oxide composite material. Chem. Eng. J. 306: 269-279. https://doi.org/10.1016/j.cej.2016.07.037.

Largitte, L. and Pasquier, R. 2016. A review of the kinetics adsorption models and their application to the adsorption of lead by an activated carbon. Chem. Eng. Res. Des. 109: 495-504. https://doi.org/10.1016/j.cherd.2016.02.006.

Li, X., Zhao, R., Sun, B., Lu, X., Zhang, C., Wang, Z., et al. 2014. Fabrication of α-Fe_2O_3-γ-Al_2O_3 core-shell nanofibers and their Cr(vi) adsorptive properties. RSC Adv. 4: 42376-42382. https://doi.org/10.1039/c4ra03692a.

Li, Y., Yang, X.Y., Feng, Y., Yuan, Z.Y., and Su, B.L. 2012. One-dimensional metal oxide nanotubes, nanowires, nanoribbons, and nanorods: Synthesis, characterizations, properties and applications. Crit. Rev. Solid State Mater. Sci. 37: 1-74. https://doi.org/10.1080/10408436.2011.606512.

Liu, M., Cai, N., Chan, V. and Yu, F. 2019. Development and applications of MOFs derivative one-dimensional nanofibers via electrospinning: A mini-review. Nanomaterials 9: 1-22. https://doi.org/10.3390/nano9091306.

Liu, P., Zhu, Y., Ma, J., Yang, S., Gong, J. and Xu, J. 2013. Preparation of continuous porous alumina nanofibers with hollow structure by single capillary electrospinning. Colloids Surf. A Physicochem. Eng. Asp. 436: 489-494. https://doi.org/10.1016/j.colsurfa.2013.07.023.

Mahapatra, A., Mishra, B.G. and Hota, G. 2011. Synthesis of ultra-fine α-Al_2O_3 fibers via electrospinning method. Ceram. Int. 37: 2329-2333. https://doi.org/10.1016/j.ceramint.2011.03.028.

Mahapatra, A., Mishra, B.G. and Hota, G. 2013a. Studies on electrospun alumina nanofibers for the removal of chromium(vi) and fluoride toxic ions from an aqueous system. Ind. Eng. Chem. Res. 52: 1554-1561. https://doi.org/10.1021/ie301586j.

Mahapatra, A., Mishra, B.G. and Hota, G. 2013b. Electrospun Fe_2O_3-Al_2O_3 nanocomposite fibers as efficient adsorbent for removal of heavy metal ions from aqueous solution. J. Hazard. Mater. 258-259: 116-123. https://doi.org/10.1016/j.jhazmat.2013.04.045.

Panda, A.P., Rout, P., Jena, K.K., Alhassan, S.M., Kumar, S.A., Jha, U., et al. 2019. Core-shell structured zero-valent manganese (ZVM): A novel nanoadsorbent for efficient removal of As(iii) and As(v) from drinking water. J. Mater. Chem. A 7: 9933-9947. https://doi.org/10.1039/c9ta00428a.

Peng, C., Zhang, J., Xiong, Z., Zhao, B. and Liu, P. 2015. Fabrication of porous hollow γ-Al_2O_3 nanofibers by facile electrospinning and its application for water remediation. Microporous Mesoporous Mater. 215: 133-142. https://doi.org/10.1016/j.micromeso.2015.05.026.

Pereao, O.K., Bode-Aluko, C., Ndayambaje, G., Fatoba, O. and Petrik, L.F. 2017. Electrospinning: Polymer nanofibre adsorbent applications for metal ion removal. J. Polym. Environ. 25: 1175-1189. https://doi.org/10.1007/s10924-016-0896-y.

Pham, T.D., Do, T.U., Pham, T.T., Nguyen, T.A.H., Nguyen, T.K.T., Vu, N.D., et al. 2019. Adsorption of poly(styrenesulfonate) onto different-sized alumina particles: Characteristics and mechanisms. Colloid Polym. Sci. 297: 13-22. https://doi.org/10.1007/s00396-018-4433-5.

Qin, H., Guo, W. and Xiao, H. 2019. Preparation of γ-Al_2O_3 membranes for ultrafiltration by reverse micelles-mediated sol-gel process. Ceram. Int. 45: 22783-22792. https://doi.org/10.1016/j.ceramint.2019.07.320.

Roque-Ruiz, J.H., Medellín-Castillo, N.A. and Reyes-López, S.Y. 2019a. Fabrication of α-alumina fibers by sol-gel and electrospinning of aluminum nitrate precursor solutions. Results Phys. 12: 193-204. https://doi.org/10.1016/j.rinp.2018.11.068.

Roque-Ruiz, J.H., Medellín-Castillo, N.A. and Reyes-López, S.Y. 2019b. Fabrication of α-alumina fibers by sol-gel and electrospinning of aluminum nitrate precursor solutions. Results Phys. 12: 193-204. https://doi.org/10.1016/j.rinp.2018.11.068.

Roque-Ruiz, J.H., Medellín-Castillo, N.A. and Reyes-López, S.Y. 2019c. Fabrication of α-alumina fibers by sol-gel and electrospinning of aluminum nitrate precursor solutions. Results Phys. 12: 193-204 https://doi.org/10.1016/j.rinp.2018.11.068.

Sahoo, S.K., Tripathy, M. and Hota, G. 2019. *In situ* functionalization of GO sheets with AlOOH-FeOOH composite nanorods: An eco-friendly nanoadsorbent for removal of toxic arsenate ions from water. J. Environ. Chem. Eng. 7: 103357. https://doi.org/10.1016/j.jece.2019.103357.

Saxena, R., Saxena, M. and Lochab, A. 2020. Recent progress in nanomaterials for adsorptive removal of organic contaminants from wastewater. ChemistrySelect 5: 335-353. https://doi.org/10.1002/slct.201903542.

Selim, M.S., Mo, P.J., Zhang, Y.P., Hao, Z. and Wen, H. 2020. Controlled-surfactant-directed solvothermal synthesis of γ-Al_2O_3 nanorods through a boehmite precursor route. Ceram. Int. 46: 9289-9296. https://doi.org/10.1016/j.ceramint.2019.12.183.

Shokati Poursani, A., Nilchi, A., Hassani, A.H., Shariat, M. and Nouri, J. 2015. A novel method for synthesis of nano-γ-Al_2O_3: Study of adsorption behavior of chromium, nickel, cadmium and lead ions. Int. J. Environ. Sci. Technol. 12: 2003-2014. https://doi.org/10.1007/s13762-014-0740-7.

Song, X., Zhang, K., Song, Y., Duan, Z., Liu, Q. and Liu, Y. 2020. Morphology, microstructure and mechanical properties of electrospun alumina nanofibers prepared using different polymer templates: A comparative study. J. Alloys Compd. 829: 154502. https://doi.org/10.1016/j.jallcom.2020.154502.

TC, P., Sharma, S.K. and Kennedy, M. 2018. Application of iron oxide and iron oxide/alumina nanocomposites for arsenic and fluoride removal: A comparative study. Int. J. Theor. Appl. Nanotechnol. 6: 7-10. https://doi.org/10.11159/ijtan.2018.001.

Teoh, G.L., Liew, K.Y. and Mahmood, W.A.K. 2007. Synthesis and characterization of sol-gel alumina nanofibers. J. Sol-Gel Sci. Technol. 44: 177-186. https://doi.org/10.1007/s10971-007-1631-x.

Tripathy, M. and Hota, G. 2020. Maghemite and graphene oxide embedded polyacrylonitrile electrospun nanofiber matrix for remediation of arsenate ions. ACS Appl. Polym. Mater. 2: 604-617. https://doi.org/10.1021/acsapm.9b00982.

Tripathy, M., Padhiari, S. and Hota, G. 2020. L-cysteine-functionalized mesoporous magnetite nanospheres: Synthesis and adsorptive application toward arsenic remediation. J. Chem. Eng. Data 65: 3906-3919. https://doi.org/10.1021/acs.jced.0c00250.

Vidal, R.R.L. and Moraes, J.S. 2019. Removal of organic pollutants from wastewater using chitosan: A literature review. Int. J. Environ. Sci. Technol. 16: 1741-1754. https://doi.org/10.1007/s13762-018-2061-8.

Wadhawan, S., Jain, A., Nayyar, J. and Mehta, S.K. 2020. Role of nanomaterials as adsorbents in heavy metal ion removal from waste water: A review. J. Water Process Eng. 33: 101038. https://doi.org/10.1016/j.jwpe.2019.101038.

Wang, L., Yang, G., Peng, S., Wang, J., Yan, W. and Ramakrishna, S. 2020. One-dimensional nanomaterials toward electrochemical sodium-ion storage applications via electrospinning. Energy Storage Mater. 25: 443-476. https://doi.org/10.1016/j.ensm.2019.09.036.

WHO/UNICEF. 2017. Progress on Drinking Water, Sanitation and Hygiene – Joint Monitoring Programme 2017 Update and SDG Baselines. Who 66. https://doi.org/10.1111 / tmi.12329.

Xu, J., Cao, Z., Zhang, Y., Yuan, Z., Lou, Z., Xu, X., et al. 2018. A review of functionalized carbon nanotubes and graphene for heavy metal adsorption from water: Preparation, application, and mechanism. Chemosphere 195: 351-364. https://doi.org/10.1016/j.chemosphere.2017.12.061.

Xu, X., Yu, Q., Lv, Z., Song, J. and He, M. 2018. Synthesis of high-surface-area rod-like alumina materials with enhanced Cr(VI) removal efficiency. Microporous Mesoporous Mater. 262: 140-147. https://doi.org/10.1016/j.micromeso.2016.12.002.

Yousefi, T., Mohsen, M.A., Mahmudian, H.R., Torab-mostaedi, M., Moosavian, M.A. and Aghayan, H. 2018. Removal of Pb(II) by modified natural adsorbent: Thermodynamics and kinetics studies. J. Water Environ. Nanotechnol. 3: 265-272. https://doi.org/10.22090/jwent.2018.03.007.

Yu, L., Ruan, S., Xu, X., Zou, R. and Hu, J. 2017. One-dimensional nanomaterial-assembled macroscopic membranes for water treatment. Nano Today 17: 79-95. https://doi.org/10.1016/j.nantod.2017.10.012.

Yuan, X., Zhu, J., Tang, K., Cheng, Y., Xu, Z. and Yang, W. 2015. Formation and properties of 1-D alumina nanostructures prepared via a template-free thermal reaction. Procedia Eng. 102: 602-609. https://doi.org/10.1016/j.proeng.2015.01.135.

Zhao, S., Wen, Y., Du, C., Tang, T. and Kang, D. 2020. Introduction of vacancy capture mechanism into defective alumina microspheres for enhanced adsorption of organic dyes. Chem. Eng. J. 402: 126180. https://doi.org/10.1016/j.cej.2020.126180.

Zou, Y., Wang, X., Khan, A., Wang, P., Liu, Y., Alsaedi, A., et al. 2016. Environmental remediation and application of nanoscale zero-valent iron and its composites for the removal of heavy metal ions: A review. Environ. Sci. Technol. 50: 7290-7304. https://doi.org/10.1021/acs.est.6b01897.

Adsorption and Desorption Aspects of Carbon-Based Nanomaterials: Recent Applications for Water Treatments and Toxic Effects

Patricia Prediger,[1,*] Melissa Gurgel Adeodato Vieira,[2] Natália Gabriele Camparotto,[1] Tauany de Figueiredo Neves,[1] Paula Mayara Morais da Silva[1] and Giani de Vargas Brião[2]

1. INTRODUCTION

In recent decades, water contamination has become a worldwide concern mainly due to rapid urbanization, population growth, and industrialization. The contamination of water comes from the discharge of untreated effluents, both domestic and industrial (cosmetics, pharmaceutical, textiles, paper, electronic devices, food processing, oil, etc.), from agricultural activities (fertilizers, pesticides, and animal waste), and storm water runoff (salts, oil, grease, chemicals, etc.) (Gusain et al. 2020). It was estimated that 80% of the global wastewater is discharged into the environment without any treatment, and this value can reach 95% in the least developed countries (Ryder 2017). Thus, large amounts of chemical compounds are released into the environment, such as polycyclic aromatic hydrocarbons (PAH), dyes, drugs, phenols, amines, phthalate, surfactants, and heavy metals.

When present in the environment, these chemical compounds pose risk to human health and also to terrestrial and aquatic ecosystems (Igiri et al. 2018). Thus, several methods have been developed aiming at water purification, including adsorption, coagulation, advanced oxidation processes, biological treatments, flocculation, desalination, membrane separation, ion exchange, and photocatalytic and electrochemical approaches (Singh et al. 2018). Among these techniques, adsorption is the most studied, low-cost, rapid, and industrially-applied method.

Activated carbon is the most popular adsorbent due to its porosity and surface area decorated with different chemical groups. However, the high cost and limited reusability of activated carbon limit its broad application. Thus, it is necessary to develop new low-cost adsorbents that have a larger surface area and a superior performance than activated carbon. Nanomaterials are suitable candidates to

[1] School of Technology, University of Campinas – UNICAMP, Limeira, São Paulo, Brazil.

[2] Process and Product Development Department, School of Chemical Engineering, University of Campinas – UNICAMP, Campinas, São Paulo, Brazil.

[*] Corresponding author: prediger@unicamp.br.

adsorb several pollutants from water. These materials have at least one of their dimensions at a nanometric scale, and their key property is the exponential increase in specific surface area which enhances the ratio between surface/internal atoms. Because of that several nanomaterials have been used as adsorbents for water purification, such as metallic and metal oxides nanoparticles, MXenes, h-boron nitride, layered double hydroxides, silicon nanomaterials, nanofibers, nanoclays, polymer-based nanomaterials, zeolites, nanogels, and carbon-based nanomaterials (CBNs) (Elessawy et al. 2020a). In the last decade, CBNs emerged as efficient adsorbents due to their outstanding properties, high abundance, low cost, ease of preparation, regeneration and reusability, and benign nature. CBNs include a series of materials, but this chapter is focused on crystalline framework materials including graphene derivatives, carbon nanotubes (CNT), nanoporous carbon (NPC), nanodiamond (ND), fullerenes, and graphitic carbon nitride (CN) and their composites.

CBNs possess different dimensions, i.e. zero-dimensional (0D) fullerenes and NDs, one-dimensional (1D) single-walled carbon nanotubes (SWCNT), bi-dimensional (2D) graphene derivatives, and tridimensional structures (3D) formed by assemblies of graphene and CNT, porous CN and NPCs. In regard to CBN structure, the crystallite size and the presence of defects in their lattice will influence their mechanical properties and thermal stability and consequently affect their performance as adsorbents for water and wastewater treatments. The quality of crystalline CBNs is directly affected by the production method and any post-production treatments, such as removal of amorphous carbon, oxidation, reduction, grafting, sonication, and thermal annealing. CBNs can be synthetically produced on a large laboratory scale, but some of them also occur naturally, i.e. fullerene (Becker et al. 1994) and CNT (Murr et al. 2004). CBNs can also be adventitiously produced as a result of human activities, including car exhausts, industrial processes, and biomass burning (Baalousha et al. 2014). Recently, more attention has been given to the effects and potential risks of CBNs on the environment and human health after the exposition.

This chapter features recent technologies aiming at water purification using CBNs, dealing with adsorption/desorption aspects and related mechanisms. The occurrence of CBNs in the environment, their toxic effects, and their ability to modulate the toxicity of other contaminants are discussed.

2. ADSORPTION AND DESORPTION ASPECTS

2.1 Adsorption

Adsorption involves the transference of substances from the bulk solution (gaseous or liquid) to the adsorbent (solid). Adsorptions processes can be carried out in batch experiments, continuous moving and fixed-bed filters, and continuous fluidized and pulsed bed processes (Patel 2019). The driving force responsible for the adsorbate adhering onto the adsorbent surface is based on physisorption or chemisorption. Physisorption is a reversible process that occurs in mono- or multilayers and involves weak electrostatic attractions, including van der Waals, dipole-dipole and π-π interactions, London forces, and hydrogen bonding (Sims et al. 2019). In turn, chemisorption is an irreversible process that only occurs in monolayers and is characterized by strong chemical interactions between adsorbent/adsorbate based on complex formation/chelation, covalent bonding, and oxidation/reduction reactions. If the energy of the interactions is lower than 8 kJ/mol, the process takes place physically; while for higher values, adsorption proceeds chemically (Anayurt et al. 2009). Metal ions and organic pollutants undergo chemisorption or physisorption when attached to CBNs.

Adsorption efficiency and the type of interactions are usually investigated by isothermal and kinetic studies. An isothermal curve is obtained by plotting the amount of adsorbed molecules versus the concentration of the pollutant in the bulk solution at a constant temperature. To forecast the operating adsorption mechanism, different models have been described to fit the experimental data, such as Langmuir, Freundlich, Temkin, Redlich-Peterson, Frumkin, and Dubinin-Radushkevich. Concerning CBNs, Langmuir and Freundlich are the most widely applied models and suggest an adsorption process with monolayer and multilayer formation, respectively.

Adsorption kinetics is a curve that represents the rate of adsorption of adsorbate from a bulk of solution to the adsorbent surface at a determined adsorbent loading, flow rate, temperature, and pH. Adsorption kinetics involves three steps: external mass transfer or film diffusion, diffusion on the pore, and mass retention of pollutant molecules onto the adsorbent surface by complexation or precipitation (Ijagbemi et al. 2009). Aiming to calculate the rate of adsorption and the rate-determining step, pseudo-first order (PFO), pseudo-second order (PSO), intraparticle diffusion, and Elovich models are usually employed to fit the experimental data of the adsorption onto CBNs. Studying how adsorption takes place and which steps most influence the velocity of the process is the key to understanding the dynamics of the interactions and determining the amount of adsorbate that it takes up and the time it requires. For large-scale applications, optimized parameters of adsorption are indispensable.

The influence of temperature on the adsorption of pollutants is examined by submitting the removal process to a series of isotherms. Thus, thermodynamic parameters, including Gibbs free energy change (ΔG), enthalpy change (ΔH), and entropy change (ΔS), can be calculated and also indicate whether the system is endothermic/exothermic and favorable. Chemisorption can be endo- or exothermic, whereas physisorption is exclusively exothermic.

2.2 Desorption and Adsorbent Recycling

Desorption is often described as a surface phenomenon whereby the adsorbed molecules are detached from a solid surface. It can occur by the release of molecules, without chemical modifications, merely by breaking the bonds with surface atoms or by the detachment of new molecules that underwent chemical modifications, such as association or decomposition. A successful desorption step requires optimized conditions whereby the best elutants are chosen, which strongly depends on the nature of the adsorbent and adsorbate and the related adsorption mechanism. The elutant must not damage the adsorbent and should be cost-effective and eco-friendly (Das 2010). The most widely adopted desorption protocols involve the use of strong acids or bases. However, some desorption procedures apply chelating or oxidizing agents, salts, ultrasound or irradiation-assistance, thermal treatments, microbial activities, or simple washing with organic solvents or water (Hu et al. 2017). Low pH may favor the desorption and/or dissolution of metal cations and organic molecules. There is a strong competition between H^+ ions and adsorbates for adsorption sites causing the displacement of adsorbed ions/molecules into the acid solution. At basic conditions, adsorbents usually became negatively charged and the attraction interactions are replaced by electrostatic repulsions, favoring the desorption process (Hou et al. 2020). Desorption is crucial to adsorbent reuse that reduces operating costs.

The following sections present and discuss adsorption/desorption aspects when pristine and modified CBNs, as well their composites, were applied to remove hazardous compounds from water.

3. RECENT ADVANCES IN THE REMOVAL OF METAL IONS AND ORGANIC COMPOUNDS BY CARBON-BASED NANOMATERIALS

3.1 Carbon Nanotubes

Described as graphite sheets rolled into cylindrical shapes, having from 1 to 15 micrometer in length and diameters from 2 to 100 nanometers, CNTs have generated and still generate a number of studies in many areas of engineering and science (Kumar et al. 2020). They are classified into two types according to the number of carbon layers. Single-walled CNTs (SWCNTs) consist of a single layer of graphene, usually occurring as hexagonally packaged bales, with a diameter ranging between 0.4 and 2 nm. Multiwalled CNTs (MWCNTs) are formed by two or several layers of carbon concentric cylinders, all held together by Van der Waals forces between the layers with a diameter ranging from 2 to 100 nm.

TABLE 1 SWCNTs and MWCNT adsorbents: settings for adsorption equilibrium experiments, q_{max}, and performance after successive regeneration processes.

Adsorbent Matrix	Pollutants	Experimental Conditions	q_{max} (mg/g)	Regeneration	References
DP/MWCNT	Cr(VI)	pH 2, $M_{ads} = 0.02$ g, $V = 0.015$ L, $C_0 = 20\text{-}140$ mg/L, 303 K, 360 min.	55.55	1 cycle, NaOH, NaCl, Above 20%	(Kumar et al. 2013)
MWCNTs/Fe_3O_4	Hg(II)	pH 2, $M_{ads} = 0.01$ g, $V = 0.1$ L, $C_0 = 10\text{-}50$ mg/L, RT, 60 min.	238.78	–	(Sadegh et al. 2018)
SWCNTs/SH	Hg(II)	pH 5, $M_{ads} = 0.0025$ g, $V = 0.01$ L, $C_0 = 10\text{-}80$ mg/L, RT, 60 min.	131	5 cycles, HCl, 88 to 80%	(Bandaru et al. 2013)
CNT	MB	pH 10, $M_{ads} = 0.01$ g, $V = 0.02$ L, $C_0 = 10\text{-}50$ mg/L, RT, 1400 min.	62.5		(Tabrizi and Yavari 2015)
MWCNTS	Anionic and cationic dyes	pH 7-8, $M_{ads} = 0.04$ g, $V = 0.1$ L, $C_0 = 10$ mg/L, RT, 60 min.	24.67, 24.3	1 cycle 98.7 to 98%	(Shabaan et al. 2020)
MWCNTs-PEG	Phenol	pH 6, $M_{ads} = 0.02$ g, $V = 0.05$ L, $C_0 = 20$ mg/L, RT, 30 min.	21.23	–	(Bin-Dahman and Saleh 2020)
SWCNTs, MWCNTS	Phenol	pH 6-7, 4-5, $M_{ads} = 0.1$ g, $V = 0.05$ L, $C_0 = 50$ mg/L, RT, 56 min.	50.4, 64.6	–	(Dehghani et al. 2016)

*RT = Room temperature.

CNTs can be synthesized by three different methods: arc discharge, laser ablation, and chemical vapor deposition. This material exhibits excellent properties of resistance, elasticity, electrical conductivity, thermal conductivity, expansion, electron emission, and high proportion. Due to their excellent chemical and physical properties, CNTs are used in several applications including adsorbents to remove contaminants from water (Shabaan et al. 2020). Table 1 compiles recent works dealing with the uptake of hazardous substances by CNTs.

Dehghani et al. (2015) reported the ability of SWCNTs to remove Cr(VI) ions from water and compared it to MWCNTs. A q_{max} of 1.26 and 2.35 mg/g was achieved by MWCNTs and SWCNTs, respectively. The Langmuir isothermal and the PSO kinetic models best fitted the experimental data, indicating that the interaction sites on the adsorbent surface were identical and energetically equivalent. When applied to the adsorption of contaminants from water, MWCNTs normally have higher adsorption potential than SWCNTs (Dehghani et al. 2016) mainly due to the greater number of graphene sheets. However, in the study presented by Dehghani et al. (2015), better performance was achieved by SWCNTs. When compared to polyaniline/MWCNTs in the removal of Cr(VI), SWCNTs showed lower performance (Kumar et al. 2013).

In relation to MWCNTs, several studies have reported their application to remove contaminants from water. High efficiency of diazinon pesticide uptake by MWCNTs (3 mg/g) was achieved in 15 min (Dehghani et al. 2019). However, the composite based on chitosan/MWCNT reached a higher q_{max} of 8.2 mg/g in pesticide removal (Firozjaee et al. 2017).

Shabaan et al. (2020) evaluated the potential of MWCNTs in removing cationic Auramine O (AO), Basic Violet 2B crystal (BVC) dyes, and anionic Acid Scarlet 3R (AS3) dye from wastewater. The q_{max} of 24.7 mg/g and 24.3 mg/g was reached for a mixture of anionic and cationic dyes, respectively. MWCNTs functionalized with carboxylic groups (MWCNTs-COOH) were used in the removal of Rhodamine B (RhB) and Crystal Violet (CV) dyes, and q_{max} up to 300 mg/g was obtained for both cationic dyes (Li et al. 2020c). The cationic dyes used in the above-mentioned studies were not the same, but both studies indicate the presence of electrostatic interactions between CNTs and dyes. Avcı et al. (2020) analyzed the uptake of the antibiotic ciprofloxacin (CP) from water by MWCNTs, and the q_{max} was 2 mg/g. The removal of CP by MWCNT was mainly due to electrostatic attractions, and the best isothermal and kinetics models were Freundlich and PSO, respectively.

Although several studies have reported the great efficiency of CNTs as an adsorbent material for the removal of organic pollutants, researchers from all over the world are developing protocols to insert new coordinating groups on CNTs to improve their adsorbent ability (Maazinejad et al. 2020). SWCNTs magnetized and functionalized with polydopamine (PDA) (SWCNTs/Fe_3O_4@PDA) were applied to the removal of Hg(II) ions from water (Ghasemi et al. 2019). The tendency to remove Hg(II) ions was enhanced when pH increased. The q_{max} reached by SWCNTs/Fe_3O_4@PDA was 249.07 mg/g. The Freundlich isothermal model and PSO kinetics were better adjusted to the experimental data. The adsorbent was reused in five cycles, and a drop of 23% in Hg (II) adsorption was observed. SWCNTs/Fe_3O_4@PDA showed a greater removal potential compared to pristine SWCNTs (Alijani and Shariatinia 2018), thiol-derivative SWCNTs (Bandaru et al. 2013), and magnetized MWCNTs (Sadegh et al. 2018).

With regard to modified MWCNTs, it was found that the increase in oxygen content in oxidized MWCNTs (MSCNTs-O) enhanced the q_{max} of CP uptake to 192 mg/g (Yu et al. 2016). Oxidized MWCNTs showed greater adsorption potential than pristine MWCNTs (Avcı et al. 2020). Both studies suggested that electrostatic interactions govern adsorption, and the PSO kinetic model was better fitted to the experimental data.

Analyzing the studies on CNTs applied to water purification, it is possible to observe that although CNTs (single- or multiwalled) functionalized or not, they still have a removal potential for a range of organic compounds and heavy metals, and their adsorption capacity is influenced by environmental factors, such as pH, contact time, temperature, and adsorbent dose. Overall, the adsorption potential of CNTs increases with the number of sheets that make up the material and their functionalization

(SWCNTs < MWCNTs < functionalized CNTs) with Freundlich's isotherm and PSO kinetics being the models that best fit most of the experimental results. These studies agree that electrostatic and π-π interactions govern CNT/pollutant interactions. Some studies claim that the mechanisms of the interaction occur via chemisorption, while other studies indicate that they occur via physisorption.

3.2 Graphene Oxide Derivatives

Since its discovery by Novoselov and co-workers (Novoselov et al. 2016), graphene and derivatives have attracted exceptional interest over the last decade due to their outstanding properties, including high specific surface area and high electrical conductivity (Kyzas et al. 2018). Graphene is the first 2D material that proved to be stable in crystalline form and the fundamental building block for graphite, CNTs, and fullerenes. Due to its poor solubility and dispersibility in water, graphene has few applications for the adsorption of contaminants from water.

Graphene derivatives can be formed by the incorporation of abundant oxygen-containing functional groups to obtain graphene oxide (GO) and reduced graphene oxide (rGO) (Khurana et al. 2017), which have better dispersibility compared to graphene. GO has a similar structure to graphene but contains functional groups, such as hydroxyl, alkoxy, carbonyl, and carboxylic acid. GO is the most explored graphene derivative due to its easier large-scale and lower-cost production (Hiew et al. 2018), higher solubility, and easy functionalization to form modified materials when compared to graphene. One of the most attractive properties of GO is that it can be transformed into pristine graphene-like nanosheets through a reduction process that removes the oxygenated groups and partially restores the sp^2-hybridized carbonaceous domain.

For the above-mentioned reasons, GO and rGO have been applied to the uptake of pollutants from water. The oxygen-containing groups, mainly in GO, allow the adsorption of metal ions and organic molecules via electrostatic interactions. The presence of a non-oxidized aromatic domain in GO and especially in rGO enables other interactions, such as π-π, n-π, and hydrogen-π, between adsorbent/adsorbate. Concerning the adsorption of pollutants by GO-based materials, it depends on the degree of oxidation and size of the nanosheets, pH, dispersion quality, nature of the pollutant, adsorbent/adsorbate ratio, presence of interfering agents, etc. The following studies address the investigation of GO properties and their influences on the adsorption of pollutants. Table 2 presents the most recent research studies that applied graphene derivatives to adsorb organic and inorganic contaminants.

It was found that pristine GO was able to remove the non-aromatic dimethoate (DMT) pesticide, while the aromatic pesticide chlorpyrifos (CPF) was not adsorbed onto its surface (Lazarević-Pašti et al. 2018). On the other hand, CPF was removed by graphene nanoplatelets (GNPs), which have a lower oxidation degree and preserved the highly ordered carbonaceous sp^2 domain, suggesting that π-π interactions play a crucial role in CPF adsorption. The experimental data obtained for DMT and CPF adsorption were best fitted to the Langmuir and Freundlich isotherm models.

Le et al. (2019) reported that sonication treatment removed oxygenated groups from GO, increasing disorder and defects on the material which, in turn, weakens π-π interactions and decreases removal of Cs(I), MB, and Methyl Orange (MO) dyes. The q_{max} for CS(I), MB, and MO were 113, 139, and 64 mg/g before sonication, and it decreased up to 75% after ultrasound treatment. The effect of pH variation on the adsorption of benzoic acid (BA), bisphenol A (BPA), bisphenol S (BPS), nitrobenzene (NB), and salicylic acid (SA) on GO revealed the q_{max} of 25, 275, 252, 298, and 40 mg/g, respectively, achieved at acidic conditions (pH 4) (Tang et al. 2020). At lower pH, GO self-aggregation was observed, which improved pollutant uptake since new and favorable adsorption sites were created. No aggregation was verified at basic conditions, but electrostatic repulsion dominated the interactions.

TABLE 2 GO derivatives: settings for adsorption equilibrium experiments, q_{max}, and performance after successive regeneration processes.

Adsorbent Matrix	Pollutants	Experimental Conditions	q_{max} (mg/g)	Regeneration	References
GO-EDA-CAC-BPED	Cu(II)	pH 7, M_{ads} = 0.1 g, V = 0.1 L, C_0 = 50-500 mg/L, 293 K, 240 min.	202.12	10 cycles EDTA	(Chaabane et al. 2020).
	Ni(II)		190.97		
	Co(II)		179.98		
GO/montmorillonite 100/1	TC	pH 7, V = 0.025 L, C_0 = 50-300 mg/L, NaCl 0.01 mol/L RT, 200 min.	262.82	–	(Li et al. 2020a)
Magnetic rGO	TC	pH 5, M_{ads} = 0.001g, V = 0.008 L, C_0 = 5-200 mg/L, RT, 24h.	252	–	(Huang et al. 2019)
	Cd(II)		234		
	As(V)		14		
GO	MG	pH 7, M_{ads} = 0.005g, V = 0.01 L, C_0 = 50-150 mg/L, RT, 1 min.	3523.5	–	(Lv et al. 2018)
GO/1-OA			2687.5	5 cycles ethanol 85%	
12CE4@Fe_3O_4-rGO	Li(I)	pH 7, M_{ads} = 0.03 g, V = 0.04 L, C_0 = 5-70 mg/L, 303 K, 24h.	7.14	5 cycles HCl ~100%	(Nisola et al. 2020).
GO	BA	pH 4, M_{ads} = 0.0002 g, V = 0.01 L, C_0 = 10-500 mg/L, 298 K, 24h.	25	–	(Tang et al. 2020)
	BPA		275		
	BPS		252		
	NB		298		
	SA		40		

Also, the presence of common dispersed materials, including natural and engineered particles, which are originated from cosmetics, drugs, coatings, paints, advanced materials, electronic devices, or humic substances (HS) (Hartland et al. 2013), affect adsorption processes, especially considering real applications. In this way, Li et al. (2020a) found that the presence of HS enhanced tetracycline (TC) uptake by GO and extended the adsorption equilibrium time, which can be attributed to TC adsorption onto HS. The data obtained was best fitted to the Freundlich isothermal and PSO kinetic models. The q_{max} for TC was 262.82 mg/g, obtained with the use of GO/montmorillonite 100/1, indicating a synergetic effect. Huang et al. (2019) evaluated magnetic GO (MGO), magnetic chemically-reduced graphene (MCRG), and magnetic annealing-reduced graphene (MARG) as adsorbents for TC, Cd(II), and As(V), and the q_{max} were 252 mg/g, 234 mg/g, and 14 mg/g, respectively, obtained by MGO due its higher oxygen content and better dispersibility.

GO functionalization, which is carried out to bring some special characteristics and prevent its agglomeration, improves the surface-area-to-volume ratio, allowing novel interactions and strengthens existing ones (Yu et al. 2020). GO can be modified via covalent bonding, non-covalent bonds, and element doping. Covalent bonds normally occur through the functionalization of oxygen-containing groups or in the sp^2 structure of the graphene lattice. Non-covalent functionalization, in turn, does not alter the structure and electronic properties of graphene sheets and simultaneously attaches new chemical structures to their surface. On the other hand, doping GO with heteroatoms such as boron, nitrogen, and sulfur gives rise to useful properties allowing a fine-tuning of the electric structure and intrinsic characteristics of the material.

Modified-GO derivatives have been applied to water purification, i.e. GO treated with ethylenediamine, chloroacetyl chloride, and *N,N*-bis(2-pyridylmethyl)ethylenediamine (GO-EDA-CAC-BPED) was applied as an adsorbent and reached a q_{max} of 202.12 mg/g for Cu(II), 190.97 mg/g for Ni(II), and 179.98 mg/g for Co(II) (Chaabane et al. 2020). The experimental data were best fitted to PSO kinetics and the Jossens isotherm model, and all adsorbents could be recycled and maintained at high removal capacities even after 10 adsorption cycles. Co(II) removal was greater than that verified for pristine and N-doped GO (Wang et al. 2018).

Lv et al. (2018) applied a new GO functionalized with tetrazolyl derivative and octadecylamine (GO/1-OA) as an adsorbent to a series of organic pollutants. Removal of Malachite Green (MG), BPA, CP, and Cu(II) by GO/1-OA in a ternary system reached 87.4%, 80.2%, 81.1%, and 98.3% of removal, respectively. In single systems GO and GO/1-OA achieved a q_{max} of 3,523.5 and 2,687.5 mg/g for MG, respectively. The adsorbent was reused after washing with ethanol and the MG removal percentage decreased from 97% to 85% after five cycles. When rGO hydrogel was used as the MG adsorbent, lower adsorption efficiency reached 740.7 mg/g (Shi et al. 2016).

Chemically modified rGO adsorbents containing crown ethers (CE) with tailorable ion recognition were used to remove metallic ions from seawater by Nisola et al. (2020). The magnetic CE@Fe_3O_4-rGO materials showed high selectivity to capture metal ions depending on the size of the attached CE, its cavity conformation, metal hardness, and dehydration energy. Polyether-modified GO was prepared via grafting of arenediazonium salt and applied as an adsorbent to a polyether (TX-100) and cationic (dodecyl trimethylammonium bromide - DTAB) surfactants (Cheminski et al. 2019). The insertion of phenyl tetraethyleneglycol (PTEG) moieties onto GO enhanced the material's dispersibility in water and improved TX-100 uptake from 1,342 mg/g (by pristine GO) (Prediger et al. 2018) to 1,690 mg/g (by GO-PTEG). Regarding DTAB surfactant, GO-PTEG removal capacity reached 730 mg/g, while pristine GO achieved 329 mg/g (Neves et al. 2020a).

3.3 Nanoporous Carbon

Nanoporous carbon (NPC) is an adsorbent material that received great attention in the last decade due to its surface properties, ease of preparation, low-cost production, and a large abundance of precursor materials. NPC can be classified according to its porous size, i.e. microporous (pore size

< 2 nm), mesoporous (pore size 2-50 nm), and macroporous (pore size 50-1,000 nm) (Zhao et al. 2019b). NPC is widely applicable to the removal of organic and inorganic pollutants from water due to its high surface area and pore volume in addition to the presence of carbonyl and hydroxyl groups. Several studies reported NPC production from cheaper and renewable sources, including tea waste (Akbayrak et al. 2020), bamboo (Mistar et al. 2020), pyrolysis char of sewage sludge (Li et al. 2020b), and others. Table 3 features recent studies on the application of NPC to the removal of pollutants from water.

TABLE 3 NPC adsorbents: settings for adsorption equilibrium experiments, q_{max}, and performance after successive regeneration processes.

Adsorbent	Pollutant	Experimental Equilibrium Conditions	q_{max} (mg/g)	Regeneration (Number of Cycles)	Reference
NPC modified with polyaniline	Resorcinol Phenol p-Cresol	m_{ads} = 1 mg, V = 25 mL, C_0 = 100-500 mg/L, RT, 4 h	714.28 476.19 238.09	–	(Anbia and Ghaffari 2009)
Oxidized NPC functionalized with Zn	Pb(II)	m_{ads} = 1 g, V = 500 mL, C_0 = 10-400 mg/L	522.8	3	(Zolfaghari et al. 2013)
NPC from *Ceratonia siliqua*	RB5	m_{ads} = 0.1 g, V = 50 mL, C_0 = 25-300 mg/L, 2 h	36.90	5	(Güzel et al. 2015)
NPC from fruit hulls	MB	m_{ads} = 0.20 g, V = 200 mL, C_0 = 25-400 mg/L, 303 K, 0-28 h	239.4	–	(Islam et al. 2017)
NPC from MOF	Cu(II)	m_{ads} = 2 mg, V = 10 mL, C_0 = 100 mg/L, pH = 5, 6 h	33.44	–	(Bakhtiari et al. 2015)
Hybrid Fe–Ce–Ni NPC	Fluoride	m_{ads} = 0.1 g, V = 250 mL, 303 K, 30 min	285.7	5	(Dhillon and Kumar 2015)
NPC from *Enteromorpha prolifera*	MB	m_{ads} = 25 mg, V = 50 mL, C_0 = 40-120 mg/L, 5 h, RT*	270.27	–	(Li et al. 2013)

RT* = room temperature

Shokry et al. (2019) reported the preparation of NPC by NaOH impregnation using a new raw Egyptian coal source. The adsorbent had a spherical shape and showed a total pore volume of 0.183 cm^3/g and a specific surface area equal to 49 m^2/g. The prepared material showed MB uptake with a q_{max} of 28.09 mg/g, and its recycling was evaluated for 10 cycles of adsorption/desorption. Also, NPC prepared from a MOF showed a high surface area of 1,054 m^2/g, a total pore volume of 2.49 cm^3/g, and the q_{max} of 415 mg/g was found for MB uptake (Shi et al. 2019). Gupta et al. (2019) synthesized a graphene-like NPC material from the Bengal gram bean husk, named BGBH-C-K. The surface area of BGBH-C-K was 1,710 m^2/g, the total pore volume was 0.834 cm^3/g, and the q_{max} of 469 mg/g to MB removal. These indicate that the efficiency in removing the dye increases with increasing surface area and pore volume.

NPC synthesized from anthracite powder, which had a remarkable surface area of 3,143 m^2/g and was used in the removal of 25 aromatic compounds, including phenols, anilines, and nitrobenzene (Yang et al. 2018). The q_{max} for phenol, aniline, and nitrobenzene was 1,030 mg/g, 1,180 mg/g, and 1,794 mg/g, respectively, and the Polanyi Dubinin-Ashtakhov model best represented the adsorption equilibrium. The adsorption mechanism was based on π-π interactions and hydrogen bonding, and the hydrophobic effect derived from the van der Waals forces.

Li et al. (2017) synthesized a zeolitic imidazolate framework-8 derived NPC carbonizated at 700°C (NPC-700) and applied it to the removal of CP. The prepared NPC-700 reached higher q_{max} (416.7 mg/g), and the obtained data were best fitted to the Freundlich isotherm model and suggested that adsorption occurred in multilayers. Adsorbent recycling was evaluated, and after seven cycles the CIP removal percentage remained above 80%.

3.4 Nanodiamond, Fullerene and Carbon Nitride

CBNs, such as NDs, fullerenes, and CNs, have been exploited as adsorbents. NDs are promising carbon-based particles that possess a tetrahedral network sp^3 at nanometer scale prepared by detonation technique (Pichot et al. 2015), high-pressure and high-temperature conditions (Liang et al. 2020b), laser shocking (Motlag et al. 2020), and other protocols. NDs are applied to water purification due to their large specific surface area and mechanical, thermal, and biological properties. Table 4 gathers recent studies of the adsorption on hazardous substances by NDs, fullerenes, and CNs.

TABLE 4 ND, fullerene and CN adsorbents: settings for adsorption equilibrium experiments, q_{max}, and performance after successive regeneration processes.

Adsorbent	Pollutant	Experimental Equilibrium Conditions	q_{max} (mg/g)	Regeneration (Number of Cycles)	Reference
Thermal oxidized ND	Methotrexate	m_{ads} = 20 mg, V = 50 mL, C_0 = 0.24-0.96 mg/L, 2 h	650.82	–	(Zamani et al. 2019)
C_{60}	TCC	m_{ads} = 1 mg, V = 100 mL, C_0= 10-50 mg/L	~70	–	(Ion et al. 2019)
Mesoporous CN	Perfluorooctane sulfonate	m_{ads} = 15 mg, V = 25 mL, T = 298 K, 24 h	625	–	(Yan et al. 2013)
Mesoporous CN	Cr(VI)	m_{ads} = 0.5 g, V = 500 mL, T = 298 K	48.31	1	(Chen et al. 2014)
Mesoporous CN	MB	m_{ads} = 10 mg, C_0= 5-400 mg/L	360.8	5	(Peng et al. 2017)
Ultrathin CN	TC	C_0 = 20-500 mg/L, m_{ads} = 0.02 g, pH = 6.5.7.5, 1 h	85	–	(Tian et al. 2020)

Molavi et al. (2018) reported that unoxidized ND (UND) was selective to MO uptake and had a q_{max} of 41.7 mg/g. However, oxidized ND (OND) exhibited selectivity to MB dye adsorption with a q_{max} of 47.6 mg/g. MB removal was due to electrostatic interactions between OND and the dye, while MO uptake was based on hydrogen bonding between sulfonate groups in MO and oxygenated groups on the UND surface. UND and OND desorption and recycling showed high desorption rates for both dyes. NDs modified with benzene-sulfonic were synthesized in a one-step method based on the aryl diazonium chemistry by Lei et al. (2020) and applied to the removal of MB dye from water. The q_{max} was 385.27 mg/g, and the adsorption mechanism occurred in multilayers, according to the Freundlich model.

Fullerenes, also called C_{60} or Buckyballs, are the third allotropic modification of carbon, after graphite and diamond and were discovered by chemists in 1985 (Kroto et al. 1985). Fullerenes have a spherical shape composed of sp^2 carbon atoms arranged into 12 pentagons and hexagons whose quantity depends on the total number of carbon atoms. The preparation and related properties of C_{20}, C_{70}, C_{84}, and C_{100}, along with others fullerenes, have been reported (Scott 2004). However, C_{60} is the most studied fullerene, and its potential as a filtering agent has been evaluated (Gupta and Saleh 2013). The removal of organic pollutants by fullerenes is based on hydrophobic interactions, low polar forces, and dominant π–π interactions (Yang and Xing 2010). The potential of fullerenes to remove organic pollutants has been evaluated, and they were able to adsorb naphthalene, phenanthrene, pyrene (Cheng et al. 2004, Yang et al. 2006), carbamates, phenols, and amines (Ballesteros et al. 2000).

Ion et al. (2019) prepared fullerene C_{60} and evaluated its ability to remove triclocarban pesticide (TCC) under simulated environmental conditions, including saline media, the presence of HA, and irradiation. Both irradiated- and non-irradiated fullerene C_{60} remained stable during TCC removal at mild pH values, ionic strengths, and amount of HA. As the saline concentration increased, so did the size of the fullerene aggregates, drastically diminishing pesticide removal. The presence of HA at high values of ionic strength moderately diminished TCC removal since the attached HA prevented fullerene aggregation. The experimental data were best fitted to the Langmuir isothermal model and π-stacking interactions were the main adsorption mechanism.

Graphitic carbon nitride (CN) is a structure based on g-C_3N_4 nanosheets, which have graphite-like planes of sp^2 hybridized domain composed of carbon and nitrogen elements arranged in *s*-triazine and tri-*s*-triazine moieties (Darkwah and Ao 2018). CN can be prepared from a simple thermal polycondensation of cheap *N*-rich precursors, such as melamine (Ba et al. 2020), cyanamide (Kim et al. 2020), dicyanamide (Lei et al. 2007), and urea (Tian et al. 2019).

CN has stood out in recent years due to numerous advantages related to its structure, such as chemical and thermal stability, low cost, non-toxicity, and easy preparation. CN has received great attention as a photocatalyst due to outstanding properties, such as low energy, tunable band gap, and high organic transformations (Wang et al. 2020). Despite its properties, efforts have been made to overcome some limitations of CN materials, such as low surface area, poor visible light absorption, and the use of centrifugation, or filtration after its use. Few studies on its adsorptive performance are described due to adsorption limiting properties, such as low surface area. To overcome this limitation, hard, and soft templates have been developed for the replication of mesoporous matrices, which provide CN materials with high surface areas. Table 4 presents studies on the application of CN to the adsorption of pollutants from water.

The hard-template technique was applied by Azimi et al. (2019) to synthesize boron-modified carbon nitride (BMCN). BMCN had a high surface area of 431 m^2/g, a pore size of 4.9 nm, and it showed MG dye adsorption capacity (310 mg/g). The insertion of boron atoms increased the acidic sites on the BMCN surface, which enhanced interactions with amine groups of the MG dye. BMCN was used in six consecutive cycles of adsorption-desorption, and the material showed excellent stability as removal efficiency remained above 90%.

CN doping with elements including boron, sulfur, carbon, and metals has been a technique to improve its performance as an adsorbent for water purification. CN doped with *s*-block metals was synthesized by Fronczak et al. (2018) and used to remove Cu(II) and MB. The presence of the *s*-block metals Li, Na, K, Ca, Sr, and Ba enhanced CN performance in MB and Cu(II) uptake. Cu(II) removal increased by ~30 times, from 10 mg/g for pristine CN to 324-339 mg/g for all doped CN. MB adsorption capacity increased by ~680 times from 11 mg/g for pristine CN to 7,500 mg/g for Mg-doped CN.

4. RECENT ADVANCES IN METAL IONS AND ORGANIC COMPOUNDS REMOVAL BY CARBON-BASED NANOMATERIAL COMPOSITES

A composite is a multi-phase material formed by the combination of two or more substances. The continuous phase is called matrix and holds the dispersed phase (or phases). For being stronger than the matrix, the dispersed phase is also called the reinforcing phase. Composite properties are usually superior to those of individual substances, which justified the application of these materials (Paul and Dai 2018).

CBN composites can be formed with several polymers, resins, and metals to improve the physical, chemical, and thermal properties of carbon or the compositing agent. High surface area and facilitated functionalization make carbon-based nanocomposites promising adsorbents to remove different contaminants from water.

In this section, we present the recent applications of CNT, GO derivatives, NPC, ND, fullerene, and CN composites as adsorbents to remove organic pollutants and toxic metals from water.

4.1 Carbon Nanotubes Composites

Polymer/CNT composites are commonly formed by solvent casting, melt mixing (thermoplastic polymers), and the *in situ* polymerization method (Choudhary et al. 2013). CNT ceramic composites are prepared for mixing in wet media by co-milling of CNT and ceramic powders. Also, they can be efficiently obtained by the synthesis of CNT *in situ* within the ceramic powders or the synthesis of the ceramic *in situ* around the CNT (Peigney and Laurent 2006). The most common method to prepare CNT metal composites is powder metallurgy, followed by electrodeposition and electroless deposition and melting and solidification.

CNT composites from diverse matrices have been applied to remediate water via adsorption. Table 5 compiles the relevant results of recent works that applied CNT composites as adsorbents.

TABLE 5 CNTs composites-based adsorbents: settings for adsorption equilibrium experiments, q_{max}, and performance after successive regeneration processes.

Adsorbent Matrix	Pollutant	Experimental Conditions	q_{max} (mg/g)	Regeneration*	Reference
Silver nanoparticle/ modified magnetic MWCNT composites	Sulfamethoxazole	S/L = 0.5 g/L, C_0 = 10-100 mg/l, pH = 5.0, room temperature, 4 h	118.58	4 cycles Methanol Above to 80%	(Song et al. 2016)
CNT/ $CoFe_2O_4$ composites	Sulfamethoxazole and E_2	m_{ads} = 5 mg, V = 0.05 L, 120 rpm C_0 = 0.4-2.4 mg/L	–	5 cycles Heat at 573 K	(Wang et al. 2015)
$CaCO_3$/CNT composites	2-Naphthol	m_{ads} = 0.05 g, V = 8 mL, C_0 = 1-10 mg/L, 4 h, 298 K, 150 rpm, pH = 7.0	–	4 cycles Ethanol 90%	(Xu et al. 2015)
$CaCO_3$/CNT composites	2-Naphthol	Teflon column, diameter = 1.6 cm, length = 8.3 cm, m_{ads} = 0.5 g, flow = 20 mL/h	**0.286	–	(Xu et al. 2018)
MOF/CNT composites	Phenol	m_{ads} = 10 mg, V = 10 mL, T = 298 K	350	5 cycles Methanol	(Han et al. 2015)

* Adsorption efficiency along of cycles of adsorption/desorption.

** Adsorption capacity achieved in a dynamic mode in a fixed-bed column with an inlet concentration of 20 mg/L.

Polymeric CNT composites have been applied to adsorb toxic metals (Nyairo et al. 2018) and organic pollutants (Jadhav et al. 2015) due to their highly porous structure, presence of different functional groups, large specific surface area, and the possibility of π interactions with the adsorbate.

Abdeen and Akl (2015) proposed poly(vinyl alcohol)/CNT composites to remove/adsorb uranium. Two films were prepared and their adsorptive capacities were compared with a cross-linked poly(vinyl alcohol) (PVA) adsorbent. The equilibrium isotherm of uranyl ions was better represented by the Langmuir model, and the q_{max} achieved by the polymer/MWCNT containing sodium dodecyl sulfate (SDS) surfactant was 232.55 mg/g, indicating that the adsorbent had greater dispersion and culminating in increased adsorption capacity. The reusability of the PVA/MWCNTs was attested by the maintenance of high adsorption capacity even after three runs of the adsorption/regeneration process.

Nyairo et al. (2018) compared the adsorptive capacity of three adsorbents: pristine MWCNT, oxidized MWCNT (O-MWCNT), and the composite O-MWCNT/polypyrrole. These adsorbents were employed to remove Pb(II) and Cu(II) from an aqueous solution. Langmuir represented the adsorption equilibrium, and the composite achieved a q_{max} of 26.32 and 24.39 for Pb(II) and Cu(II), respectively, which corresponds to an increase of about 50% compared to O-MWCNT. The reusability of the composite was confirmed by high adsorption efficiencies, up to 90%, along with five successive processes of adsorption/desorption.

Ceramic CNT composites have also been investigated as adsorbents mainly because reinforcement of the ceramic increases fracture toughness of the CNT-ceramic composites (Peigney and Laurent 2006). Esfandiyari et al. (2017) proposed a composite formed by a sol-gel technique to remove the cationic dye basic yellow 28 (BY28). Adsorption equilibrium was best represented by the Freundlich model, indicating that adsorption occurs in multilayers and on heterogeneous surfaces. However, Langmuir also fitted the data satisfactorily, informing a q_{max} of 22.5 mg/g.

In contrast, Tofighy and Mohammadi (2015) prepared a ceramic adsorbent (CNT-mullite composite) to remove a toxic metal from an aqueous solution. Mullite is a relevant ceramic material with high-temperature stability (Ilić et al. 2014). The composite was also submitted to acid, chitosan, and acid-chitosan treatments. The composite treated with acid-chitosan presented the highest adsorption capacity of Langmuir, 60.34 mg/g, which can be explained by the increase of functional groups with affinity with copper ions by the acid (–COOH and –OH) and chitosan treatment ($-NH_2$ and –OH).

Aiming at water remediation, metal-CNT composites can be a promising alternative. In particular, CNT composites with magnetic properties have been studied due to their easy separation from an aqueous solution by exposure to an external magnetic field.

Song et al. (2016) proposed a silver nanoparticle-modified magnetic MWCNT composite (Ag-FMWCNTs) and employed it to remove sulfamethoxazole (SMZ). From the Langmuir isotherm model, the adsorbent containing 5% silver in its composition had a q_{max} of 118.58 mg/g, which was much higher than that achieved with lower proportions of Ag. The process could be classified as chemical sorption. Regarding reusability, methanol regenerates the adsorbent promoting adsorption efficiency superior to 80% even in the fourth cycle.

Considering adsorption processes, there is recurrent loss of CBN during water purification, which apart from posing health risks to the environment and human health, increases the operational costs. To overcome this problem, several protocols have been reported to efficiently fix CBNs using filtering membranes, beads, continuous-flow techniques, and aerogels. CNT composites based on $CaCO_3$ core-shell with loaded CNT to remove organic pollutants were evaluated in batch mode (Xu et al. 2015) and dynamic mode through a fixed-bed column (Xu et al. 2018). In batch mode, for the adsorption of 2-naphthol, the Freundlich constant increased from 128 to 222 (mg/kg)/(mg/L)^n with the increase of CNT content on the surface, i.e. with the increase of active sites on the surface of the carbonate core-shell. The adsorbent was also regenerated by using ethanol, which provided an adsorption rate above 90% in four cycles. In the continuous mode, Xu et al. (2018) studied the effect of flow rate, inlet concentration, and adsorbent mass on the capacity to adsorb 2-naphthol. The total efficiency range in the experiments was 60-70%, and the equilibrium adsorption capacity reached 286.6 mg/kg with an inlet concentration of 20 mg/L. The Thomas model indicated that external mass transfer dominates the overall kinetic mechanism.

4.2 Graphene Oxide Derivatives Composites

Graphene composites are described as all graphene-based materials, which have been grafted with reactive groups and functionalized with polymers (Kyzas et al. 2018). The high surface area of graphene makes graphene composites potential adsorbents for the removal of pollutants (Kyzas et al. 2018). Table 6 compiles recent works involving the application of graphene-based nanocomposites to the adsorption of hazardous substances.

TABLE 6 Graphene composites-based adsorbents: settings for adsorption equilibrium experiments, q_{max}, and performance after successive regeneration processes.

Adsorbent Matrix	Pollutant	Experimental Equilibrium Conditions	q_{max} (mg/g)	Regeneration*	Reference
GO polysulfone nanocomposite pellets	CP	pH 5.0, m_{ads} = 7 mg, V = 15 mL, C_0 = 13-130 mg/L	82.78 21.49	10 cycles ethanol	(Indherjith et al. 2019)
GO/dopamine	U(VI)	pH 5.0, m_{ads} = 0.01 g, V = 20 mL	717	5 cycles NaCl 0.8 mol/L ~85-75%	(Li et al. 2019a)
GO/dopamine/ polyethylene imine			734	5 cycles NaCl 0.8 mol/L 91-80%	
Chitosan grafted GO	Cr(VI)	pH 2.0, V = 10 mL, 600 min, C_0 = 10-125 mg/L	104.16	10 cycles NaOH 96 to 82%	(Samuel et al. 2019)
Poly(4-vinylpyridine)–GO–Fe_3O_4	MB	m_{ads} = 0.02 g V = 50 mL, 10 h, 180 rpm, T = 283 K, C_0 = 10-100 mg/L	164.20	5 cycles Ethyl alcohol 38.65-29.5 mg/g	(Li et al. 2019b)
Polyacrylamide-Fe_3O_4–rGO	CR	m_{ads} = 10 mg, V = 20 mL, 500 rpm, T = 298 K	166.7	5 cycles Acetone ~ 50-46 mg/g	(Pourjavadi et al. 2019)
Chitosan-Fe_3O_4-GO	As(III)	C_0 = 10 mg/L, S/L = 5 g/L, pH 7.3, 298 K, 250 rpm, 240 min	2.34	5 cycles NaOH 0.1 mol/L 61-47.8%	(Sherlala et al. 2019)
Polyglycerol-Fe_3O_4-GO	TC	m_{ads} = 4.5 mg, V = 5 mL, 150 rpm, 24 h, pH = 7.0, C_0 = 0-500 mg/L	684.93	5 cycles NaOH 0.01 mol/L 92-68 mg/g	(Yu et al. 2019)

* Adsorption efficiency or adsorption capacity along of cycles of adsorption/desorption

According to Mohan et al. (2018), polymer nanocomposites based on graphene have currently attracted the attention of the scientific community due to the great dispersibility of graphene derivatives in polymeric matrices and their action as a reinforcing agent.

Indherjith et al. (2019) evaluated the capacity of adsorbents based on GO and reduced GO polysulfone composites to remove the antibiotic CP. The adsorption equilibrium isotherm followed the Freundlich model and therefore adsorption of CP on both adsorbents is considered a multilayer adsorption process on energetically heterogeneous surfaces. The composite formed with GO achieved higher adsorption capacity than the rGO composite due to the increase of hydrogen bonding and π-π interactions between the polymeric nanomaterial and the antibiotic. Despite the satisfactory original adsorption capacities, both adsorbents had their adsorption capacity reduced by about 40% after 10 cycles of adsorption/desorption.

Li et al. (2019a) compared the adsorption capacity to remove uranium of two GO-based adsorbents composited by dopamine and dopamine/polyethyleneimine. Adsorption at equilibrium was represented by the Langmuir isotherm, i.e. the adsorbent surface had energetically equaled active sites and adsorbate uptake occurred by monolayer formation. The composite with two polymeric matrices had a slightly higher adsorption capacity at equilibrium at 298 K, 734 mg/g, while that composed with dopamine only presented a q_{max} of 717 mg/g. The authors affirmed that the increase in adsorption capacity is due to the chelation of U(VI) with the carboxyl and amino groups of the GO/dopamine/polyethyleneimine composite. The reusability of the absorbents was also validated for five cycles in which the adsorption efficiency remained up 75%.

There has been a recent increase in the number of papers studying the adsorptive capacity of magnetic graphene and CNT nanocomposites to retain hazardous materials. These kinds of composites include those based on iron oxides and elemental metals (Fe, Co, Ni, and Mn) among others. Bao et al. (2018) prepared a magnetic rGO composite by co-precipitating Fe(III) and Mn(II) on GO nanosheets reduced with N_2H_4. This material was assessed for its efficiency to adsorb TC. The q_{max} was 870, 1,131, and 1,326 mg/g, at 283, 298, and 313 K, respectively, which indicates the endothermic nature of the adsorption process. The authors affirmed that the multilayer assumption of the Freundlich model can be explained by the fact the TC had four aromatic rings that could be easily adsorbed on the adsorbent by π-π interactions and attract additional adsorbate molecules. The desorption experiment shows that the material could be reused four times reducing its efficiency from 86 to 71% using hydrochloric acid as eluent.

Lei et al. (2019) investigated the adsorptive performance of a magnetic GO composite to remove dyes from water. The cationic dyes evaluated were Rose Red B (RRB), CV, and MG, and the anionic ones were MO and CR. Due to the different nature of the dyes, the pH of maximum adsorption efficiency operation was very distinct for each dye. The q_{max} was 144.93, 105.15, 59.92, 77.94, and 208.33 mg/g for MG, MO, CV, RRB, and CR, respectively. These values are slightly higher than those obtained using GO as adsorbent, except for CV, which had a low q_{max} compared to the non-composite material. Regarding reusability, the adsorption capacity remained up to 42% even on the fifth cycle.

Through hydrothermal methods, Sahoo and Hota (2018) synthesized a magnetic GO adsorbent based on MgO-$MgFe_2O_4$ for fluoride removal. The adsorption equilibrium indicated that the process occurs by monolayer formation over the homogenous surface of the composite, according to the Langmuir model assumptions. The q_{max} values for 293, 303, and 313 K were 31.54, 34.24, and 40.98 mg/g, which were higher than those obtained with the non-carbon adsorbent (MgO-$MgFe_2O_4$). The mechanism proposed by the authors is electrostatic interactions, complexation-forming Mg-F^- and hydrogen bonding between OH-F^-. The regeneration of the adsorbent with sodium hydroxide guaranteed an adsorption efficiency of 57% on the fifth cycle.

As exposed above, polymer nanocomposites reinforced with graphene structure have been explored extensively for the adsorption of hazardous substances. In addition, magnetic graphene nanocomposites have been investigated as adsorbents with the advantage of magnetic solid-phase separation. Combining the advantages of these two approaches, some studies have been performed to investigate the improvement of the adsorptive properties of graphene or graphene derivatives by composite formation with polymers and the loading of magnetic particles.

Li et al. (2019b) studied the adsorption of MB on poly(4-vinylpyridine)–GO–Fe_3O_4 nano adsorbents. In this investigation, the authors showed that GO, ferrite, and the polymer had a synergic effect to improve the adsorption capacity to remove methylene blue. In addition, the adsorbent achieved, experimentally, the adsorption capacity of 152.51 mg/g at equilibrium conditions. The Freundlich model fitted better the experimental data, indicating MB forms multilayers on the adsorbent surface. Regarding reusability, the adsorbent was regenerated using ethanol, and after five cycles the adsorption capacity decreased from 38.65 to 29.59 mg/g.

Pourjavadi et al. (2019) evaluated the adsorption of CR dye on GO decorated with ferrite nanoparticles, reduced and functionalized with 3-(trimethoxysilyl)propyl methacrylate, and then grafted with polyacrylamide. The adsorption of this anionic dye on the proposed adsorbent followed the Langmuir model which assumes the monolayer formation with a q_{max} of 166.7 mg/g. The authors indicated that this high value was achieved thanks to the dispersion facility provided by the polymer insertion on the composite matrix. The adsorbent had a slight loss of the adsorption capacity after five cycles of adsorption.

Graphene adsorbents prepared by the insertion of surface-active agents have been reported in the literature (Mahmoodi et al. 2017). Cationic surfactant-loaded GO sponge was prepared and used to simultaneously remove Cu(II) and BPA from an aqueous solution (Kuang et al. 2019). When the loading of the cationic surfactant hexadecyltrimethylammonium bromide (HDTMA) was between 6.4 and 11.5%, GO nanosheets were able to form arranged 3D sponge. This can be attributed to the decrease of repulsive forces between electron-rich GO nanosheets in the presence of cationic HDTMA and the increase of hydrophobic interactions among the alkyl chains of the surfactant molecules adsorbed on the GO surface. PBA q_{max} by GO-HDTMA sponge was 141.0 mg/g, while pristine GO and GO/chitosan sponge provided lower efficiencies (1.5 and 1.6 mg/g, respectively). Regarding Cu(II) uptake, pristine GO showed the best performance (84.3 mg/g) and GO/chitosan and GO/HDTAM sponges were able to remove Cu(II) with a q_{max} of 66.1 and 59.7 mg/g, respectively. GO modified by the insertion of cationic surfactant-like moieties (GO-QAS) was synthesized by grafting arenediazonium salt and used in the preparation of a magnetic composite with chitosan (Neves et al. 2020b). The prepared mCS/GO-QAS was then applied to the removal of basic brown 4 dye (BB4) and reached a q_{max} of 650 mg/g, a performance superior to MWCNTs (200 mg/g) (Mahalingam et al. 2013).

4.3 Nanoporous Carbon Composites

NPC is a very porous material used to remove contaminant species from water-based mainly on surface adsorption (Khan et al. 2018). NPC has been applied to decolorization, drinking water and wastewater treatment, gas absorption, and solvent recovery systems. NPC composites have been employed to remediate water. Table 7 presents the most recent research works that applied NPC composites to remove organic and inorganic pollutants.

Dutta and Nath (2018) produced SiO_2-nano carbon from corn cobs and used it in the adsorption of cationic dye (MB) and toxic metals (uranium and chromium). The Freundlich isotherm model better described the adsorption equilibrium of all adsorbates, indicating the formation of multilayers on the adsorbent surface. The nanocomposite with silica had greater adsorption capacity than nanocarbon alone due to the high surface area. SiO_2-nano carbon presented a specific surface area of 715.22 m^2/g against 430.17 m^2/g to nanocarbon. For the desorption of MB, pH 1 was able to regenerate the adsorbent which in a subsequent adsorption process achieved an efficiency of 74%.

Huang et al. (2017) synthesized magnetic *N*-doped carbon composites via mixture and carbonization of zinc oxide nanoparticles, methylimidazole, and ferrite-silica nanoparticles. Comparing the adsorptive capacity to uptake mercury (Hg(II)) of the adsorbents made by different means of synthesis, those produced at 700°C and with 3% of nitrogen had an increased adsorption capacity (429 mg/g) due to the high surface area, and the binding sites also presented proper magnetism thanks to nitrogen content. The adsorbent produced was stable during ten cycles of adsorption/desorption using HCl as an eluent and maintained the original adsorption capacity.

Liang et al. (2020a), similar to Huang et al. (2017), proposed an *N*-doped hollow NPC composite using a zeolitic imidazolate framework as a template through different routes and compared their capacity to adsorb TC. The Langmuir model, indicating the monolayer formation, better represented the equilibrium data for all systems. The q_{max} NPC, the NPC formed by using resorcinol and formaldehyde solutions, and the NPC produced by using tannin acid solutions were 180.2, 284.9, and 518.1 mg/g, respectively. The excellent adsorption capacity of the composite formed by using tannin is due to the adsorbent' porosity, which permits the diffusion and fixation of TC, and to the high nitrogen content which implies a high number of binding sites. Regarding reusability, the q_{max} fell to 85% after the last (eight) regeneration cycles.

TABLE 7 NPC composites-based adsorbents: settings for adsorption equilibrium experiments, q_{max}, and performance after successive regeneration processes.

Adsorbent Matrix	Pollutant	Experimental Equilibrium Conditions	q_{max} (mg/g)	Regeneration*	Reference
SiO_2/NPC from corn cobs	U(VI)	m_{ads} = 30 mg, V = 50 mL, 2 h, pH = 5	255.12	–	(Dutta and Nath 2018)
	Cr(VI)	m_{ads} = 10 mg, V = 50 mL, 4 h, pH = 2	90.01	–	
	MB	m_{ads} = 10 mg, V = 25 mL,	–	1 cycle, pH 1, 74.24%	
Magnetic nitrogen doped NPC composite	Hg(II)	m_{ads} = 10 mg, V = 10 mL, C_0 = 50-500 mg/L, 60 min	429	10 cycles, HCl 429 mg/g (1st -10th)	(Huang et al. 2017)
Nitrogen doped hollow NPC	TC	m_{ads} = 2 mg, V = 50 mL, C_0 = 3-100 mg/L, 60 min, T = 298 K, 16 h	518.1	8 cycles NaOH/ethanol ~ 240-180 mg/g	(Liang et al. 2020a)
MOF-derived NPC	Vanadium	m_{ads} = 0.01 g, V = 20 mL, C_0 = 50-250 mg/L, T = 298 K, 15 min	285.71	3 cycles ~198-192 mg/g	(Salehi et al. 2020)
Attapulgite/NPC composites	Chlortetracycline	m_{ads} = 20 mg, V = 20 mL, C_0 = 50-1,000 mg/L, 4 h, T = 298 K	336.4	10 cycles Heat at 573 K	(Tang et al. 2018)
	TC		297.9		

* Adsorption efficiency or adsorption capacity along of cycles of adsorption/desorption

Salehi et al. (2020) synthesized a nanoadsorbent through carbonization of the crystalline metal-organic framework (MOF-199) in the presence or not of carbon precursors, which includes CNTs. These materials were put in contact with vanadium solutions and their adsorptive efficiency was compared. The Langmuir isotherm model fitted well the equilibrium data and indicated that the composite of CNTs as precursors had the best adsorptive performance, achieving a q_{max} of 285.71 mg/g. Desorption experiments, carried out in 3 runs, demonstrated that the adsorbent could be effectively regenerated by using NaOH.

Tang et al. (2018) studied the adsorption of the antibiotics chlortetracycline and TC on attapulgite/NPC produced through direct calcination of spent bleaching earth derived from soybean oil bleaching. The q_{max} for chlortetracycline and TC was 336.4 and 297.9 mg/g, respectively. In terms of reusability, the adsorbent maintained 67.3% and 62.9% of the original adsorption capacity for chlortetracycline and TC in the tenth cycle.

4.4 Nanodiamond, Fullerene, and Carbon Nitride Composites

Carbon materials can exist in different structural arrangements of carbon atoms within their unit cells and consequently have variable properties. Innovative nanocomposites based on these carbon materials (NDs, fullerene, and CN) have been recently applied to remediate water, according to Table 8.

TABLE 8 NDs, fullerene, and CN composites-based adsorbents: settings for adsorption equilibrium experiments, q_{max}, and performance after successive regeneration processes.

Adsorbent Matrix	Pollutant	Experimental Equilibrium Conditions	q_{max} (mg/g)	Regeneration*	Reference
ND-containing polyethyleneimine	Pb(II)	m_{ads} = 0.1 g, V = 15 mL, 250 rpm, C_0 = 5-300 mg/L, pH = 5.0, 7 h	17.12	3 cycles HCl 0.5 mol/L 16.5-14.1 mg/g	(Çiğil et al. 2019)
MOF-oxidized ND composite	MR	m_{ads} = 20 mg, V = 100 mL, T = 298 K, 2 h	500	3 cycles Ethanol ~94-88 mg/g	(Molavi et al. 2020)
ND-filled chitosan	MO	V = 100 mL, T = 298 K	454.5	5 cycles Ethanol ~88-70%	(Raeiszadeh et al. 2018)
Magnetic fullerene nanocomposite	MB	C_0 = 200 mg/L, S/L = 0.2 mg/mL, 298 K	833.3	10 cycles HCl 1 mol/L ~ 94-90%	(Elessawy et al. 2020a)
	Acid Blue 25 (AB25)		806.5	10 cycles NaOH 1 mol/L ~92-88%	
	CP	V = 50 mL, T = 303, 150 rpm, C_0 = 50-200 mg/L	356	5 cycles Methanol and NaOH (3%) ~100-98%	(Elessawy et al. 2020b)
Three-dimensional magnetic CN composite	Pb(II)	V = 100 mL, T = 298 K	423.73	5 cycles HNO_3 2 mol/L ~192-175 mg/g	(Guo et al. 2018)

* Adsorption efficiency or adsorption capacity along of cycles of adsorption/desorption

According to Al-Tamimi et al. (2019), ND is a kind of carbon-based nanoparticle with diamond properties in the nanometer scale. The unique geometry and the variety of functional groups on the surface, such as carboxyl, hydroxyl, ketones, and ethers, make ND an interesting material to create nanocomposites with exceptional adsorptive properties.

Çiğil et al. (2019) evaluated the adsorption of lead on a ND/polyethyleneimine composite. The q_{max} from the Langmuir fit was 17.12 mg/g. The authors suggested that the adsorption mechanism is based on the formation of chelates between Pb(II) ions and amine groups on the adsorbent surface. Desorption of loaded lead by using HCl in three runs occurred with high elution rates, up to 92% even in the last cycle, with adsorption capacity varying from 16.5 (1st) to 14.1 mg/g (3rd).

Molavi et al. (2020) synthesized a hybrid nanoparticle based on thermally oxidized ND and UiO-66 MOF to treat water contaminated with Methyl Red (MR) dye. The equilibrium data were described by the Langmuir isotherm model with q_{max} of 500.00, 555.56, 560.25 mg/g, for 298, 308, and 318 K, respectively, which indicates the endothermic nature of the process. The desorption experiment was performed three times using ethanol with ultrasonic assistance in which the adsorbent lost 7% of the original adsorption capacity.

Raeiszadeh et al. (2018) studied the adsorption of the anionic dye MO onto a composite based on the biopolymer (chitosan) and NDs with high carboxyl contents. The equilibrium results revealed that the chitosan achieved a q_{max} of 167 mg/g, while the composite reached 454 mg/g which indicates that the incorporation of NDs on chitosan considerably favored the adsorption of MO. The authors indicated that the increase in adsorption capacity is due to the oxygen-containing groups on the external surface of NDs, which form hydrogen bonding and electrostatic interactions with the dye molecules. Over five cycles of adsorption/desorption, the polymer/ND adsorbent lost fifteen percent of adsorption efficiency.

Fullerene is the first nanocarbon material to feature a closed-cage structure. Fullerene is chemically reactive due to the presence of electron-deficient polyalkene, which favors the formation of fullerene-based composites (Chen et al. 2019).

Elessawy et al. (2020a) proposed obtaining a functionalized magnetic fullerene nanoadsorbent by catalytic thermal decomposition of recycled bottles of polyethylene terephthalate (PET). The adsorbent was applied to remove cationic (MB) and anionic (AB25) dyes and an antibiotic (CP). The q_{max} for MB and AB25 was 833.3 and 806.5 mg/g, respectively. After ten cycles of adsorption/desorption, the adsorbent was effectively regenerated using HCl or NaOH. However, the investigation of Elessawy et al. (2020b) indicates that the Langmuir model also represented well the equilibrium of CP adsorption, whose q_{max} was 365 mg/g. The adsorbent was satisfactorily reused in five processes of regeneration using methanolic and basic solutions. In both works, the excellent adsorption ability of the composites is due to the possibility of different adsorbate/adsorbent bindings, such as electrostatic and π-π interactions, and hydrogen bonding to uptake the dye molecules.

Graphitic carbon nitride (CN), the most stable allotrope of CN under ambient conditions, presents an exclusive assembly comprised of a 2D sheet of tri-s-triazine connected via tertiary amines (Chen et al. 2019). As with 0D carbon nanomaterials (fullerene), carbon 2D nanomaterials such as CN composites have their adsorptive proprieties improved to purify water solutions.

Guo et al. (2018) proposed a composite formed by magnetic particles (Fe_3O_4) and CN to remove toxic metal lead from water. The Langmuir model predicted the q_{max} of ferrite, CN and the nanocomposite as 80.13, 251.89, and 423.73 mg/g, respectively. The nanocomposite was regenerated by using HNO_3 and maintained 89.3% of the original adsorption capacity in the fifth cycle.

In general, carbon-based composites used as adsorbents for the removal of adsorbates with different chemical natures showed improved adsorptive properties compared to their non-composite form. Structure parameters, insertion of new and more selective functional groups, and stability were relevant attributes gained by carbon-based composite adsorbents. The high adsorption capacities of these materials indicate a potential for industrial applications; however, studies on dynamic mode, pilot scale, or with real effluents are still lacking to make this technology mature enough for large-scale use.

5. CARBON-BASED NANOMATERIALS AND TOXIC EFFECTS

Advances in nanoscale science and nanoengineering lead us to believe that many problems related to water purification will be remedied or at least minimized. With high production, the discharge of nanomaterials into the environment will be inevitable. Thus, there is growing concern regarding the effects caused by the presence of nanomaterials in the environment, which has already been reported (Sanchís et al. 2018, 2020, Richardson and Kimura 2020). Regarding CBN toxicity, it may be inherent to the material itself or stem from synergistic interactions with other pollutants or common NOM. On the other hand, the presence of CBNs can be beneficial and decrease the toxicity of other co-existent pollutants (antagonistic effect) mainly through the adsorption mechanism. The associated toxicity of CBN has been studied for several organisms, such as bacteria (Mortimer et al. 2020), aquatic organisms (Medeiros et al. 2020), plants (Jia et al. 2019), insects (Dziewięcka et al. 2020), and mammals (Mrózek et al. 2020).

Toxicities of GO and thermally reduced GO at 200°C (rGO_{200}), 500°C (rGO_{500}), and 800°C (rGO_{800}) were analyzed for *Escherichia coli* (Barrios et al. 2019). GO was the least toxic CBN probably due to its lower carbon/oxygen, indicating that hydrophobic materials interact better with organisms. GO had the highest risk for oxidative stress, but rGO_{500}/rGO_{800} had higher wrapping capacity that decreased membrane permeability and caused higher toxicity by physical mechanism.

GO, ascorbic acid, and hydrazine-reduced GO, A-rGO, and H-rGO, respectively, were evaluated for their cytotoxicity in epithelial lung cells (A549) (Dervin et al. 2018). High toxicity for the mitochondrial function was caused by rGO, while ArGO compromised the membrane integrity. The lower biocompatibility of ArGO was probably due to its irregular and sharp surface that could cause cell apoptosis/necrosis. HrGO becomes coarse, but not irregular, indicating a physical membrane rupture as the main toxicity route. These toxicity mechanisms were also observed for *S. obliquus* and *D. magna* in the presence of GO-HA (Zhang et al. 2019). *S. obliquus* cells indicated that GO aggregated on its surface, damaging it, and GO was ingested and accumulated in *D. magna* digestive tract, indicating a long-term concentration process that affected the first brood birth.

Recent studies have addressed the modulation of the toxicity of co-contaminants in the presence of CBNs in terms of bioavailability and transport. If the effect caused by simultaneous exposure is equal to the sum of the toxicities of a single component, the relation is named additive, whereas when a greater effect is verified, the relationship is named synergistic. However, the antagonistic effect is observed when the toxicity in simultaneous exposure is less than that predicted by the additive relation. Ouyang et al. (2020) studied the influence of HA and natural nanocolloids (NCs) on GO phytotoxicity for *Chlorella vulgaris*. The presence of HA mitigated GO phytotoxicity. The aggregates of GO and GO-NCs trapped algae cells and NCs, GO, and GO-NCs were able to enter algal cells, damaging the cell wall and consequently caused plasmolysis. A similar effect of HA on GO toxicity was reported for zebrafish (Chen et al. 2015) and *D. magna* (Zhang et al. 2019).

GO, G, and amine-modified graphene (GNH) toxicities to *Scenedesmus obliquus* were studied in the presence of Cd(II) ions (Zhang et al. 2020). Cd toxicity increased in the presence of G and GO and decreased with GNH. The higher reactive oxygen species (ROS) in the presence of Cd/nanomaterials could cause intracellular oxidative stress as a toxicity route. Cell analysis indicated the coating of nanomaterials on the cell surface, and they penetrate it due to an increase in nanomaterial structure disorder at higher CBN concentrations and smaller particle production when interacting with Cd ions, suggesting physical damage as another toxicity route.

The modulation of Cd(II)/Zn(II) toxicities to the fresh water fish *Geophagus iporangensis* caused by the presence of GO was evaluated (Medeiros et al. 2020). GO-Cd (4 mg/L) reduced the metabolic rate of the fish by 33.7%, whereas GO-Zn at 10 mg/L increased it around three times when compared to the control group. It was verified that the GO-Zn complex showed an increment in its hydrodynamic diameter, indicating the formation of aggregates. A low GO effect on *E. coli* was also found by Sun (Sun et al. 2018), who studied levofloxacin (LV) and CP removal by GO. These antibiotics affected less the bacteria after GO sorption with survival rates increasing to 67.6% and 70% for LV and CP, respectively.

CNTs, like GO, are widely used in several areas, and therefore their related toxicities have been investigated, ranging from cultured cells (Mohanta et al. 2019) to model organisms (Lam et al. 2006). Zhao et al. (2019a) investigated the diameter-dependent toxicity of MWCNTs to human umbilical vein endothelial cells (HUVECs). Using similar CBN doses, toxicity tests indicated that the smallest diameter MWCNTs (XFM4) induced significantly higher cytotoxicity than the other evaluated MWCNTs. The modulation of the stress of endoplasmic reticulum autophagy by chemicals resulted in a significant increase in the cytotoxicity of MWCNT. Zhou et al. (2017) reported how the functionalization of MWCNTs could influence their cytotoxicity to human lung cancer cells. Taking into account a dose-response effect and matrix dispersion, the results obtained suggest that pristine

MWCNTs induce more cell death than functionalized MWCNTs, while functionalized MWCNTs are more genotoxic.

The accumulation and toxicity of MWCNTs in *Xenopus tropicalis* were evaluated by Zhao et al. (2020). A high survival rate of the organisms (97%) was observed in a 24h period of exposure to MWCNTs, which decreased considerably over time. LC_{50} of 2.53 mg/L for tadpoles exposed for 72 hours to MWCNTs was found, indicating that the nanomaterial accumulated in the gills and digestive tract of the model organism.

On the other hand, Sun et al. (2020) evaluated the effects of CNTs on the toxicity of the metals Cu, Cd, and Zn in relation to the freshwater microalgae *Scenedesmus obliquus.* When co-exposed with Cu, Cd, and Zn, CNTs alleviated the adverse effects initially observed for the metals.

The high potential of fullerene (C_{60}) material motivated the development of research regarding its toxicity. Biby et al. (2020) studied the associated neurotoxicity of dextran-stabilized C_{60} nanoparticles (Dex-C_{60}) to glial cells C_6. The interactions between C_6-Dex-C_{60} and glial cells demonstrated that the Dex-C_{60} nanoparticles penetrate deeper into the cells and cause dose-dependent cellular toxic responses.

Chronic and acute toxicity of CBN to *Daphnia magna* indicates that a reduction in the organism's reproduction occurred when nC_{60} was 1 mg/L (Wang et al. 2019). The ecotoxicity of C_{60} to *Lumbriculus variegatus* and *Nitzschia palea* was also evaluated by Ponte et al. (2019). The C_{60} inhibited *N. paela* photosynthesis at all concentrations tested (7-100 mg/L), and for *L. variegatus* population growth was affected at a higher C_{60} dose.

Regarding NDs, due to the wide application of this nanomaterial in the biomedical therapeutic and diagnostic areas, a series of studies in cultured cells and model organisms were evaluated in order to demonstrate the low toxicity of NDs in mammals (Kumar et al. 2019). However, there are few studies that address the toxic potential of this nanomaterial in biota.

The combined acute toxicity of NDs and thiabendazole (TBZ) in relation to *Daphia magna* were conducted in the presence and absence of the green algae *Raphidocelis subcapitata* to assess the influence of different feeding conditions (Martín-de-Lucía et al. 2019). The immobilization of *D. magna* exposed to NDs in the presence of *R. subcapitata* mitigated the toxicity of the nanomaterial. The results of the binary mixtures of NDs and TBZ showed synergistic interactions at low concentrations and at high doses of NDs, an antagonistic effect was observed.

Due to the recent trend in applying CN materials, few studies in the literature report their toxicity. Dong et al. (2018) evaluated the toxicity of CNs for lung cancerous epithelial cells (A549). The results obtained indicated that the toxicological effects of CN depended on the dose and the N/C ratio of the materials. In addition, CNs showed less toxicity when compared to GO, indicating the potential application of CN to biomedicine.

CBNs toxicity mechanisms are often grouped into two main categories, which are physical (based on cell membrane damage) and chemical (due to the generation of ROS). Regarding the physical mechanism, it can happen due to a dependence on the structure and size of the nanoparticles, and the loss of cell membrane integrity is caused by microorganism cell contact with the sharp and irregular edge of nanomaterials. CBNs are hydrophobic and can easily form aggregates (Su et al. 2017), but when interacting with other compounds, surface/charge can improve their dispersity.

In relation to chemical mechanisms, CBN can be internalized by target organisms or model cells, through material coating/adsorption on the cell surface (Ouyang et al. 2020, Zhang et al. 2020). It can prevent cell growth, cause shading effect and consequently limit nutrient/light access (Markovic et al. 2020), causing structure damage (Zhang et al. 2019, 2020). The internalization process of CBNs can also lead to oxidative stress of cells, triggering lipid peroxidation of cell membranes (Zhang et al. 2020).

Taking into account what was discussed about CBNs toxicity, the presence of CBNs in the environment caught the attention of the scientific community due to uncertainties regarding its disposal, mobility, and possible toxic effects. However, few studies have experimentally determined their concentration levels in aquatic systems. Sanchís et al. (2018) determined the presence of fullerenes in the river Sava, located in southeastern Europe. C_{60} was the ubiquitous fullerene in the analyzed samples with concentrations ranging from 8 to 59 ng/L in water. In other studies, also developed by Sanchís et al. (2014, 2015), fullerene C_{60} was detected in concentrations ranging from 31 pg/L to 4.5 ng/L in surface waters of Spain and Brazil, respectively. No data related to the presence of other members of the CBN family were found mainly due to the difficulty of quantifying these materials in environmental samples.

Conclusions

CBNs have shown promising use as filtering agents in batch and continuous flow experiments. Chemical and physical modification of CBNs' surfaces can improve some properties, enhancing their ability to adsorb hazardous compounds, such as metallic ions, dyes, drugs, surfactants, pesticides, and PAH. The generation of composites through the mixing of CBNs with other materials, including MOF, metallic ions, metal oxides, polymer, and ceramics, enhance the surface area of the adsorbents and facilitates their functionalization. The use of CBNs composites as adsorbents has many advantages, mainly relating to their high performance, recycling, and easy extraction of adsorbents from the water after use. Despite the enormous potential of CBNs as filtering agents, most studies are batch-based; however, it is known that flow adsorption systems are required for real applications. Thus, a significant investment in research to pass batch-mode/laboratory scale to real large-scale applications is required. It is also important to highlight the toxicity of CBNs to human health and the environment as well as the effect of the presence of common interfering agents and the ability of CBNs to modulate the toxicity of other contaminants. Therefore, technologies based on CBNs for water purification must take into account aspects of process efficiency, cost-effectiveness, disposal after use as well as safety for human health and the environment.

Acknowledgments

We thank the Research Supporting Foundation of the State of São Paulo (FAPESP, proposal nº. 2017/18236-1, 2019/11353-8, 2019/07822-2 and 2019/25228-0), Brazilian National Research Council (CNPq, proposal nº. 406193/2018-5 and 308046/2019-6), Coordination for the Improvement of Higher Education Personnel (CAPES, Financial code – 001), and to Espaço da Escrita – Coordenadoria Geral da Universidade – UNICAMP – for the language services provided.

References

Abdeen, Z. and Akl, Z.F. 2015. Uranium(VI) adsorption from aqueous solutions using poly(vinyl alcohol)/carbon nanotube composites. RSC Adv. 5: 74220-74229.

Akbayrak, S., Özçifçi, Z. and Tabak, A. 2020. Activated carbon derived from tea waste: A promising supporting material for metal nanoparticles used as catalysts in hydrolysis of ammonia borane. Biomass Bioenerg. 138: 105589.

Al-Tamimi, B.H., Jabbar, I.I. and Al-Tamimi, H.M. 2019. Synthesis and characterization of nanocrystalline diamond from graphite flakes via a cavitation-promoted process. Heliyon 5: e01682.

Alijani, H. and Shariatinia, Z. 2018. Synthesis of high growth rate SWCNTs and their magnetite cobalt sulfide nanohybrid as super-adsorbent for mercury removal. Chem. Eng. Res. Des. 129: 132-149.

Anayurt, R.A., Sari, A. and Tuzen, M. 2009. Equilibrium, thermodynamic and kinetic studies on biosorption of Pb(II) and Cd(II) from aqueous solution by macrofungus (Lactarius scrobiculatus) biomass. Chem. Eng. J. 151: 255-261.

Anbia, M. and Ghaffari, A. 2009. Adsorption of phenolic compounds from aqueous solutions using carbon nanoporous adsorbent coated with polymer. Appl. Surf. Sci. 255: 9487-9492.

Avcı, A., İnci, İ. and Baylan, N. 2020. Adsorption of ciprofloxacin hydrochloride on multiwall carbon nanotube. J. Mol. Struct. 1206: 1-7.

Azimi, B.E., Badiei, A. and Ghasemi, J.B. 2019. Efficient removal of malachite green from wastewater by using boron-doped mesoporous carbon nitride. Appl. Surf. Sci. 469: 236-245.

Ba, G., Liang, Z., Li, H., Deng, Q., Du, N. and Hou, W. 2020. Synthesis of hierarchically mesoporous polymeric carbon nitride with mesoporous melamine as a precursor for enhanced photocatalytic performance. Chem. Eng. J. 380: 122535.

Baalousha, M., How, W., Valsami-Jones, E. and Lead, J.R. 2014. Chapter 1 – Overview of environmental nanoscience. *In*: Lead, Jamie R. and Valsami-Jones, Eugenia. [eds.]. Nanoscience and the Environment. Front. Nanosci. 7: 1-54.

Bakhtiari, N., Azizian, S., Alshehri, S.M., Torad, N.L., Malgras, V. and Yamauchi, Y. 2015. Study on adsorption of copper ion from aqueous solution by MOF-derived nanoporous carbon. Microporous Mesoporous Mater. 217: 173-177.

Ballesteros, E., Gallego, M. and Valcárcel, M. 2000. Analytical potential of fullerene as adsorbent for organic and organometallic compounds from aqueous solutions. J. Chromatogr. A 869: 101-110.

Bandaru, N.M., Reta, N., Dalal, H., Ellis, A.V., Shapter, J. and Voelcker, N.H. 2013. Enhanced adsorption of mercury ions on thiol derivatized single wall carbon nanotubes. J. Hazard. Mater. 261C: 534-541.

Bao, J., Zhu, Y., Yuan, S., Wang, F., Tang, H., Bao, Z., et al. 2018. Adsorption of tetracycline with reduced graphene oxide decorated with $MnFe_2O_4$ nanoparticles. Nanoscale Res. Lett. 13: 1-8.

Barrios, A.C., Wang, Y. and Gilbertson, L.M. 2019. Structure–property–toxicity relationships of graphene oxide: Role of surface chemistry on the mechanisms of interaction with bacteria. Environ. Sci. Technol. 53: 14679-14687.

Becker, L., Bada, J., Winans, R., Hunt, J., Bunch, T. and French, B. 1994. Fullerenes in the 1.85-billion-year-old Sudbury impact structure. Science 265(5172): 642-645. doi: 10.1126/science.11536660.

Biby, T.E., Prajitha, N., Ashtami, J., Sakthikumar, D., Maekawa, T. and Mohanan, P.V. 2020. Toxicity of dextran stabilized fullerene C60 against C6 Glial cells. Brain Res. Bull. 155: 191-201.

Bin-Dahman, O.A. and Saleh, T.A. 2020. Synthesis of carbon nanotubes grafted with PEG and its efficiency for the removal of phenol from industrial wastewater. Environ. Nanotechnology, Monit. Manag. 13: 100286.

Chaabane, L., Beyou, E., El Ghali, A. and Baouab, M.H.V. 2020. Comparative studies on the adsorption of metal ions from aqueous solutions using various functionalized graphene oxide sheets as supported adsorbents. J. Hazard. Mater. 389: 121839.

Cheminski, T., de Figueiredo Neves, T., Silva, P.M., Guimarães, C.H. and Prediger, P. 2019. Insertion of phenyl ethyleneglycol units on graphene oxide as stabilizers and its application for surfactant removal. J. Environ. Chem. Eng. 7: 102976.

Chen, H., Yan, T. and Jiang, F. 2014. Adsorption of Cr(VI) from aqueous solution on mesoporous carbon nitride. J. Taiwan Inst. Chem. Eng. 45: 1842-1849.

Chen, M., Guan, R. and Yang, S. 2019. Hybrids of fullerenes and 2D nanomaterials. Adv. Sci. 6: 1800941.

Chen, Y., Ren, C., Ouyang, S., Hu, X. and Zhou, Q. 2015. Mitigation in multiple effects of graphene oxide toxicity in zebrafish embryogenesis driven by humic acid. Environ. Sci. Technol. 49: 10147-10154.

Cheng, X., Kan, A.T. and Tomson, M.B. 2004. Naphthalene adsorption and desorption from aqueous C_{60} fullerene. J. Chem. Eng. Data 49: 675-683.

Choudhary, V., Singh, B.P. and Mathur, R.B. 2013. Chapter 9 – Carbon nanotubes and their composites. pp. 52897. *In*: Satoru, Suzuki [ed.]. Syntheses and Applications of Carbon Nanotubes and Their Composites. IntechOpen.

Çiğil, A.B., Urucu, O.A. and Kahraman, M.V. 2019. Nanodiamond-containing polyethyleneimine hybrid materials for lead adsorption from aqueous media. J. Appl. Polym. Sci. 136: 1-9.

Darkwah, W.K. and Ao, Y. 2018. Mini review on the structure and properties (photocatalysis), and preparation techniques of graphitic carbon nitride nano-based particle, and its applications. Nanoscale Res. Lett. 13: 388.

Das, N. 2010. Recovery of precious metals through biosorption: A review. Hydrometallurgy 103: 180-189.

Dehghani, M.H., Taher, M.M., Bajpai, A.K., Heibati, B., Tyagi, I., Asif, M., et al. 2015. Removal of noxious Cr(VI) ions using single-walled carbon nanotubes and multi-walled carbon nanotubes. Chem. Eng. J. 279: 344-352.

Dehghani, M.H., Mostofi, M., Alimohammadi, M., McKay, G., Yetilmezsoy, K., Albadarin A.B., et al. 2016. High-performance removal of toxic phenol by single-walled and multi-walled carbon nanotubes: Kinetics, adsorption, mechanism and optimization studies. J. Ind. Eng. Chem. 35: 63-74.

Dehghani, M.H., Kamalian, S., Shayeghi, M., Yousefi, M., Heidarinejad, Z., Agarwal, S., et al. 2019. High-performance removal of diazinon pesticide from water using multi-walled carbon nanotubes. Microchem. J. 145: 486-491.

Dervin, S., Murphy, J., Aviles, R., Pillai, S.C. and Garvey, M. 2018. An *in vitro* cytotoxicity assessment of graphene nanosheets on alveolar cells. Appl. Surf. Sci. 434: 1274-1284.

Dhillon, A. and Kumar, D. 2015. Development of a nanoporous adsorbent for the removal of health-hazardous fluoride ions from aqueous systems. J. Mater. Chem. A 3: 4215-4228.

Dong, Q., Mohamad Latiff, N., Mazánek, V., Rosli, N.F., Chia, H.L., Sofer, Z., et al. 2018. Triazine- and heptazine-based carbon nitrides: Toxicity. ACS Appl. Nano Mater. 1: 4442-4449.

Dutta, D.P. and Nath, S. 2018. Low cost synthesis of SiO_2/C nanocomposite from corn cobs and its adsorption of uranium(VI), chromium(VI) and cationic dyes from wastewater. J. Mol. Liq. 269: 140-151.

Dziewięcka, M., Flasz, B., Rost-Roszkowska, M., Kędziorski, A., Kochanowicz, A. and Augustyniak, M. 2020. Graphene oxide as a new anthropogenic stress factor-multigenerational study at the molecular, cellular, individual and population level of Acheta domesticus. J. Hazard. Mater. 396: 122775.

Elessawy, N.A., El-Sayed, E.M., Ali, S., Elkady, M.F., Elnouby, M. and Hamad, H.A. 2020a. One-pot green synthesis of magnetic fullerene nanocomposite for adsorption characteristics. J. Water Process Eng. 34: 101047.

Elessawy, N.A., Elnouby, M., Gouda, M.H., Hamad, H.A., Taha, N.A., Gouda, M., et al. 2020b. Ciprofloxacin removal using magnetic fullerene nanocomposite obtained from sustainable PET bottle wastes: Adsorption process optimization, kinetics, isotherm, regeneration and recycling studies. Chemosphere 239: 124728.

Esfandiyari, T., Nasirizadeh, N., Ehrampoosh, M.H. and Tabatabaee, M. 2017. Characterization and absorption studies of cationic dye on multi walled carbon nanotube–carbon ceramic composite. J. Ind. Eng. Chem. 46: 35-43.

Firozjaee, T.T., Mehrdadi, N., Baghdadi, M. and Nabi Bidhendi, G.R.N. 2017. The removal of diazinon from aqueous solution by chitosan/carbon nanotube adsorbent. Desalin. Water Treat. 79: 291-300.

Fronczak, M., Demby, K., Strachowski, P., Strawski, M. and Bystrzejewski, M. 2018. Graphitic carbon nitride doped with the s-block metals: Adsorbent for the removal of methyl blue and copper(II) ions. Langmuir 34: 7272-7283.

Ghasemi, S.S., Hadavifar, M., Maleki, B. and Mohammadnia, E. 2019. Adsorption of mercury ions from synthetic aqueous solution using polydopamine decorated SWCNTs. J. Water Process Eng. 32: 100965.

Guo, S., Duan, N., Dan, Z., Xu, F., Zhang, C., Shi, F., et al. 2018. Three-dimensional magnetic graphitic carbon nitride composites as high-performance adsorbent for removal Pb^{2+} from aqueous solution. J. Taiwan Inst. Chem. Eng. 89: 169-182.

Gupta, K., Gupta, D. and Khatri, O.P. 2019. Graphene-like porous carbon nanostructure from Bengal gram bean husk and its application for fast and efficient adsorption of organic dyes. Appl. Surf. Sci. 476: 647-657.

Gupta, V.K. and Saleh, T.A. 2013. Sorption of pollutants by porous carbon, carbon nanotubes and fullerene: An overview. Environ. Sci. Pollut. Res. 20: 2828-2843.

Gusain, R., Kumar, N. and Ray, S.S. 2020. Recent advances in carbon nanomaterial-based adsorbents for water purification. Coord. Chem. Rev. 405: 213111.

Güzel, F., Sayłili, H., Akkaya Sayłili, G. and Koyuncu, F. 2015. New low-cost nanoporous carbonaceous adsorbent developed from carob (Ceratonia siliqua) processing industry waste for the adsorption of anionic textile dye: Characterization, equilibrium and kinetic modeling. J. Mol. Liq. 206: 244-255.

Han, T., Xiao, Y., Tong, M., Huang, H., Liu, D., Wang, L., et al. 2015. Synthesis of CNT@MIL-68(Al) composites with improved adsorption capacity for phenol in aqueous solution. Chem. Eng. J. 275: 134-141.

Hartland, A., Lead, J.R., Slaveykova, V.I., O'Carroll, D. and Valsami-Jones, E. 2013. The environmental significance of natural nanoparticles. Nat. Educ. Knowl. 4: 7.

Hiew, B.Y.Z., Lee, L.Y., Lee, X.J., Thangalazhy-Gopakumar, S., Gan, S., Lim, S.S., et al. 2018. Review on synthesis of 3D graphene-based configurations and their adsorption performance for hazardous water pollutants. Process Saf. Environ. Prot. 116: 262-286.

Hou, L., Liang, Q. and Wang, F. 2020. Mechanisms that control the adsorption–desorption behavior of phosphate on magnetite nanoparticles: The role of particle size and surface chemistry characteristics. RSC Adv. 10: 2378-2388.

Hu, Z., Qin, S., Huang, Z., Zhu, Y., Xi, L. and Li, Z. 2017. Recyclable graphene oxide-covalently encapsulated magnetic composite for highly efficient Pb(II) removal. J. Environ. Chem. Eng. 5: 4630-4638.

Huang, D., Wu, J., Wang, L., Liu, X., Meng, J., Tang, X., et al. 2019. Novel insight into adsorption and co-adsorption of heavy metal ions and an organic pollutant by magnetic graphene nanomaterials in water. Chem. Eng. J. 358: 1399-1409.

Huang, L., He, M., Chen, B.B., Cheng, Q. and Hu, B. 2017. Highly efficient magnetic nitrogen-doped porous carbon prepared by one-step carbonization strategy for Hg^{2+} removal from water. ACS Appl. Mater. Interfaces 9: 2550-2559.

Igiri, B.E., Okoduwa, S.I.R., Idoko, G.O., Akabuogu, E.P., Adeyi, A.O. and Ejiogu, I.K. 2018. Toxicity and bioremediation of heavy metals contaminated ecosystem from tannery wastewater: A review. J. Toxicol. 2018: 1-16.

Ijagbemi, C.O., Baek, M.-H. and Kim, D.-S. 2009. Montmorillonite surface properties and sorption characteristics for heavy metal removal from aqueous solutions. J. Hazard. Mater. 166: 538-546.

Ilić, S., Zec, S., Miljković, M., Poleti, D., Pošarac-Marković, M., Janaćković, D., et al. 2014. Sol-gel synthesis and characterization of iron doped mullite. J. Alloys Compd. 612: 259-264.

Indherjith, S., Karthikeyan, S., Monica, J.H.R. and Krishna Kumar, K. 2019. Graphene oxide & reduced graphene oxide polysulfone nanocomposite pellets: An alternative adsorbent of antibiotic pollutant-ciprofloxacin. Sep. Sci. Technol. 54: 667-674.

Ion, I., Ivan, G.R., Senin, R.M., Doncea, S.M., Capra, L., Modrogan, C., et al. 2019. Adsorption of triclocarban (TCC) onto fullerene C60 in simulated environmental aqueous conditions. Sep. Sci. Technol. 54: 2759-2772.

Islam, M.A., Sabar, S., Benhouria, A., Khanday, W.A., Asif, M. and Hameed, B.H. 2017. Nanoporous activated carbon prepared from karanj (Pongamia pinnata) fruit hulls for methylene blue adsorption. J. Taiwan Inst. Chem. Eng. 74: 96-104.

Jadhav, A.H., Mai, X.T., Ofori, F.A. and Kim, H. 2015. Preparation, characterization, and kinetic study of end opened carbon nanotubes incorporated polyacrylonitrile electrospun nanofibers for the adsorption of pyrene from aqueous solution. Chem. Eng. J. 259: 348-356.

Jia, W., Zhai, S., Ma, C., Cao, H., Wang, C. and Sun, H. 2019. The role of different fractions of humic acid in the physiological response of amaranth treated with magnetic carbon nanotubes. Ecotoxicol. Environ. Saf. 169: 848-855.

Khan, A.A.P., Khan, A. and Asiri, A.M. 2018. Chapter 24 – Nanocarbon and its Composites for Water Purification, in Nanocarbon and its Composites Preparation, Properties and Applications. Woodhead Publishing Series in Composites Science and Engineering, 711-731. https://doi.org/10.1016/B978-0-08-102509-3.00024-9.

Khurana, I., Saxena, A., Bharti, Khurana, J.M. and Rai, P.K. 2017. Removal of dyes using graphene-based composites: A review. Water. Air. Soil Pollut. 228: 1-17.

Kim, Y.H., Park, B.H., Choi, Y.J., Lee, G.W., Kim, H.K. and Kim, K.B. 2020. Compact graphene powders with high volumetric capacitance: Microspherical assembly of graphene via surface modification using cyanamide. Energy Storage Mater. 24: 351-361.

Kroto, H.W., Heath, J.R., O'Brien, S.C., Curl, R.F. and Smalley, R.E. 1985. C60: Buckminsterfullerene. Nature 318: 162-163.

Kuang, Y., Yang, R., Zhang, Z., Fang, J., Xing, M. and Wu, D. 2019. Surfactant-loaded graphene oxide sponge for the simultaneous removal of Cu^{2+} and bisphenol A from water. Chemosphere 236: 124416.

Kumar, R., Ansari, M.O. and Barakat, M.A. 2013. DBSA doped polyaniline/multi-walled carbon nanotubes composite for high efficiency removal of Cr(VI) from aqueous solution. Chem. Eng. J. 228: 748-755.

Kumar, S., Nehra, M., Kedia, D., Dilbaghi, N., Tankeshwar, K. and Kim, K.H. 2019. Nanodiamonds: Emerging face of future nanotechnology. Carbon N. Y. 143: 678-699.

Kumar, S.D., Ravichandran, M., Alagarsamy, S.V., Chanakyan, C., Meignanamoorthy, M. and Sakthivelu, S. 2020. Processing and properties of carbon nanotube reinforced composites: A review. Mater. Today Proc. 27: 1152-1156.

Kyzas, G.Z., Deliyanni, E.A., Bikiaris, D.N. and Mitropoulos, A.C. 2018. Graphene composites as dye adsorbents: Review. Chem. Eng. Res. Des. 129: 75-88.

Lam, C.W., James, J.T., McCluskey, R., Arepalli, S. and Hunter, R.L. 2006. A review of carbon nanotube toxicity and assessment of potential occupational and environmental health risks. Crit. Rev. Toxicol. 36: 189-217.

Lazarević-Pašti, T., Anićijević, V., Baljozović, M., Anićijević, D.V., Gutić, S., Vasić, V., et al. 2018. The impact of the structure of graphene-based materials on the removal of organophosphorus pesticides from water. Environ. Sci. Nano 5: 1482-1494.

Le, G.T.T., Chanlek, N., Manyam, J., Opaprakasit, P., Grisdanurak, N. and Sreearunothai, P. 2019. Insight into the ultrasonication of graphene oxide with strong changes in its properties and performance for adsorption applications. Chem. Eng. J. 373: 1212-1222.

Lei, L., Kang, X., Zhang, Y., Jia, C., Yang, Q. and Chen, Z. 2019. Highly efficient magnetic separation of organic dyes from liquid water utilizing Fe_3O_4@graphene composites. Mater. Res. Express 6: 106106.

Lei, M., Zhao, H.Z., Yang, H., Li, P.G., Tang, H.L., Song, B., et al. 2007. Ammonium dicyanamide as a precursor for the synthesis of metal nitride and carbide nanoparticles. Diam. Relat. Mater. 16: 1974-1981.

Lei, Y., Huang, Q., Gan, D., Huang, H., Chen, J., Deng, F., et al. 2020. A novel one-step method for preparation of sulfonate functionalized nanodiamonds and their utilization for ultrafast removal of organic dyes with high efficiency: Kinetic and isotherm studies. J. Environ. Chem. Eng. 8: 103780.

Li, M., Liu, Y., Yang, C., Liu, S., Tan, X., He, Y., et al. 2020a. Effects of heteroaggregation with metal oxides and clays on tetracycline adsorption by graphene oxide. Sci. Total Environ. 719: 137283.

Li, S., Yang, P., Liu, X., Zhang, J., Xie, W., Wang, C., et al. 2019a. Graphene oxide based dopamine mussel-like cross-linked polyethylene imine nanocomposite coating with enhanced hexavalent uranium adsorption. J. Mater. Chem. A 7: 16902-16911.

Li, S., Zhang, X. and Huang, Y. 2017. Zeolitic imidazolate framework-8 derived nanoporous carbon as an effective and recyclable adsorbent for removal of ciprofloxacin antibiotics from water. J. Hazard. Mater. 321: 711-719.

Li, Y., Du, Q., Liu, T., Peng, X., Wang, J., Sun, J., et al. 2013. Comparative study of methylene blue dye adsorption onto activated carbon, graphene oxide, and carbon nanotubes. Chem. Eng. Res. Des. 91: 361-368.

Li, Y., Lu, H., Wang, Y., Zhao, Y. and Li, X. 2019b. Efficient removal of methyl blue from aqueous solution by using poly(4-vinylpyridine)–graphene oxide–Fe_3O_4 magnetic nanocomposites. J. Mater. Sci. 54: 7603-7616.

Li, Y.H., Chang, F.M., Huang, B., Song, Y.P., Zhao, H.Y. and Wang, K.J. 2020b. Activated carbon preparation from pyrolysis char of sewage sludge and its adsorption performance for organic compounds in sewage. Fuel 266: 117053.

Li, Z., Sellaoui, L., Franco, D., Netto, M.S., Georgin, J., Dotto, G.L., et al. 2020c. Adsorption of hazardous dyes on functionalized multiwalled carbon nanotubes in single and binary systems: Experimental study and physicochemical interpretation of the adsorption mechanism. Chem. Eng. J. 389: 124467.

Liang, C., Tang, Y., Zhang, X., Chai, H., Huang, Y. and Feng, P. 2020a. ZIF-mediated N-doped hollow porous carbon as a high performance adsorbent for tetracycline removal from water with wide pH range. Environ. Res. 182: 109059.

Liang, J., Ender, C.P., Zapata, T., Ermakova, A., Wagner, M. and Weil, T. 2020b. Germanium iodide mediated synthesis of nanodiamonds from adamantane "seeds" under moderate high-pressure high-temperature conditions. Diam. Relat. Mater. 108: 108000.

Lv, M., Yan, L., Liu, C., Su, C., Zhou, Q., Zhang, X., et al. 2018. Non-covalent functionalized graphene oxide (GO) adsorbent with an organic gelator for co-adsorption of dye, endocrine-disruptor, pharmaceutical and metal ion. Chem. Eng. J. 349: 791-799.

Maazinejad, B., Mohammadnia, O., Ali, G.A.M., Makhlouf, A.S.H., Nadagouda, M.N., Sillanpää, M., et al. 2020. Taguchi L9 (34) orthogonal array study based on methylene blue removal by single-walled carbon nanotubes-amine: Adsorption optimization using the experimental design method, kinetics, equilibrium and thermodynamics. J. Mol. Liq. 298: 112001.

Mahalingam, P., Maiyalagan, T., Manikandan, E., Syed Shabudeen, P.S. and Karthikeyan, S. 2013. Dynamic and equilibrium studies on the sorption of basic dye (Basic brown 4) onto multi-walled carbon nanotubes prepared from renewable carbon precursors. J. Environ. Nanotechnol. 2: 43-62.

Mahmoodi, N.M., Maroofi, S.M., Mazarji, M. and Nabi-bidhendi, G. 2017. Preparation of modified reduced graphene oxide nanosheet with cationic surfactant and its dye adsorption ability from colored wastewater. J. Surfactants Deterg. 20: 1085-1093.

Markovic, M., Andelkovic, I., Shuster, J., Janik, L., Kumar, A., Losic, D., et al. 2020. Addressing challenges in providing a reliable ecotoxicology data for graphene-oxide (GO) using an algae (Raphidocelis subcapitata), and the trophic transfer consequence of GO-algae aggregates. Chemosphere 245: 125640.

Martín-de-Lucía, I., Gonçalves, S.F., Leganés, F., Fernández-Piñas, F., Rosal, R. and Loureiro, S. 2019. Combined toxicity of graphite-diamond nanoparticles and thiabendazole to Daphnia magna. Sci. Total Environ. 688: 1145-1154.

Medeiros, A.M.Z. de, Côa, F., Alves, O.L., Teodoro Martinez, D.S. and Barbieri, E. 2020. Metabolic effects in the freshwater fish Geophagus iporangensis in response to single and combined exposure to graphene oxide and trace elements. Chemosphere 243: 1-8.

Mistar, E.M., Alfatah, T. and Supardan, M.D. 2020. Synthesis and characterization of activated carbon from Bambusa vulgaris striata using two-step KOH activation. J. Mater. Res. Technol. 20: 6278-6286.

Mohan, V.B., tak Lau, K., Hui, D. and Bhattacharyya, D. 2018. Graphene-based materials and their composites: A review on production, applications and product limitations. Compos. Part B Eng. 142: 200-220.

Mohanta, D., Patnaik, S., Sood, S. and Das, N. 2019. Carbon nanotubes: Evaluation of toxicity at biointerfaces. J. Pharm. Anal. 9: 293-300.

Molavi, H., Shojaei, A. and Pourghaderi, A. 2018. Rapid and tunable selective adsorption of dyes using thermally oxidized nanodiamond. J. Colloid Interface Sci. 524: 52-64.

Molavi, H., Neshastehgar, M., Shojaei, A. and Ghashghaeinejad, H. 2020. Ultrafast and simultaneous removal of anionic and cationic dyes by nanodiamond/UiO-66 hybrid nanocomposite. Chemosphere 247: 125882.

Mortimer, M., Li, D., Wang, Y. and Holden, P.A. 2020. Physical properties of carbon nanomaterials and nanoceria affect pathways important to the nodulation competitiveness of the symbiotic N_2-fixing bacterium bradyrhizobium diazoefficiens. Small 16: 190605.

Motlag, M., Liu, X., Nurmalasari, N.P.D., Jin, S., Nian, Q., Park, C., et al. 2020. Molecular-scale nanodiamond with high-density color centers fabricated from graphite by laser shocking. Cell Reports Phys. Sci. 1: 100054.

Mrózek, O., Melounková, L., Smržová, D., Machálková, A., Vinklárek, J., Němečková, Z., et al. 2020. Salt-washed graphene oxide and its cytotoxicity. J. Hazard. Mater. 398: 1-9.

Murr, L.E., Esquivel, E.V., Bang, J.J., de la Rosa, G. and Gardea-Torresdey, J.L. 2004. Chemistry and nanoparticulate compositions of a 10,000 year-old ice core melt water. Water Res. 38: 4282-4296.

Neves, T.F., Assano, P.K., Sabino, L.R., Nunes, W.B. and Prediger, P. 2020a. Influence of adsorbent/adsorbate interactions on the removal of cationic surfactants from water by graphene oxide. Water. Air. Soil Pollut. 231: 1-6.

Neves, T.F., Dalarme, N.B., da Silva, P.M.M., Landers, R., Picone, C.S.F. and Prediger, P. 2020b. Novel magnetic chitosan/quaternary ammonium salt graphene oxide composite applied to dye removal. J. Environ. Chem. Eng. 8: 103820.

Nisola, G.M., Parohinog, K.J., Cho, M.K., Burnea, F.K.B., Lee, J.Y., Seo, J.G., et al. 2020. Covalently decorated crown ethers on magnetic graphene oxides as bi-functional adsorbents with tailorable ion recognition properties for selective metal ion capture in water. Chem. Eng. J. 389: 123421.

Novoselov, K.S., Mishchenko, A., Carvalho, A. and Castro Neto, A.H. 2016. 2D materials and van der waals heterostructures. Science 353: aac9439. doi: 10.1126/science.aac9439.

Nyairo, W.N., Eker, Y.R., Kowenje, C., Akin, I., Bingol, H., Tor, A., et al. 2018. Efficient adsorption of lead(II) and copper(II) from aqueous phase using oxidized multiwalled carbon nanotubes/polypyrrole composite. Sep. Sci. Technol. 53: 1498-1510.

Ouyang, S., Zhou, Q., Zeng, H., Wang, Y. and Hu, X. 2020. Natural nanocolloids mediate the phytotoxicity of graphene oxide. Environ. Sci. Technol. 54: 4865-4875.

Patel, H. 2019. Fixed-bed column adsorption study: A comprehensive review. Appl. Water Sci. 9: 45.

Paul, R. and Dai, L. 2018. Interfacial aspects of carbon composites. Compos. Interfaces 25: 539-605.

Peigney, A. and Laurent, C.H. 2006. Carbon nanotubes-ceramic composites. *In*: Low, I.M. [ed.]. Ceramic-Matrix Composites Microstructure, Properties and Applications. 148: 309-333.

Peng, J., Zhang, W., Liu, Y., Jiang, Y., Ni, L. and Qiu, J. 2017. Superior adsorption performance of mesoporous carbon nitride for methylene blue and the effect of investigation of different modifications on adsorption capacity. Water. Air. Soil Pollut. 228: 1-16.

Pichot, V., Comet, M., Risse, B. and Spitzer, D. 2015. Detonation of nanosized explosive: New mechanistic model for nanodiamond formation. Diam. Relat. Mater. 54: 59-63.

Ponte, S., Moore, E.A., Border, C.T., Babbitt, C.W. and Tyler, A.C. 2019. Fullerene toxicity in the benthos with implications for freshwater ecosystem services. Sci. Total Environ. 687: 451-459.

Pourjavadi, A., Nazari, M., Kohestanian, M. and Hosseini, S.H. 2019. Polyacrylamide-grafted magnetic reduced graphene oxide nanocomposite: Preparation and adsorption properties. Colloid Polym. Sci. 297: 917-926.

Prediger, P., Cheminski, T., Neves, T.F., Nunes, W.B., Sabino, L., Picone, C.S.F., et al. 2018. Graphene oxide nanomaterials for the removal of non-ionic surfactant from water. J. Environ. Chem. Eng. 6: 1536-1545.

Raeiszadeh, M., Hakimian, A., Shojaei, A. and Molavi, H. 2018. Nanodiamond-filled chitosan as an efficient adsorbent for anionic dye removal from aqueous solutions. J. Environ. Chem. Eng. 6: 3283-3294.

Richardson, S.D. and Kimura, S.Y. 2020. Water analysis: Emerging contaminants and current issues. Anal. Chem. 92: 473-505.

Ryder, G. 2017. The United Nations World Water Development Report. Wastewater: The Untapped Resource. Paris, UNESCO.

Sadegh, H., Ali, G.A.M., Makhlouf, A.S.H., Chong, K.F., Alharbi, N.S., Agarwal, S., et al. 2018. MWCNTs-Fe_3O_4 nanocomposite for Hg(II) high adsorption efficiency. J. Mol. Liq. 258: 345-353.

Sahoo, S.K. and Hota, G. 2018. Surface functionalization of GO with MgO/$MgFe_2O_4$ binary oxides: A novel magnetic nanoadsorbent for removal of fluoride ions. J. Environ. Chem. Eng. 6: 2918-2931.

Salehi, S., Mandegarzad, S. and Anbia, M. 2020. Preparation and characterization of metal organic framework-derived nanoporous carbons for highly efficient removal of vanadium from aqueous solution. J. Alloys Compd. 812: 152051.

Samuel, M.S., Bhattacharya, J., Raj, S., Santhanam, N., Singh, H. and Pradeep Singh, N.D. 2019. Efficient removal of Chromium(VI) from aqueous solution using chitosan grafted graphene oxide (CS-GO) nanocomposite. Int. J. Biol. Macromol. 121: 285-292.

Sanchís, J., Bosch-Orea, C., Farré, M. and Barceló, D. 2014. Nanoparticle tracking analysis characterisation and parts-per-quadrillion determination of fullerenes in river samples from Barcelona catchment area. Anal. Bioanal. Chem. 407: 4261-4275.

Sanchís, J., Silva, L.F.O., de Leão, F.B., Farré, M. and Barceló, D. 2015. Liquid chromatography-atmospheric pressure photoionization-orbitrap analysis of fullerene aggregates on surface soils and river sediments from Santa Catarina (Brazil). Sci. Total Environ. 505: 172-179.

Sanchís, J., Milačič, R., Zuliani, T., Vidmar, J., Abad, E., Farré, M., et al. 2018. Occurrence of C_{60} and related fullerenes in the Sava River under different hydrologic conditions. Sci. Total Environ. 643: 1108-1116.

Sanchís, J., Jiménez-Lamana, J., Abad, E., Szpunar, J. and Farré, M. 2020. Occurrence of cerium-, titanium-, and silver-bearing nanoparticles in the Besòs and Ebro Rivers. Environ. Sci. Technol. 54: 3969-3978.

Scott, L.T. 2004. Methods for the chemical synthesis of fullerenes. Angew. Chemie Int. Ed. 43: 4994-5007.

Shabaan, O.A., Jahin, H.S. and Mohamed, G.G. 2020. Removal of anionic and cationic dyes from wastewater by adsorption using multiwall carbon nanotubes. Arab. J. Chem. 13: 4797-4810.

Sherlala, A.I.A., Raman, A.A.A., Bello, M.M. and Buthiyappan, A. 2019. Adsorption of arsenic using chitosan magnetic graphene oxide nanocomposite. J. Environ. Manage. 246: 547-556.

Shi, X., Zhang, S., Chen, X. and Mijowska, E. 2019. Evaluation of nanoporous carbon synthesized from direct carbonization of a metal-organic complex as a highly effective dye adsorbent and supercapacitor. Nanomaterials 9: 601.

Shi, Y.C., Wang, A.J., Wu, X.L., Chen, J.R. and Feng, J.J. 2016. Green-assembly of three-dimensional porous graphene hydrogels for efficient removal of organic dyes. J. Colloid Interface Sci. 484: 254-262.

Shokry, H., Elkady, M. and Hamad, H. 2019. Nano activated carbon from industrial mine coal as adsorbents for removal of dye from simulated textile wastewater: Operational parameters and mechanism study. J. Mater. Res. Technol. 8: 4477-4488.

Sims, R., Harmer, S. and Quinton, J. 2019. The role of physisorption and chemisorption in the oscillatory adsorption of organosilanes on aluminium oxide. Polymers (Basel). 11: 410.

Singh, N.B., Nagpal, G., Agrawal, S. and Rachna. 2018. Water purification by using adsorbents: A review. Environ. Technol. Innov. 11: 187-240.

Song, Q., Wang, H., Yang, B., Wang, F. and Sun, X. 2016. A novel adsorbent of Ag-FMWCNTs for the removal of SMX from aqueous solution. RSC Adv. 6: 75855-75861.

Su, Y., Yang, G., Lu, K., Petersen, E.J. and Mao, L. 2017. Colloidal properties and stability of aqueous suspensions of few-layer graphene: Importance of graphene concentration. Environ. Pollut. 220: 469-477.

Sun, C., Li, W., Xu, Y., Hu, N., Ma, J., Cao, W., et al. 2020. Effects of carbon nanotubes on the toxicities of copper, cadmium and zinc toward the freshwater microalgae Scenedesmus obliquus. Aquat. Toxicol. 224: 105504.

Sun, K., Dong, S., Sun, Y., Gao, B., Du, W. and Xu, H. 2018. Graphene oxide-facilitated transport of levofloxacin and ciprofloxacin in saturated and unsaturated porous media 348: 92-99.

Tabrizi, N.S. and Yavari, M. 2015. Methylene blue removal by carbon nanotube-based aerogels. Chem. Eng. Res. Des. 94: 516-523.

Tang, H., Zhang, S., Huang, T., Cui, F. and Xing, B. 2020. pH-Dependent adsorption of aromatic compounds on graphene oxide: An experimental, molecular dynamics simulation and density functional theory investigation. J. Hazard. Mater. 395: 122680.

Tang, J., Zong, L., Mu, B., Kang, Y. and Wang, A. 2018. Attapulgite/carbon composites as a recyclable adsorbent for antibiotics removal. Korean J. Chem. Eng. 35: 1650-1661.

Tian, C., Zhao, H., Sun, H., Xiao, K. and Keung Wong, P. 2020. Enhanced adsorption and photocatalytic activities of ultrathin graphitic carbon nitride nanosheets: Kinetics and mechanism. Chem. Eng. J. 381: 122760.

Tian, J., Wu, T., Wang, D., Pei, Y., Qiao, M. and Zong, B. 2019. One-pot synthesis of potassium and phosphorus-doped carbon nitride catalyst derived from urea for highly efficient visible light-driven hydrogen peroxide production. Catal. Today 330: 171-178.

Tofighy, M.A. and Mohammadi, T. 2015. Copper ions removal from water using functionalized carbon nanotubes-mullite composite as adsorbent. Mater. Res. Bull. 68: 54-59.

Wang, F., Sun, W., Pan, W. and Xu, N. 2015. Adsorption of sulfamethoxazole and 17*β*-estradiol by carbon nanotubes/$CoFe_2O_4$ composites. Chem. Eng. J. 274: 17-29.

Wang, P., Huang, B., Chen, Z., Lv, X., Qian, W., Zhu, X., et al. 2019. Behavioural and chronic toxicity of fullerene to Daphnia magna: Mechanisms revealed by transcriptomic analysis. Environ. Pollut. 255: 113181.

Wang, X., Liu, Y., Pang, H., Yu, S., Ai, Y., Ma, X., et al. 2018. Effect of graphene oxide surface modification on the elimination of Co(II) from aqueous solutions. Chem. Eng. J. 344: 380-390.

Wang, X., Li, X., Wang, J. and Zhu, H. 2020. Recent advances in carbon nitride-based nanomaterials for the removal of heavy metal ions from aqueous solution. J. Inorg. Mater. 35: 260-270.

Xu, L., Li, J. and Zhang, M. 2015. Adsorption characteristics of a novel carbon-nanotube-based composite adsorbent toward organic pollutants. Ind. Eng. Chem. Res. 54: 2379-2384.

Xu, L., Wang, S., Zhou, J., Deng, H. and Frost, R.L. 2018. Column adsorption of 2-naphthol from aqueous solution using carbon nanotube-based composite adsorbent. Chem. Eng. J. 335: 450-457.

Yan, T., Chen, H., Wang, X. and Jiang, F. 2013. Adsorption of perfluorooctane sulfonate (PFOS) on mesoporous carbon nitride. RSC Adv. 3: 22480-22489.

Yang, K., Zhu, L. and Xing, B. 2006. Adsorption of polycyclic aromatic hydrocarbons by carbon nanomaterials. Environ. Sci. Technol. 40: 1855-1861.

Yang, K. and Xing, B. 2010. Adsorption of organic compounds by carbon nanomaterials in aqueous phase: Polanyi theory and its application. Chem. Rev. 110: 5989-6008.

Yang, K., Zhu, L., Yang, J. and Lin, D. 2018. Adsorption and correlations of selected aromatic compounds on a KOH-activated carbon with large surface area. Sci. Total Environ. 618: 1677-1684.

Yu, B., Wang, J., Yang, X., Wang, W. and Cai, X. 2019. Preparation of polyglycerol mediated superparamagnetic graphene oxide nanocomposite and evaluation of its adsorption properties on tetracycline. Environ. Sci. Pollut. Res. 26: 32345-32359.

Yu, F., Sun, S., Han, S., Zheng, J. and Ma, J. 2016. Adsorption removal of ciprofloxacin by multi-walled carbon nanotubes with different oxygen contents from aqueous solutions. Chem. Eng. J. 285: 588-595.

Yu, W., Sisi, L., Haiyan, Y. and Jie, L. 2020. Progress in the functional modification of graphene/graphene oxide: A review. RSC Adv. 10: 15328-15345.

Zamani, M., Aghajanzadeh, M., Molavi, H., Danafar, H. and Shojaei, A. 2019. Thermally oxidized nanodiamond: An effective sorbent for separation of methotrexate from aqueous media: Synthesis, characterization, *in vivo* and *in vitro* biocompatibility study. J. Inorg. Organomet. Polym. Mater. 29: 701-709.

Zhang, Y., Meng, T., Shi, L., Guo, X., Si, X., Yang, R., et al. 2019. The effects of humic acid on the toxicity of graphene oxide to Scenedesmus obliquus and Daphnia magna. Sci. Total Environ. 649: 163-171.

Zhang, Y., Duan, X., Bai, L. and Quan, X. 2020. Effects of nanomaterials on metal toxicity: Case study of graphene family on Cd. Ecotoxicol. Environ. Saf. 194: 110448.

Zhao, J., Xie, G., Xu, Y., Zheng, L. and Ling, J. 2020. Accumulation and toxicity of multi-walled carbon nanotubes in Xenopus tropicalis tadpoles. Chemosphere 257: 127205.

Zhao, X., Chang, S., Long, J., Li, J., Li, X. and Cao, Y. 2019a. The toxicity of multi-walled carbon nanotubes (MWCNTs) to human endothelial cells: The influence of diameters of MWCNTs. Food Chem. Toxicol. 126: 169-177.

Zhao, X., Chen, H., Kong, F., Zhang, Y., Wang, S., Liu, S., et al. 2019b. Fabrication, characteristics and applications of carbon materials with different morphologies and porous structures produced from wood liquefaction: A review. Chem. Eng. J. 364: 226-243.

Zhou, L., Jay, H., Ge, Y. and Lunec, J. 2017. Multi-walled carbon nanotubes: A cytotoxicity study in relation to functionalization, dose and dispersion. Toxicol. Vitr. 42: 292-298.

Zolfaghari, G., Esmaili-Sari, A., Anbia, M., Younesi, H. and Ghasemian, M.B. 2012. A zinc oxide-coated nanoporous carbon adsorbent for lead removal from water: Optimization, equilibrium modeling, and kinetics studies. Int. J. Environ. Sci. Technol. 10: 325-340.

Recent Development in Industrial Scale Fabrication of Nanoparticles and Their Applications

Sandeep Kumar[1], Bandna Bharti[2], Xiaoxiong Zha[2], Feng Ouyang[2] and Peng Ren[1,*]

1. INTRODUCTION

Nanotechnology has been booming worldwide in many areas, which include materials science, electronics, optics, medicine, plastics, energy and aerospace (Yetisen et al. 2016, Kamat and Schatz 2009). Nanotechnology aims at building nanomaterials which are of countless scientific attention in these materials because of their unusual chemical and physical properties that serve as an association between atomic and bulk structures. The new scientific technologies are the result of human dreams and imagination that can provide tremendous facilities to the world. The emergence of nanotechnology, a twenty-first-century frontier is the outcome of such dreams. The word 'nano' originally came out from the Greek word, which meant 'dwarf'. Nanotechnology is based on super small dimensions (1 and 100 nm) particles prepared mainly from silicon and is capable of a variety of tasks in some fields (Wong et al. 2020). As particles become very small, they begin to behave in a way that is not defined by the principles of classical mechanics (movement, energy, etc.). The materials on the nanometer scale may, therefore, reveal physical characteristics that are completely different from the bulk. These can be recognized as nanoscale quantum structures that begin to emerge as a result of the significant reduction of permissible states in the tiny particles, which increase the band gap (Li et al. 2020). Furthermore, the properties of nanomaterials, are decided not just by the size of the particles, but also through many other factors, such as the particles' structure, shape and surfaces. In semiconductors, the electric and optical properties, dependent on the size of nanoparticles derived from the spatial confining of electronic motion to a length scale that is equivalent or smaller than the length of the e-Bohr radius.

For the realization of better properties of nanomaterials, significant attention has been given to the attenuation of devices in the nanoscale regime. In this respect, as an evolving interdisciplinary technology, nanotechnology has come to meet the growing energy demands for the realization of new materials with unique functionality and selectivity. Multidisciplinary teams from different fields of research, including physicists, chemists, molecular biologists, material scientists and engineers, are therefore working together on the synthesis and processing of nanomaterials, the

[1] School of Science, Harbin Institute of Technology, Shenzhen 518055, PR China.
[2] School of Civil and Environmental Engineering, Harbin Institute of Technology, Shenzhen 518055, PR China.
[*] Corresponding author: renpeng@hit.edu.cn

design and manufacture of nano-devices as building blocks of nanomaterials and the development of novel instruments for the characterization of these materials (Kamat and Schatz 2009). The next technological revolution for exponential progress is believed to bring nanotechnology into the fastest-growing field. By supplying society with tremendous benefits, the field of nanotechnology has an immense influence on human life (Yuan et al. 2020).

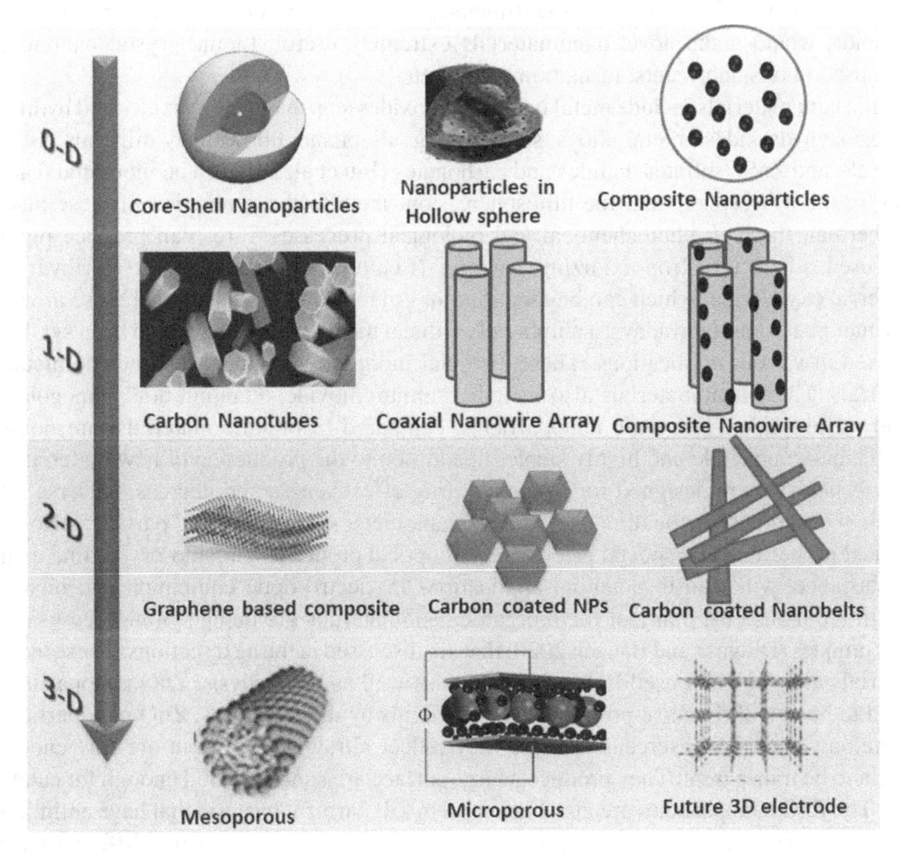

FIGURE 1 Classification of nanomaterials on the basis of dimensions

Based on the properties and structures, nanoparticles have been classified in different phrases, which are based on dimensionality (0, 1, 2 and 3D), and on the basis of origin (natural and artificial) (Kufer and Konstantatos 2016, Shao et al. 2020). Furthermore, natural nanoparticles have been divided into two parts which are organic and inorganic nanoparticles, while artificial nanoparticles may divide into four parts, such as metal, carbon-based, composites and dendrimers (Zhang et al. 2020). Classification of the nanomaterials has been shown in Figure 1, which is based on the dimensionality of nanoparticles. Because of dimensionality changes, the surface area, as well as properties of the materials, also changes. Nanomaterial systems in fluid crystals have been attracting growing interest in recent years because of the ability to add functionalities to liquid crystals through the properties of the scattered particles, and also the self-organization of fluid crystals can be used to form ordered nanomaterial structures. A nanocrystalline material has a significant fraction of crystal grains in the nanoscale. The need for high surface area, excellent electrical conductivity, high sensitivity, and catalytic activity was the key behind the extent of its usefulness and numerous applications. They are classified as nanocomposites, nanofoam, nanoporous and nanocrystalline materials on the basis of the phases of matter contained by the nanostructured materials. Nanocomposites are called solid materials as they contain one physically or chemically distinct region with at least one nanoscale-dimensioned region. A liquid or solid matrix is contained in nanofoams that are filled with a gaseous phase, and one of the two phases has nanoscale dimensions. Nanoporous materials, cavities with nanoscale dimensions, are regarded as nanoporous. At the nanoscale, nanocrystalline materials have crystal grains. Despite these obstacles, a controlled assembly of crystals using soft materials, such as proteins, DNA, nanoparticles and other colloids, would make novel nanomaterials extremely useful. Liquid crystal nanomaterials have been used in oils, lubricants, formation grease, etc.

Inorganic nanomaterials include metal oxides/hydroxides (e.g. manganese oxides and hydroxides, iron oxides/oxyhydroxides), metal alloys, silicates (e.g. allophane, fibrous clay minerals, asbestos), sulfides (FeS_2 and ZnS), sulfates, halides and carbonates (Liu et al. 2017a). The inorganic reactions, occurring in the hydrosphere and the lithosphere, contribute to the generation of these materials via non-thermal, thermal, photochemical and biological processes. Fires can produce pigments, cement, fused silica, etc. Iron-oxidizing bacteria (Leptothrix, Gallionela) of ferrihydrite can produce ferric oxyhydride, which can be identified in ground water and soils. All these are natural inorganic nanomaterials. Nowadays, a number of artificial nanomaterials have also been synthesized and are used in various applications. These artificial inorganic nanomaterials include metal base nanomaterials. These nanomaterials also include titanium dioxide, quantum dots, nanogold, nano silver and metal oxides. Inorganic nanoparticles, compared to organic materials, are non-toxic, hydrophilic, bio-compatible and highly stable. In addition to the production of new materials, drug delivery mechanisms are designed for improved drug effectiveness and decrease adverse effects. When their size approaches the dimensions of the nanometer scale, inorganic particles also exhibit new physical properties. The special electronic and optical properties of nano-crystalline quantum dots, for instance, will lead to potential applications in electro-optic equipment and biomedical imaging. In the industries, many of the inorganic nanomaterials are being synthesized using flow reactor techniques (Długosz and Banach 2020) that are discussed in the next sections. These inorganic nanomaterials are also been used in heterogeneous as well as biocatalysis. ZnO nanoparticles are considered to be one of the most produced nanomaterials by the industries. ZnO nanoparticles are most commonly used in sunscreens because they reflect ultraviolet light but are tiny enough for visible light to be transparent. They produce a large surface area and are good enough for catalytical activities. The ZnO nanoparticles are also been used to kill harmful bacteria that have antimicrobial activities. These nanomaterials can be synthesized at an industrial scale using different methods, such as hydrothermal, sol-gel, electrodepositing, co-precipitation, etc. Although ZnO is a very promising and important material but also has some disadvantages to human health as it is a very tiny particle so can penetrate via skin to the human body. They can enter into the body also via food and water and can show adverse effects (Gomez-Gonzalez et al. 2019). Iron oxide nanomaterials have a size range from 1-100 nm, and they have gained attention due to their superparamagneteic properties. They are using in various fields, including catalysis and degradation of various organic dyes from wastewater. Many magnetic devices have also been made using iron oxide nanoparticles and are also used for imaging, sensing and medical diagnosis. Although there are many methods for the synthesis of iron oxide, coprecipitation is one of the most important methods at an industrial scale. The following reactions take place during synthesis.

$$2Fe^{3+} + Fe^{2+} + 8OH^{-} \rightarrow Fe_3O_4 \downarrow + 4H_2O \quad \text{and} \quad 2Fe_3O_4 + O_2 \rightarrow 2\gamma Fe_2O_3$$

The surface area and the size of the nanomaterials can be tunned by using proper reaction conditions by changing temperature, pH, solvent and reaction time. These nanoparticles are mostly used in biomedical applications as drug delivery agents. They can carry anti-cancer drugs to the tumor cells using electron releasing properties (Hu et al. 2018).

Organic nanomaterials mainly include carbon-based materials. Nature has given the first carbon nanotube and the first buckyball (carbon fullerene) produced from fossil fuel combustion. In this case, there is no metal contained and may act as toxic in nature in some cases. The volcanic ashes that are released during eruptions contain naturally occurring bucky balls. Petroleum and natural gas contain organic nanostructures which deposit as nanoscale diamond structures. Although artificial nanomaterials are often made of carbon, the hollow spheres, ellipses or tubes most commonly take shape. Spherical or ellipsoidal-shaped carbon materials are referred to as fullerenes, whereas cylindrical-shaped are called nanotubes (carbon nanotubes (CNTs). They are of much interest in the scientific and engineering communities because of their unusual chemical, physical, mechanical, optical and thermal properties. Out of all this graphene is the 2D substance that is the thinnest substance in a single-atom-thick planar layer, which consists of carbon atoms with sp^2-hybridization. On the other hand, carbon nanotubes often have a cylindrical-shaped nanostructure (Kamat and Schatz 2009). Graphene is the fundamental structure of carbon nanotubes, graphite, and fullerene. Carbon tube is basically known as carbon nanotubes and can be classified into two categories: (1) single-walled carbon nanotubes (SWCNTs) and (2) carbon nanotubes with multiple walls (MWCNTs) (Skarmoutsos et al. 2020). This carbon-based nanoparticle has been widely studied in various applications. It is well recognized that by adding carbon nanoparticles, the electrical conductivity of polymeric substances can be enhanced significantly. Carbon-based nanoparticles can also be used in the treatment of human cancer (Zhang et al. 2020).

Carbon nanoparticles have exceptional electrical conductivity, mechanical properties and heat conductivity. Therefore, they are formed by pure carbon that has high stability, conductivity, and very low toxicity. Since a huge portion of the human body is made up of carbon-based materials, so it is commonly considered a biocompatible material. The excellent conductivity, linear geometry and high surface area make them highly usable materials for different applications. Nanomaterials based on carbon have a high thermal conductivity in an anisotropic manner, which enables advanced computing electronics to use carbon-based nanomaterials where the temperature of electronic chips can reach around 100°C. As fossil fuel supplies are diminishing, the introduction of realistic alternative energy systems is becoming increasingly necessary. One of the most possible sources of energy could be fuel cells. Carbon-based nanomaterials have been widely studied for the storage of different gases, such as hydrogen, methane, etc., even though they are used as a catalyst for various organic transformations. Because of their inherent properties, this type of material is being used for making electrodes in capacitors and different types of batteries. The extraordinary electrical conductivity, surface area, thermal stability and conductivity of carbon-based materials also allow the CNT to be used in gas diffusion layers and current collectors.

In many imaging applications, carbon-based materials have long been studied. These nanomaterials are used in various imaging applications based on research, diagnosis, medical or treatment of diseases. Different techniques include fluorescence imaging (FL), magnetic resonance imaging (MRI), Raman imaging, tomography (CT), two-photon FL, photoacoustic imaging (PAI), computed tomography with positron emission/computed tomography with single-photon emission (PET/SPECT) and multimodal imaging. In its bioimaging applications, carbon-based nanomaterial, such as carbon quantum dots (CQDs), has recently attracted considerable interest. CQDs have been extensively studied for their fluorescent properties since the initial study in 2006. CQDs have size-dependent tunable emission, similar to conventional QDs, which makes them a good imaging material. Moreover, the dangerous toxicity of typical QDs because of consisting of heavy metals, such as cadmium is also overcome by using CQDs (Ghosal and Sarkar 2018). Therefore, its uses for biological marking and bio-imaging in both human and animal cells have been thoroughly studied. In case drug and gene delivery, some research studies have shown that CNTs can act as great drugs carrier because they can make a bond with DNA and can then be inserted into a cell successfully (Stando et al. 2019, Kokarneswaran et al. 2020).

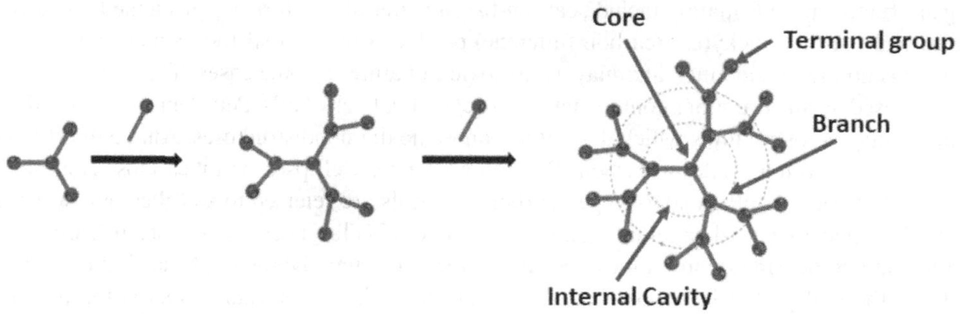

FIGURE 2 Representation of formation of dendrimer

The nanoparticles based on dendrimers are nano-sized polymers constructed from branched groups. There are multiple-chain ends in the dendrimer's surface that can be customized to perform particular chemical functions. This can also be useful for catalytic research. Furthermore, they might be useful for medicine because the three-dimensional dendrimers contain cavities inside which other molecules may be positioned. Dendrimers are human-made compounds with special properties, making them useful for the health and pharmaceutical industries in both developing current and completely new products (Sowinska and Urbanczyk-Lipkowska 2014). Because of the successive layers of branching groups, dendrimers are constructed and used for different applications. Recent applications are drug delivery, gene transfection, catalysis, energy processing, imaging, determination of molecular weight and size, alteration of rheology and science and technology of nanoscale. Dendrimer nanostructures have enhanced different properties including chemical, physical and biological, such as solubility, durability, the ability to act as supply networks, etc. The huge potential of nanotechnology is for cancer, viral and bacterial diseases that can be prevented, identified, diagnosed and treated (Chakravarty et al. 2019). Many examples address different forms of dendrimers used as nanomaterials. One of the structures of the dendrimer is shown in Figure 2, which is used in the human body to improve the immune system.

This technology has now been used on an industrial scale for many days to manufacture various nanomaterials used in the food, water treatment, electronics, medicine, and coating industries. Because of nanotechnology, a great improvement in textiles has been observed and has become smart in terms of novel production of electronics, cosmetic and fiber products. Nowadays, nanomaterials-based industries are reaching heights because of the production of smart and valuable products that makes human life enjoyable as well as healthier. When one thinks of nanotechnology, the first thing that comes to mind is developments in different kinds of materials and protective coatings. Nanotechnology applications, ranging from fabrics and sporting gear to lenses and computer and camera screens, offer numerous possibilities. Finally, nanotechnology showed the importance in a variety of environments as well as human health-based applications, including treatment of water and air beyond enhancing solar cells. Furthermore, to enhance air quality, nanotechnology is being used. Another use of silk and its nano-fibrils has been identified by MIT researchers in the filtration systems, while ABB (NYSE: ABB) uses nanotechnology for air filtration to effectively block dust. In the battle against COVID-19, nanotechnology applications for air treatment can also prove helpful. Researchers at the Negev University of Ben-Gurion are creating a new form of air filter in Israel that can self-sterilize the decontaminated air. This method of air filtration is based on laser-induced graphene filters that eliminate waterborne viruses and bacteria (McGuinness et al. 2015).

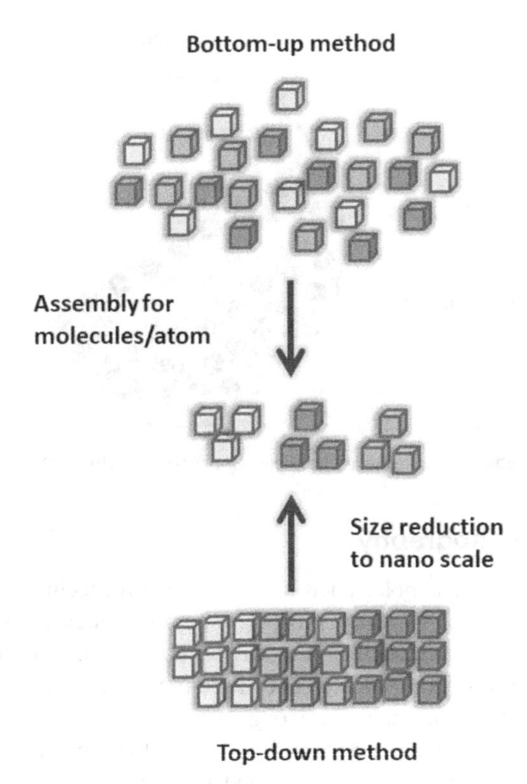

FIGURE 3 Bottom-up and Top-down approaches for the synthesis of various sized nanoparticles

2. INDUSTRIAL SCALE FABRICATION OF NANOMATERIALS

Different techniques were used by the researchers on a small scale from time to time to produce the nanoparticles (NPs). Basically, two methods were used for the synthesis of nanoparticles; one of them is the bottom-up approach while another is the top-down approach (Figure 3). The bottom-up technique includes the miniaturization of constituents of the material (up to the atomic level) followed by the process of self-assembly to create nanoparticles. Different physical forces or interactions operate at the nanoscale to combine the fundamental components into large and stable structures during the self-assembly process. It has been known that this method is more favorable for the synthesis and development of nanoparticles that have different sizes. Different techniques have been developed to use this approach to synthesize nanomaterials. The formation of nanoparticles from colloidal dispersions is a typical example. While the top-down approach requires the creation of nanosized particles by splitting or sculpting a large block of material using physical ball-milling techniques, lithography, mask etching and extreme plastic deformation application. We provide a brief introduction of some of these techniques in the next section, which is followed by the method used at the industrial scale to synthesize metal NPs.

2.1 High Energy Ball Milling

Benzamin and his co-workers at the Multinational Nickel Corporation first invented this method in the late 1960s. The bulk material (i.e., powder mixture) is put in the ball mill with high-energy spinning balls that compress the solid material into nanocrystallites in this process (Chen et al. 2013). This technique, referred to as mechanical alloying, could effectively generate small, uniform oxide particle dispersions (Al_2O_3, Y_2O_3 and ThO_2) in nickel-based superalloys, which could not be developed by more traditional methods of powder metallurgy (Baláž et al. 2013) (Figure 4).

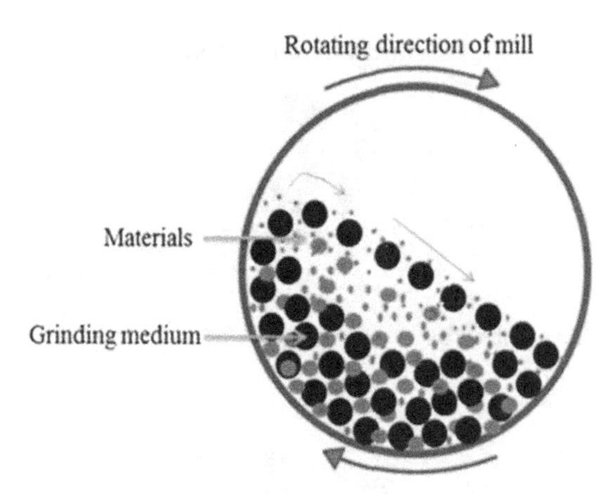

FIGURE 4 Representation of high energy Ball-Milling process for the synthesis of nanoparticles

2.2 Electron Beam Lithography

In the 1980s, the boom of nanotechnology led to a very effective technique called electron beam lithography (Manfrinato et al. 2013) that constructs nanostructures or mesoscopic systems by scanning a highly concentrated electron beam to write a pattern on a surface covered by a resist called an electron sensitive film. An improvement in the molecular structure and solubility of the resistance film is caused by the electron beam (Hong et al. 2018). For the selective removal of either the exposed or the non-exposed areas, the resist film may further be submerged in a solvent. This technique provides high-resolution advantages (as good as sub 10 nm), high versatility that can work with a variety of materials, high processing efficiency and high positioning/alignment accuracy.

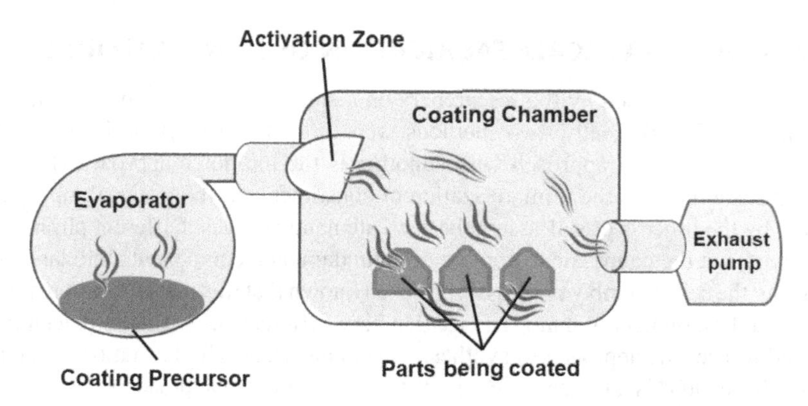

FIGURE 5 Schematic representation of chemical vapor deposition method for synthesis of nanoparticles

2.3 Chemical Vapor Deposition (CVD)

It is one of the best procedures for the solid-state processing of nanomaterials (Huang et al. 2018a). For the development of thin films, this method was used and acquired with great success. Intermetallic compounds may be coated using this chemical process on the surface of various types of metallic or ceramic compounds, metals and their alloys. In this technique, one or more volatile precursors are transported to the reaction chamber through the vapor process, whereby they decompose on a heated substrate (Gaur et al. 2014). The volatile products are then eliminated in the reaction chamber by gas flow (Figure 5). For this reason, a wide range of volatile materials has been used for deposition of the surfaces of thin layer supports (Feng et al. 2015). These materials include precursors of industrial importance, metal hydrides (AiH_4, AsH_3), metal halides (WF_6, $TiCl_4$) and metal alkyls ($GaEt_3$, $AliBu_3$). An example would be TiB_2 preparation where melting point 3,325°C may be deposited at 1,000°C by CVD, that is:

$$TiCl_4 + 2BCl_3 + 5H_2 \rightarrow TiB_2 + 10\ HCl$$

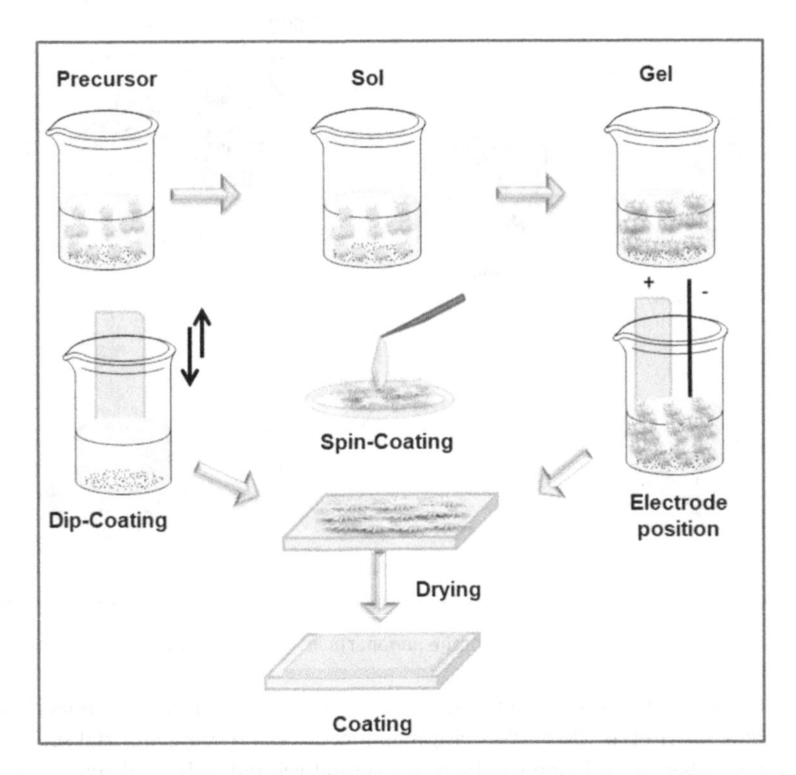

FIGURE 6 Schematic representation of Sol-Gel method for the fabrication of nanoparticles

2.4 Sol-Gel Method

The sol-gel phase may be considered as the formation of the network by polycondensation of the precursor in the liquid and used for the synthesis of nanoparticles of different shapes and sizes. The sol-gel process (Hitihami-Mudiyanselage et al. 2013) consists of several steps, including hydrolysis, condensation and drying. The metal precursor initial undergoes rapid hydrolysis, leading to the creation and immediate condensation of the metal hydroxide solution leading to the formation of three-dimensional gel (Deshmukh et al. 2020). The drying process is then subjected to the obtained gel, which results in a rapid conversion into Xerogel or Aerogel using a drying system (Figure 6).

2.5 Hydrothermal/Solvothermal Method

This is a solution-based approach to synthesize different materials that can withstand high temperatures and high autogenous pressure for a long time using pressurized vessels called autoclaves (Nicolae et al. 2020). Substances that under normal conditions are virtually insoluble include oxides, silicates and sulfides that can be synthesized using this technique (Figure 7). By controlling certain reaction parameters, such as temperature profile, solvents, time and additives, materials with desired shapes, morphologies and sizes can be generated. This procedure also has many benefits over other growth methods, such as the use of simple equipment, low process temperature, low cost, high purity, environmentally friendly and less dangerous catalyst-free growth. The name switches from hydrothermal to solvothermal by using some solvent other than water (Kuo and Huang 2008). In most industrial scale fabrication, continuous flow hydrothermal reactors are being used to synthesize at a very high scale (Naik et al. 2015).

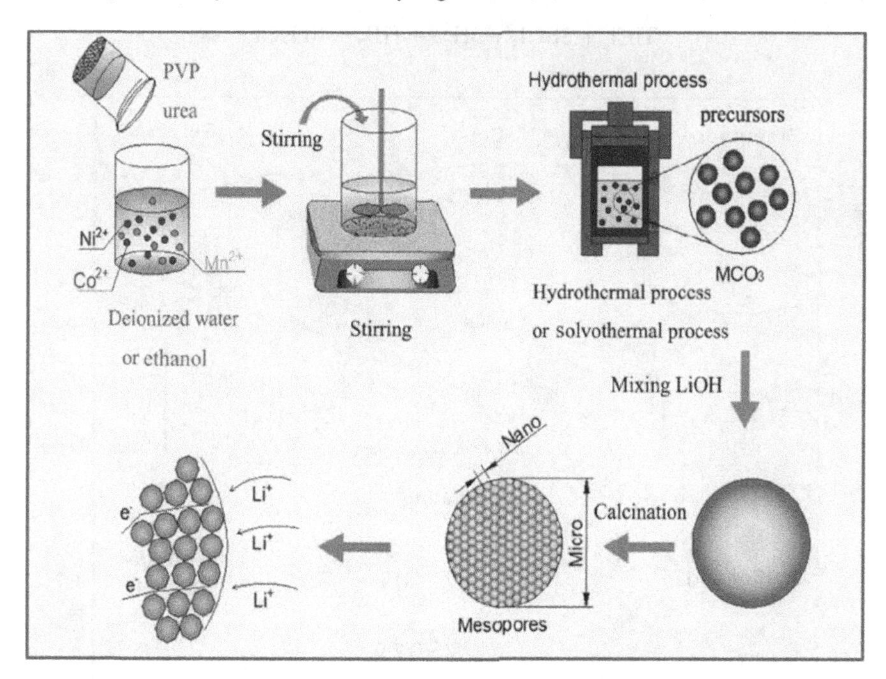

FIGURE 7 Schematic representation of the nanoparticles via hydro/solvothermal methods

All these methods have been employed by the industries for the synthesis of nanoparticles at large or very large scales (Darr et al. 2017). We have mentioned some of the industrial scale fabrication in the next section. There is a difference between laboratory and industrial scale formulation of nanoparticles. On a laboratory scale, we have to use small apparatus, while a big plant has to be used at industrial scale. For that, it should be in mind that the process should be continuous, so the flow process is mostly preferred for that.

2.6 Industrial Scale Synthesis of CaP Nanoparticles Using Flow Process

Nanoparticles of calcium phosphate are an important class of nanomaterials used primarily for biomedical applications, but also very promising in other sectors, such as cosmetics, catalytic, water repair and agriculture (CaP NPs) (Esposti et al. 2020). Unfortunately, their wide-ranging use, as with other nanomaterials, is hindered by the complexity of translating from laboratory to manufacturing on an industrial scale, usually performing in batch or semi-batch systems, responsible for size, morphology, purity and degree of particle inclusion. The use of continuous flux synthesis in this regard would help to solve the problem, providing more homogeneous and highly reproducible conditions for the reaction. They investigate the impact on the phytochemical properties of the nanoparticles of some of the most critical process variables (some reactant flow rate, sonic amplitude and ripening time), and the precipitation studies of citrate functionalized CaP nanoparticles assisted by sonication by continuous wet flow precipitation (Figure 8). The dimension of CaP nanoparticles was determined by the statistical data analysis by the reactor flow rate, while the maturation step was influenced by the dimensions of the crystalline domain and the purity of the substance. This work offers an overview and a valuable contribution to the scale-up of CaP nanoparticles in nano medics or other reaction factors and CaP applications.

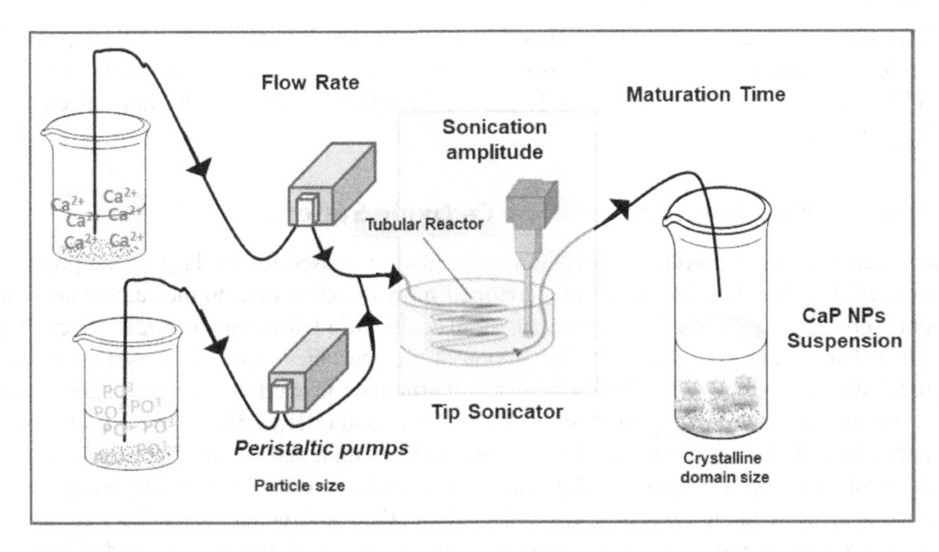

FIGURE 8 Industrial scale synthesis of CaP nanoparticles using flow process at industries

2.7 Industrial Scale Synthesis of Silver Nanoparticles

Due to their wide range of antimicrobial and antiviral properties, silver nanoparticles, if available on a broad (industrial) scale, would have excellent applications in wound dressings and as an antiseptic and disinfectant in various consumer goods. This patent deals with the biosynthesis of silver nanoparticles by using a biological fungus called T Reesei, which is environmentally friendly and industrially abundant. This method of synthesis is the only route for the production of silver and other metallic nanoparticles on an industrial scale, which will significantly reduce the cost of their production (Vahabi et al. 2011). According to the flow chart of the pilot plant at industry, first $AgNO_3$ and T-Reesei are mixed in a big reactor and the color of the mixture is noted down. Furthermore, the mixture can be mixed for a particular time using a rotating shaker and is followed by transfer to another reactor when the color is changed. The product mixture is then used for purification by separating solid and dissolved impurities and the resulting pure substances will be then loaded for the market (Figure 9).

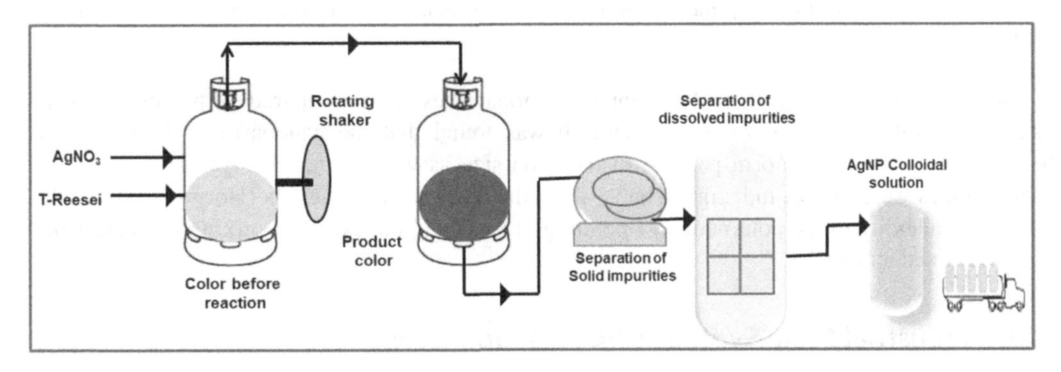

FIGURE 9 Pilot-scale synthesis of silver nanoparticles at industries using flow process

2.8 Industrial Scale Fabrication of Radio-labeled ^{64}CuS Nanoparticles

Industrial synthesis was performed in the hot cell with remote operating tongs for ^{64}CuS-CTAC and ^{64}CuS-PEG nanoparticles. For the synthesis of non-radioactive nanoparticles, the radio label formulation was transferred to a centrifugal 50 kDa filter tube and was centrifuged for 15 minutes at 5,000 rpm to remove unreacted PEG-SH and $^{64}CuCl_2$ after modification to PEG. This procedure has been repeated once to complete conversion of the unreacted reactant (Chakravarty et al. 2016. ^{64}CuS-PEG nanoparticles were subsequently passed through the 0.22 μm filtration injection and send for quality check.

2.9 Industrial Scale Synthesis of Zn-Ce Oxide System

For the synthesis of Zn-Ce oxide, a continuous pilot plant was used for the large-scale processing of nanomaterials. The pilot plant mixes in a confined spray mixer, similar to the current laboratory continuous process, aqueous salt solutions at room temperature with supercritical water (Tighe et al. 2013). This results in a continuous formation of nanoparticles. The Zn-Ce system was used as a model system to distinguish the difference in particulate properties because of the physical expansion of the mixer. In the typical flow KOH, $Zn(NO_3)_2$ and $(NH_4)CeNO_3)_6$ were transferred to the reactor and mixed for a particular time. During this constant supply, the deionized water was provided through an electric heater. In the end, the suspension was cooled down by using a cooler and the nanoparticle-containing slurry was collected and the resulting nanoparticles were used for different applications (Figure 10). In this method, flow rate affect the particle size and dimensions.

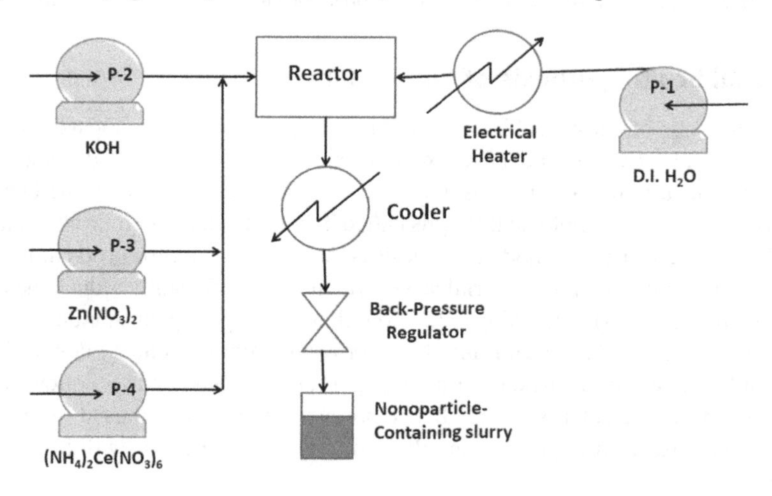

FIGURE 10 Flow diagram for the synthesis of Zn-Ce oxide NPs at industrial scale using Continuous Hydrothermal Flow Synthesis (CHFS)

The data obtained from the pilot plant for nanoparticles were compared with previous work using a laboratory-scale continuous reactor. It was found that the nanoparticles formed using the flow process have uniform particle distribution size as well as morphology. The pilot plant nanoparticles were almost indistinguishable from the ones made at the laboratory because of the inherent scalability of the continuously operating process and the excellent mixing characteristics of the confined jet blender.

2.10 Industrial Scale Synthesis VO_2 Nanoparticles

Because of the thermochromic properties of the material, monoclinic VO_2 Nanoparticles are of interest because the high formulation temperatures and extensive reaction times are necessary for the direct pathways to VO_2 nanoparticles that are often not accessible (Powell et al. 2015). The process is completed in two steps and is followed by a heat treatment phase to prepare monoclinic VO_2 nanoparticles using Continuous hydrothermal-Flow Synthesis (CHFS) (Ma et al. 2015). A number of particle sizes and particle size depending on the synthesis are produced by different reaction temperatures and residence times (average particle size range 50 to 200 nm). Techniques such as Powder X-ray diffraction (PXRD), UV/Vis spectroscopy, transmission electron microscopy, electron scanning microscope, differential calorimetry scanning and Raman spectroscopy are the main characteristics for the nanoparticles (Gruar et al. 2015).

The nanoparticles are very crystalline and are highly dependent on both temperature and time of reaction. Scanning electron microscopy images confirmed the uniform nature of the bulk sample that was produced by using the CHFS process. To confirm the pure monoclinic VO_2 phase, a heat treatment was given to the samples that resulted in them displaying a large and reversible switch in optical properties. The CHFS path allows for an even greater synthesis of such particles in future industrial applications. Moreover, the procedure offers the ability, among other parameters, to tailor the size and morphology of the synthesized particles by independent temperature control and time of residence that allows modifications to be made to suit a particular process while maintaining the desired properties. Not only these but various metal-containing nanoparticles such as Ni-nanoparticles (Roberts et al, 2017), tin oxide-based nanoparticles (Roberts et al. 2017), conjugated gold nanoparticles (Chakravarty et al. 2018), cobalt oxide (Denis et al. 2015), etc., have been synthesized using continuous flow methods.

3. APPLICATIONS OF NANOMATERIALS

3.1 Use of Nanomaterials in Textile Dyeing Industry

The textile dyeing industry is one of the main causes of water pollution. The effluents released from the textile dyeing units degrade the quality of the water resources. Synthetic dye released from the industrial waste water has been used to give color to the fabric, which produces a major hazard to the environment due to their carcinogenic nature. Because of uneconomic and ineffective dyeing methods, up to 50% of the dyes were discharged directly into the water (He et al. 2012). Dyes discharged in wastewater may undergo incomplete degradation because of their complex structure. Furthermore, the intensive color of the dyes decreases the diffusion of the sunlight and dissolution of the oxygen in the water that is a major threat to the aquatic ecosystem (Rauf et al. 2011). Therefore, efficient, ecofriendly and greener methods are required to eradicate the dyes. The conventional methods include ozonation, coagulation, adsorption, etc., but these methods are expensive, time-consuming and are not efficient as the dyes remain in the water. In recent years, various types of nanoparticles are considered to be effective that are prepared by various synthesis methods and applied in various applications, such as waste water treatment, nanosensors, biosensors, solar cells, photocatalysis, food packaging, optoelectronics, energy storages devices, etc. In this section, different types of nanoparticles and nanocomposites were illustrated for the degradation of dyes with different techniques.

Fe/Al/Ti trimetallic nanocomposite was synthesized by a flexible tuned chemical method, which provided a novel route for water treatment. The grain size and average particle size of the trimetallic nanocomposite were found to be 155 and 42 nm, respectively, as confirmed by scanning electron microscopy and transmission electron microscopy. For various concentrations of methylene blue solution, trimetallic nanocomposite shows greater than 90% photodegradation efficiency. Further, it was confirmed that the removal efficiency was not retarded even after five-set of experiments (Mukherjee et al. 2019). In another study, graphene oxide interconnected and reduced graphene oxide, and fluorinated graphene oxide was fabricated and consistently studied for the removal of rhodamine B and methylene blue dye from their aqueous solution. The experimental data obtained from both the dyes were fitted using Freundlich and Langmuir isotherm models. The maximum adsorption capacity for methylene blue and rhodamine B on graphene oxide was found to be 403.3 mg/g and 686.6 mg/g revealed by the Langmuir model. This study also illustrated the importance of graphene oxide as an auspicious adsorbent for the treatment of wastewater which efficiently separates the cationic species from the ionic compounds (Mao et al. 2020).

A composite nanofiber of amidoximated polyacrylonitrile/magnetite nanoparticles was synthesized with the help of the electrospinning technique which was further followed by chemical

TABLE 1 Various types of nanomaterials used in the textile dyeing industry.

Nanoparticles	Method used for the Preparation	Dye used for the Degradation	Shape/Size	Process used for the Degradation of Dye	Performance	References
Ni-Ethylene glycol	Solvothermal method	Congo Red	Spherical/600nm	Adsorption	A high surface area of 222 m^2g^{-1}, a maximum adsorption capacity of 440 mg g^{-1} and promising materials for waste water treatment.	Zhu et al. 2012
Phytogenic Magnetic Nanoparticles	Green method	Crystal Violet	Cube, spherical/30-55nm	Adsorption	98.57% of dye was removed within 120 minutes, pH 6.0-12.0, 88.65 mg/g adsorption capacity at 25 °C, and it can be used in the removal of cationic dyes from textile wastewater.	Ali et al. 2018
TiO_2	Modified sol-gel method	Indigo Carmine, methylene blue, methyl orange, Rhodamine B and Eriochrome black T	Spherical/ ~15nm for anatase TiO_2 and ~37nm for rutile TiO_2	Photocatalysis	For rutile TiO_2 the degradation efficiency was found to be 62%, 50.63%, 42.98%, 29.67% and 64.93%, respectively, for methylene blue, methyl orange, Rhodamine B, indigo carmine and Eriochrome Black T, whereas the degradation efficiency for anatase TiO_2 was 64%, 56.36%, 54.85%, 54.66% and 65.09%, respectively.	Gautam et al. 2016
$BiFeO_3$	Hydrothermal method	Rhodamine B	Spherical/20nm	Photocatalysis	The photocatalytic degradation efficiency of $BiFeO_3$ nanoparticles synthesized at 160 °C, 140 °C and 120 °C was 79%, 53% and 42%, respectively, and the as-synthesized nanoparticles can be used for improving the production of solar H_2.	Basith et al. 2018
Au-Cu_3N nano heterostructures	A simple one-pot synthesis	Methylene blue and methyl orange	-----/10 ± 5 nm	Photocatalysis	Nearly 93% of dye was degraded by Au-Cu_3N (Au~ 10 nm), Similarly, in the case of MO degradation, 84% of dye was degraded by ACN2, and this photocatalyst at the same interval of time effective for better photocatalytic dye degradation in the presence of solar light as well as green light-emitting diodes.	Barman et al. 2019
MFe_2O_4: M=Co, Ni, Cu, Zn	CTAB-mediated co-precipitation-oxidation method	Methylene blue, methyl orange, bromo green, and methyl red	Cubic/27-36	Photocatalysis	In the optimized condition, NiF was found to degrade 89%, 92%, 93%, and 78% of methylene blue, methyl orange, bromo green, and methyl red, respectively, within 1 minute of UV-irradiation; these photocatalysts are highly suitable for the remediation of dye-contaminated wastewater.	Gupta et al. 2020
ZnO/Cellulose Nanofibre	*In situ* precipitation	Methylene blue	–	Adsorption/ Photocatalysis	After 10 minutes: >99%.	Dehghani et al. 2020

Contd.

GO–ZnO–Cu nanocomposite and GO–ZnO–Ag	Simple one-pot method	Methylene blue	Nanorods and nanosheets/–	Photocatalysis	100% degradation efficiency was shown by GO-ZnO-Ag within 40 minutes of sunlight irradiation, whereas the degradation efficiency was decreased by 50% as shown by GO-ZnO-Cu nanocomposites, GO-ZnO-Ag nanocomposites can be considered as effective photocatalysts for the degradation of organic dyes released from industrial wastewater.	Al-Rawashdeh et al. 2020
Ce:SnO_2	Hydrothermal chemical precipitation method	Methyl orange	Spherical/–	Photocatalysis	4% Ce:SnO_2 photocatalyst shows 94.5% degradation of methylene orange within 100 minutes, under UV–Vis illumination, long-term robustness after four repeatable cycles.	Baig et al. 2020
AuNPs	Plant extracted green synthesis method	Methyl Orange and Rhodamine B	Spherical/21.52nm	Photocatalysis	83.25% and 87.64% degradation efficiency were shown by as prepared green synthesized AuNPs for methyl orange and Rhodamine B dye in the presence of sunlight, the rate constant for methyl orange and Rhodamine of were found to be 1.443 $\times 10^{-2}$ min^{-1} and 1.494 $\times 10^{-2}$ min^{-1}, respectively.	Baruah et al. 2018
Iron nanoparticles	Facile borohydride reduction method	Congo red	Spherical/35.5±15.7nm, irregular/ 16.7±5.2nm	Adsorption	A maximum adsorption capacity of 1,735 mg g^{-1} was shown by amorphous iron, pseudosecond-order model.	Kim and Choi 2017
Cotton fiber–graphene oxide	Hummers method	Methylene Blue (MB) and Crystal Violet (CV)	–	Adsorption	The obtained linear correlation coefficient values (R^2) for the adsorption of MB and CV from the binary system were 0.996 and 0.999, respectively.	Nayl et al. 2020
Fe_3O_4	Chemical precipitation method	Acridine orange (cationic dye), Coomassie Brilliant Blue R-250 (anionic dye) and Congo red (azo dye)	Spherical/23 ± 2 nm	Adsorption	Adsorption process follows the pseudosecond-order reaction and the rate constant for acridine, coomassie brilliant blueR-250 and congo red dye is found to be 6.82×10^{-6}, 8.70×10^{-4} & 1.01×10^{-4}, respectively.	Chaudhary et al. 2013
Layered porous C/ Fe3C/γ-Fe2O3	Carbonization (carbonizing Fe^{3+} treated bagasse)	Methylene blue, congo red and commercial hair dye	–/15-80nm	Adsorption	For congo red and methylene blue > 96% adsorption was observed at pH 2 and pH 8 within 24 minutes. For commercial hair dye, 92.6% adsorption was achieved within 20 minutes; maximum adsorption capacity of 531.9 mg/g for CR, 185.2 mg/g for MB, respectively.	Manippady et al. 2020

cross-linking. The as-prepared composite nanofiber was estimated for the degradation of indigo carmine dye from its aqueous solutions. The effects of different parameters, such as solution pH, adsorption equilibrium isotherms, initial dye concentration and contact time, were also studied. At pH of 5, it was observed that the maximum loading capacity was 154.5 mg g^{-1}. The adsorption studies of indigo carmine dye obey the Langmuir model. From the reusability experiment of composite nanofiber, it was confirmed that within 5 minutes, more than 90% of indigo carmine was recovered. The adsorption efficiency of the composite nanofiber can remain almost constant after the five consecutive cycles as revealed by the results of the stability studies (Yazdi et al. 2018).

In another study, nanocomposites of lead and zinc ferrite were synthesized by the co-precipitation method and characterized by various techniques. A simple and common batch adsorption technique was used in this study. UV-Visible spectrophotometer was used for the confirmation of the removal of dye from its aqueous solution. The as-prepared nanocomposites show an adsorption efficiency of 1,042 mg g^{-1}, which confirmed the excellent performance of the nanocomposite as compared to the available adsorbents. The removal of 96.49% of congo red dye has been reported by lead@ zinc ferrite within 90 minutes. A CCD (central composite design) is applied to evaluate the role of adsorption variables. Therefore, based on these findings, lead@zinc ferrite was considered as an effective candidate for the production of efficient adsorbent for environmental application (Jethave et al. 2019). Table 1 illustrated the degradation of various dyes used in the textile dyeing industry with the help of several nanoparticles.

3.2 Use of Nanomaterials in Construction Industry

Various types of nanocomposites and nanomaterials are being progressively considered for several uses in the construction industry because of their exceptional chemical and physical properties. These nanomaterials behave differently at the nanoscale, which generates stimulating innovative scenarios in a variety of construction applications. Different types of manufacturing nanomaterials are incorporated into the construction materials to increase its durability, lightness, strength (Ge and Gao 2008, Li 2004, Sobolev and Gutiérrez 2005) and also provides some useful properties like self-cleaning, heat-insulating and anti-fogging (Irie et al. 2004, Kumar et al. 2008). Recent construction materials used for the civil structure and infrastructure include steel, bricks, timber and concrete and among them, steel and concrete are the most important materials for large-scale projects. Carbon nanotubes, carbon nanofibers, nano clay, alumina, nanosilica, zirconium dioxide, titanium dioxide and so on are the most widely used nanomaterials in the center of cementitious materials (Srinivas 2014). Different research groups have reported the various uses of nanomaterials in the field of construction industries. For the building materials matrices, nanosilica is the most frequently used additive. It increases the resistance to strength, workability and water penetration and also speeds up the hydration reaction. Nanosilica concrete plays an important role in the dispersion of reagents and the paste of the cement during the synthesis process (Sanchez and Sobolev 2010).

The effect of nanosilica has been examined on the mechanical properties of polymer-cement composites. In this study, the influence of the quantity of nanosilica (1%, 3%, 5% by weight of cement) and diameter (100 nm and 250 nm) on the mechanical properties of polymer cement mortars were investigated. The X-ray diffraction was used to study the hydration of cement compounds. The authors revealed that the addition of nanosilica has an excessive probability to increase the speed of the pozzolanic reaction (Sikora et al. 2015). In another study, flexible, high strength and lighter concrete were produced by examined the mechanical properties of a concrete composite, including nanosilica and polyethylene terephthalate. The mechanical properties of the composites were weakened with the polyethylene terephthalate aggregates, but these properties were significantly enhanced by adding nanosilica. With the optimum composition of polyethylene terephthalate (10 wt%) and nanosilica (3 wt%), the flexural, tensile and compressive strength was increased to 9%, 27% and 30% respectively, as compared to original concrete (Behzadian and Shahrajabian 2019).

Some studies have reported the importance of photocatalytic cementitious materials which were used in the infrastructure and outdoor of the buildings for air purifying and self-cleaning application. The effect of nanosilica was thoroughly investigated on the leaching attack upon the photocatalytic cement mortar. A varying amount of nanosilica (0-2%) and 3% of TiO_2 were used for the preparation of photocatalytic mortar. The as-prepared mortar was immersed in a 6 M ammonium nitrate solution in order to form an accelerated leaching environment. From the results, it was observed that leaching attack creates some microstructural and mechanical damages in the mortars, however, these damages were decreased with the addition of nanosilica. Nanosilica can also enhance the resistance of mortars to leaching attacks (Qudoos et al. 2019).

Alumina is also an important additive for improving the mechanical properties of cement. The properties of the concrete were examined with the addition of nano alumina. After the addition of nano alumina, the mechanical strength of the cement composite was investigated by preparing the specimens with a varied volume percentage of nano alumina i.e., 0.5%, 0.75% and 1%, respectively. From these three percentages, 58 specimens were prepared and cured for 28 days and then they were tested. It was confirmed from the experiments that the compressive strength of the concrete cubes was enhanced with the addition of nano alumina into the matrix and the compressive strength was increased to 33.14% within 28 days (Jaishankar and Karthikeyan 2017). In another study, nano alumina was used as a cement replacement in mortar mix. The mechanical properties of cement mortar were studied after the addition of nano alumina (1%, 3% and 5%). Various parameters, such as a change in workability, fire resistance, hardened and fresh properties of the mortar, were observed. It has been obtained from the results that sustainable increase in the fire resistance and a moderate increase in the compressive strength of cement mortar. However, the workability of the cement mortar was decreased (Gowda et al. 2017).

Besides, carbon nanofibers and carbon nanotubes have revealed exceptional strength and have moduli of elasticity in the order of terapascals and tensile strength in the range of gigapascal. These nanomaterials have extraordinary chemical and electronic properties (Sanchez and Sobolev 2010) along with high thermal and electrical conductivity. Graphene oxide/carbon nanotubes composites have been used to improve the mechanical properties of cement paste. The optical and ultraviolet-visible microscopy results confirmed the better dispersion of carbon nanotubes in the solution of graphene oxide because of its greater electrostatic repulsion. These results allow a novel method for the dispersion of carbon nanotubes by eliminating the use of any other dispersant. The prepared composite plays a crucial role in the enhancement of flexural and compressive strength by 24.21% and 21.21%, respectively, which was much greater as compared to carbon nanotubes (10.14% and 6.40%) and graphene oxide (16.20% and 11.05%). The authors also proposed the mechanism for the space interlocking structure, which plays a major role in the improvement of load transfer efficiency, i.e. from cement matrix to the graphene oxide/carbon nanotubes (Lu et al. 2015).

The tensile, compressive and bond strength of reinforced concrete have been investigated with and without the addition of carbon nanotubes. The microstructure of the prepared carbon nanotube-reinforced concrete was examined by scanning electron microscopy. Four samples were prepared with different percentages of carbon nanotubes i.e., 0.01%, 0.02% and 0.03%, respectively, along with a reference sample. Carbon nanotubes with 0.01% and 0.03% increase the tensile strength from 10 to 20% and compressive strength from 7 to 20%. From the scanning electron microscopy, it was confirmed that carbon nanotubes acted as a bridge across the micro-cracks, which illustrated the enhancement in the mechanical properties (Hassan et al. 2019). Apart from these nanomaterials, a variety of manufacturing nanomaterials have superior applications in construction that include functional coating and paints, high resolution actuating/sensing devices as describes in Table 2.

3.3 Use of Nanomaterials in the Food Industry

In the food industry, various types of nanomaterials have been used in numerous applications (Figure 11), such as TiO_2 is used as a food additive in various food products to give a whiting effect. Similarly, aluminum silicate is used as an additive in milk powder and cereals. The nanoparticles used in the production of nano enhancers are zinc, nanocapsules with vitamins, iron, coenzyme

TABLE 2 Nanomaterials used in construction.

Manufacturing Nanomaterials	Construction Materials	Method	Nanomaterials Size (nm)	Remarks	References
Nano-ferrite	Concrete	Citrate-gel method	29.6	2% nano-ferrites show a compressive strength of 71% and 66%, respectively, as compared to the control mix, the gamma-ray attenuation coefficient, splitting strength and density enhanced to 138.5%,101% and 2.5%, respectively, after the aging of 28.	Tobbala 2019
Iron oxide	Cement mortar composites	Sand and cement were mixed homogeneously for 5 minutes at a low speed of 50 rpm, iron oxide was added to it and mixed for 5 minutes before the mixture was added to the prismatic molds with a 4 × 4 × 16 cm dimension. For vibrating the mixture, a vibration machine was used to eliminate the air bubbles. After 24 hours at 25°C and 95% of relative humidity demolding the sample and pour into the water bath till the testing day.	>30	The flexural and compressive strengths of iron oxide cement-mortar nanocomposite increased to 2% and 12% as compared to cement mortar, iron oxide prevents the creation of microcracks and enhance the accessibility of H_2O molecules to the cement C-S-H and iron oxide oxygen group.	Kiamahalleh et al. 2020
Zinc oxide	White Portland Cement	Ball- milling method	24, 27, 37, and 70	The setting time and the normal consistency of the white Portland cement paste were enhanced by ZnO with the optimized parameter, such as average particle size-37 nm, 0.4% weight percentage of ZnO improved the compressive strength to 28% at 28 days via filler effect.	Shafeek et al. 2020
Silver	Cementitious composites	Experimental design methods	11.2, 52.39 and 72.58	Higher splitting tensile strength, ultrasonic pulse velocity and conductivity were shown by smaller silver nanoparticles, and also form homogeneous structure with cement	Ceran et al. 2019

Contd.

SiO_2 TiO_2 Fe_2O_3	Fly ash blended cement mortars	A suspension of nanoparticles was formed by mechanically mixing nanoparticles with water, cementitious materials were mixed to the above suspension to produce a homogeneous paste, sample was placed in molds for one day after casting at relative humidity 65 ± 5% and temperature of 23 ± 2 °C, after demolded the sample, place it in a water tank for further processing.	20-30 20-40 15	Enhancing the flexural and compressive strengths of the mortar.	Ng et al. 2020
Nano ZrO_2	Cement paste	Mechanical dispersion of ZrO_2 in the mixture of water+ polycarboxylate admixture for 5 min, the suspension was poured into the cement and thermal expansion coefficient was completed with different molds, covered with plastic wrap and moistened for 24 hours, samples were demolded and cured at 25 ± 2°C for 7 days in saturated alkali solution.	20	Improved volume heat capacity and residual thermal conductivity with the addition of 1 wt% of ZrO_2, composite paste displayed heat capacity of 1946 kJ m^3 after heating at 350°C.	Yuan et al. 2013
Graphene oxide nanosheets	Cement	Mobasher–Tixier model		Enhanced corrosion resistance coefficient and compressive strength by 40.9% and 68.7%, addition of 0.03 wt% graphene oxide into mortars, inhibits the microcracks, refines the microstructure, improve the sulfate resistance.	Zeng et al. 2020
TiO_2	Portland cement paste	TiO_2 nanoparticles+ water ultrasonically vibrated for 30 minutes for better dispersion, stirring the water +cement mixture, after mixing the mixture was transferred into a plastic Petri dish, after 24 hours demolded and cured at room temperature, relative humidity of 95% for 28 days	20-50	Increased 28-day three-point bend strength by 4.52%, 8.00%, 8.26%, and 6.71%, respectively, with different amount of TiO_2 nanoparticles by mass of cement, were added, enhancement in the microstructure of the paste, enlarged the amount of cementitious phase in the paste, reduced the microporosity.	Feng et al. 2013

Q10 and omega-3 fatty acids (Jampilek and Kralova 2020). The nanoparticles easily interact with the components of food products during the preparation and packaging for protection and security. Nanoparticles formed by dispersion are superior as compared to the conventional polymers owing to their enhanced properties, including stability, antimicrobial, durability, nutritional and flexibility properties (Assis et al. 2012). Food processing and food packaging are the main application of nanoparticles in the food industry as described below.

FIGURE 11 Pictorial representation of different applications of nanomaterials used in food industry

3.4 Food Processing

The progress of nanotechnology in food processing has attracted much attention to the production of spontaneous foodstuffs in respect of the texture and the taste of the components present in the food products. Supervising the release of flavors, encapsulating the food additives or components and improving the bioavailability of nutritional components producing food more hygienic and nutritive for consumers (Lamabam and Thangjam 2018). Therefore, according to the requirements of the consumers, food processing undergoes several changes with time.

Antimicrobial, vitamins, enzymes, preservatives, flavorings, colorants and antioxidants are utilized as functional food ingredients in the food industry. During the different phases of food processing, these functional materials should be secured from any kind of deterioration. Currently, nanocarriers are used to adding food additives in the product as a delivery system without changing their morphology. Some research groups have reported the importance of particle size that directly influences the distribution of any bioactive compound in the body (Ezhilarasi et al. 2013). It was also observed that microparticles cannot be absorbed in the cell line, but submicron nanoparticles can easily get absorbed. An idealistic delivery system should have the following characteristics, that should be (i) suitable for the transfer of the active compounds accurately to the target, (ii) should assure the accessibility at the target time and the specific rate and (iii) in the short duration, it should be effective to keep the active compounds for a long time at a suitable level. Nanotechnology has been used in the creation of biopolymer matrices, simple solutions, encapsulations, colloids and emulsions that provide effective delivery systems. As compared to the traditional encapsulation process, nanoparticles have improved properties for encapsulation. Nanoencapsulated food supplements, ingredients and additives can enhance the taste of the food, increase the dispersibility of the fat-soluble additives, facilitate hygienic food storage and reduce the utilization of preservatives, sugar, salt and fat. It also helps in masking the uninvited taste of some additives like a synthetic form of ascorbic acid, citric acid, benzoic acid, tomato carotenoid lycopene certain supplements, like isoflavones, lutein, vitamins A and E, coenzyme Q10 and omega-3 fatty acid (Chaudhry and Castle 2011). Nanoemulsions are the most important types of liquid-based nanocarriers which are prepared by biopolymer/surfactants, water and oil in several forms of water/oil emulsion or oil/water nanoemulsions, structural nanoemulsions and pickering nanoemulsions (Jafari 2017, Akhavan et al. 2018).

Nanoemulsions have been used in order to increase the bioavailability and solubility of a bioactive polyphenolic compound such as curcumin with anti-inflammatory, antioxidant and anticancer activity. A nanoemulsion of curcumin was prepared by ultrasonication method. Small droplet sizes and a greater load of curcumin were used to form the nanoemulsions. In this study, glycerol was added to stimulate the trapping efficiency of curcumin and the prepared nanoemulsions revealed physical stability during the storage (Ochoa et al. 2016). In another study, nanoemulsions droplets loaded with essential oil i.e., garlic, sunflower, cinnamon and curcumin have been prepared without or with pectin coating. The influence of the prepared nanoemulsions has been checked on the shelf life of chilled chicken fillets. Nanoemulsions with chicken fillets had greater water holding capacity, texture scores and lower total volatile basic nitrogen values as compared to the chicken fillets. This study revealed the enhancement in the nutritional profile, quality and texture of the chicken fillets (Abdou et al. 2018). In beef patties, chitosan-cinnamon oil nanoemulsions were used to restrict the oxidation of lipid, which was mainly due to the antioxidant activity of trans cinnamaldehyde (Ghaderi-Ghahfarokhi et al. 2017).

A thin-film-hydration sonication method was used to prepared vitamin A palmitate-loaded nanoliposomes. The encapsulation efficiency of vitamin A palmitate was decreased with the greater level of cholesterol, however, with the implementation of the lecithin-cholesterol mixture the efficiency was extended to 15.8% (Pezeshky et al. 2016). Vitamin C connecting with multi-layered nanoliposomes by deposing the negative sodium alginate and positive chitosan on anionic nanoliposomes, respectively, and activated it to mandarin juice. The surface characteristic of the nanoliposomes was enhanced by the coated structure of nanoliposomes and decreased the rate of lipid peroxidation and the protection of vitamin C after 90 days of storing (Liu et al. 2017b). A controlled ionic gelation technique was used for the preparation of $FeSO_4$ nanoparticles, where Na-alginate was used as a wall material for $FeSO_4$. The 15-30 nm of alginate loaded with 0.06% w/v Fe discovered an extended *in vitro* release kinetics for four days and the rate of release of Fe was dependent on the pH values. The release of Fe at pH 2.0 and 6.0 and 7.4 were 20 and 65% to 70%, respectively (Katuwavila et al. 2016). Besides nanoemulsions, eugenol nanocapsules were reported for the preservation of chilled pork by creating the hindrance in the increase in the total volatile basic nitrogen value, pH and lipid oxidation. Higher water holding capacity and cohesiveness were obtained in the treated chilled pork sample as compared to the untreated one (Wang et al. 2020a).

In the food industry, flavors are the most important part of food systems. 2-Isobutyl-3 methoxypyrazine in green pepper, allylpyrazine in the roasted nut, acetyl-L-pyrazines in popcorn, methoxypyrazines in vegetables, terpenoids in citrus and piney, phenolics in smoked products and aldehydes in fruits are some examples of flavors that are present in the food products (Jafari et al. 2017). To reduce the instability of the flavor structure, nanoencapsulation was used to preserve the properties of the bioactive compounds. Here, the main advantage of encapsulation was the conversion of liquids flavors into powder and enhanced their stability when they were exposed to light, oxygen, and higher temperature. The influence of the flavor composition was studied on the preparation and the properties of oil/water nanoemulsions.

3.5 Food Packaging

Packaging is essentials for all the food items to protect them from physical changes, preserve their nutritional value, ensure the proper handling and accelerate the storage and transport to the supply chain from producer to consumer. The materials used in packaging must be mechanically and physically resistant and will not add any odor to the packed product. Laminated paper, glass, plastic film, biodegradable film and metal were used as a material in the packaging of the food products. Traditional packaging materials have been replaced by synthetic polymer owing to its low density, resistance to microbial growth, thermoplastic, inertia, transparency and low cost. But synthetic polymer is not completely biodegradable creating some environmental problems and their use is being moderately limited. Therefore, some biodegradable polymer, such as protein-based film, poly-β-hydroxyalkanoates, starch and polylactic acid, has been developed. The mechanical, functional and physical properties of biodegradable polymers were enhanced by adding inorganic and organic nanoparticles, such as chitosan, zinc, titanium, cellulose, zein silica, starch and clay (Samanta et al. 2016). An environment-friendly starch/TiO_2 bio-nanocomposites was prepared using various amounts of nano TiO_2 (1, 3 and 5%). The prepared films showed improved thermal and hydrophobicity properties and decreased water vapor permeability. Young's modulus and tensile strength were decreased with the increase in the TiO_2 content while increased the tensile energy to break and elongation at break. Starch/TiO_2 bio-nanocomposite effectively protects the materials against ultraviolet light and could be used as an ultraviolet shielding packaging material (Goudarzi et al. 2017).

A coating of chitosan and chitosan/ZnO nanocomposite on polyethylene films was used to study the antimicrobial properties. The polyethylene films were treated with oxygen plasma in order to increases the adhesion by 2% of chitosan and the coating solution to the packaging films. The addition of ZnO nanoparticles into the matrix of chitosan resulted in an increase in the water contact angle from 60° to 95°, an increase in the solubility up to 42% and a decrease in swelling by 80%. Chitosan/ZnO nanocomposites coated on polyethylene completely prevented or inactivated the growth of food pathogens such as *Staphylococcus aureus*, *Salmonella enterica* and *Escherichia coli* (Al-Naamani et al. 2016). In another study, an innovative *in situ* method has been reported for the preparation of magnesium oxide nanoparticles with spherical shape by heat-treating magnesium carbonate/polymethyl methacrylate composite precursor. Magnesium oxide nanoparticles were added in the chitosan nanocomposite thin film to enhance its physical properties for appropriate packaging applications. The elastic modulus and tensile stress were enhanced by 38% and 86%, respectively. The prepared films displayed exceptional flame-retardant properties, thermal stability moisture barrier and ultraviolet shielding properties (De Silva et al. 2017).

The influence of high hydrostatic pressure treatment and nano TiO_2 on water vapor, microstructure, gas barrier, mechanical and antibacterial properties of polyvinyl alcohol-chitosan biodegradable films were investigated. The author also studied the migration behavior of the nanoparticles from the film to the food simulants. TiO_2 nanoparticles also enhanced the antibacterial activity of the films which plays a crucial role as a plasticizer in the films. The interaction between chitosan and polyvinyl alcohol molecules was enhanced by the high hydrostatic pressure treatment. There was a reduction in the migration of TiO_2 nanoparticles from the films (Lian et al. 2016). With the production of new materials, various ways were formed to create active packaging that delivers the functional properties to increase the shelf-life and enhanced the safety of the materials used in the food packaging. These include moisture absorber to reduce microbial growth (Gaikwad et al. 2019), ethylene scavenger to decrease fruits and vegetables ripening (Sadeghi et al. 2019) and oxygen scavenger to reduce fat oxidation (Gaikwad et al. 2018). Silver/hyperbranched polyamide-amine nanoparticles enclosed cellulose films have been reported to study the antibacterial effect on *Staphylococcus aureus* and *Escherichia coli*. For the preparation of *in situ* silver nanoparticles, hyperbranched polyamide-amine was used as a stabilizer and reducing agent which embedded silver nanoparticles onto the oxidized cellulose to form a cellulose film with less silver leakage. The prepared films show better antibacterial activity when the leakage of silver was less than 10% and expand the storage life of cherry tomatoes as food packaging (Gu et al. 2019).

A green method has been reported for the fabrication of zinc oxide-silver nanocomposites. In this method leaf extract of Thymus vulgaris was used as a reducing agent and the nanocomposite was mixed into poly(3-hydroxybutyrate-co-3hydroxyvalerate)-chitosan with a specific ratio of solvent by ultrasonication. The prepared nanocomposite increases the shelf-life of poultry items and preserve the food safety (Zare et al. 2019).

3.6 Use of Nanomaterials in Semiconductor Industry

Semiconductors are extensively used in various types of industries ranging from high-tech industries to aerospace, aviation and health care. They are the driving forces for growth and technological developments. The semiconductor industry, which comprises numerous companies that fabricate or supply semiconductor materials and equipment is predictable to experience sustainable growth with increasing technological advances. Currently, the semiconductor industry attaining much attention in the field of energy harvesting, the auspicious field that used semiconductors is solar energy harvesting. Nanotechnology is used to fabricate photovoltaic materials or solar cells with increasing photon trapping efficiency. Semiconductor nanoparticles are the propitious materials used in several devices like an optical detector, light-emitting diode, telecommunication relays, clean conversion cell and electronic (Table 3). InP, InAs, GaN and GaP from group III-V, Si and Ge from group IV and ZnS, CdTe, CdSe, CdS, ZnO from group II-VI are some examples of semiconductor nanoparticles.

A hybrid nanostructure of phenothiazine with two-dimensional CdSe nanoplatelets was designed by varying the thickness of CdSe nanoplatelets. The synthesized nanostructure was used in a high-performance photodetector. The shortening of the decay time and the considerable photoluminescence quenching of CdSe nanoplatelets with phenothiazine explain the process of charge transfer. A photocurrent of 4.7×10^3 fold (photo-to-dark current ratio) was displayed by the optimized CdSe nanoplatelets-phenothiazine at the voltage of 1.5 V. The maximum detecti, response time, responsivity value and quantum efficiency for hybrid were found to be 4×10^{11} jones, 107 ms, 160 mA/W and ~40%, respectively. Therefore, from these findings, it was suggested that the synthesized hybrid system can be used as an alternate for an ultrasensitive photodetector (Dutta et al. 2020). All-inkjet printing technology was used first time for the deposition of inkjet printing silver nanoparticles on the inkjet printing ZnO UV photodetector. The surface defects of ZnO were passivated by inkjet printing silver nanoparticles. Normalized detectivity of 1.45×10^{10} jones at 0.715 mW incident light power was exhibited by silver nanoparticles modified detector, which was greater than the ZnO photodetector (Wang et al. 2020b). Another study reported the enhanced photoresponse of UV photodetectors by the insertion of several plasmonic nanoparticles in the detector architecture. A solid-state dewetting method was used to fabricate silver, gold and silver/gold bimetallic nanoparticles on GaN (0001). Among all these nanoparticles, silver/gold nanoparticles with a greater silver percentage at a low bias of 0.1 V displayed the improved detectivity of 2.4×10^{12} jones, the quantum efficiency of 3.6×10^4% and photo responsibility of 112 A/W (Kunwar et al. 2020).

A transparent electrode, insulating layer of SiO_2 and a thin film of AlGaN were used for the fabrication of graphene-insulator-semiconductor ultraviolet light-emitting diode. An outstanding ultraviolet-detecting capability was shown by a graphene-insulator-semiconductor ultraviolet light-emitting diode which can be used in the fabrication of multifunctional optoelectronic devices. This approach avoids the doping of both *n*-type and *p*-type semiconductors as compared to the conventional light-emitting diodes (Wu et al. 2020). In another study, a sol-gel process was used to prepare zinc oxide nanoparticles which act as an electron transport layer in quantum dots light-emitting diode. With the decrease in the size of the nanoparticles, the defect density was reduced and the bandgap changed from 3.44 to 3.66 eV, and it also avoided the exciton quenching at the interface of zinc oxide nanoparticles/emitting quantum dots. There was an enhancement in the radiative recombination rates and the electron-hole balance in the emissive quantum dot layer in quantum dots light-emitting diode as the conductivity of the smaller nanoparticles were reduced (Moyen et al. 2020).

TABLE 3 Potential industrial applications of various types of nanomaterials.

Material Used	Method	Use	Remarks	References
PbS Quantum dots	Hot injection method, Layer by layer method	Photodiode	At zero voltage bias, rise/fall time as short as 0.33 μs, specific detectivity value up to 3.2×10^{11} Jones at 1,125 nm.	Xu et al. 2020
PbS/CdS Quantum dot	Spin coating process	Photodetector	4.0×10^{12} Jones detectivity value, showing enhancement in these values more than 10 times as compared to PbS QDs.	Kwon et al. 2020
Cu@graphene nanowire	Chemical vapor deposition	Light-emitting diode	Strong antioxidant stability, broad transparency range (200~3,000 nm), the enhanced optoelectronic performance of 33 Ω/sq.	Huang et al. 2018b
CdSe/ZnS quantum dots	Step controlled mixing process	Blue quantum dot light-emitting diode	Quantum dot light-emitting diodes show a maximum quantum efficiency of 19.8%, current efficiency of 14.1 cd A^{-1}	Wang et al. 2017
Core-Shell ZnO@ SnO_2 nanoparticles	Solvothermal method	Inorganic perovskite solar cells	The core-shell ZnO@ SnO_2 nanoparticles act as an electron transfer layer, attained 14.35% power conversion efficiency with 79% fill factor, 16.45 mA cm^{-2} short circuit current density and 1.11 V open circuit voltage	Li et al. 2020
CdTe or CdTe/CdS quantum dots	One-step linker assisted chemical bath deposition method	Quantum dot sensitized solar cells	Enhanced light absorptivity of CdTe, CdS shell protect the sensitive CdTe quantum dots and important for the photo-generated charges separation, the power conversion efficiency of 3.8% and 5.25% in the presence of AM 1.5 G one sun and 0,12 sun illumination	Yu et al. 2011
InZnSnO semiconductors	Atomic layer deposition	Thin-film transistor	Superior mobility higher than 20 $cm^2 V^{-1} s^{-1}$, promising materials for large area displays and high resolution, retained the electrical characteristics of thin-film transistors	Sheng et al. 2019

Contd.

p-type $CuCrO_2$ semiconductor	Spin coating process	Thin-film transistor	A hole mobility of 0.59 $cm^2 V^{-1} s^{-1}$, on/off current ratio of ~105, used in low-cost *p-n* junctions	Nie et al. 2018
In-Zn-O thin-film	Sol-gel synthesis, electron beam evaporator deposition	Thin-film transistor	Improved thin-film transistors mobilities, 6.5-time higher mobilities as compared to the device fabricated by single solution-processed layer, increase in the on/off current ratio by 2 order of magnitude.	Bang et al. 2020
AgNPPANI-graphene/ CFP	Spray-coating process	Supercapacitors	Attained 828 Fg^{-1} specific capacitance, at an applied current density of 1.5 A^{-1} AgNPPANI-graphene/CFP shows capacity retention of 97.5%.	Sawangphruk et al. 2013
Sb_2S_3-M@S-C Sb_2S_3-P@S-C nanoparticles	Microwave-assisted synthesis	Supercapacitors	A higher specific capacitance value of 1,179 and 1,380 Fg^{-1} was shown by Sb_2S_3-M@S-C and Sb_2S_3-P@S-C, respectively, with a current density of 1 Ag^{-1}, a hybrid device was designed with Sb_2S_3-M@S-C/Sb_2S_3-P@S-C showing an outstanding energy density of 49 $Whkg^{-1}$.	Sahoo et al. 2019

Titania/benzoic acid-fullerene bishell decorated with silver nanoparticles have been reported for the improved performance of organic and perovskite solar cell. Coating of silver/titanium dioxide nanoparticles with fullerene shell can remove the accumulation of the charge, activate the effective plasma-exciton coupling, decreasing the monomolecular recombination and encourage exciton dissociation. Enhanced light absorption and improved carrier extraction of the device with the prepared nanoparticles were responsible for the increased fill factor and short-circuit current. Plasmonic photovoltaic solar cells and organic solar cells displayed a power conversion efficiency of 20.2% and 13.0%, respectively, because of the plasmon coupling effect (Yao et al. 2019). Pt–$Cu_3InSnSe_5$ and Au–Cu_3 In $SnSe_5$, heteronanostructures were synthesized by using the seed growth approach. These heterostructures in the form of thin films were transferred to the counter electrode of dye-sensitized solar cells to evaluate their power conversion efficiency. The power conversion efficiency of $Cu_3InSnSe_5$ nanoparticle-based dye-sensitized solar cell was 5.8%, which was further enhanced to 7.6% and 6.5%, respectively, with the addition of Pt-/Au-$Cu_3InSnSe_5$ heterojunction counterpart (Lou et al. 2017).

Apart from these applications, thin films of semiconductors can also be used in thin-film transistors. A low-cost ultrasonic spray pyrolysis method has been explored for the fabrication of zirconium oxide thin films. The synthesized thin films were optically transparent with a bandgap of 5.35 eV, uniform, smooth and amorphous. Thin films displayed a leakage current of 3.56×10^{-6} A cm^{-2} at a bias of 1 V and relative permittivity of 22.7 at 50 Hz. Zinc tin oxide/silicon dioxide thin-film transistors showed saturation mobility of 0.5 $cm^2 V^{-1} s^{-1}$. A full solution-processed and cost-effective device was formed by replacing silicon dioxide gate dielectric with zirconium oxide dielectric. At a low threshold voltage of 0.03 V, greater saturation mobility was achieved by zinc tin oxide/zirconium oxide thin-film transistors (Oluwabi et al. 2020). In another study, the spin coating process was used to synthesized $CuCrO_2$ semiconductor thin films. After the synthesis process, these films were integrated as a channel layer in thin-film transistors. With the increasing annealing temperature from 500°C to 800°C, the electrical performance of the thin film transistors on silicon dioxide dielectric was also enhanced. The optimized thin-film transistors show hole mobility of 0.59 $cm^2 V^{-1} s^{-1}$ and an on/off current ratio of ~105 (Nie et al. 2018). Atomic Layer Deposition (ALD) process was used to fabricate ZnO/graphene composite thin film by depositing ZnO on graphene. The graphene layer was prepared using a chemical vapor deposition process. ZnO/graphene thin film transistor revealed improved carrier mobility of TFT displayed enhanced carrier mobility of 1–14 cm^2/Vs as compared to ZnO thin-film transistors (Liu et al. 2016).

4. CONCLUSIONS AND FUTURE RECOMMENDATIONS

A general introduction of various types of nanomaterials, their industrial-scale fabrication and their uses in different industrial applications were discussed in this chapter. The rapid progress in nanotechnology has enabled the conversion of different concepts from research laboratories to consumer uses with low cost and wide availability. At an industrial scale that is different from the laboratory synthesis methods was reported for the preparation of various types of nanoparticles, such as CaP NPs, Ag NPs, ^{64}CuS NPs, Zn-Ce oxide, VO_2 NPs, Ni NPs, Sn NPs and conjugated Au NPs. There is a difference between laboratory and industrial-scale formulation of nanoparticles. At laboratory scale, we have to use small apparatus, while a big plant to be used at industrial scale. Among all the accessible methods for the preparation of nanoparticles, some of them are relatively efficient for the control of size distribution and particle size, producing high-quality products. Different industrial applications of nanoparticles were reviewed in this chapter. In the textile dyeing industry, nanoparticles such as magnetic, TiO_2, $BiFeO_3$, Au–Cu_3N nano heterostructures, ZnO/Cellulose nanofibre, GO–ZnO–Cu nanocomposite, Au NPs, Fe_3O_4 NPs, etc., were used for the degradation of dyes discharged from several industries. The nanoparticles were used as a photocatalyst and nanoadsorbent for the degradation of various types of dyes. Nanoparticles play a significant role in the food industry together with food processing and food packaging. The shelf life of food products was enhanced by protecting them from lipids, gases and moisture. They also provide a superior delivery system for the transfer of bioactive compounds. The application of nanoparticles in the construction industry presents numerous chances for increasing materials' properties and functions. Nanoparticles of silica, TiO_2, iron oxide, zinc oxide, silver, zirconium oxide, graphene oxide, etc. were used to enhance the durability of the structure like their tensile and compressive strength, increased the resistance to corrosion, fatigue, abrasion and wear. These nanoparticles were used in several types of construction materials and improved their mechanical properties. Semiconductor nanoparticles like PbS Quantum dots, PbS/CdS quantum dots, CdTe or CdTe/CdS quantum dots, *p*-type $CuCrO_2$ semiconductor, zirconium oxide thin films, $CuCrO_2$ semiconductor thin films, etc., were used in the semiconductor industry. These nanoparticles were used in different applications such as photovoltaic solar cell, organic solar cell, dye-sensitized solar cell, photodetector, photodiodes, light-emitting diode, thin-film transistors and supercapacitor, etc. In supercapacitors, nanoparticles increase the specific capacitance and energy density of the prepared device. In the case of the solar cell, the power conversion efficiency of the solar cells was increased by using these nanoparticles.

Therefore, in the future, the demands of nanomaterials tools and devices will be increased day by day in all industries. But there are some factors that should be kept in mind while fabricating the nanomaterials devices, like toxicology, bioavailability, transport, health and environmental impact. Environmental safety must be a priority in research which deal with threats associated with the nanodevices in the industry. Therefore, the protection and security problems will be prerequisites that need to be wisely addressed and considered in the future (Li et al. 2019).

References

Abdou, E.S., Galhoum, G.F. and Mohamed, E.N. 2018. Curcumin loaded nanoemulsions/pectin coatings for refrigerated chicken fillets. Food Hydrocoll. 83: 445-453.

Akhavan, S., Assadpour, E., Katouzian, I. and Jafari, S.M. 2018. Lipid nano scale cargos for the protection and delivery of food bioactive ingredients and nutraceuticals. Trends Food Sci. Technol. 74: 132-146.

Al-Naamani, L., Dobretsov, S. and Dutta, J. 2016. Chitosan-zinc oxide nanoparticle composite coating for active food packaging applications. Innov. Food Sci. Emerg. Technol. 38: 231-237.

Al-Rawashdeh, N.A., Allabadi, O. and Aljarrah, M.T. 2020. Photocatalytic activity of graphene oxide/zinc oxide nanocomposites with embedded metal nanoparticles for the degradation of organic dyes. ACS Omega 5(43): 28046-28055.

Ali, I., Peng, C., Khan, Z.M., Sultan, M. and Naz, I. 2018. Green synthesis of phytogenic magnetic nanoparticles and their applications in the adsorptive removal of crystal violet from aqueous solution. Arab. J. Sci. Eng. 43(11): 6245-6259.

Assis, L.M.D., Zavareze, E.D.R., Prentice-Hernández, C. and Souza-Soares, L.A.D. 2012. Revisão: Características de nanopartículas e potenciais aplicações em alimentos. Braz. J. Food Technol. 15(2): 99-109.

Baig, A.B.A., Rathinam, V. and Palaninathan, J. 2020. Facile synthesis of Ce-doped SnO_2 nanoparticles with enhanced performance for photocatalytic degradation of organic dye. J. Iran. Chem. Soc. 18: 13-27.

Baláž, P., Achimovičová, M., Baláž, M., Billik, P., Cherkezova-Zheleva, Z., Criado, J.M., et al. 2013. Hallmarks of mechanochemistry: From nanoparticles to technology. Chem. Soc. Rev. 42(18): 7571.

Bang, S.Y., Mocanu, F.C., Lee, T.H., Yang, J., Zhan, S., Jung, S.M., et al. 2020. Robust In-Zn-O thin-film transistors with a bilayer heterostructure design and a low-temperature fabrication process using vacuum and solution deposited layers. ACS Omega 5(34): 21593-21601.

Barman, D., Paul, S., Ghosh, S. and De, S.K. 2019. Cu_3N nanocrystals decorated with Au nanoparticles for photocatalytic degradation of organic dyes. ACS Appl. Nano Mater. 2(8): 5009-5019.

Baruah, D., Goswami, M., Yadav, R.N.S., Yadav, A. and Das, A.M. 2018. Biogenic synthesis of gold nanoparticles and their application in photocatalytic degradation of toxic dyes. J. Photochem. Photobiol. B, Biol. 186: 51-58.

Basith, M.A., Yesmin, N. and Hossain, R. 2018. Low temperature synthesis of $BiFeO_3$ nanoparticles with enhanced magnetization and promising photocatalytic performance in dye degradation and hydrogen evolution. RSC Adv. 8(52): 29613-29627.

Behzadian, R. and Shahrajabian, H. 2019. Experimental study of the effect of nano-silica on the mechanical properties of concrete/PET composites. KSCE J. Civ. Eng. 23(8): 3660-3668.

Ceran, Ö.B., Şimşek, B., Doruk, S., Uygunoğlu, T. and Şara, O.N. 2019. Effects of dispersed and powdered silver nanoparticles on the mechanical, thermal, electrical and durability properties of cementitious composites. Constr. Build Mater. 222: 152-167.

Chakravarty, R., Chakraborty, S., Ningthoujam, R.S., Vimalnath Nair, K.V., Shitaljit Sharma, K., Ballal, A., et al. 2016. Industrial-scale synthesis of intrinsically radiolabeled64CuS nanoparticles for use in positron emission tomography (PET) imaging of cancer. Ind. amp; Eng. Chem. Res. 55(48): 12407-12419.

Chakravarty, R., Chakraborty, S., Guleria, A., Shukla, R., Kumar, C., Vimalnath Nair, K.V., et al. 2018. Facile one-pot synthesis of intrinsically radiolabeled and cyclic RGD conjugated 199Au nanoparticles for potential use in nanoscale brachytherapy. Ind. amp; Eng. Chem. Res. 57(43): 14337-14346.

Chakravarty, R., Chakraborty, S., Guleria, A., Kumar, C., Kunwar, A., Nair, K.V.V., et al. 2019. Clinical scale synthesis of intrinsically radiolabeled and cyclic RGD peptide functionalized 198Au nanoparticles for targeted cancer therapy. Nucl. Med. Biol. 72-73: 1-10.

Chaudhary, G.R., Saharan, P., Kumar, A., Mehta, S.K., Mor, S. and Umar, A. 2013. Adsorption studies of cationic, anionic and azo-dyes via monodispersed Fe_3O_4 nanoparticles. J. Nanosci. Nanotechnol. 13(5): 3240-3245.

Chaudhry, Q. and Castle, L. 2011. Food applications of nanotechnologies: An overview of opportunities and challenges for developing countries. Trends Food Sci. Technol. 22(11): 595-603.

Chen, D., Zhang, Y., Chen, B. and Kang, Z. 2013. Coupling effect of microwave and mechanical forces during the synthesis of ferrite nanoparticles by microwave-assisted ball milling. Ind. amp; Eng. Chem. Res. 52(39): 14179-14184.

Darr, J.A., Zhang, J., Makwana, N.M. and Weng, X. 2017. Continuous hydrothermal synthesis of inorganic nanoparticles: Applications and future directions. Chem. Rev. 117(17): 11125-11238.

Dehghani, M., Nadeem, H., Singh Raghuwanshi, V., Mahdavi, H., Banaszak Holl, M.M. and Batchelor, W. 2020. ZnO/cellulose nanofiber composites for sustainable sunlight-driven dye degradation. ACS Appl. Nano Mater. 3(10): 10284-10295.

Denis, C.J., Tighe, C.J., Gruar, R.I., Makwana, N.M. and Darr, J.A. 2015. Nucleation and growth of cobalt oxide nanoparticles in a continuous hydrothermal reactor under laminar and turbulent flow. Cryst. Growth Des. 15(9): 4256-4265.

Deshmukh, K., Kovářík, T., Křenek, T., Docheva, D., Stich, T. and Pola, J. 2020. Recent advances and future perspectives of sol-gel derived porous bioactive glasses: A review. RSC Adv. 10(56): 33782-33835.

De Silva, R.T., Mantilaka, M.M.M.G.P.G., Ratnayake, S.P., Amaratunga, G.A.J. and de Silva, K.N. 2017. Nano-MgO reinforced chitosan nanocomposites for high performance packaging applications with improved mechanical, thermal and barrier properties. Carbohydr. Polym. 157: 739-747.

Dutta, A., Medda, A., Bera, R., Sarkar, K., Sain, S., Kumar, P., et al. 2020. Hybrid nanostructures of 2D CdSe nanoplatelets for high-performance photodetector using charge transfer process. ACS Appl. Nano Mater. 3(5): 4717-4727.

Długosz, O. and Banach, M. 2020. Inorganic nanoparticle synthesis in flow reactors-applications and future directions. React. Chem. Eng. 5(9): 1619-1641.

Esposti, L.D., Dotti, A., Adamiano, A., Fabbi, C., Quarta, E., Colombo, P., et al. 2020. Calcium phosphate nanoparticle precipitation by a continuous flow process: A design of an experiment approach. Crystals 10(10): 1-17.

Ezhilarasi, P.N., Karthik, P., Chhanwal, N. and Anandharamakrishnan, C. 2013. Nanoencapsulation techniques for food bioactive components: A review. Food Bioproc. Tech. 6(3): 628-647.

Feng, D., Xie, N., Gong, C., Leng, Z., Xiao, H., Li, H., et al. 2013. Portland cement paste modified by TiO_2 nanoparticles: A microstructure perspective. Ind. amp; Eng. Chem. Res. 52(33): 11575-11582.

Feng, J., Biskos, G. and Schmidt-Ott, A. 2015. Toward industrial scale synthesis of ultrapure singlet nanoparticles with controllable sizes in a continuous gas-phase process. Sci. Rep. 5: 15788.

Gaikwad, K.K., Singh, S. and Lee, Y.S. 2018. Oxygen scavenging films in food packaging. Environ. Chem. Lett. 16(2): 523-538.

Gaikwad, K.K., Singh, S. and Ajji, A. 2019. Moisture absorbers for food packaging applications. Environ. Chem. Lett. 17(2): 609-628.

Gaur, R., Mishra, L., Siddiqi, M.A. and Atakan, B. 2014. Ruthenium complexes as precursors for chemical vapor-deposition (CVD). RSC Adv. 4(64): 33785-33805.

Gautam, A., Kshirsagar, A., Biswas, R., Banerjee, S. and Khanna, P.K. 2016. Photodegradation of organic dyes based on anatase and rutile TiO_2 nanoparticles. RSC Adv. 6(4): 2746-2759.

Ge, Z. and Gao, Z. 2008. Applications of Nnanotechnology and Nanomaterials in Construction. First International Conference on Construction in Developing Countries. 235-240.

Ghaderi-Ghahfarokhi, M., Barzegar, M., Sahari, M.A., Gavlighi, H.A. and Gardini, F. 2017. Chitosan-cinnamon essential oil nano-formulation: Application as a novel additive for controlled release and shelf life extension of beef patties. Int. J. Biol. Macromol. 102: 19-28.

Ghosal, K. and Sarkar, K. 2018. Biomedical applications of graphene nanomaterials and beyond. ACS Biomater. Sci. Eng. 4(8): 2653-2703.

Gomez-Gonzalez, M.A., Koronfel, M.A., Goode, A.E., Al-Ejji, M., Voulvoulis, N., Parker, J.E., et al. 2019. Spatially resolved dissolution and speciation changes of ZnO nanorods during short-term *in situ* incubation in a simulated wastewater environment. ACS Nano 13(10): 11049-11061.

Goudarzi, V., Shahabi-Ghahfarrokhi, I. and Babaei-Ghazvini, A. 2017. Preparation of ecofriendly UV-protective food packaging material by starch/TiO_2 bio-nanocomposite: Characterization. Int. J. Biol. Macromol. 95: 306-313.

Gowda, R., Narendra, H., Rangappa, D. and Prabhakar, R. 2017. Effect of nano-alumina on workability, compressive strength and residual strength at elevated temperature of Cement Mortar. Mater. Today 4(11): 12152-12156.

Gruar, R.I., Tighe, C.J., Southern, P., Pankhurst, Q.A. and Darr, J.A. 2015. A direct and continuous supercritical water process for the synthesis of surface-functionalized nanoparticles. Ind. amp; Eng. Chem. Res. 54(30): 7436-7451.

Gu, R., Yun, H., Chen, L., Wang, Q. and Huang, X. 2019. Regenerated cellulose films with amino-terminated hyperbranched polyamic anchored nano silver for active food packaging. ACS Appl. Bio Mater. 3(1): 602-610.

Gupta, N.K., Ghaffari, Y., Kim, S., Bae, J., Kim, K.S. and Saifuddin, M. 2020. Photocatalytic degradation of organic pollutants over MFe_2O_4 (M = Co, Ni, Cu, Zn) nanoparticles at neutral pH. Sci. Rep. 10(1): 1-11.

Hassan, A., Elkady, H. and Shaaban, I.G. 2019. Effect of adding carbon nanotubes on corrosion rates and steel-concrete bond. Sci. Rep. 9(1): 1-12.

He, Y., Gao, J.F., Feng, F.Q., Liu, C., Peng, Y.Z. and Wang, S.Y. 2012. The comparative study on the rapid decolorization of azo, anthraquinone and triphenylmethane dyes by zero-valent iron. Chem. Eng. J. 179: 8-18.

Hitihami-Mudiyanselage, A., Senevirathne, K. and Brock, S.L. 2013. Assembly of phosphide nanocrystals into porous networks: Formation of InP gels and aerogels. ACS Nano 7(2): 1163-1170.

Hong, Y., Zhao, D., Liu, D., Ma, B., Yao, G., Li, Q., et al. 2018. Three-dimensional *in situ* electron-beam lithography using water ice. Nano Lett. 18(8): 5036-5041.

Hu, Y., Hu, H., Yan, J., Zhang, C., Li, Y., Wang, M., et al. 2018. Multifunctional porous iron oxide nanoagents for MRI and photothermal/chemo synergistic therapy. Bioconjug. Chem. 29(4): 1283-1290.

Huang, H., Xu, W., Chen, T., Chang, R.-J., Sheng, Y., Zhang, Q., et al. 2018a. High-performance two-dimensional schottky diodes utilizing chemical vapor deposition-grown graphene–MoS_2 heterojunctions. ACS Appl. Mater. Interfaces 10(43): 37258-37266.

Huang, Y., Huang, Z., Zhong, Z., Yang, X., Hong, Q., Wang, H., et al. 2018b. Highly transparent light emitting diodes on graphene encapsulated Cu nanowires network. Sci. Rep. 8(1): 1-11.

Irie, H., Sunada, K. and Hashimoto, K. 2004. Recent developments in TiO_2 photocatalysis: Novel applications to interior ecology materials and energy saving systems. Electrochemistry 72(12): 807-812.

Jafari, S.M. [ed.]. 2017. Nanoencapsulation of Food Bioactive Ingredients: Principles and Applications. Academic Press. pp. 183-221. ISBN: 9780128097403.

Jafari, S.M., Paximada, P., Mandala, I., Assadpour, E. and Mehrnia, M.A. 2017. Encapsulation by nanoemulsions. pp. 36-73. *In*: Jafari, S.M. (ed.). Nanoencapsulation Technologies for the Food and Nutraceutical Industries. Academic Press. ISBN: 978-0-12-809436-5.

Jaishankar, P. and Karthikeyan, C. 2017. Characteristics of cement concrete with nano alumina particles. IOP Conf. Ser. Earth Environ. Sci. 80: 012005.

Jampilek, J. and Kralova, K. 2020. Potential of nanonutraceuticals in increasing immunity. Nanomaterials 10(11): 2224.

Jethave, G., Fegade, U., Attarde, S., Ingle, S., Ghaedi, M. and Sabzehmeidani, M.M. 2019. Exploration of the adsorption capability by doping Pb@$ZnFe_2O_4$ nanocomposites (NCs) for decontamination of dye from textile wastewater. Heliyon 5(9): e02412.

Kamat, P.V. and Schatz, G.C. [eds.]. 2009. Nanotechnology for next generation solar cells. J. Phys. Chem. C 113(35): 15473-15475.

Katuwavila, N.P., Perera, A.D.L.C., Dahanayake, D., Karunaratne, V., Amaratunga, G.A. and Karunaratne, D.N. 2016. Alginate nanoparticles protect ferrous from oxidation: Potential iron delivery system. Int. J. Pharm. 513(1-2): 404-409.

Kiamahalleh, M.V., Alishah, A., Yousefi, F., Astani, S.H., Gholampour, A. and Kiamahalleh, M.V. 2020. Iron oxide nanoparticle incorporated cement mortar composite: Correlation between physico-chemical and physico-mechanical properties. Adv. Mater. 1(6): 1835-1840.

Kim, S.H. and Choi, P.P. 2017. Enhanced Congo red dye removal from aqueous solutions using iron nanoparticles: Adsorption, kinetics, and equilibrium studies. Dalton Trans. 46(44): 15470-15479.

Kokarneswaran, M., Selvaraj, P., Ashokan, T., Perumal, S., Sellappan, P., Murugan, K.D., et al. 2020. Discovery of carbon nanotubes in sixth century BC potteries from Keeladi, India. Sci. Rep. 10(1): 19786.

Kufer, D. and Konstantatos, G. 2016. Photo-FETs: Phototransistors enabled by 2D and 0D nanomaterials. ACS Photonics 3(12): 2197-2210.

Kumar, A., Vemula, P.K., Ajayan, P.M. and John, G. 2008. Silver-nanoparticle-embedded antimicrobial paints based on vegetable oil. Nat. Mater. 7(3): 236-241.

Kunwar, S., Pandit, S., Jeong, J.H. and Lee, J. 2020. Improved photoresponse of UV photodetectors by the incorporation of plasmonic nanoparticles on GaN through the resonant coupling of localized surface plasmon resonance. Nano-Micro Lett. 12(1): 91.

Kuo, C.-L. and Huang, M.H. 2008. Hydrothermal synthesis of free-floating Au_2S nanoparticle superstructures. J. Phys. Chem. C 112(31): 11661-11666.

Kwon, J.B., Kim, S.W., Kang, B.H., Yeom, S.H., Lee, W.H., Kwon, D.H., et al. 2020. Air-stable and ultrasensitive solution-cast SWIR photodetectors utilizing modified core/shell colloidal quantum dots. Nano Converg. 7(1): 1-10.

Lamabam, S.D. and Thangjam, R. 2018. Progress and challenges of nanotechnology in food engineering. pp. 87-112. *In*: Grumezescu, A. and Holban, A.M. (eds.). Impact of Nanoscience in the Food Industry. Academic Press, Elsevier publications.

Li, G. 2004. Properties of high-volume fly ash concrete incorporating nano-SiO_2. Cem. Concr. Res. 34(6): 1043-1049.

Li, Y., Wang, X., Zhang, Y. and Nie, G. 2020. Recent advances in nanomaterials with inherent optical and magnetic properties for bioimaging and imaging-guided nucleic acid therapy. Bioconjug. Chem. 31(5): 1234-1246.

Li, Z., Wang, R., Xue, J., Xing, X., Yu, C., Huang, T., et al. 2019. Core–shell ZnO@SnO_2 nanoparticles for efficient inorganic perovskite solar cells. J. Am. Chem. Soc. 141(44): 17610-17616.

Lian, Z., Zhang, Y. and Zhao, Y. 2016. Nano-TiO_2 particles and high hydrostatic pressure treatment for improving functionality of polyvinyl alcohol and chitosan composite films and nano-TiO_2 migration from film matrix in food simulants. Innov. Food Sci. Emerg Technol. 33: 145-153.

Liu, Q., Zhan, C. and Kohane, D.S. 2017a. Photo triggered drug delivery using inorganic nanomaterials. Bioconjug. Chem. 28(1): 98-104.

Liu, R., Peng, M., Zhang, H., Wan, X. and Shen, M. 2016. Atomic layer deposition of ZnO on graphene for thin film transistor. Mater. Sci. Semicond. Process. 56: 324-328.

Liu, W., Tian, M., Kong, Y., Lu, J., Li, N. and Han, J. 2017b. Multilayered vitamin C nanoliposomes by self-assembly of alginate and chitosan: Long-term stability and feasibility application in mandarin juice. LWT 75: 608-615.

Lou, Y., Zhao, W., Li, C., Huang, H., Bai, T., Chen, C., et al. 2017. Application of $Cu_3InSnSe_5$ heteronanostructures as counter electrodes for dye-sensitized solar cells. ACS Appl. Mater. Interfaces 9(21): 18046-18053.

Lu, Z., Hou, D., Meng, L., Sun, G., Lu, C. and Li, Z. 2015. Mechanism of cement paste reinforced by graphene oxide/carbon nanotubes composites with enhanced mechanical properties. RSC Adv. 5(122): 100598-100605.

Ma, C.Y., Liu, J.J., Zhang, Y. and Wang, X.Z. 2015. Simulation for scale-up of a confined jet mixer for continuous hydrothermal flow synthesis of nanomaterials. J. Supercrit. Fluids 98: 211-221.

Manfrinato, V.R., Zhang, L., Su, D., Duan, H., Hobbs, R.G., Stach, E.A., et al. 2013. Resolution limits of electron-beam lithography toward the atomic scale. Nano Lett. 13(4): 1555-1558.

Manippady, S.R., Singh, A., Basavaraja, B.M., Samal, A.K., Srivastava, S. and Saxena, M. 2020. Iron–carbon hybrid magnetic nanosheets for adsorption-removal of organic dyes and 4-nitrophenol from aqueous solution. ACS Appl. Nano Mater. 3(2): 1571-1582.

Mao, B., Sidhureddy, B., Thiruppathi, A.R., Wood, P.C. and Chen, A. 2020. Efficient dye removal and separation based on graphene oxide nanomaterials. New J. Chem. 44(11): 4519-4528.

McGuinness, N.B., Garvey, M., Whelan, A., John, H., Zhao, C., Zhang, G., et al. 2015. Nanotechnology Solutions for Global Water Challenges. ACS Symposium Series 1206: 375-411. ISBN: 9780841231061.

Moyen, E., Kim, J.H., Kim, J. and Jang, J. 2020. ZnO nanoparticles for quantum dot-based light-emitting diodes. ACS Appl. Nano Mater. 3(6): 5203-5211.

Mukherjee, A., Adak, M.K., Upadhyay, S., Khatun, J., Dhak, P., Khawas, S., et al. 2019. Efficient fluoride removal and dye degradation of contaminated water using Fe/Al/Ti oxide nanocomposite. ACS Omega 4(6): 9686-9696.

Naik, A.J.T., Gruar, R., Tighe, C.J., Parkin, I.P., Darr, J.A. and Binions, R. 2015. Environmental sensing semiconducting nanoceramics made using a continuous hydrothermal synthesis pilot plant. Sens. Actuators B Chem. 217: 136-145.

Nayl, A.A., Abd-Elhamid, A.I., Abu-Saied, M.A., El-Shanshory, A.A., Soliman, H.M., Akl, M.A., et al. 2020. A novel method for highly effective removal and determination of binary cationic dyes in aqueous media using a cotton–graphene oxide composite. RSC Adv. 10(13): 7791-7802.

Ng, D.S., Paul, S.C., Anggraini, V., Kong, S.Y., Qureshi, T.S., Rodriguez, C.R., et al. 2020. Influence of SiO_2, TiO_2 and Fe_2O_3 nanoparticles on the properties of fly ash blended cement mortars. Constr. Build Mater. 258: 119627.

Nicolae, S.A., Au, H., Modugno, P., Luo, H., Szego, A.E., Qiao, M., et al. 2020. Recent advances in hydrothermal carbonisation: From tailored carbon materials and biochemicals to applications and bioenergy. Green Chem. 22(15): 4747-4800.

Nie, S., Liu, A., Meng, Y., Shin, B., Liu, G. and Shan, F. 2018. Solution-processed ternary p-type $CuCrO_2$ semiconductor thin films and their application in transistors. J. Mater. Chem. C 6(6): 1393-1398.

Ochoa, A.A., Hernández-Becerra, J.A., Cavazos-Garduño, A., Vernon-Carter, E.J. and García, H.S. 2016. Preparación ycaracterización de nanoemulsiones de curcumina obtenidas por los métodos de emulsificación por hidratación de capa fina y ultrasonicación. Rev. Mex. Ing. Quim. 15(1): 79-90.

Oluwabi, A.T., Katerski, A., Carlos, E., Branquinho, R., Mere, A., Krunks, M., et al. 2020. Application of ultrasonic sprayed zirconium oxide dielectric in zinc tin oxide-based thin film transistor. J. Mater. Chem. C 8(11): 3730-3739.

Pezeshky, A., Ghanbarzadeh, B., Hamishehkar, H., Moghadam, M. and Babazadeh, A. 2016. Vitamin A palmitate-bearing nanoliposomes: Preparation and characterization. Food Biosci. 13: 49-55.

Powell, M.J., Marchand, P., Denis, C.J., Bear, J.C., Darr, J.A. and Parkin, I.P. 2015. Direct and continuous synthesis of VO_2 nanoparticles. Nanoscale 7(44): 18686-18693.

Qudoos, A., Jakhrani, S.H., Kim, H.G. and Ryou, J.S. 2019. Influence of nano-silica on the leaching attack upon photocatalytic cement mortars. Int. J. Concr. Struct. Mater. 13(1): 35.

Rauf, M.A., Meetani, M.A. and Hisaindee, S. 2011. An overview on the photocatalytic degradation of azo dyes in the presence of TiO_2 doped with selective transition metals. Desalination 276(1-3): 13-27.

Roberts, E.J., Habas, S.E., Wang, L., Ruddy, D.A., White, E.A., Baddour, F.G., et al. 2017. High-throughput continuous flow synthesis of nickel nanoparticles for the catalytic hydrodeoxygenation of guaiacol. ACS Sustain. Chem. Eng. 5(1): 632-639.

Sadeghi, K., Lee, Y. and Seo, J. 2019. Ethylene scavenging systems in packaging of fresh produce: A review. Food Rev. Int. 37(2): 155-176.

Sahoo, R.K., Singh, S., Yun, J.M., Kwon, S.H. and Kim, K.H. 2019. Sb_2S_3 nanoparticles anchored or encapsulated by the sulfur-doped carbon sheet for high-performance supercapacitors. ACS Appl. Mater. Interfaces 11(37): 33966-33977.

Samanta, K.K., Basak, S. and Chattopadhyay, S.K. 2016. Potentials of fibrous and nonfibrous materials in biodegradable packaging. pp. 75-113. *In*: Environmental Footprints of Packaging. Springer, Singapore. ISBN: 978-981-287-911-0.

Sanchez, F. and Sobolev, K. 2010. Nanotechnology in concrete: A review. Constr. Build Mater. 24(11): 2060-2071.

Sawangphruk, M., Suksomboon, M., Kongsupornsak, K., Khuntilo, J., Srimuk, P., Sanguansak, Y., et al. 2013. High-performance supercapacitors based on silver nanoparticle–polyaniline–graphene nanocomposites coated on flexible carbon fiber paper. J. Mater. Chem. A 1(34): 9630-9636.

Shafeek, A.M., Khedr, M.H., El-Dek, S. and Shehata, N. 2020. Influence of ZnO nanoparticle ratio and size on mechanical properties and whiteness of White Portland Cement. Appl. Nanosci. 10(1): 3603-3615.

Shao, X., Wu, Y., Jiang, S., Li, B., Zhang, T. and Yan, Y. 2020. Chiral 3D CdSe nanotetrapods. Inorg. Chem. 59(19): 14382-14388.

Sheng, J., Hong, T., Kang, D., Yi, Y., Lim, J.H. and Park, J.S. 2019. Design of InZnSnO semiconductor alloys synthesized by supercycle atomic layer deposition and their rollable applications. ACS Appl. Mater. Interfaces 11(13): 12683-12692.

Sikora, P., Łukowski, P., Cendrowski, K., Horszczaruk, E. and Mijowska, E. 2015. The effect of nanosilica on the mechanical properties of polymer-cement composites (PCC). Procedia. Eng. 108: 139-145.

Skarmoutsos, I., Koukaras, E.N., Froudakis, G.E., Maurin, G. and Klontzas, E. 2020. Confinement effects on the properties of polar hydrogen-bonded fluids: A showcase on methanol adsorbed in three-dimensional pillared graphene and carbon nanotube networks. J. Mater. Chem. C 124(42): 22959-22971.

Sobolev, K. and Gutiérrez, M.F. 2005. How nanotechnology can change the concrete world: Part two of a two-part series. Bull. Am. Ceram. Soc. 84(11): 16-19.

Sowinska, M. and Urbanczyk-Lipkowska, Z. 2014. Advances in the chemistry of dendrimers. New J. Chem. 38(6): 2168.

Srinivas, K. 2014. Nanomaterials for concrete technology. International Journal of Civil, Structural, Environmental and Infrastructure Engineering Research and Development (IJCSEIERD) 1(4): 79-90.

Stando, G., Łukawski, D., Lisiecki, F. and Janas, D. 2019. Intrinsic hydrophilic character of carbon nanotube networks. Appl. Surf. Sci. 463: 227-233.

Tighe, C.J., Cabrera, R.Q., Gruar, R.I. and Darr, J.A. 2013. Scale up production of nanoparticles: Continuous supercritical water synthesis of Ce-Zn oxides. Ind. amp; Eng. Chem. Res. 52(16): 5522-5528.

Tobbala, D.E. 2019. Effect of nano-ferrite addition on mechanical properties and gamma ray attenuation coefficient of steel fiber reinforced heavy weight concrete. Constr. Build Mater. 207: 48-58.

Vahabi, K., Mansoori, G.A. and Karimi, S. 2011. Biosynthesis of silver nanoparticles by fungus trichoderma reesei (A route for large-scale production of AgNPs). Insciences J. 1(1): 65-79.

Wang, H.C., Hong, Y., Chen, Z., Lao, C., Lu, Y., Yang, Z., et al. 2020a. ZnO UV photodetectors modified by Ag nanoparticles using all-inkjet-printing. Nanoscale Res. Lett. 15(1): 1-8.

Wang, L., Lin, J., Hu, Y., Guo, X., Lv, Y., Tang, Z., et al. 2017. Blue quantum dot light-emitting diodes with high electroluminescent efficiency. ACS Appl. Mater. Interfaces 9(44): 38755-38760.

Wang, Q., Zhang, L. and Ding, W. 2020b. Eugenol nanocapsules embedded with gelatin-chitosan for chilled pork preservation. Int. J. Biol. Macromol. 158: 837-844.

Wong, X.Y., Sena-Torralba, A., Álvarez-Diduk, R., Muthoosamy, K. and Merkoçi, A. 2020. Nanomaterials for nanotheranostics: Tuning their properties according to disease needs. ACS Nano 14(3): 2585-2627.

Wu, M.J., Wang, Y., Wu, S.C., Chien, Y.C., Chang, C.Y., Chen, J.H., et al. 2020. Graphene–insulator–semiconductor ultraviolet light-responsive nitride LEDs for multi-applications. ACS Appl. Electron. Mater. 2(7): 2104-2112.

Xu, Q., Meng, L., Sinha, K., Chowdhury, F.I., Hu, J. and Wang, X. 2020. Ultrafast colloidal quantum dot infrared photodiode. ACS Photonics 7(5): 1297-1303.

Yao, K., Zhong, H., Liu, Z., Xiong, M., Leng, S., Zhang, J., et al. 2019. Plasmonic metal nanoparticles with core–bishell structure for high-performance organic and perovskite solar cells. ACS Nano 13(5): 5397-5409.

Yazdi, M.G., Ivanic, M., Mohamed, A. and Uheida, A. 2018. Surface modified composite nanofibers for the removal of indigo carmine dye from polluted water. RSC Adv. 8(43): 24588-24598.

Yetisen, A.K., Qu, H., Manbachi, A., Butt, H., Dokmeci, M.R., Hinestroza, J.P., et al. 2016. Nanotechnology in textiles. ACS Nano 10(3): 3042-3068.

Yu, X.Y., Lei, B.X., Kuang, D.B. and Su, C.Y. 2011. Highly efficient CdTe/CdS quantum dot sensitized solar cells fabricated by a one-step linker assisted chemical bath deposition. Chem. Sci. 2(7): 1396-1400.

Yuan, C., Liu, Y., Wang, T., Sun, M. and Chen, X. 2020. Nanomaterials as smart immunomodulator delivery system for enhanced cancer therapy. ACS Biomater. Sci. Eng. 6(9): 4774-4798.

Yuan, H., Shi, Y., Xu, Z., Lu, C., Ni, Y. and Lan, X. 2013. Influence of nano-ZrO_2 on the mechanical and thermal properties of high temperature cementitious thermal energy storage materials. Constr. Build Mater. 48: 6-10.

Zare, M., Namratha, K., Ilyas, S., Hezam, A., Mathur, S. and Byrappa, K. 2019. Smart fortified PHBV-CS biopolymer with ZnO–Ag nanocomposites for enhanced shelf life of food packaging. ACS Appl. Mater. Interfaces 11(51): 48309-48320.

Zeng, H., Lai, Y., Qu, S. and Qin, Y. 2020. Graphene oxide-enhanced cementitious materials under external sulfate attack: Implications for long structural life. ACS Appl. Nano Mater. 3(10): 9784-9795.

Zhang, Y., Fang, F., Li, L. and Zhang, J. 2020. Self-assembled organic nanomaterials for drug delivery, bioimaging, and cancer therapy. ACS Biomater. Sci. Eng. 6(9): 4816-4833.

Zhu, T., Chen, J.S. and Lou, X.W. 2012. Highly efficient removal of organic dyes from waste water using hierarchical NiO spheres with high surface area. J. Phys. Chem. C 116(12): 6873-6878.

Carbon Nanomaterials and Biopolymers Derived Aerogels for Wastewater Remediation

Kanika Gupta[1, 2], Pratiksha Joshi[1, 2] and Om P Khatri[1, 2, *]

1. INTRODUCTION

Aerogels are dried gels or 3D structured scaffolds enriched with air pouches and capillaries. They are known as a particular class of open-cell structures and possess fascinating properties, like ultralow densities, high accessible surface area, channelized porous network, and excellent elasticity. The constituent materials or building blocks of aerogels are interconnected in a 3D network and furnishes a structural scaffold of the desired shape. The aerogels are gaining increasing interest for a wide range of applications as they are made by interconnecting the nanoscopic materials into macroscopic scaffolds. Most of the desired properties of nanoscopic materials are retained in the aerogels. They can be either transparent or black, conductive or insulating, chemically inert or efficient catalyst, luminescent or colorful, magnetic or non-magnetic, and can be excellent absorbers based on the composition and structural features of constituent materials (Ziegler et al. 2017).

The aerogels are usually prepared by the sol-gel technique, which involves transforming molecular precursors into cross-linked organic or inorganic gels that can be dried via different methods, viz. supercritical drying, freeze-drying, etc., to retain the porous and solid 3D network. The first aerogel, i.e. silica gel, was synthesized by S.S. Kistler in 1931. However, silica gel could not find much practical value until 1974, when Cantin et al. demonstrated the potential of silica aerogels for Cherenkov radiation detection. Subsequently, the polymers-based carbon aerogels were prepared and shown their applicability as electrode materials (Araby et al. 2016, Biener et al. 2011, Fricke and Emmerling 1992, Pierre and Pajonk 2002). In the past, silica aerogels have been widely explored and commercially accepted for several applications. The organic and inorganic aerogels viz. clay, metal oxide, mixed metal oxide, graphene-derived, carbon-based, biomass-derived carbon, and polymeric aerogels, etc., are gaining increasing attention for fundamental studies to a wide range of applications. The synthetic and natural polymers-based aerogels exhibit good elastic properties than silica aerogels and have been explored extensively (S. Zhao et al. 2018, Zhu 2019). The mechanical strength, hydrophobicity, elasticity, and catalytic features of silica and polymeric aerogels can be enhanced by incorporating specific components/fillers to broaden their potential for a wide range of applications (Maleki, 2016).

The carbon aerogels having 3D interconnected structural building blocks, exhibit outstanding physical properties and promises their potential in energy storage, electromechanical sensing,

[1] Chemical and Material Sciences Division, CSIR – Indian Institute of Petroleum, Dehradun – 248005, India.
[2] Academy of Scientific and Innovative Research (AcSIR), Ghaziabad – 201002, India.
[*] Corresponding author: opkhatri@iip.res.in.

biomedical, oil-water separation, electrocatalysis, wastewater treatment, and chemical adsorption (T. Chen et al. 2020, Lee and Park 2020, C. Wang et al. 2020). Carbon aerogels are primarily composed of graphene, carbon nanotubes, carbon nanofibers, biomass-derived carbon, polymeric carbon, carbide, and carbonitride-based precursors (T. Chen et al. 2020, Lee and Park 2020). The properties of carbon aerogels are governed by preparation conditions, type of carbon materials, and binders. The microporous structure of aerogels is governed by the intraparticle/building block framework, whereas the interparticle structure furnishes meso- and macroporosity (Lee and Park 2020). The conventional route for carbon aerogels synthesis involves sol-gel/hydro-gel chemistry by transforming molecular precursors into cross-linked organic gels. Subsequently, the carbonization in an inert atmosphere generates carbon aerogels. Graphene and carbon nanotubes are the emerging precursors for carbon aerogels, and their individual properties are translated into 3D free-standing macroscopic assemblies. However, the preparation cost and tedious process of these aerogels impede their use in many areas, mainly for energy and environmental applications. Therefore, efficient and economical routes to fabricate carbon aerogels based on renewable resources are gaining increasing interest (S. C. Li et al. 2018).

Biopolymers have gained increasing interest in preparing aerogels as they are environmentally sustainable materials and provide remarkable physical, mechanical, and chemical features. The alginate, pectin, proteins, lignin, chitosan, cellulose, etc., are primarily used as precursors for the production of biopolymers-based aerogels (Raman et al. 2015, Subrahmanyam et al. 2015). The presence of surface-active functional groups, such as carboxylic acids and amines in pectin, alginate, chitosan, etc., unlock their usability for developing a wide range of aerogels. The non-toxic nature of biopolymers makes them suitable candidates for developing aerogels for food and health-relevant applications (S. Zhao et al. 2018). These aerogels have shown potential for tissue engineering, regenerative medicines, pharmaceuticals, nutraceuticals, catalysis, sensors, adsorption, functional foods, oil-water separation, and as supporting/sacrificial materials for the development of porous metal oxides and carbon aerogels (Raman et al. 2015, Subrahmanyam et al. 2015). The environmental applications, particularly the use of aerogels for the adsorption of variable pollutants from wastewater and air, have shown immense interest. These aerogels have been explored to remove air pollutants, such as CO_2 and volatile organic compounds, adsorption of hazardous organic and heavy metal contaminants from wastewater, sorption of organic solvents and oils, etc. The textural properties, hydrophobicity/hydrophilicity, and surface chemistry primarily govern their potential for environmental applications, including their selective use for the targeted pollutants (Maleki 2016).

The efficient removal of organic pollutants (oils, solvents, pesticides, dyes, active pharmaceuticals ingredients, etc.) and inorganic pollutants (heavy metals, metalloids, and oxides) from industrial effluents, domestic wastewater, and agricultural runoff has attracted wide attention as a global challenge. Several efforts have been made to eradicate the pollutants for providing sustainable and clean water. The unique and tunable characteristics, viz. low density, high accessible surface area, tailorable hydrophobicity, channelized porosity, and ease of separation, make the aerogels promising candidates for wastewater remediation (Maleki 2016). The activated carbon and graphene-based aerogels have been explored for fast, selective, and facile wastewater treatment with excellent recyclability (Lee and Park 2020). However, their high cost and poor mechanical strength compromise their potential for real-time applications. The introduction of polymeric materials enhances mechanical strength and elasticity. The use of biopolymers or waste biomass for the production of aerogels makes them economically viable. Recently, carbon aerogels derived from renewable resources and waste biomass show good potential for adsorptive separation. Over the recent past, immense research has been directed to meet the various challenges, viz. selectivity for variable pollutants, good mechanical strength, elasticity, reusability, etc. The present chapter covers the role of carbon and biopolymers-based aerogels for wastewater remediation. It emphasizes the structural and morphological features of aerogels and their applicability for wastewater cleaning applications. The recent advances for the adsorptive removal of organic dyes, APIs, pesticides, and heavy metal ions from the wastewater and the oil-water separation using variable aerogels are further discussed, emphasizing their potential for real-time applications and associated challenges.

2. AEROGELS FOR WASTEWATER REMEDIATION

The discharge of industrial effluents, domestic wastewater, and agricultural runoff are primarily accountable for water pollution. The significant volumes of industrial oils, hazardous organic compounds, particularly dyes, pesticides, pharmaceuticals, detergents, and metal/metalloids ions (Cr, Co, As, Cd, Hg, Pb, etc.), are discharged into aquatic bodies without proper treatment (Reza and Singh 2010). Therefore, effective and sustainable remediation approaches are urgently required to prevent the deterioration of the water bodies. Adsorption has been established as an economical and practical approach for wastewater remediation, and the adsorbent materials can be reused for multiple cycles (Gusain et al. 2019). The activated carbon (Gupta et al. 2019, Gupta and Khatri 2019), aerogels (S. Zhao et al. 2018), polymers (Ngah et al. 2011), metal oxides (Joshi et al. 2020), graphene-based materials (Gupta and Khatri 2017), and clays and composite materials (Sani et al. 2017) have been explored as efficient adsorbents for wastewater remediation. Aerogels have drawn increasing interest as sorbent materials over the recent past because of their high accessible surface area, low density, versatile surface chemistry, and tunable porosity. The critical benefits of aerogels are their macroscale scaffolds made by the interconnection of nanostructural units, ease of separation, controlled porosity for fast and effective diffusion of liquid and gaseous media (Gan et al. 2019). A wide range of aerogels prepared by variable precursors like 2D nanostructured materials, activated carbon (Araby et al. 2016), biomass (Rudaz et al. 2014), metal oxides (Aegerter et al. 2011), polymers (Salimian et al. 2018), and so on have been used for the removal of variable contaminants from wastewater. The aerogels prepared from carbon nanomaterials, like activated carbon, graphene oxide (GO), carbon nanotubes, etc., and biopolymers like cellulose, chitosan, sodium alginate, starch, etc., are demonstrated as effective adsorbents for the removal of a broad range of biological, inorganic, and organic contaminants from wastewater (Maleki 2016, S. Zhao et al. 2018). The carbon nanomaterials and biopolymers-derived aerogels are discussed in the following section to remove organic and inorganic pollutants.

3. REMOVAL OF ORGANIC POLLUTANTS FROM AEROGELS

3.1 Adsorptive Removal of Organic Dyes

Dyes are complex organic species, ubiquitously used in textile, paper, leather, and tanning industries. The dyes and organic pollutants-enriched industrial effluents impart color to the water, lead to foul smell, high BOD, and disturb the aquatic environment by blocking the sunlight penetration, which retards the aquatic plant growth by inhibiting the photosynthesis activity (Ray et al. 2020). Even at ppm level, the dye-contaminated water may lead to health hazards viz. carcinogenesis, mutagenesis, teratogenesis, impairment of nervous system, lungs, genital system, etc. (Gusain et al. 2019). Therefore, the removal of organic pollutants from industrial wastewater is of great importance.

Aerogels are demonstrated for the fast and effective removal of organic dyes from polluted water. F. Liu et al. 2012 explored the adsorptive removal of methyl violet (MV) and methylene blue (MB) dyes by a 3D graphene-based sponge. It adsorbed 99% of these cationic dyes within 2 minutes, driven by π-π and ionic interactions. However, the mechanical strength, accessible porosity, and surface area of graphene-based aerogels suffer from few drawbacks, which limit their use for real-time applications. The GO layers agglomerate with each other in the GO-based aerogels, which eventually decreases the accessible surface area and lessen the adsorption capability of GO-based aerogels. Moreover, GO-based aerogels are fragile. The major challenge is to maintain the intactness of the 3D framework while using it for adsorption applications (Ha et al. 2015). The multiple studies are focused on cross-linking methods to strengthen graphene-based aerogels. The incorporation of chitosan in graphene sponge provides both positive and negative charged species and improves the strength. The composite aerogel successfully removed both anionic (eosin Y) and cationic (MB) dyes based on the ion-complex formation and electrostatic interaction (Yunqiang Chen et al. 2013). The hydrothermally-reduced graphene oxide (rGO)/PVA aerogels prepared by a hydrothermal route adsorb the cationic, anionic, and non-ionic dyes. The PVA macromolecule serves as a spacer and inhibits the aggregation of GO sheets, which results in a high accessible surface area for enhanced removal capacity for organic dyes (Xiao et al. 2017). The intercalation of hollow carbon spheres in graphene aerogel prevents the restacking of graphene sheets. The hydroxyl and carboxyl groups over the surface of carbon spheres and oxygen functionalities of GO sheets enhanced the adsorptive removal capacity of these aerogels (Hou et al. 2019).

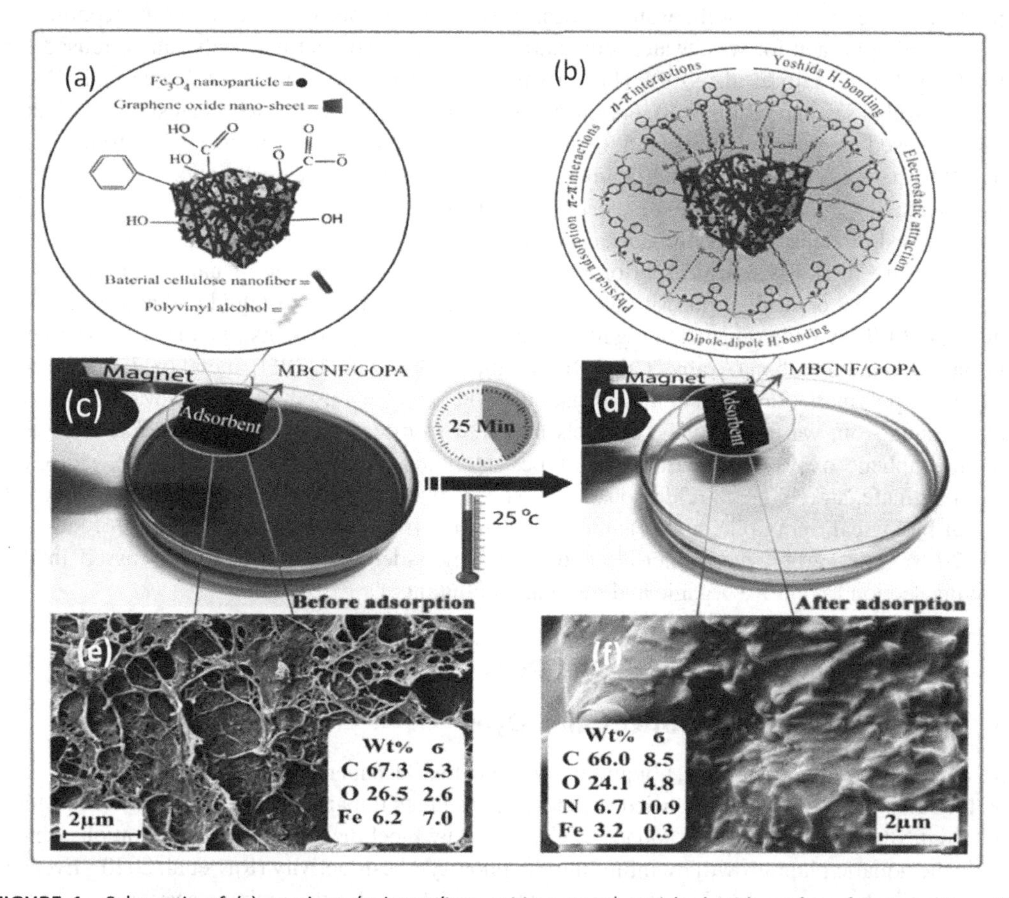

FIGURE 1 Schematic of (a) graphene/polymer/iron-oxide aerogel enriched with surface functionalities, (b) plausible interactions between the aerogel and dye molecule. (c, d) Adsorptive removal of dye by aerogel and then separation of foam driven by external magnetic effect. (e, f) morphological images of aerogels before and after the adsorption (Arabkhani and Asfaram 2020). Reprinted with permission from Copyright (2020) Elsevier.

Cellulosic and carbon aerogels are promising adsorbents for the removal of dyes. The cellulose nanofibrils with their honeycomb morphology adsorbed the malachite green (MG) dye with an adsorption capacity of 212.7 mg.g^{-1} (Jiang et al. 2017). Li et al. prepared a fiber-based carbon aerogel derived from cotton waste; demonstrating the adsorption of MB dye with a capacity of 102.2 mg.g^{-1} (Z. Li et al. 2017). The calcium alginate-based carbon aerogel acts as an excellent adsorbent for triphenylmethane group-based dyes, especially for MG with a removal capacity of 7,059 mg.g^{-1} (Tian et al. 2020). The Ni^{2+} ions elevate the cross-linking between resorcinol-formaldehyde and the graphene aerogel. The metal ions create active sites for the alignment of the aerogel layer, resulting in grooves with interconnected porous systems ranging in pore size from sub-micrometer to tens of micrometers (Wei et al. 2013). The 3D scaffold prepared using the GO, PVA, bacterial cellulose, and iron oxide nanoparticles as building blocks could efficiently remove the MG dye from the simulated and real-time polluted water samples. The carboxylic, hydroxyl, and π-electron-rich systems are primarily accountable for the adsorptive removal of dye via hydrogen bonding, Yoshida, π-π, dipole-dipole, and electrostatic attractions (Figure 1). The aerogel could attain an equilibrium within 25 minutes with an adsorption efficiency of 82% (Arabkhani and Asfaram 2020). Table 1 presents variable aerogels for the adsorptive removal of organic dyes.

TABLE 1 Graphene and biopolymer-derived aerogels for the adsorptive removal of organic dyes.

S. No.	Adsorbent	Targeted Pollutant	Adsorption Capacity, mg g^{-1}	Reference
1	Cellulose aerogel	MG	212.7	(Jiang et al. 2017)
2	Carbon aerogel	MG	249.6 245.3	(M. Yu et al. 2017)
3	Starch-graphene aerogel	Rh B CV	539.0 318.0	(Yun Chen et al. 2019)
4	Graphene oxide/aminated lignin aerogel	MG	113.5	(H. Chen et al. 2020)
5	Cellulose/chitosan aerogel	MB	785.0	(H. Yang et al. 2016)
6	N-doped carbon aerogel	MB MG	230.4 238.2	(Yu et al. 2018)
7	Calcium alginate carbon aerogel	MG CV FA	7,059 2,390 6,964	(Tian et al. 2020)
8	Magnetic cellulose GO aerogel	CR MB	282.0 346.0	(Xiong et al. 2020)
9	GO/poly(ethyleneimine)	MB MO	249.6 331.0	(Q. Zhao et al. 2018)
10	Xanthan gum–GO aerogel	Rh B MB	244.4 290.6	(S. Liu et al. 2017)
11	PDA-c-GO aerogel	MB	633.0	(T. Huang et al. 2018)
12	Polyacrylamide/GO aerogel	BF	1,034	(X. Yang et al. 2015)
13	Propylene glycol adipate/cellulose aerogel	CR	120.0	(Tang et al. 2020)
14	Chitosan/GO/lignosulfonate	MB	1,024	(Yan et al. 2019)
15	Amino functionalized graphene aerogel	MO	3,059	(Shu et al. 2017)
16	Agar/graphene oxide aerogel	MB	578.0	(Long Chen et al. 2017)
17	β-Cyclodextrin/activated carbon aerogel	MB	166.7	(K. Zhou et al. 2018)

3.2 Adsorptive Removal of Active Pharmaceutical Ingredients (APIs)

The APIs are non-biodegradable, and their ingestion via polluted water has detrimental and hazardous impacts on living creatures, including human health. The consumption of APIs-enriched water leads to several health hazards, viz. fertility and respiratory complications, chronic anxiety, mental retardation, cancer, and physical anomalies (Chander et al. 2016). The aerogels have been explored as effective adsorbents for the removal of APIs from wastewater. The GO-cellulose aerogels have shown efficient adsorption of 21 antibiotics of six different categories, driven by electrostatic, π-π interactions, and hydrogen bonding, as shown in Figure 2. The adsorption efficiency for these antibiotic classes are found in following order; tetracycline (82-90%) > quinolones (81-82%) > sulphonamides (80-81%) > chloramphenicol (77-78%) > β-lactams (70-78%) > macrolides (74-75%). The fast and effective adsorption was attributed to an interconnected 3D porous network of aerogels, which enabled the fast diffusion of antibiotics, leading to a higher accessible surface area for the adsorption. The conjugated π electron network and the presence of the hydroxyl group furnished the highest adsorption efficiency for the tetracycline (Yao et al. 2017). The magnetic GO sponge enriched with Fe_3O_4 showed excellent removal of tetracycline with a maximum adsorption capacity of 473 $mg.g^{-1}$, which is 50% higher than GO-based aerogel. The electrostatic, π-π interactions, and hydrogen bonding are suggested as primary interactive pathways that enabled effective adsorption of tetracycline by magnetic GO-based aerogels (B. Yu et al. 2017).

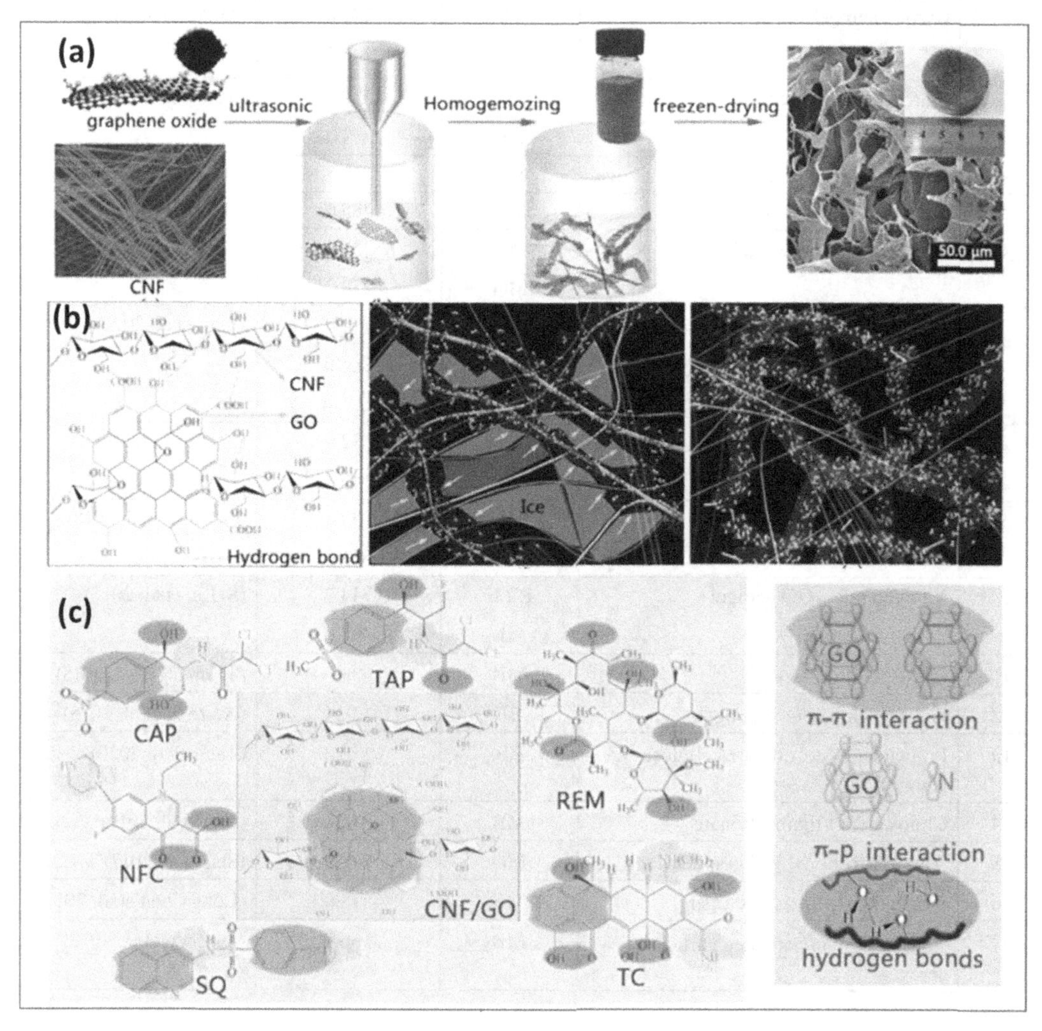

FIGURE 2 (a) The synthesis of CNF/GO aerogel, (b) Interactions between GO and cellulosic nanofibrils, (c) Mechanism of antibiotic adsorption on CNF/GO aerogel surface (Yao et al. 2017). Reprinted with permission from Copyright (2017) Springer Nature.

Carbon aerogels prepared via the carbonization of chitosan successfully removed the phenicol antibiotics (chloramphenicol, thiamphenicol, and florfenicol) from the polluted water, driven by π-π interaction, hydrogen bonding, pore-filling effect, π-π/n-π EDA (electron-donor-acceptor), and electrostatic interactions (H. Liu et al. 2019). The sodium alginate (SA) enriched with hydroxyl and carboxylic groups showed excellent cross-linking with GO. The GO-SA, prepared by encapsulating the GO into SA, displayed the excellent removal of ciprofloxacin. Moreover, the GO-SA aerogel showed high resistance to pH shifts and good porosity, revealing the potential of GO-SA aerogel for adsorptive removal of ciprofloxacin (Fei et al. 2016). The modification of κ-carrageenan foam with sodium alginate removed the ciprofloxacin hydrochloride effectively with a removal capacity of 291.6 $mg.g^{-1}$ (L. Li et al. 2019). Table 2 presents the variable aerogels for the removal of pharmaceutical ingredients.

TABLE 2 Carbon and biopolymer-derived aerogels for the adsorptive removal of APIs.

S. No.	Adsorbent	Pharmaceutical Ingredients	Adsorption Capacity, $mg.g^{-1}$	Reference
1	Chitosan carbon aerogel	Chloramphenicol	786.1	(H. Liu et al. 2019)
		Florfenicol	751.5	
		Thiamphenicol	691.9	
2	Carbon aerogel	Tetracycline	1030	(K. Li et al. 2019)
		Oxytetracycline	813.8	
		Doxycycline	922.9	
		Sulfamethazine	834.9	
3	Carbon aerogel	Hygromycin B	104.2	(Aylaz et al. 2020)
		Gentamicin	81.30	
		Vancomycin	107.5	
4	ZIF-67/PANI/cellulose	Tetracycline	409.6	(Q. Liu et al. 2020)
5	Alginate-graphene-ZIF67	Tetracycline	456.6	(Y. Kong et al. 2020)
6	Peanut shell/GO aerogel	Norfloxacin	228.8	(Dan et al. 2020)
7	Graphene–soy protein aerogel	Tetracycline	164.0	(Zhuang et al. 2016)
8	Sodium alginate/κ-carrageenan	Ciprofloxacin.HCl	291.6	(L. Li et al. 2019)
9	TiO_2–graphene sponge	Tetracycline	1805	(Zhao et al. 2015)
10	UiO-66/polydopamine/ bacterial cellulose	Aspirin	149.0	(Cui et al. 2020)

3.3 Adsorptive Removal of Pesticides

Pesticides are used as pest controllers for agricultural or domestic uses. The fungicides, insecticides, herbicides, etc., are used to combat bugs, insects (e.g., rats, mice, ticks, etc.), fungi, and herbs. They control weeds, insects, and fungus infections (Ray et al. 2020). The residual contents of pesticides in domestic wastewater and agricultural runoff gradually mixed with mainstream water bodies. The removal of pesticides from water has been a major challenge because of their good solubility. The permissible limit for single and multiple pesticides in water is 0.1 and 0.5 $\mu g.L^{-1}$, respectively, as per the WHO guidelines (Joshi et al. 2020). Their occurrence in water above the permissible limit negatively affects human health. Therefore, the use of pesticides without proper handling leads to several challenges and threats to human health, marine system, aquatic lives, soil, etc. The removal of pesticides from wastewater by adsorption process has shown considerable interest. The use of aerogels for the adsorptive removal of pesticides is still in its infancy stage.

The amyloid nanofibrils synthesized from β-lactoglobulin efficiently removed the herbicide (bentazone) and fungicide (chlorothalonil) from the polluted water with efficiencies of 92 and 94%, respectively. The adsorption process is driven by electrostatic interaction between the amino acid functionalities of aerogel and the pollutants (Peydayesh et al. 2020). The functionalized GO aerogel prepared by GO, PVA, and carboxylated chitosan removed the glyphosate from the contaminated water with an adsorption capacity of 578 $mg.g^{-1}$. The aerogel could be regenerated by NaOH treatment and used for multiple cycles without a significant loss in its efficiency (Ding et al. 2018). The carboxylated CNT-GO aerogel showed a removal capacity of 546 $mg.g^{-1}$ and the aerogels could be reused up to 20 cycles of adsorption-desorption (H. Liu et al. 2020). The metal-organic frameworks (UiO66-NH_2 and ZIF-8) over the polydopamine modified multi-walled carbon nanotubes-based composite aerogels have been demonstrated to remove alachlor and chipton herbicides from the polluted water. The hydrogen bonding, π-π stacking, and electrostatic interaction between the herbicides and composite aerogels facilitated the adsorption events (Liang et al. 2021).

3.4 Adsorptive Separation of Oils and Organic Solvents From Wastewater

Adsorptive removal of floating fluids (oils/solvents) from the water surface by variable aerogels has drawn significant attention. The hydrophobic nature of carbon aerogels makes them suitable materials for oil-water separation. The carbon aerogels prepared from carbon nanotubes, graphene, carbon nanofiber, and activated carbon derived from natural resources show excellent sorption capacity for the sorption of oils and petroleum products (W.-J. Yang et al. 2019). The ultra-light 3D aerogels prepared by the interconnected network of graphene via chemical approach exhibited high oil sorption capacity, i.e. 350 times to its weight with excellent reusability (F. Wang et al. 2017). The graphene aerogels of variable pore sizes and volumes prepared through the Pickering emulsion method showed a sorption capacity of 106 $g{\cdot}g^{-1}$ for diesel oil in 50 s, whereas it can hardly adsorb water (2.39 $g.g^{-1}$) within the same time. The internal and external parts of graphene aerogels showed a difference in the adsorption behavior. The rough internal part displayed lower adsorption than the smooth exterior surface. The rough interior surface establishes lesser contact between aerogel and emulsified oil. The increase in the reduction degree of graphene aerogels decreased the number of oxygen-containing functional groups, resulting in higher hydrophobicity which enhanced the equilibrium adsorption capacity of graphene aerogels for emulsified oils in water (Diao et al. 2020).

The shrinkage volume in aerogels at higher temperatures limits their potential for high-temperature applications. The vapor-liquid deposition process prepares the superhydrophobic and superoleophilic graphene aerogels to eradicate these limitations. They exhibit superior characteristics for the oil-water separation and displayed an adsorption capacity of 109 $g.g^{-1}$ for commonly used oils and organic solvents, suggesting their potential for oil sorption and oil-water separation applications (S. Yang et al. 2018). The renewable Enteromorpha was introduced into the graphene aerogel via hydrothermal freeze casting technique to resolve the volume shrinkage and form the ultra-light and amphiphilic adsorbent for water remediation application. The Enteromorpha in graphene aerogels (EGA) increased the oil uptake capacity compared to the pristine graphene aerogels, possessing 68-160 $g.g^{-1}$ uptake capacity for a number of oils and organic solvents within 10 seconds (Figure 3). They also showed a good regenerability for a wide range of oils and organic solvents (Ji et al. 2021).

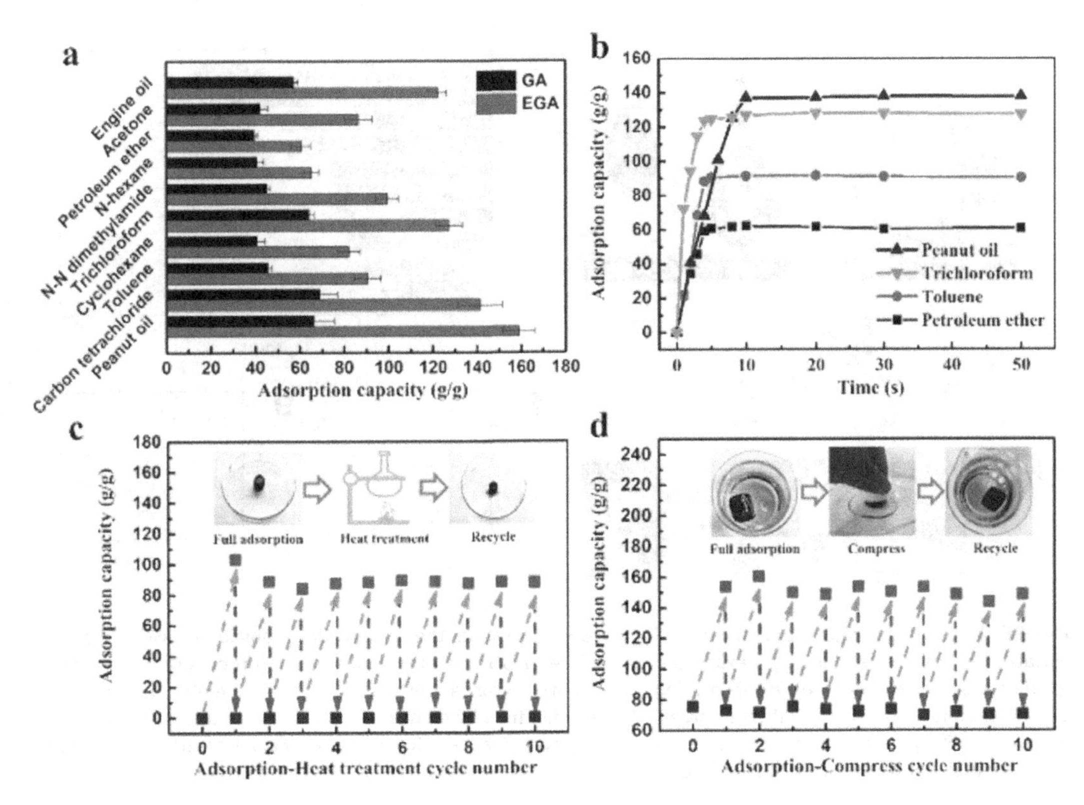

FIGURE 3 (a) Absorption capacities of GA and EGA aerogels for various solvents and oils. (b) Adsorption rate of EGA for multiple solvents and oils. (c) Recyclability of the EGA by heat treatment. (d) Recyclability of the EGA by compression (Ji et al. 2021). Reprinted with permission from Copyright (2021) Elsevier.

The poor structural integration of graphene aerogels compromises long-term stability, particularly the regenerative ability. Recently, N-doped rGO aerogels are prepared by introducing cellulosic biomass and carbon fibers to improve mechanical strength. These aerogels showed high surface area (3,29.6 $m^2.g^{-1}$), ultra-low density, superhydrophobicity, and high adsorptive capacity (206.4 $g.g^{-1}$) for tetrachloroethylene. Moreover, > 90% adsorption efficiency is retained after multiple cycles of heating or squeezing (Qin et al. 2021). The graphene nanoribbons aerogels modified with polydimethylsiloxane displayed adsorption capacities of 302 and 121 times for phenixin and *n*-hexane to their weight, respectively (Liang Chen et al. 2015). The cellulose acetate nanofibers are introduced in graphene for preparing the GO/nanofiber aerogels (GNA) to prevent overstacking and enhance the interconnectivity of cell walls. The deposition of polydopamine and polyethyleneimine makes the GNA foams superhydrophilic and superoleophobic materials. The cross-linking of GNA with hexadecyltrimethoxysilane furnished the superhydrophobic and elastic properties, which effectively adsorbed variable organic liquids with a sorption capacity of 230-734 $g{\cdot}g^{-1}$ (Figure 4). The GNA-based foams showed excellent regenerability (Xiao et al. 2018).

The incorporation of CNTs enhanced the porosity, robustness, and hydrophobicity of graphene-based aerogels and furnished superior oleophilic properties. These aerogels are demonstrated as suitable materials for the adsorptive removal of fats, petroleum products, and organic solvents under the continuous vacuum regime with an adsorption capacity of 28 $L{\cdot}g^{-1}$ (Kabiri et al. 2014). The superhydrophobic and superoleophilic aerogels of graphene/polyvinylidene fluoride are prepared through a solvothermal reduction of graphene and PVDF dispersion. These aerogels are used to absorb a vast number of oils and organic solvents, revealing their excellent performance for oil-spill cleaning (Li et al. 2014). The compressible and light-weight 3D graphene-carbon nitride/PVA-based aerogels of high porosity and hydrophobicity showed sorption of a number of oils and organic

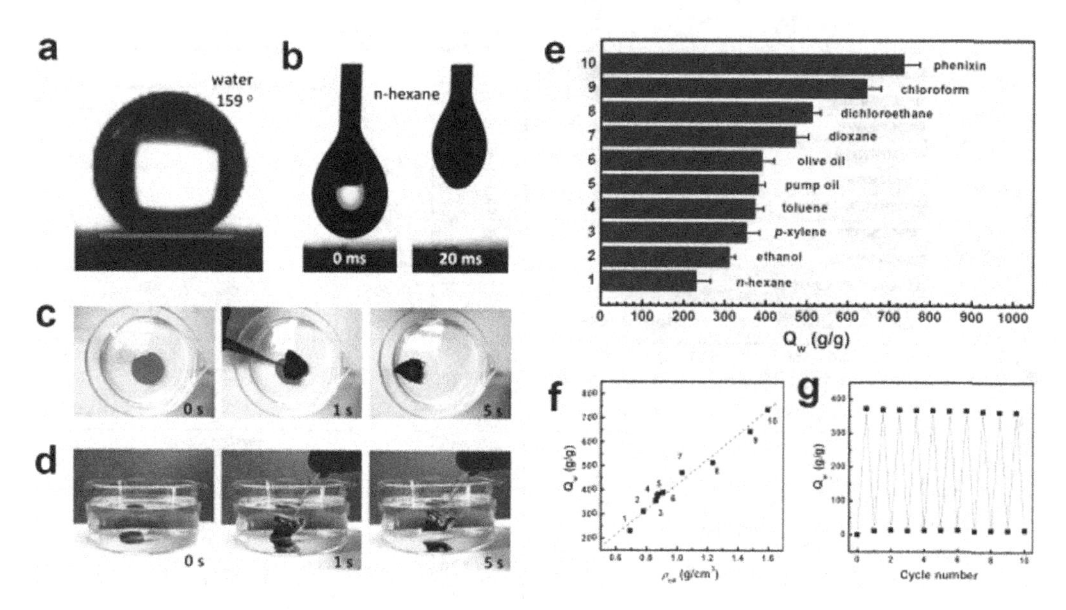

FIGURE 4 (a) High water contact angle (159°) demonstrating the superhydrophobicity of GNA aerogel surface. (b) Fast permeation of *n*-hexane on the surface of GNA. (c-d). The absorption process of toluene and 1,2-dichloroethane by GNA. (e) The absorption capacity of GNA for various oils and organic solvents. (f) Absorption capacity as a function of ρ for different organic liquids identified by numbers being the same as those marked in graph 'e'. (g) Recyclability of GNA for the adsorptive removal of toluene for ten cycles (Xiao et al. 2018). Reprinted with permission from Copyright (2018) Elsevier.

solvents with a removal capacity of 228-695 times to own weight. Moreover, the extraordinary reusable properties, retaining core structural features, and high sorption performance even after 40 cycles, i.e. 95, 93, and 95% removal performance by distillation, squeezing, and combustion, revealed their outstanding oil-water separation properties (Song et al. 2020). The chitosan/reduced graphene oxide composite aerogels (CGA) are prepared by self-polymerization of dopamine and chitosan as reinforcement and interconnecting materials. The CGA exhibits super amphiphilicity in air and super oleophobicity in water because of the rough surface and hydrophilic functionalities ($-NH_2$ and -OH) of chitosan. These foams were further modified with 1H, 1H, 2H, 2H perflourodecanethiol to introduce superhydrophobicity to extend their applications in oily wastewater treatment (N. Cao et al. 2017). The carbon aerogels prepared via 2,2,6,6-tetramethyl-1-piperidinyloxy (TEMPO)-oxidized cellulose nanofibers (TOCN) exhibited 99.5% porosity and adsorbed variety of oils and organic solvents with sorption capacity of 110-260 $g.g^{-1}$, variable with the density of carbon aerogels (M. Wang et al. 2017).

Nevertheless, expensive chemicals, instruments, or complicated fabrication processes restrict the large-scale production of aerogels for industrialization. Over the recent past, the development of new and facile approaches to prepare economically and environmentally viable carbon-based aerogels has been gaining considerable interest. The rich source, low cost, sustainability, and non-toxicity of waste biomass make them suitable alternatives for fabricating carbon-based aerogels. The highly elastic carbon aerogels prepared by waste newspaper displayed excellent hydrophobicity as revealed by high water contact angle (132°) and adsorbed the variable oils and solvents with a capacity of 29-51 $g.g^{-1}$. The recyclability results promise its potential for oil sorption applications. These aerogels showed 63.9, 97, and 100% sorption capacity for pump oil, ethanol, and gasoline even after five consecutive cycles (Han et al. 2016). Kapok, a natural hollow fiber recently considered as a new sustainable resource for oil uptake. The micro-fibrillated aerogels of Kapok exhibited hierarchically porous structures and are considered cost-effective adsorbents. The pictorial representation of kapok-based micro-fibrillated aerogels preparation, along with wettability and microscopic images, are presented in Figure 5. The ultralow density, porosity, and hydrophobicity of these aerogels selectively adsorbed the waste oils with a sorption capacity of 104-190 $g.g^{-1}$ (Zhang et al. 2021).

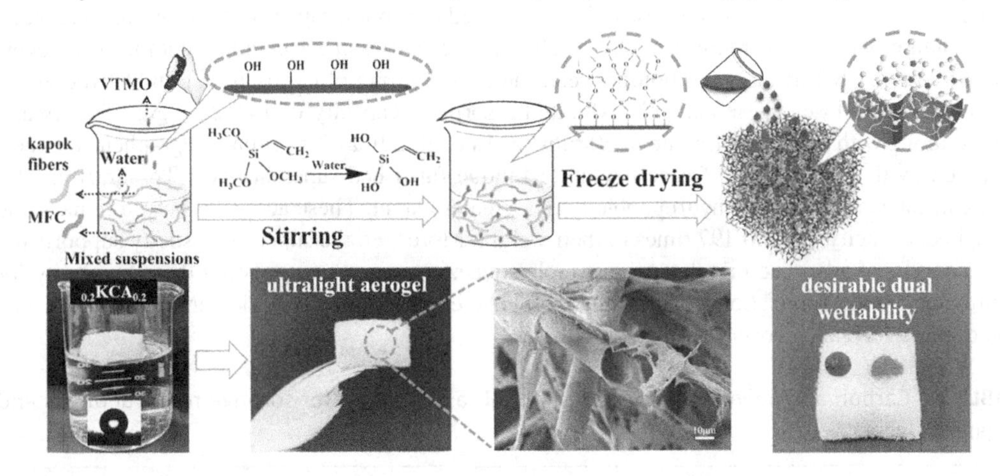

FIGURE 5 (a) Schematic presentation for the preparation of Kapok/micro-fibrillated aerogels (KCA). (b) Photograph of a KCA aerogel floating over the water. (c) SEM images of KCA. (d) Water (dyed with methyl blue) and vegetable oil (dyed with Susan III) drops over the surface of KCA aerogel (Zhang et al. 2021). Reprinted with permission from Copyright (2021) Elsevier.

The rice straw, an agricultural waste, was utilized to prepare ecofriendly aerogels and used for a wide range of applications, including cleaning oil-spill, thermal insulation, and acoustic absorption. The PVA and cationic starch were used as binders to induce cross-linking and reinforcement to the aerogel structure. The aerogels showed a sorption capacity of 13 $g.g^{-1}$ for oils and can be reused for up to 3 cycles. Moreover, the hydrophobic nature of aerogels is restored even after a long time of exposure to the ambient environment, suggesting their potential for oil-water separation applications (S. T. Nguyen et al. 2020). The multifunctional low-cost carbon fiber-based aerogels are prepared from disposable bamboo chopsticks. These compressive aerogels exhibit efficient separation of oil droplets from water and displayed high sorption for various organic solvents and oils with good recyclability (S. Yang et al. 2015). The high carbonization temperature has been a major challenge for the preparation of economically viable carbon-based aerogels. The cellulose nanofibers modified with glucose and choline chloride-based deep eutectic solvents are carbonized at significantly low-temperature (350°C) for preparing the carbon aerogels. These aerogels displayed 74 to 95% sorption efficiency for a variety of organic pollutants and oils. Moreover, carbon aerogels could be used for more than 100 cycles of adsorption-desorption because of their excellent mechanical and compressive properties, which promises their potential for oil-water separation applications (S. Yang et al. 2015).

The bio-polymers have shown an increasing demand for aerogel production for oily wastewater treatment. The cellulose is used as an adsorbent for oil-spill cleanup and treatment of wastewater. However, hydrophilicity and recyclability after adsorption events is a tedious process and hamper the application of cellulosic materials for wastewater remediation. The oleic acid-doped magnetic nano cellulose aerogels exhibited high hydrophobicity and porous network. The oleic acid enhanced the surface area with hydrophobic characteristics, thereby increased the oil sorption performance. These magnetic nano-cellulosic aerogels adsorbed 56.3, 68.1, and 33.2 $g.g^{-1}$ of ethyl acetate, cyclohexane, and vacuum pump oil, respectively. Moreover, the magnetic features enable the easy separation of spent aerogels by applying an external magnetic field (Gu et al. 2020). The magnetic chitosan aerogel is fabricated through electrostatic interaction between chitosan particles, itaconic acid, and iron nanoparticles. These hydrophobic magnetic chitosan aerogels showed a sorption capacity of 17.7-43.8 $g.g^{-1}$ for various oils and solvents. The biodegradability and eco-friendliness of magnetic chitosan aerogels make them suitable candidates as oil scavengers for wastewater treatment applications (Yin et al. 2020).

The composite aerogels of cellulose/chitosan showed a sorption capacity of 10 $g.g^{-1}$ for oil uptake from contaminated water (Meng et al. 2017). The oxidized chitosan and cellulose composite aerogels are prepared by Schiff base reaction between the amino group of chitosan and dialdehyde group of oxidized-cellulose. These aerogels exhibited a sorption capacity of 13.8-28.2 $g.g^{-1}$ for organic liquids along with cyclic stability up to 50 runs (Z. Li et al. 2018a). The cellulose/graphene aerogels prepared by the bidirectional freeze-drying technique furnished superelasticity. The grafting of a long carbon chain increased hydrophobicity to a greater extent. These aerogels showed remarkable oil uptake capacity, i.e. 180-197 times to their weight. Moreover, the superior elasticity supports the fast recovery of adsorbed oils through a simple squeezing process, suggesting their suitability for commercial-scale applications. (Mi et al. 2018). The carbon and polymeric aerogels used for oils and organic solvents recovery are listed in Table 3.

TABLE 3 Carbon and polymeric aerogels and their applications for sorptive removal of oils and organic solvents.

S. No.	Adsorbent	Organic Solvents and Oils	Absorption Capacity	Reference
1	Cellulosic graphene aerogel	Chloroform, benzene, etc.	80-197 $g.g^{-1}$	(Mi et al. 2018)
2	Carbon nanotubes sponge	Various oils and organic solvents	80-180 $g.g^{-1}$	(Gui et al. 2010)
3	Graphene/CNT foam	Compressor oil, organic solvents	80-140 $g.g^{-1}$	(Dong et al. 2012)
4	Carbon nanofiber aerogels	Oils and organic solvents	106-312 $g.g^{-1}$	(Z. Y. Wu et al. 2013)
5	Carbonized graphene aerogel	Chloroform, pump oil, etc.	180-350 $g.g^{-1}$	(F. Wang et al. 2017)
6	Carbon@SiO_2@MnO_2 aerogel	Carbon tetrachloride, toluene, etc.	60-120 $g.g^{-1}$	(Yuan et al. 2018)
7	PAA/rGO aerogel	Oil	120 $g.g^{-1}$	(Ha et al. 2015)
8	Seashell derived graphene foam	Solvents/Oils	200-250 $g.g^{-1}$	(Shi et al. 2016)
9	PANI/n-dodecylthiol coated melamine sponge	Pump oil, vegetable oil, petroleum ether, etc.	51-122 $g.g^{-1}$	(Q. Liu et al. 2015)
10	Trimethoxysilane/GO-PU sponge	Lubrication oil, n-hexane, crude oil, diesel oil	26-44 $g.g^{-1}$	(S. Zhou et al. 2016)
11	Graphene-based sponge	Oils and organic solvents	60-160 $g.g^{-1}$	(D. D. Nguyen et al. 2012)
12	Stearic acid/PVA/rGO aerogel	Chloroform, hexane, etc.	105-250 $g.g^{-1}$	(J. Cao et al. 2018)
13	Carbon aerogel	Oils and organic solvents	80-181 $g.g^{-1}$	(L. Li et al. 2017)
14	Cellulose/PVA/rGO aerogel	DMF, pump oil, etc.	57-97 $g.g^{-1}$	(Xu et al. 2018)
15	Nano-cellulosic sponge	DCM, Silicon oil, toluene, ethanol, etc.	25-55 $g.g^{-1}$	(Phanthong et al. 2018)

4. ADSORPTIVE REMOVAL OF HEAVY METALS AND METALLOIDS FROM AEROGELS

Inorganic impurities, like anionic species, heavy metals, metalloids, and metal oxides, are major environmental challenges as the cumulative contamination caused by these pollutants is far greater than organic and nuclear debris (Antoniou et al. 2016). The aggregation of inorganic compounds in manufacturing sludge and drainage represents the wastage of utilized feed substances and disturbing the ecological equilibrium. Inorganic contaminants, such as arsenic/arsenate, nitrates, cadmium, phosphates, mercury, and fluorides, are serious contaminants in water (Joshi et al. 2020). Their permissible limits in potable water are quite stringent due to their toxic and health hazards. Especially, heavy metals are considered highly hazardous and carcinogenic to human health, and they can persist in the atmosphere for a long time because of their non-biodegradability. The water contamination by these pollutants, particularly in coastal areas, nourishes algal bloom and decreases dissolved oxygen in water bodies, resulting in dead zones (Sizmur et al. 2017). Therefore, removing inorganic pollutants from the polluted water is mandatory before discharging them into water bodies.

A diversified variety of aerogels, primarily carbon and biopolymers-derived scaffolds, with controlled porosity and surface characteristics are investigated to remove the inorganic contaminants from wastewater. The adsorption mechanisms are mainly governed by physical interactions. The presence of heteroatoms like N, S, P, and O in the aerogels and hydrophilicity influences the adsorption potential for inorganic pollutants. The rGO aerogels fabricated via a hydrothermal approach effectively removed the Pb^{2+} ions, and it was ascribed to an excellent porous network and accessible surface area of aerogel (Gao et al. 2020). The adsorption of heavy metal ions by sulfonated graphene aerogels (3D-SRGO) is enabled by effective interactions between sulfur and heavy metal ions. The adsorptive removal of Cd^{2+} ions by 3D-SRGO aerogels is attributed to cation exchange reaction and electrostatic interaction. The adsorbent can be effectively regenerated by treating with HNO_3 solution and restored for subsequent utilization (S. Wu et al. 2015). The graphene aerogel modified with chitosan successfully adsorbed Cu^{2+} and Pb^{2+} ions with removal capacities of 70 and 90 $mg.g^{-1}$, respectively, driven by electrostatic interactions and ion complex formation (Yunqiang Chen et al. 2013).

The carbon aerogel prepared via carbonization of resorcinol-formaldehyde showed effective adsorption of Ni, Cd, Cu, Pb, Mn, Zn, and Hg ions. The adsorption potential is governed by pH, adsorbent dosage, concentration, temperature, and time. The Cd^{2+} showed maximum adsorption (400.8 $mg{\cdot}g^{-1}$), whereas Pb^{2+} displayed least adsorption (0.7 $mg{\cdot}g^{-1}$) by carbon aerogels. The difference is ascribed to the variance in chemical affinity and ion-exchange ability of the particular metal ions toward binding sites of carbon aerogel (Meena et al. 2005). The adsorption efficiency of aerogels for these heavy metal ions increases by modifying the surfaces, incorporating active sites, and increasing the porosity of carbon aerogels. The capturing of heavy metal ions via chelating them with specific functional moieties comprising N, P, S, and O, furnished the coordination complex with metal ions. The N-doped carbon aerogel prepared by employing the globular protein adsorbed the Pb^{2+} and Cr^{6+} ions from wastewater driven by electrostatic interaction and reduction as major adsorption events. The N-doped carbon aerogel showed maximum removal capacities of 68 and 240 $mg{\cdot}g^{-1}$ for Cr^{6+} and Pb^{2+} ions (Alatalo et al. 2015). The N-doped carbon aerogels prepared via NH_3-assisted pyrolysis are employed for adsorptive removal of Pb^{2+} and Cu^{2+} ions. It is found that the oxygen content of N-doped samples was higher than that of non-doped samples, thus increasing their adsorption capacity by 2-6 times than that of non-doped samples (Veselá and Slovák 2014). Kong et al. have synthesized a graphene-based aerogel co-doped with nitrogen and sulfur, which can simultaneously adsorb Cd^{2+} and organic dyes from the polluted water. The adsorbed dye enhanced the adsorption of Cd^{2+} ions. The surface and intraparticle diffusions regulate the rate of adsorption (Q. Kong et al. 2018).

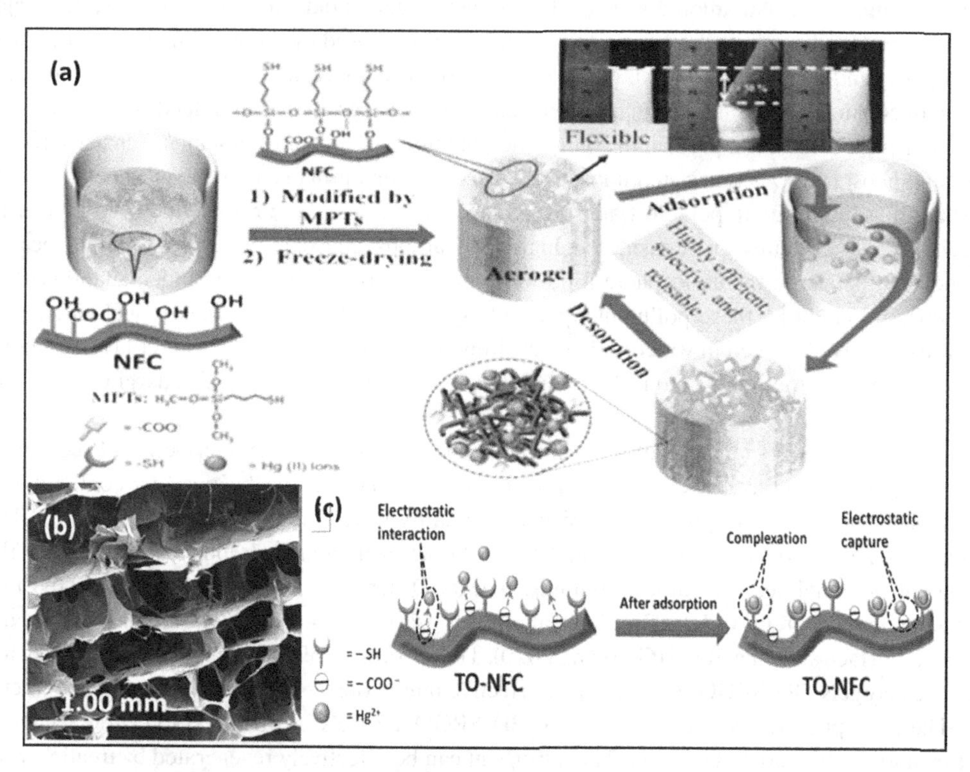

FIGURE 6 (a) Nano cellulose aerogel modified with thiol groups for effective removal of Hg^{2+} ions. (b) SEM micrograph of cellulose aerogel. (c) Adsorption of Hg^{2+} ion by cellulose aerogel (Geng et al. 2017). Reprinted with permission from Copyright (2017) American Chemical Society.

The aerogels prepared by biopolymers, like cellulose, starch, and chitosan, have recently been used to remove heavy metal ions. The oriented micro-channeled cellulose/chitosan aerogel efficiently adsorbed the Pb^{2+} ions from polluted water with an adsorption capacity of 252.6 $mg.g^{-1}$. The prepared aerogel maintained the adsorption efficiency up to 85% for five adsorption-desorption cycles (Y. Li et al. 2019). The incorporation of ethylenediamine in sodium alginate aerogel provides –CO-NH, –OH, NH_2 functionalities to successfully remove Cu^{2+} and Pb^{2+} metal ions via chelation. The XPS studies revealed that cation exchange and complexation are the primary adsorptive pathways for removing metal ions (Y. Huang and Wang 2018). The flexible nano cellulosic aerogels enriched with surface thiol groups exhibited interconnected porous structures (Figure 6b), excellent selectivity, and adsorption capacity for Hg^{2+} ions from the polluted water. The negatively charged surface of the aerogel due to carboxylate ions electrostatically attracted the Hg^{2+} ions, which gradually led to complexation between the thiol and mercury ions, as illustrated in Figure 6c (Geng et al. 2017). Table 4 presents the variable aerogels for the adsorption of heavy metal ions along with their maximum adsorption capacity. Adsorbent regeneration is an imperative activity in removing heavy metal ions. There have been studies on the usage of acids, ethanol, alkalis, distilled water, buffer solutions, and chelating agents as regenerative agents (Lata et al. 2015).

TABLE 4 Aerogels for the adsorptive removal of heavy metal ions.

S. No.	Adsorbent	Targeted Pollutant	Maximum Capacity, $mg.g^{-1}$	Reference
1	Resorcinol-formaldehyde-derived carbon aerogel	Cd^{2+} Hg^{2+} Cu^{2+}	400.8 45.6 561.7	(Meena et al. 2005)
2	Nano cellulose aerogel	Hg^{2+}	718.5	(Geng et al. 2017)
3	Alginate aerogel beads	Cd^{2+} Cu^{2+}	244.6 126.8	(Deze et al. 2012)
4	Calcium-alginate/GO aerogel	Pb^{2+} Cu^{2+} Cd^{2+}	275.0 60.0 110.0	(Pan et al. 2018)
5	Cellulose/polyethyleneimine aerogel	Pb^{2+} Cu^{2+}	250.0 123.0	(J. Li et al. 2018)
6	Polydopamine@chitosan aerogel	Cr^{6+} Pb^{2+}	181.0 188.0	(Guo et al. 2018a)
7	Calcium alginate aerogel	Pb^{2+}	390.0	(Z. Wang et al. 2016)
8	Waste paper/chitosan aerogel	Cu^{2+}	165.0	(Z. Li et al. 2018b)
9	GO/chitosan aerogel	Pb^{2+} Cu^{2+}	249.4 95.4	(Luo et al. 2019)
10	Chitosan/polydopamine	Cr^{6+} Pb^{2+}	374.4 441.2	(Guo et al. 2018b)

5. SUMMARY AND FUTURE PERSPECTIVE

Highly porous 3D aerogels gained significant interest over the last few decades due to their fascinating features including, ultralow density, high porosity, tailorable structural and chemical characteristics, which make them suitable materials to bridge the properties of nanoscale building blocks to macroscale scaffolds. These unique features make them suitable candidates for a wide range of applications, like energy storage, gas separation, adsorbents, oil/water separation, catalyst, thermal insulation, etc. A number of organic, inorganic, and carbon materials-based aerogels have shown interest in high-performance applications. The present chapter discussed different types of aerogels, their characteristics, and applications, primarily focused on wastewater remediation. It also emphasizes the adsorptive removal of organic pollutants (dyes, APIs, pesticides), inorganic contaminants, oils, and organic solvents from wastewater. The carbon-based aerogels are used as sorbents for wastewater treatment applications, including the separation of oily products, dyes, and heavy metal ions. However, the tedious and expensive synthetic procedure, poor mechanical strength, structural instability, and the low number of repetitive cycles remained as obstacles for its indulgence in commercial applications. The development of carbon aerogels using waste biomass has decreased the cost and made them environmentally and economically viable. The incorporation of polymeric materials in carbon aerogels enhances their structural stability and compressibility, providing essential characteristics for their facile regeneration.

Moreover, it emphasizes the recent progress in the development of biopolymer-based aerogels for wastewater remediation applications. Biopolymeric aerogels with high structural stability and excellent absorptive capacity enlarge their demand as adsorbent materials for environmental remediation applications. However, few drawbacks related to the facile preparation of 3D scaffolds of highly porous, superhydrophobic, and super oleophilic aerogels need to be addressed for their effective utilization. The high cost, low mechanical or structural stability, less adsorptive removal efficiency, environmental or economical non-viability are the major challenges that hamper their large-scale production and application for industrialization. These drawbacks received extensive attention, and there is a growing need for the emergence of many new productive ways, implying the involvement of low-cost waste biomass as raw precursors and other natural biopolymeric materials.

References

Aegerter, M.A., Leventis, N. and Koebel, M.M. 2011. Aerogels Handbook. Springer Science & Business Media.

Alatalo, S.-M., Pileidis, F., Mäkilä, E., Sevilla, M., Repo, E., Salonen, J., et al. 2015. Versatile cellulose-based carbon aerogel for the removal of both cationic and anionic metal contaminants from water. ACS Appl. Mater. Interfaces 7(46): 25875-25883.

Antoniou, M., Litter, M., Ibanez, P.F., Marugan, J., Lim, T.T., Malato, S., et al. 2016. Photocatalysis: Applications. Royal Society of Chemistry. U.K. (BOOK).

Arabkhani, P. and Asfaram, A. 2020. Development of a novel three-dimensional magnetic polymer aerogel as an efficient adsorbent for malachite green removal. J. Hazard. Mater. 384: 121394.

Araby, S., Qiu, A., Wang, R., Zhao, Z., Wang, C.-H. and Ma, J. 2016. Aerogels based on carbon nanomaterials. J. Mater. Sci. 51(20): 9157-9189.

Aylaz, G., Okan, M., Duman, M. and Aydin, H.M. 2020. Study on cost-efficient carbon aerogel to remove antibiotics from water resources. ACS Omega 5(27): 16635-16644.

Biener, J., Stadermann, M., Suss, M., Worsley, M.A., Biener, M.M., Rose, K.A., et al. 2011. Advanced carbon aerogels for energy applications. Energy Environ. Sci. 4(3): 656-667.

Cao, J., Wang, Z., Yang, X., Tu, J., Wu, R. and Wang, W. 2018. Green synthesis of amphipathic graphene aerogel constructed by using the framework of polymer-surfactant complex for water remediation. Appl. Surf. Sci. 444: 399-406.

Cao, N., Lyu, Q., Li, J., Wang, Y., Yang, B., Szunerits, S., et al. 2017. Facile synthesis of fluorinated polydopamine/chitosan/reduced graphene oxide composite aerogel for efficient oil/water separation. Chem. Eng. J. 326: 17-28.

Chander, V., Sharma, B., Negi, V., Aswal, R., Singh, P., Singh, R., et al. 2016. Pharmaceutical compounds in drinking water. J. Xenobiot. 6(1): 1-7.

Chen, H., Liu, T., Meng, Y., Cheng, Y., Lu, J. and Wang, H. 2020. Novel graphene oxide/aminated lignin aerogels for enhanced adsorption of malachite green in wastewater. Colloids Surf, A Physicochem. Eng. Asp. 603: 125281.

Chen, L., Du, R., Zhang, J. and Yi, T. 2015. Density controlled oil uptake and beyond: From carbon nanotubes to graphene nanoribbon aerogels. J. Mater. Chem. A 3(41): 20547-20553.

Chen, L., Li, Y., Du, Q., Wang, Z., Xia, Y., Yedinak, E., et al. 2017. High performance agar/graphene oxide composite aerogel for methylene blue removal. Carbohydr. Polym. 155: 345-353.

Chen, T., Li, M., Zhou, L., Ding, X., Lin, D., Duan, T., et al. 2020. Bio-inspired biomass-derived carbon aerogels with superior mechanical property for oil–water separation. ACS Sustain. Chem. Eng. 8(16): 6458-6465.

Chen, Y., Chen, L., Bai, H. and Li, L. 2013. Graphene oxide–chitosan composite hydrogels as broad-spectrum adsorbents for water purification [10.1039/C2TA00406B]. J. Mater. Chem. A 1(6): 1992-2001.

Chen, Y., Dai, G. and Gao, Q. 2019. Starch nanoparticles–graphene aerogels with high supercapacitor performance and efficient adsorption. ACS Sustain. Chem. Eng. 7(16): 14064-14073.

Cui, J., Xu, X., Yang, L., Chen, C., Qian, J., Chen, X., et al. 2020. Soft foam-like UiO-66/polydopamine/ bacterial cellulose composite for the removal of aspirin and tetracycline hydrochloride. Chem. Eng. J. 395: 125174.

Dan, H., Li, N., Xu, X., Gao, Y., Huang, Y., Akram, M., et al. 2020. Mechanism of sonication time on structure and adsorption properties of 3D peanut shell/graphene oxide aerogel. Sci. Total Environ. 739: 139983.

Deze, E.G., Papageorgiou, S.K., Favvas, E.P. and Katsaros, F.K. 2012. Porous alginate aerogel beads for effective and rapid heavy metal sorption from aqueous solutions: Effect of porosity in Cu^{2+} and Cd^{2+} ion sorption. Chem. Eng. J. 209: 537-546.

Diao, S., Liu, H., Chen, S., Xu, W. and Yu, A. 2020. Oil adsorption performance of graphene aerogels. J. Mater. Sci. 55(11): 4578-4591.

Ding, C., Wang, X., Liu, H., Li, Y., Sun, Y., Lin, Y., et al. 2018. Glyphosate removal from water by functional three-dimensional graphene aerogels. Environ. Chem. 15(6): 325-335.

Dong, X., Chen, J., Ma, Y., Wang, J., Chan-Park, M.B., Liu, X., et al. 2012. Superhydrophobic and superoleophilic hybrid foam of graphene and carbon nanotube for selective removal of oils or organic solvents from the surface of water. Chem. Commun. 48(86): 10660-10662.

Fei, Y., Li, Y., Han, S. and Ma, J. 2016. Adsorptive removal of ciprofloxacin by sodium alginate/graphene oxide composite beads from aqueous solution. J. Colloid Interface Sci. 484: 196-204.

Fricke, J. and Emmerling, A. 1992. Aerogels—Preparation, properties, applications. pp. 37-87. *In*: Jochen Fricke and Andreas Emmerling. [eds.]. Chemistry, Spectroscopy and Applications of Sol-Gel Glasses. Springer Berlin Heidelberg.

Gan, G., Li, X., Fan, S., Wang, L., Qin, M., Yin, Z., et al. 2019. Carbon aerogels for environmental clean-up. Eur. J. Inorg. Chem. 2019(27): 3126-3141.

Gao, C., Dong, Z., Hao, X., Yao, Y. and Guo, S. 2020. Preparation of reduced graphene oxide aerogel and its adsorption for Pb(II). ACS Omega 5(17): 9903-9911.

Geng, B., Wang, H., Wu, S., Ru, J., Tong, C., Chen, Y., et al. 2017. Surface-tailored nanocellulose aerogels with thiol-functional moieties for highly efficient and selective removal of Hg(II) ions from water. ACS Sustain. Chem. Eng. 5(12): 11715-11726.

Gu, H., Zhou, X., Lyu, S., Pan, D., Dong, M., Wu, S., et al. 2020. Magnetic nanocellulose-magnetite aerogel for easy oil adsorption. J. Colloid Interface Sci. 560: 849-856.

Gui, X., Wei, J., Wang, K., Cao, A., Zhu, H., Jia, Y., et al. 2010. Carbon nanotube sponges. Adv. Mater. 22(5): 617-621.

Guo, D.-M., An, Q.-D., Xiao, Z.-Y., Zhai, S.-R. and Yang, D.-J. 2018a. Efficient removal of Pb(II), Cr(VI) and organic dyes by polydopamine modified chitosan aerogels. Carbohydr. Polym. 202: 306-314.

Guo, D.-M., An, Q.-D., Xiao, Z.-Y., Zhai, S.-R. and Yang, D.-J. 2018b. Efficient removal of Pb(II), Cr(VI) and organic dyes by polydopamine modified chitosan aerogels. Carbohydr. Polym. 202: 306-314.

Gupta, K. and Khatri, O.P. 2017. Reduced graphene oxide as an effective adsorbent for removal of malachite green dye: Plausible adsorption pathways. J. Colloid Interface Sci. 501: 11-21.

Gupta, K. and Khatri, O.P. 2019. Fast and efficient adsorptive removal of organic dyes and active pharmaceutical ingredient by microporous carbon: Effect of molecular size and charge. Chem. Eng. J. 378: 122218.

Gupta, K., Gupta, D. and Khatri, O.P. 2019. Graphene-like porous carbon nanostructure from Bengal gram bean husk and its application for fast and efficient adsorption of organic dyes. Appl. Surf. Sci. 476: 647-657.

Gusain, R., Gupta, K., Joshi, P. and Khatri, O.P. 2019. Adsorptive removal and photocatalytic degradation of organic pollutants using metal oxides and their composites: A comprehensive review. Adv. Colloid Interface Sci. 272: 102009.

Ha, H., Shanmuganathan, K. and Ellison, C.J. 2015. Mechanically stable thermally cross-linked poly (acrylic acid)/reduced graphene oxide aerogels. ACS Appl. Mater. Interfaces 7(11): 6220-6229.

Han, S., Sun, Q., Zheng, H., Li, J. and Jin, C. 2016. Green and facile fabrication of carbon aerogels from cellulose-based waste newspaper for solving organic pollution. Carbohydr. Polym. 136: 95-100.

Hou, P., Xing, G., Tian, L., Zhang, G., Wang, H., Yu, C., et al. 2019. Hollow carbon spheres/graphene hybrid aerogels as high-performance adsorbents for organic pollution. Sep. Purif. Technol. 213: 524-532.

Huang, T., Dai, J., Yang, J.-h., Zhang, N., Wang, Y. and Zhou, Z.-w. 2018. Polydopamine coated graphene oxide aerogels and their ultrahigh adsorption ability. Diam. Relat. Mater. 86: 117-127.

Huang, Y. and Wang, Z. 2018. Preparation of composite aerogels based on sodium alginate, and its application in removal of Pb^{2+} and Cu^{2+} from water. Int. J. Biol. Macromol. 107: 741-747.

Ji, K., Gao, Y., Zhang, L., Wang, S., Yue, Q., Xu, X., et al. 2021. A tunable amphiphilic enteromorpha-modified graphene aerogel for oil/water separation. Sci. Total Environ. 763: 142958.

Jiang, F., Dinh, D.M. and Hsieh, Y.-L. 2017. Adsorption and desorption of cationic malachite green dye on cellulose nanofibril aerogels. Carbohydr. Polym. 173: 286-294.

Joshi, P., Gupta, K., Gusain, R. and Khatri, O.P. 2020. Metal oxide nanocomposites for wastewater treatment. pp. 361-397. *In*: Raneesh, B. and Visakh, P.M. [eds.]. Metal Oxide Nanocomposites: Synthesis and Applications. John Wiley & Sons, Inc.

Kabiri, S., Tran, D.N., Altalhi, T. and Losic, D. 2014. Outstanding adsorption performance of graphene–carbon nanotube aerogels for continuous oil removal. Carbon 80: 523-533.

Kong, Q., Wei, C., Preis, S., Hu, Y. and Wang, F. 2018. Facile preparation of nitrogen and sulfur co-doped graphene-based aerogel for simultaneous removal of Cd^{2+} and organic dyes. Environ. Sci. Pollut. Res. 25(21): 21164-21175.

Kong, Y., Zhuang, Y., Han, K. and Shi, B. 2020. Enhanced tetracycline adsorption using alginate-graphene-ZIF67 aerogel. Colloids Surf, A Physicochem Eng. Asp. 588: 124360.

Lata, S., Singh, P. and Samadder, S. 2015. Regeneration of adsorbents and recovery of heavy metals: A review. Int. J. Environ. Sci. Technol. 12(4): 1461-1478.

Lee, J.-H. and Park, S.-J. 2020. Recent advances in preparations and applications of carbon aerogels: A review. Carbon 163: 1-18.

Li, J., Zuo, K., Wu, W., Xu, Z., Yi, Y., Jing, Y., et al. 2018. Shape memory aerogels from nanocellulose and polyethyleneimine as a novel adsorbent for removal of Cu(II) and Pb(II). Carbohydr. Polym. 196: 376-384.

Li, K., Zhou, M., Liang, L., Jiang, L. and Wang, W. 2019. Ultrahigh-surface-area activated carbon aerogels derived from glucose for high-performance organic pollutants adsorption. J. Colloid Interface Sci. 546: 333-343.

Li, L., Hu, T., Sun, H., Zhang, J. and Wang, A. 2017. Pressure-sensitive and conductive carbon aerogels from poplars catkins for selective oil absorption and oil/water separation. ACS Appl. Mater. Interfaces 9(21): 18001-18007.

Li, L., Zhao, J., Sun, Y., Yu, F. and Ma, J. 2019. Ionically cross-linked sodium alginate/κ-carrageenan double-network gel beads with low-swelling, enhanced mechanical properties, and excellent adsorption performance. Chem. Eng. J. 372: 1091-1103.

Li, R., Chen, C., Li, J., Xu, L., Xiao, G. and Yan, D. 2014. A facile approach to superhydrophobic and superoleophilic graphene/polymer aerogels. J. Mater. Chem. A 2(9): 3057-3064.

Li, S.C., Hu, B.C., Ding, Y.W., Liang, H.W., Li, C., Yu, Z.Y., et al. 2018. Wood-derived ultrathin carbon nanofiber aerogels. Angew. Chem. Int. Ed. 57(24): 7085-7090.

Li, Y., Guo, C., Shi, R., Zhang, H., Gong, L. and Dai, L. 2019. Chitosan/ nanofibrillated cellulose aerogel with highly oriented microchannel structure for rapid removal of Pb(II) ions from aqueous solution. Carbohydr. Polym. 223: 115048.

Li, Z., Jia, Z., Ni, T. and Li, S. 2017. Adsorption of methylene blue on natural cotton based flexible carbon fiber aerogels activated by novel air-limited carbonization method. J. Mol. Liq. 242: 747-756.

Li, Z., Shao, L., Hu, W., Zheng, T., Lu, L., Cao, Y., et al. 2018a. Excellent reusable chitosan/cellulose aerogel as an oil and organic solvent absorbent. Carbohydr. Polym. 191: 183-190.

Li, Z., Shao, L., Ruan, Z., Hu, W., Lu, L. and Chen, Y. 2018b. Converting untreated waste office paper and chitosan into aerogel adsorbent for the removal of heavy metal ions. Carbohydr. Polym. 193: 221-227.

Liang, W., Wang, B., Cheng, J., Xiao, D., Xie, Z. and Zhao, J. 2021. 3D, eco-friendly metal-organic frameworks@ carbon nanotube aerogels composite materials for removal of pesticides in water. J. Hazard. Mater. 401: 123718.

Liu, F., Chung, S., Oh, G. and Seo, T.S. 2012. Three-dimensional graphene oxide nanostructure for fast and efficient water-soluble dye removal. ACS Appl. Mater. Interfaces 4(2): 922-927.

Liu, H., Wei, Y., Luo, J., Li, T., Wang, D., Luo, S., et al. 2019. 3D hierarchical porous-structured biochar aerogel for rapid and efficient phenicol antibiotics removal from water. Chem. Eng. J. 368: 639-648.

Liu, H., Wang, X., Ding, C., Dai, Y., Sun, Y., Lin, Y., et al. 2020. Carboxylated carbon nanotubes-graphene oxide aerogels as ultralight and renewable high performance adsorbents for efficient adsorption of glyphosate. Environ. Chem. 17(1): 6-16.

Liu, Q., Meng, K., Ding, K. and Wang, Y. 2015. A superhydrophobic sponge with hierarchical structure as an efficient and recyclable oil absorbent. ChemPlusChem 80(9): 1435.

Liu, Q., Yu, H., Zeng, F., Li, X., Sun, J., Hu, X., et al. 2020. Polyaniline as interface layers promoting the *in situ* growth of zeolite imidazole skeleton on regenerated cellulose aerogel for efficient removal of tetracycline. J. Colloid Interface Sci. 579: 119-127.

Liu, S., Yao, F., Oderinde, O., Zhang, Z. and Fu, G. 2017. Green synthesis of oriented xanthan gum–graphene oxide hybrid aerogels for water purification. Carbohydr. Polym. 174: 392-399.

Luo, J., Fan, C., Xiao, Z., Sun, T. and Zhou, X. 2019. Novel graphene oxide/carboxymethyl chitosan aerogels via vacuum-assisted self-assembly for heavy metal adsorption capacity. Colloids Surf, A Physicochem. Eng. Asp. 578: 123584.

Maleki, H. 2016. Recent advances in aerogels for environmental remediation applications: A review. Chem. Eng. J. 300: 98-118.

Meena, A.K., Mishra, G.K., Rai, P.K., Rajagopal, C. and Nagar, P.N. 2005. Removal of heavy metal ions from aqueous solutions using carbon aerogel as an adsorbent. J. Hazard Mater. 122(1-2): 161-170.

Meng, G., Peng, H., Wu, J., Wang, Y., Wang, H., Liu, Z., et al. 2017. Fabrication of superhydrophobic cellulose/ chitosan composite aerogel for oil/water separation. Fibers Polym. 18(4): 706-712.

Mi, H.-Y., Jing, X., Politowicz, A.L., Chen, E., Huang, H.-X. and Turng, L.-S. 2018. Highly compressible ultra-light anisotropic cellulose/graphene aerogel fabricated by bidirectional freeze drying for selective oil absorption. Carbon 132: 199-209.

Ngah, W.W., Teong, L. and Hanafiah, M. 2011. Adsorption of dyes and heavy metal ions by chitosan composites: A review. Carbohydr. Polym. 83(4): 1446-1456.

Nguyen, D.D., Tai, N.-H., Lee, S.-B. and Kuo, W.-S. 2012. Superhydrophobic and superoleophilic properties of graphene-based sponges fabricated using a facile dip coating method. Energy Environ. Sci. 5(7): 7908-7912.

Nguyen, S.T., Do, N.D., Thai, N.N.T., Thai, Q.B., Huynh, H.K.P. and Phan, A.N. 2020. Green aerogels from rice straw for thermal, acoustic insulation and oil spill cleaning applications. Mater. Chem. Phys. 253: 123363.

Pan, L., Wang, Z., Yang, Q. and Huang, R. 2018. Efficient removal of lead, copper and cadmium ions from water by a porous calcium alginate/graphene oxide composite aerogel. Nanomaterials 8(11): 957.

Peydayesh, M., Suter, M.K., Bolisetty, S., Boulos, S., Handschin, S., Nyström, L., et al. 2020. Amyloid fibrils aerogel for sustainable removal of organic contaminants from water. Adv. Mater. 32(12): 1907932.

Phanthong, P., Reubroycharoen, P., Kongparakul, S., Samart, C., Wang, Z., Hao, X., et al. 2018. Fabrication and evaluation of nanocellulose sponge for oil/water separation. Carbohydr. Polym. 190: 184-189.

Pierre, A.C. and Pajonk, G.M. 2002. Chemistry of aerogels and their applications. Chem. Rev. 102(11): 4243-4266.

Qin, W., Zhu, W., Ma, J., Yang, Y. and Tang, B. 2021. Carbon fibers assisted 3D N-doped graphene aerogel on excellent adsorption capacity and mechanical property. Colloids Surf, A Physicochem. Eng. Asp. 608: 125602.

Raman, S., Gurikov, P. and Smirnova, I. 2015. Hybrid alginate based aerogels by carbon dioxide induced gelation: Novel technique for multiple applications. J. Supercrit Fluids 106: 23-33.

Ray, S.S., Gusain, R. and Kumar, N. 2020. Chapter 2 – Classification of water contaminants. pp. 11-36. *In*: Ray, S.S., Gusain, R. and Kumar, N. [eds.]. Carbon Nanomaterial-Based Adsorbents for Water Purification. Elsevier, United States.

Reza, R. and Singh, G. 2010. Heavy metal contamination and its indexing approach for river water. Int. J. Environ. Sci. Technol. 7(4): 785-792.

Rudaz, C., Courson, R.m., Bonnet, L., Calas-Etienne, S., Sallée, H.b. and Budtova, T. 2014. Aeropectin: Fully biomass-based mechanically strong and thermal superinsulating aerogel. Biomacromolecules 15(6): 2188-2195.

Salimian, S., Zadhoush, A., Naeimirad, M., Kotek, R. and Ramakrishna, S. 2018. A review on aerogel: 3D nanoporous structured fillers in polymer-based nanocomposites. Polym. Compos. 39(10): 3383-3408.

Sani, H.A., Ahmad, M.B., Hussein, M.Z., Ibrahim, N.A., Musa, A. and Saleh, T.A. 2017. Nanocomposite of ZnO with montmorillonite for removal of lead and copper ions from aqueous solutions. Process Saf. Environ. Prot. 109: 97-105.

Shi, L., Chen, K., Du, R., Bachmatiuk, A., Rümmeli, M.H., Xie, K., et al. 2016. Scalable seashell-based chemical vapor deposition growth of three-dimensional graphene foams for oil–water separation. J. Am. Chem. Soc. 138(20): 6360-6363.

Shu, D., Feng, F., Han, H. and Ma, Z. 2017. Prominent adsorption performance of amino-functionalized ultra-light graphene aerogel for methyl orange and amaranth. Chem. Eng. J. 324: 1-9.

Sizmur, T., Fresno, T., Akgül, G., Frost, H. and Moreno-Jiménez, E. 2017. Biochar modification to enhance sorption of inorganics from water. Bioresour. Technol. 246: 34-47.

Song, P., Wang, M., Di, J., Xiong, J., Zhao, S. and Li, Z. 2020. Reusable graphitic carbon nitride nanosheet-based aerogels as sorbents for oils and organic solvents. ACS Appl. Nano Mater. 3(8): 8176-8181.

Subrahmanyam, R., Gurikov, P., Dieringer, P., Sun, M. and Smirnova, I. 2015. On the road to biopolymer aerogels—Dealing with the solvent. Gels 1(2): 291-313.

Tang, M., Jia, R., Kan, H., Liu, Z., Yang, S., Sun, L., et al. 2020. Kinetic, isotherm, and thermodynamic studies of the adsorption of dye from aqueous solution by propylene glycol adipate-modified cellulose aerogel. Colloids Surf, A Physicochem. Eng. Asp. 602: 125009.

Tian, X., Zhu, H., Meng, X., Wang, J., Zheng, C., Xia, Y., et al. 2020. Amphiphilic calcium alginate carbon aerogels: Broad-spectrum adsorbents for ionic and solvent dyes with multiple functions for decolorized oil–water separation. ACS Sustain. Chem. Eng. 8(34): 12755-12767.

Veselá, P. and Slovák, V. 2014. N-doped carbon xerogels prepared by ammonia assisted pyrolysis: Surface characterisation, thermal properties and adsorption ability for heavy metal ions. J. Anal. Appl. Pyrolysis. 109: 266-271.

Wang, C., Kim, J., Tang, J., Na, J., Kang, Y.M., Kim, M., et al. 2020. Large-scale synthesis of MOF-derived superporous carbon aerogels with extraordinary adsorption capacity for organic solvents. Angew. Chem. Int. Ed. 132(5): 2082-2086.

Wang, F., Wang, Y., Zhan, W., Yu, S., Zhong, W., Sui, G., et al. 2017. Facile synthesis of ultra-light graphene aerogels with super absorption capability for organic solvents and strain-sensitive electrical conductivity. Chem. Eng. J. 320: 539-548.

Wang, M., Shao, C., Zhou, S., Yang, J. and Xu, F. 2017. Preparation of carbon aerogels from TEMPO-oxidized cellulose nanofibers for organic solvents absorption. RSC Adv. 7(61): 38220-38230.

Wang, Z., Huang, Y., Wang, M., Wu, G., Geng, T., Zhao, Y., et al. 2016. Macroporous calcium alginate aerogel as sorbent for Pb^{2+} removal from water media. J. Environ. Chem. Eng. 4(3): 3185-3192.

Wei, G., Miao, Y.-E., Zhang, C., Yang, Z., Liu, Z., Tjiu, W.W., et al. 2013. Ni-doped graphene/carbon cryogels and their applications as versatile sorbents for water purification. ACS Appl. Mater. Interfaces 5(15): 7584-7591.

Wu, S., Zhang, K., Wang, X., Jia, Y., Sun, B., Luo, T., et al. 2015. Enhanced adsorption of cadmium ions by 3D sulfonated reduced graphene oxide. Chem. Eng. J. 262: 1292-1302.

Wu, Z.Y., Li, C., Liang, H.W., Chen, J.F. and Yu, S.H. 2013. Ultralight, flexible, and fire-resistant carbon nanofiber aerogels from bacterial cellulose. Angew. Chem. Int. Ed. 125(10): 2997-3001.

Xiao, J., Zhang, J., Lv, W., Song, Y. and Zheng, Q. 2017. Multifunctional graphene/poly(vinyl alcohol) aerogels: *In situ* hydrothermal preparation and applications in broad-spectrum adsorption for dyes and oils. Carbon 123: 354-363.

Xiao, J., Lv, W., Song, Y. and Zheng, Q. 2018. Graphene/nanofiber aerogels: Performance regulation towards multiple applications in dye adsorption and oil/water separation. Chem. Eng. J. 338: 202-210.

Xiong, J., Zhang, D., Lin, H. and Chen, Y. 2020. Amphiprotic cellulose mediated graphene oxide magnetic aerogels for water remediation. Chem. Eng. J. 400: 125890.

Xu, Z., Zhou, H., Tan, S., Jiang, X., Wu, W., Shi, J., et al. 2018. Ultralight superhydrophobic carbon aerogels based on cellulose nanofibers/poly (vinyl alcohol)/graphene oxide (CNFs/PVA/GO) for highly effective oil–water separation. Beilstein J. Nanotechnol. 9(1): 508-519.

Yan, M., Huang, W. and Li, Z. 2019. Chitosan cross-linked graphene oxide/lignosulfonate composite aerogel for enhanced adsorption of methylene blue in water. Int. J. Biol. Macromol. 136: 927-935.

Yang, H., Sheikhi, A. and van de Ven, T.G.M. 2016. Reusable green aerogels from cross-linked hairy nanocrystalline cellulose and modified chitosan for dye removal. Langmuir 32(45): 11771-11779.

Yang, S., Chen, L., Mu, L., Hao, B. and Ma, P.-C. 2015. Low cost carbon fiber aerogel derived from bamboo for the adsorption of oils and organic solvents with excellent performances. RSC Adv. 5(48): 38470-38478.

Yang, S., Shen, C., Chen, L., Wang, C., Rana, M. and Lv, P. 2018. Vapor–liquid deposition strategy to prepare superhydrophobic and superoleophilic graphene aerogel for oil–water separation. ACS Appl. Nano Mater. 1(2): 531-540.

Yang, W.-J., Yuen, A.C.Y., Li, A., Lin, B., Chen, T.B.Y., Yang, W., et al. 2019. Recent progress in bio-based aerogel absorbents for oil/water separation. Cellulose 26(11): 6449-6476.

Yang, X., Li, Y., Du, Q., Sun, J., Chen, L., Hu, S., et al. 2015. Highly effective removal of basic fuchsin from aqueous solutions by anionic polyacrylamide/graphene oxide aerogels. J. Colloid Interface Sci. 453: 107-114.

Yao, Q., Fan, B., Xiong, Y., Jin, C., Sun, Q. and Sheng, C. 2017. 3D assembly based on 2D structure of cellulose nanofibril/graphene oxide hybrid aerogel for adsorptive removal of antibiotics in water. Sci. Rep. 7(1): 45914.

Yin, Z., Sun, X., Bao, M. and Li, Y. 2020. Construction of a hydrophobic magnetic aerogel based on chitosan for oil/water separation applications. Int. J. Biol. Macromol. 165: 1869-1880.

Yu, B., Bai, Y., Ming, Z., Yang, H., Chen, L., Hu, X., et al. 2017. Adsorption behaviors of tetracycline on magnetic graphene oxide sponge. Mater. Chem. Phys. 198: 283-290.

Yu, M., Li, J. and Wang, L. 2017. KOH-activated carbon aerogels derived from sodium carboxymethyl cellulose for high-performance supercapacitors and dye adsorption. Chem. Eng. J. 310: 300-306.

Yu, M., Han, Y., Li, J. and Wang, L. 2018. Magnetic N-doped carbon aerogel from sodium carboxymethyl cellulose/collagen composite aerogel for dye adsorption and electrochemical supercapacitor. Int. J. Biol. Macromol. 115: 185-193.

Yuan, D., Zhang, T., Guo, Q., Qiu, F., Yang, D. and Ou, Z. 2018. Recyclable biomass carbon@SiO_2@MnO_2 aerogel with hierarchical structures for fast and selective oil-water separation. Chem. Eng. J. 351: 622-630.

Zhang, H., Wang, J., Xu, G., Xu, Y., Wang, F. and Shen, H. 2021. Ultralight, hydrophobic, sustainable, cost-effective and floating kapok/microfibrillated cellulose aerogels as speedy and recyclable oil superabsorbents. J. Hazard. Mater. 406: 124758.

Zhao, L., Xue, F., Yu, B., Xie, J., Zhang, X., Wu, R., et al. 2015. TiO_2–graphene sponge for the removal of tetracycline. J. Nanopart Res. 17(1): 16.

Zhao, Q., Zhu, X. and Chen, B. 2018. Stable graphene oxide/poly(ethyleneimine) 3D aerogel with tunable surface charge for high performance selective removal of ionic dyes from water. Chem. Eng. J. 334: 1119-1127.

Zhao, S., Malfait, W.J., Guerrero-Alburquerque, N., Koebel, M.M. and Nyström, G. 2018. Biopolymer aerogels and foams: Chemistry, properties, and applications. Angew. Chem. Int. Ed. 57(26): 7580-7608.

Zhou, K., Li, Y., Li, Q., Du, Q., Wang, D., Sui, K., et al. 2018. Kinetic, isotherm and thermodynamic studies for removal of methylene blue using β-cyclodextrin/activated carbon aerogels. J. Polym. Environ. 26(8): 3362-3370.

Zhou, S., Hao, G., Zhou, X., Jiang, W., Wang, T., Zhang, N., et al. 2016. One-pot synthesis of robust superhydrophobic, functionalized graphene/polyurethane sponge for effective continuous oil–water separation. Chem. Eng. J. 302: 155-162.

Zhu, F. 2019. Starch based aerogels: Production, properties and applications. Trends Food Sci. Technol. 89: 1-10.

Zhuang, Y., Yu, F., Ma, J. and Chen, J. 2016. Facile synthesis of three-dimensional graphene–soy protein aerogel composites for tetracycline adsorption. Desalin. Water Treat. 57(20): 9510-9519.

Ziegler, C., Wolf, A., Liu, W., Herrmann, A.K., Gaponik, N. and Eychmüller, A. 2017. Modern inorganic aerogels. Angew. Chem. Int. Ed. 56(43): 13200-13221.

Liquid and Crystal Nanomaterials for Water Remediation: Fundamentals, Synthesis and Strategies for Improving TiO_2 Activity for Wastewater Treatment

Shipra Mital Gupta[1,*], Babita Sharma[2] and S.K. Sharma[2]

1. INTRODUCTION

About 71% of the Earth's surface is water-covered. Still, only a minor portion (about 2.5%) is available as freshwater (Karpińska and Kotowska 2019), which can be used for human consumption and industrial purposes. Clean and pure water is essential for our existence as it has several vital roles to play in the human body (Karpińska and Kotowska 2019). Apart from portable water, clean, germ-free, and chemical-free water is also required for various domestic and industrial activities. So, there is a continuously increasing demand for water due to industrialization and an immense increase in population.

Due to urbanization and industrialization, water pollution has reached its maximum in the last few decades (Akerdi and Bahrami 2019, Alshammari 2020). Improper disposal of industrial effluents and poor wastewater management, especially in developing countries, are the major factors that are responsible for intensification in water pollution at a frightening level (Verma and Sarlkar 2018, Jing et al. 2013, Liu et al. 2017). A variety of untreated or improperly treated effluents, like the release of industrial and domestic, effluent wastes, marine dumping, leakage from water tanks, atmospheric deposition, and radioactive waste, are known to be toxic and pose negative impacts on the environment (Akpor et al. 2014, Haseena et al. 2017), resulting in a scarcity of clean water (Mehndiratta et al. 2013, Zhang et al. 2017, Jing et al. 2018, Liu et al. 2017). Since water is a universal solvent, so it is considered the foremost source of infection. According to reports of WHO, about 80% of diseases are water-borne and about 3.1% of deaths occur due to unhygienic water (Pawari and Gawande 2015). It is also reported that in most of the countries, even the water used for drinking does not meet the suggested standards (Khan et al. 2013). So, effective removal of waste, i.e. water remediation, to improve the quality of water has become an immensely challenging task.

Waste-water effluents originate from various types of industries, such as pesticides, textiles, plastics, food, pharmaceuticals, acid-lead batteries, electro-plating designing of batteries, tanneries,

[1] University School of Basic and Applied Sciences, GGSIPU, New Delhi, India.
[2] University School of Chemical Technology, GGSIPU, New Delhi, India.
[*] Corresponding author: shipra.mital@gmail.com.

paper and pulp, agriculture, glass, polishing, photovoltaic cell, metallurgy process, and fabric (Bibi et al. 2016, Verma and Sarkar 2018, Akerdi and Bahrami 2019, Baby et al. 2019, Sarma et al. 2019, Ambaye and Hagos 2020). The significant contaminants in wastewater effluents are nutrients (nitrogen and phosphorus), hydrocarbons organics, microbes, endocrine disruptors, dyes, and heavy metal ions, such as lead, cadmium, zinc, nickel, arsenic, chromium, and mercury. There are several significant problems due to the accumulation of wastewater effluents in water. Eutrophication is one of the foremost sources to diminish the quality of freshwater and marine ecosystems all over the world. It is generally characterized by excess growth of plants and algal due to the increased accessibility of one or more controlling growth factors, which is required for the process of photosynthesis (Schindler 2006), such as nutrient fertilizers, carbon dioxide, and sunlight. Lehtiniemi et al. (2005) reported that algal blooms restrict the penetration of light, reduces growth, and hence cause the die-offs of plants in coastal zones. Moreover, high photosynthesis rates associated with eutrophication will diminish the dissolved inorganic carbon and raise the pH to the highest level in the daytime. The increased value of pH can, in turn, 'blind' organisms that use chemical signals for their existence by damaging their chemosensory capabilities (Turner and Chislock 2010). The contaminations present in effluents, which are stable and non-biodegradable, pose a significant risk to human health (Akerdi and Bahrami 2019, Akpor et al. 2014, Juneja and Chauhdary 2013). There are several diseases that are caused due to the presence of contaminants in water, like cancer, kidney damage, immune suppression, mental confusion, profuse sweating, hepatitis, miscarriages, anemia, encephalopathy, nephritic syndrome, reproductive organ failure, epilepsy, Alzheimer, renal damage, hypertension, infectious diseases (cholera, typhoid, and diarrhea), skin allergies, liver damage, nephritis, and nasal mucous ulcer (Samir and Ibrahim 2008, Akpor et al. 2014, Haseena et al. 2017, Akerdi and Bahrami 2019, Baby et al. 2019). The release of wastewater effluents in water bodies does not merely lead to eutrophication and human health risks. It also promotes greenhouse gas emissions significantly in the form of nitrous oxide and methane.

2. MAJOR WATER POLLUTANTS

A wide variety of organic pollutants are produced as domestic waste, municipal sewage, stormwater, wastewater, and industries like food processing, canning, slaughter-houses, pulp and paper, tanneries, breweries, and distilleries in which organic pollutants are presented in suspended or in the dissolved form (Connell et al. 2006). In sewage, the concentration of organic pollutants is small; however, its volume is large. Stormwater consists of organic pollutants released from vehicles on the surfaces of the road, which flounce into water streams through storm run-off. It can be said that vehicles are the foremost sources of dioxins, polycyclic aromatic hydrocarbons, and petroleum hydrocarbons released into the atmosphere in the form of particulate matter. Other sources of organic pollutants are agricultural activities. The harvesting of crops frequently includes the release of pesticides into the environment, resulting in the contamination of soil and watercourses. Over the years many disasters of accidental spillage of petroleum have happened to discharge thousands of tons of petroleum into the water bodies (Connell et al. 2006).

According to Zheng et al. (2013), organic pollutants present in wastewater could be classified into several types depending on their biological degradation abilities; these are (i) organometallic compounds, (ii) oxygen, nitrogen and phosphorus compounds, and (iii) hydrocarbons (Connell et al. 2006). The organic pollutants which have a simple structure and good hydrophilicity are easy to degrade in the environment. For example, methanol and polysaccharides may perhaps be degraded in the presence of algae, fungus, and bacteria. The major categories, which are persistent in nature are hydrocarbons and related compounds. The persistent organic pollutants, like polynuclear aromatic hydrocarbons, polychlorinated biphenyls, and dichlorodiphenyl trichloroethane, degrade gradually. Pesticides have been extensively utilized over a number of years, but the concentration of pesticides and toxicity in wastewater is lesser compared to the soluble organic pollutants, which isolate themselves as a deposit and exist for eras and get transported into the water streams and then into the food chain. The persistent organic pollutants are lipid-soluble and mostly neurotoxic, teratogenic, and carcinogenic in nature. As they are persistent, toxic, and transported a long way, these organic pollutants have thus attracted major concern from researchers (Zheng et al. 2013). Almost all the persistent organic pollutants show poor solubility in water and are resistant to decomposition. Hence, persistent organic pollutants remain in the environment for a longer period (Olatunji 2019, Ashraf 2017, Dai et al. 2016). Therefore, it can be said that persistent organic pollutants accumulate within the soil, segregated between sediment and water or exponentially bio-accumulate in the food chain (by factors up to 70,000 fold) because of their great affinity for organic materials (Jones and De Voogt 1999, Jing et al. 2018, Fitzgerald et al. 2007). According to the reported data, these compounds are found in human blood, breast milk, and tissues which might be introduced through the intake of dairy products, fish, and meat (Van den Berg et al. 2006). It is also reported that exposure to persistent organic pollutant compounds, even at very low concentrations, can lead to severe toxic concerns and negative effects on health, like damage to critical organs, decreased pulmonary function, damage to respiratory and immune systems, interference with the hormones leading to miscarriages, cancer, and even death in human and animals (Araújo et al. 2016, Schecter et al. 2006, Wang et al. 2016, Zheng et al. 2013, Di and Chang 2011, Mukerjee 1998). Numerous pollutants present in water that cause pollution, affect living beings, and whose removal is essential can be broadly summarized as follows:

2.1 Phenols and Substituted Phenols

Phenols are chemical compounds that consist of one or more hydroxyl groups (–OH) attached to the aromatic ring (Anku et al. 2017, Zheng et al. 2013, Chen et al. 2020). The phenolic compounds which are present in wastewater may originate from several industries, like a phenolic chemical plant, coking plant, oil refining, detergents, pharmaceuticals, petrochemicals, pesticide synthesis, textiles, refining, pulp and paper, resin manufacturing, insulation material manufacturing, plastics, paint, and phenolic resin manufacturing (Sun et al. 2015, Zheng et al. 2013, Das et al. 2018, Mohamed et al. 2020). These phenolic compounds after entering water have the tendency to undergo alterations into other moieties that can even be more harmful than the original compounds. These alterations are normally due to their interactions with physical, chemical, and biological factors in wastewater (Kulkarni and Kaware 2013). As human lungs quickly absorb phenols, so it can directly pose dangers to human beings. Mostphenols and their derivatives can cause irritation to the eyes, skin, and respiratory tract, irregular breathing, muscle weakness, tremor, and even coma in humans (Villegas et al. 2016). According to the study of Rathna and Nakkeeran (2018), exposure for a longer period to the diluted concentrations of phenol could be the source of a serious burn on the human body, which may also lead to dermatitis. Phenolic compounds also have the potential to reduce the development and generative capability of aquatic organisms (Zheng et al. 2013). Studies on animals have shown that there is a considerable fetal bodyweight reduction, growth retardation, and abnormal development in the offspring (Villegas et al. 2016). Therefore, there is an urgent need to treat wastewater affected with phenolic compounds before its discharge to water bodies.

2.1.1 Polychlorinated Biphenyls

The chemical formula of Polychlorinated biphenyls is $C_{12}H_{(0-9)}Cl_{(1-10)}$. It comprises a light yellow/deep yellow oily liquid with properties, like thermal stability, insulating ability, resistance to acids, oxidation, hydrolysis, and flame resistance (Dai et al. 2016). The unique chemical and physical properties of polychlorinated biphenyls permit them to be used in an extensive range of industrial applications. Polychlorinated biphenyls are extensively utilized as coolant and dielectric fluids in electric motors, capacitors, and transformers (Van Den Berg et al. 1998, Jing et al. 2018, Fitzgerald et al. 2007, Alcock et al. 1994).

Polychlorinated biphenyls are recognized as persistent organic pollutants that are toxic in nature and have harmful effects on human beings and the environment (Lallas 2001). So far, researchers have recognized approximately 209 polychlorinated biphenyl compounds (Jacob 2013, Dai et al. 2016) based on the placement of chlorine atoms in the benzene ring (Dai et al. 2016). Generally, polychlorinated biphenyls are released into the environment through incineration of waste, discharge from the landfills, and leak in the process of transporting and disposing of industrial waste containing polychlorinated biphenyls (Van Gerven et al. 2004, Choi et al. 2008, Kim et al. 2011, Mari et al. 2008, Xing et al. 2005).

According to the reported data, severe developmental problems can be found in children, like hearing loss, low birth-weight, and behavioral disorders at a comparatively high exposure to polychlorinated biphenyls (Gregoraszczuk and Ptak 2013, De-coster and Larebeke 2012, Wang et al. 2016, Golding et al. 2014, Ribeiro et al. 2017, Marques-Pinto and Carvalho 2013, Urban et al. 2014). Studies also indicate the adverse effect of polychlorinated biphenyls exposure to animals, like immune system suppression, liver damage, abnormalities in the development of fetal, non-Hodgkin lymphomas, sarcomas, serum lipids, and enzyme induction (Goncharov et al. 2008, Kogevinas 2001).

2.1.2 Bisphenol A

Bisphenol is a commercially important chemical product and usually produced for its extensive utilization in powder paints, adhesives, paper coating, and thermal paper (Mohapatra et al. 2011, Wang et al. 2019). Since it is utilized on a large scale, it resulted in widespread dispersal in soil, water, sediment, air, and human tissues (Koumaki et al. 2018, Wang et al. 2019). Bisphenol is found in the surroundings either through landfill leachate or wastewater treatment plants (Melcer and Klecka 2011, Wang et al. 2019). Industrial discharges of Bisphenol A into the surrounding are expected to come from the wastewater and washing deposit generated during the manufacture and treating of products (Mohaptra et al. 2011). Identification and quantification of Bisphenol A in wastewater are necessary because it has the potential to cause diseases, like diabetes, cancer, the disorder in brain functioning, obesity prevalence, reduction in sperm count, and immunodeficiency (Mohaptra et al. 2011, Melcer and Klecka 2011, Wang et al. 2019, Moreira et al. 2019). Exposure to Bisphenol A is related to the health risks of the human being, so it has directed to the formation of rules to restrict its usage and production in a few countries or areas, such as the European Union, Canada, and the United state of American (Mohaptra et al. 2011, Chen et al. 2016, Wang et al. 2019). Bisphenol has been recorded as one of the banned micro-pollutants in the drinking water quality standards of Japan and China (Wang et al. 2019).

2.1.3 Nitrophenols

Nitrophenols are broadly utilized industrial organic compounds. These are commonly utilized as intermediates or raw materials in the production of pharmaceuticals, explosives, pigments, pesticides, rubber chemicals, dyes, pigments, and wood preservatives (Shukla et al. 2009, Ahmadimoghaddam et al. 2010). Under natural conditions, the derivatives of phenol are stubborn and toxic stubborn pollutants. It is also reported that phenols are partially biodegradable and hence their elimination is not easy in biological wastewater treatment plants (Costner 1998). Nitrophenols degrade on the surface of the soil and in water; however, the process of degradation takes a longer period even at a low depth of soil and groundwater. So, it is estimated that nitrophenols stay for an extended time in the deep soil of landfill sites than to the surface of the soil and may even stay for an indefinite period (Shukla et al. 2009). Since nitrophenols are carcinogenic in nature, so they are able to pose significant health risks to human beings and animals. These compounds may enter into the surroundings through industrial effluents (dye, textile, and nitrobenzene plants), degradation products of pesticides containing 2,4-dinitrophenol moieties, and spills (Ahmadimoghaddam et al. 2010). Among all the nitrophenols, 2,4-, dinitrophenol is commonly found in industrial effluents detected in agricultural and urban wastes (Shukla et al. 2009). According to the United States Environmental Protection Agency, 4-nitrophenol, 2-nitrophenol, and 2,4-dinitrophenol (2,4-DNP) are categorized as priority pollutants. Additionally there concentration in natural water is restricted < 10 ng/L^{-1} (Ahmadimoghaddam et al. 2010).

2.1.4 Nonylphenols

Textile industry is counted under the most polluting industries, which discharges persistent and high toxic chemicals in the surroundings (Hasanbeigi and Price 2015). According to the reported data, nonylphenol ethoxylates are extensively utilized as washing, auxiliaries in wool scouring, printing, dyeing, detergents, and hydrogen peroxide bleaching in textile industries (Ho and Watanabe 2017). Brigden et al. (2013) reported the existence of nonylphenol ethoxylates in most of the textile products and its lower concentration in the products involves the high release of it into the effluent. Nonylphenols are persistent at different points in the surroundings. Meanwhile, these have a high affinity toward sewage sludge, soil, sediment (Ahel et al. 1994), and also lipids (Ademollo et al. 2008). Therefore, it was found that it mounted up in organisms (Ying and Fate 2006). According to the reported data, nonylphenol ethoxylates have severe effects on the human being as well as on animals since they are indirectly responsible for the endocrine disruptor, especially nonylphenol (Ho and Watanabe 2017). Researchers have found the adversarial effects of nonylphenol on the immune, central nervous system, and reproductive systems of human beings, birds, rats, and fish, possibly aberrations in offspring and embryos (Cosnefroy et al. 2009, Mao et al. 2010, Vosges et al. 2012). According to the study of Kim et al. (2016), Forte et al. (2016) on carcinogenic materials, it was found that nonylphenol has the potential to cause prostate cancer and breast cancer in men and women, respectively. So, in regards to the adverse effects on human beings and on the surroundings, nonylphenol has been categorized as a significant hazardous material under the Directive 2000/60/EC of the European Council and Parliament (European parliament and council 2008). According to their reports, the normal yearly level of nonylphenol must not go beyond 0.3 μg/L.

2.2 Pesticides

The major utilization of pesticides by human beings is to kill undesirable pests or say organism and for the preservation of crops (EPA 2009, Rani and Shanker 2017). On the basis of their effect on the targeted pest, the majorly used pesticides are herbicides and insecticides and tracked acaricides, fungicides, nematicides, rodenticides, and molluscicides (EPA 2009, London and Meyers 1995). Pesticides can be chemically categorized as (i) organochlorines, (ii) organophosphates, (iii) substituted urea, and (iv) carbamates. Among all the listed pesticides, organochlorines are the utmost hazardous persistent organic pollutants and pose a severe threat to human beings as well as to the environment.

2.2.1 Organochlorines

Organochlorine pesticides were utilized widely to improve agriculture yields and to control pests. However, these are classified under persistent organic pollutants (Rani et al. 2017, Rani and Shaner 2017). According to the reports of Rani and Shanker (2017), merely 10-15% of the total amount of the applied pesticide truly reached the main objective and the leftover amount was distributed into the surroundings, which further damaged its quality. Rani et al. (2017) reported that dieldrin, dichlorodiphenyltrichloroethane, methoxychlor, benzene hexachloride, lindane, and chlordane are the main represented compounds that are included under organochlorines. Among all the organochlorines, use of most is restricted all over the world because these cause adverse effects on the health of animals as well as on human beings. Rani and Shanker (2017) reported that a very low concentration of organochlorines in water may be challenged for the health of human beings. These organochlorines can mount up in biota, get biomagnified via the food chain causing toxicity with the endocrine disruptor potential (Beard 2006, Igbedioh 1991, Rani and Shanker 2017). According to Varma and Varma (2005), during the manufacturing of pesticides and their formulation, chances of health hazards are very higher. Furthermore, numerous pesticides stay for an extended period (often many years) into the environment; for example, Kafilzadeh (2015) projected that the half-life period of g-HCH in water is 191 days. Cortes and Hites (2000) reported that the average half-life of a-HCH and g-HCH nearby the Great Lakes region was found in the range of 3 to 4 years. Additionally, organochlorines display long-range transport properties, so these can be carried to long distances by means of water and air. For instance, endosulfan and HCH were mainly transported to the estuary from the farmlands under a tropical regime that eventually causes a severe risk to the aquatic biota (Leadprathom et al. 2009).

2.2.2 Organophosphorus

Organophosphorus pesticides belong to the group of chlorinated hydrocarbons, which are utilized throughout the world, especially in the agriculture sector (Priyadharshini et al. 2017, Rani et al. 2017, Rani and Shanker 2017). Organophosphorous are the derivatives of esters of phosphorothionic acid or phosphoric acid that are widely used as insecticides. Approximately 200 different chemical structured organophosphorus pesticides are commercially available, articulated in form of sprays, liquids, and powders (Singh et al. 2011, Chambers et al. 2001). Pesticides have caused severe difficulties to human health in many developing countries (Turabi et al. 2007). Various studies have described an increased risk of respiratory problems, like asthma, wheeze, and chronic bronchitis among farmers or other agricultural workers (Kimbell-Dunn et al. 2001, Hoppin et al. 2006). According to the reports, both disruptive and obstructive lung dysfunctions are reported in organophosphorus poisoning (Senanayake and Karalliedde 1987, Pope 1999). Zheng et al. (2013) reported that the wastewater of industries that manufacture organophosphorus pesticide and wastewater from the farmland contains organophosphorus and causes severe environmental pollution.

2.3 Formaldehyde

Low volatile compounds are organic chemicals having high vapor pressure at room temperature. At high vapor pressure, organic chemicals have a low boiling point, which causes a large number of molecules to evaporate from liquid/solid form and then enter into the surrounding air. Formaldehyde is universally found in the environment as it is formed largely by numerous natural sources and anthropogenic activities. The main sources of formaldehyde are organic synthesis, electroplating, petrochemical industries, pharmaceutical, municipal wastewater, synthetic fiber, dyestuff, nitrite effluent, textile dyes, agroindustries, wood processing, pulp and paper, municipal waste-water, paper and pulp waste-water, and oil mill waste-water (Zheng et al. 2013, Ezhilkumar et al. 2016). Although it is not very toxic, it has long-term compounding health effects. Its effects can range from a simple headache to nausea and chest tightness. Formaldehyde is a stimulus to skin and mucous membrane. It could enter the central nervous of humans and cause retinal damage (Zheng et al. 2013). At low concentrations up to 2.0 ppm, it causes eye irritation; at high concentrations ranging from 5 to 30 ppm, it causes pulmonary effects and carcinogenic problems (Ezhilkumar et al. 2016).

2.4 Organic Dyes

In the middle of the nineteenth century, all the significant colors were utilized in industries by means of natural dyes. Later, with the discovery of Perkin's, the improvement in the commercial colorants was so fast that it resulted in >90% of synthetic dyes being used in industries and that happened within 50 years (Gordon and Gregory 1983). According to the reported data, 10,000 kinds of dyes are produced (utilized in various industries, like textiles, tanneries, rubber plastics, paper, cosmetics, leather, and paints) with a production of >700,000 tons are annually produced worldwide (Pereira and Alves 2012, Peng et al. 2016, Hassan and Nemr 2017, Moussavi and Mahmoudi 2009). The extensive utilization of synthetic dyes is due to the fact that synthetic dyes have excellent tinctorial strength, easy availability of raw material, economical, easy preparation, capability to cover the entire shade range, and good fastness properties (Shanker et al. 2017). Generally, dyes are soluble organic compounds that are categorized based on their origin (natural and synthetic), nature of their respective chromophores (Azo dyes, Nitro and nitroso dyes, Triarylmethane dyes, Indigo dyes, and Anthraquinone dyes), and their method of final application (direct dyes, Martius yellow, disperse dyes, Mordant dyes, vat dyes, and Azoic dyes).

About 12% of the entire world production is lost through different processing operations, and around 20% of the lost dye go into the environment by the direct discharge of untreated effluent and release of toxic chemicals into the aquatic environment, which causes adulteration and poses a severe threat to aquatic living organisms (Pereira and Alves 2012, Gita et al. 2017, Hassan and Nemr 2017, Demirbas 2009, Varma et al. 2020) as it reduces the amount of oxygen in the water body due to the presence of hydrosulfides as well as reduces the penetration of light and thus adversely affect the photosynthetic activities of aquatic flora and fauna (Kant 2012, Gita et al. 2017, Shanker et al. 2017). Most of these dyes show high solubility in water, making it problematic to eliminate them from wastewater (Hassan and Carr 2018). According to Shanker et al. (2017), the extensive use of dyes and their untreated discharged dyes (even at very low concentration) imparts an intensive color to the effluents and is noticed to the primary source of water pollution. In the textile industry, the dyeing process is an important step. According to the literature, excess chemicals and dyes are found in the effluents of textile industries (Carmen and Daniela 2012). So, the textile industries' effluent has received considerable attention due to its unacceptable toxicity and appearance (Yaseen and Scholz 2019, Chen et al. 2020). Even after its breakdown, their dumping has resulted in serious contamination of surface water, sediment, and adjacent soil, which is becoming the foremost worldwide environmental pollution challenge (Hassan and Nemr 2017, Yaseen and Scholz 2019, Chen et al. 2020). Additionally, dyes contribute adverse effects on human beings as well as on animals (Pereira and Alves 2012, Shakoor and Nasar 2016). Approximately 40% of worldwide used dyes contain organically bound chlorine, which is carcinogenic in nature (Sivakumar and Palanisamy 2008, Sarma et al. 2019). In textile industries effluent, heavy metals are also present which are non-biodegradable; hence, they accumulate in primary organs in the human body and over time begin to fester, leading to different symptoms, like irritation, abdominal discomfort, diarrhea, vomiting, and nausea (Hassan and Nemr 2017, Shakoor and Nasar 2016). Dyes can sustain for a longer period in the environment (Pereira and Alves 2012) due to their high thermal stability toward heat and photo-stability to resist biodegradation (Gita et al. 2017, Shanker et al. 2017). As dyes adversely influence the surroundings via the generation of a huge amount of secondary pollution, highly toxic effluent and persistent behavior, motivated researchers to design and develop an effective treatment for the removal of dyes (Sachdeva and Kumar 2009, Tan et al. 2008, Ghaedi et al. 2011) to save the environment.

2.5 Nitrobenzene

Chemical formula of nitrobenzene is $C_6H_5NO_2$. It is not naturally produced. It is manufactured in industries for its use in the production of aniline in medications, dyes, explosives, pesticides, rubbers, synthetic resins, paint and polishes, insecticides, herbicides, perfumes, and pharmaceuticals (Wen et al. 2012, Lv et al. 2013, Pan and Guan 2010, Xie et al. 2018). According to the report of Torosyan and Simonyan (2019), the production of nitrobenzene in the world is approximately 2 million tons per year. Nitrobenzene enters the environment mainly from the wastewater of plants, wastewater aniline dye as well as insulating and glossy materials (Shuxin et al. 2004, Ersoy and Celik 2004). It is very important to treat industrial effluents before disposing them into wastewater bodies due to the presence of nitrobenzene, which is declared as a priority pollutant by many of the countries (Liang et al. 2007, Elshafei et al. 2014). Inappropriate treatment of nitrobenzene present in wastewater will seriously threaten human health (Xie et al. 2018) due to its strong mutagenic and carcinogenic effects in humans (Elshafei et al. 2014). It is reported that nitrobenzene causes a serious hazard to the environment mainly because of two reasons (1) it is recalcitrant to chemical/biological oxidation and hydrolysis due to electron-withdrawing effect of nitro-group, (2) it hinders the metabolism of micro-organisms and hence bio-degradation of nitrobenzene is usually tough to attain (Hung et al. 2000, Klausen et al. 2001, Mu et al. 2004, Xu et al. 2005). Nitrobenzene is responsible for damage to the nervous system as well as to the red blood cells via the formation of methemoglobin, hence causing hemolytic anemia, hepatotoxic, and nephrotoxic effects (Liang et al. 2007, Torosyan and Simonyan 2019). Prolonged exposure to nitrobenzene may damage vision, cause serious injury to the central nervous system, anemia, lung irritation, and produce liver or kidney damage (Ju and Parales 2010). Nitrobenzene is stable due to the presence of nitro-group, so the treatment of wastewater effluent by conventional biological methods is often inadequate to eliminate it (Xie et al. 2018). Although various methodologies like ozonation, electro-chemical reduction, Fenton oxidation, and ultrasonic irradiation have been projected for the elimination of nitrobenzene from the industrial effluent (Yang et al. 2016, Jiang et al. 2011, Subbaramaiah et al. 2014, Zhao et al. 2015), the high operating and maintenance costs and difficulty in handling the features of these methodologies condense their practice in economically concerned regions (Yang et al. 2018).

2.6 Petroleum Hydrocarbons

Raw materials and energy resources that are produced from petroleum hydrocarbons are crucial for several industries (Chen et al. 2020). Contamination due to crude oil is very common due to its widespread usage and its related dumping process and accidental spills (Srivastava et al. 2019, Chen et al. 2020). The introduction of petroleum hydrocarbons into a pristine environment directly alters its nature (Truskewycz et al. 2019). Pollutants of petroleum hydrocarbon are unmanageable compounds that are categorized as significant environmental pollutants (Costa et al. 2012) due to their persistence and recalcitrant nature (Gennadiev et al. 2015, Jesus et al. 2015, Dindar et al. 2013). Petroleum hydrocarbon forms a waterproof thin layer onto the surface of the water that further prevents the exchange of oxygen between air and water bodies and its weathering process consists of spreading, evaporation, dissolution, dispersion, and emulsification, while the higher molecular fractions sink to the bottom of the water (Mishra and Kumar 2015, Srivastava et al. 2019). The occurrence of petroleum hydrocarbon pollutants in water causes a substantial impact on the environment and poses a significant hazard to all forms of life (Sammarco et al. 2016, Hentati et al. 2013). To develop suitable, environmental-friendly and cost-effective methods that can be fast, practical, and adaptable in many physical locations for restoration and reclamation of the affected areas, researchers are studying the distribution, environmental providence, transport, and chemical properties of the different types of petroleum hydrocarbon pollutants in the affected areas (Fallgren and Jin 2008, Stroud et al. 2009, Costa et al. 2012).

3. GENERAL SCHEME OF WASTEWATER TREATMENT

The main motive of wastewater treatment plants is to remove pollutants followed by appropriate management of resources and prevention of secondary pollution. Generally, wastewater treatment plants are as shown in Figure 1, comprising five successive steps that are (1) pre-treatment or preliminary treatment which involves physical and mechanical processing of wastewater, (2) primary treatment which involves physico-chemical and chemical processing of wastewater, (3) purification or secondary treatment which involves the chemical and biological process of wastewater, (4) final or tertiary treatment which involves further chemical and physical, and (5) sludge treatment which involves recycling and incineration (Crini and Badot 2010, Anjaneyulu et al. 2005).

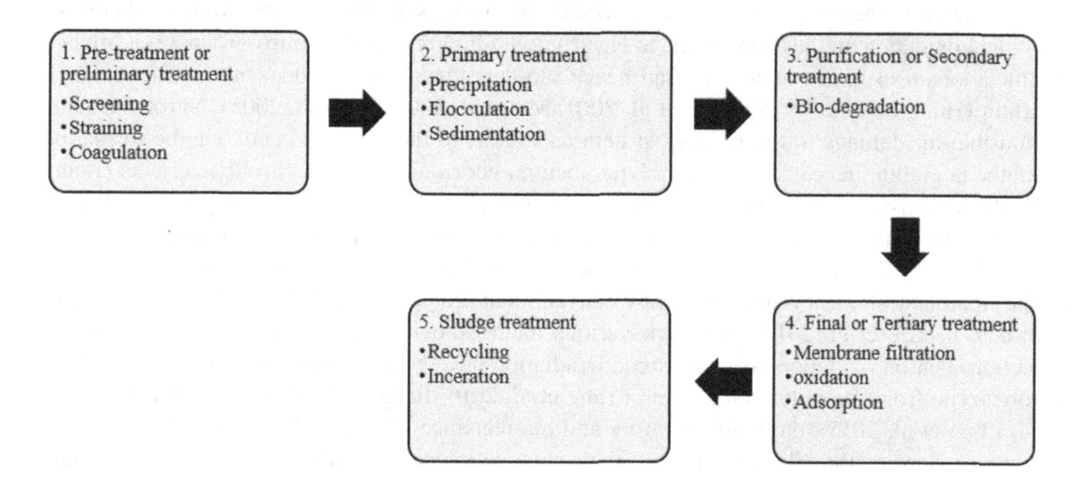

FIGURE 1 Five successive steps for the purification of wastewater.

3.1 Pre-Treatment or Preliminary Treatment

Pre-treatment is mainly utilized to protect the pumping equipment and to support the achievement of successive treatment steps. Pre-treatment devices, for example, screens for the removal of stones or pebbles are intended and employed to get rid of the big-sized floating and suspended solids or dense materials, which damage the pumping equipment. At times, the froth flotation technique is also utilized for the removal of extreme grease or oils present within the wastes.

3.2 Primary Treatment

Generally, a simple sedimentation method which is a physical process, is utilized for the removal of settleable solids from the wastewater. During this process, the horizontal velocity of wastewater is kept at a level that offers satisfactory time for the settling of solids and elimination of the floatable matter that is present on the surface of the wastewater. Thus, steps involved in primary treatment comprises settling and flotation tanks, which direct the supernatant to the microbiological treatment units and separate solids to the digestion units.

3.3 Secondary or Biological Treatment

It utilizes bacterial action under variable growth conditions to bio-chemically degrade the organic compounds present within the wastewater, which passed from the primary treatment units. For the biological treatment of wastewater, an arrangement of reactors is employed which comprise suspended biomass, biofilm, fixed-film reactors, and lagoon or pond systems.

Through the transformation of waste carbon to the new cells, most of the biological treatment processes produce an excess of biomass. So, before final treatment, such as removal of nutrients or disinfections, solids are essentially separated from the effluent of secondary treatment. This is generally accomplished through settling. Separated solids are either reprocessed back to the head of the process train or directed to the digesters for the reduction of solids and further processing, based on the type of digester.

3.4 Final or Tertiary Treatment

Tertiary or final treatment comprises processes that accomplish the higher quality of effluent compared to conventional secondary treatment techniques. These processes consist of fine steps like ion exchange, activated carbon adsorption, reverse osmosis, electro-dialysis, nutrient removal, and chemical oxidation. Even though technically not a tertiary process, actually final effluent disinfection process is frequently utilized after secondary treatment by means of ultraviolet methods, chlorination, ozonation, and other methods which are designed especially to destroy the organisms remaining in the wastewater after the application of preceding treatment steps.

3.5 Sludge Treatment

Residue which accumulates during the treatment of wastewater is called sludge. Sewage sludge in waste-water is present generally in the form of slurry, solid, and semi-solid, which is formed as a by-product of waste-water treatment techniques. The residue is usually categorized as primary or secondary sludge. Primary sludge is produced from sedimentation, chemical precipitation, or other primary processes, while secondary sludge is produced from biological treatments.

Treatment and discarding of sludge are the main concerns in designing and operating all the wastewater treatment plants. Basically, there are two major objectives of treating sludge before its final disposal that is (i) to condense its volume and (ii) to stabilize the organic materials. The lesser volume of sludge lessens the cost of storage and pumping. The stabilized sludge does not have a disagreeable odor, and it can be held without causing irritation or any health hazard. Treatment of sludge may consist of a combination of dewatering, digestion, and thickening processes.

4. TECHNOLOGIES AVAILABLE FOR REMOVAL OF WATER POLLUTANTS

Identification of pollutants and efficient elimination of pollutants from the wastewater is a significant and challenging task (Baby et al. 2019). Owing to the scarcity of pure water, special treatment and management of wastewater have been executed for several years due to its importance. Though approaches have changed and evolved with time. Numerous researchers have tried several treatment techniques, developed gradually with time comprising chemical, physical, and biological approaches (Ahammad et al. 2013, Chen et al. 2020). Physical treatment of wastewater is based on the application of physical forces, like sedimentation, mixing, floatation, filtration, and screening. On the other hand, chemical treatment of wastewater is achieved by adding some kind of chemical or by stimulating specific chemical reactions. Few examples of chemical treatment are precipitation, ion exchange, coagulation, adsorption, disinfection, and extraction (Ahammad et al. 2013, Ucankus et al. 2018). Physical and chemical methods are frequently combined and called physio-chemical processes which are utilized in industries. Apart from chemical and physical processes, removal of organic contaminants is utilized through microbiological activities, such as aerobic treatment in ponds, lagoons, and trickling filters (Tchobanoglous et al. 2003) and anaerobic treatment (Lettinga 1996) in similar reactor systems. Different wastewater treatment techniques along with their main characteristics, advantages, and disadvantages are listed in Table 1 (Barakat 2011, Mohan and Pittman 2007, Sharma and Sanghi 2012, Chen et al. 2020).

Membrane separation is a physical method to treat wastewater (Barakat 2011). Separation mechanisms usually depend on solubility, particle size, charge, and diffusivity. It poses merits like being simple, effective, and rapid removal even at high concentrations of various types of pollutants, such as solids, microorganisms, inorganic matter, and phenols (Strong and Burgess 2008). Although most of the commercial membranes are not considered semi-permeable, membranes with the smallest pore size and reverse osmosis are not proficient in restricting solute, especially for low molecular weight non-charged organics (Bellona et al. 2004). Additionally, despite high efficiency at laboratory scale, it fails at commercial scale for small and medium scale industries as it requires a high cost of investment, including maintenance and operation costs. Moreover, low throughput or low flow rates due to clogging or fouling problems may limit some applications for membrane separations at high concentrations of pollutants (Chen et al. 2020).

Conventional processes like adsorption, precipitation, coagulation, and sedimentation usually have the following disadvantages that are (a) non-destructive so it usually produces a large amount of slug which need to be treated further, (b) removal rates are slow so it requires a large storage tank (Srinivasan and Viraraghavan 2010), and (c) changing from one organic removal compartment to another can greatly reduce the efficiency (Klauck 2017). For example, the adsorption method is preferred due to its simplicity and low cost (Kalia et al. 2014). However, the adsorbents used in the adsorption technique generally have low adsorption capability along with low selectivity (Amin 2013, Dean et al. 1972). Chemical precipitation is a widely utilized technique in order to remove several types of pollutants present in wastewater. However, precipitation has the demerit of generation of a high amount of sludge, excessive usage of chemicals, such as lime, oxidants, or H_2S, which has become an issue of secondary pollution (Hena 2010, Crini et al. 2019, Sharma and Sanghi 2012).

The selection of the techniques is dependent on the characteristics of wastewater (Anjaneyulu et al. 2005). Also, each technique has its own merits, demerits, in terms of its effect on the environment, efficiency, economy, and feasibility (Crini and Lichtfouse 2019). According to the current scenario, there is not a single method that has the capability of satisfactory treatment of water, which is basically because of the complex behavior of the effluents. Generally, an arrangement of various techniques is widely utilized to attain the desired quality of water in the best cost-effective manner. Many studies have been undertaken to consider the usage of new techniques and satisfy the weaknesses of former techniques (Akerdi and Bahrami 2019). Several researchers have considered that advanced oxidation processes pose several advantages over other conventional wastewater treatment techniques because in this process there is complete mineralization of organic pollutants within a very short period, no production of secondary pollutants, a minute or negligible consumption of chemicals, rapid degradation of pollutants, even in case of industrial wastewater having low concentrations. It is environmentally friendly, low process cost, high energy savings, and possesses the ease of operation (Chen et al. 2020, Pawar et al. 2018, Varma et al. 2020). Advanced oxidation processes are a kind of destructive technique. It has gained special consideration since it can eliminate organic pollutants by converting them into harmless species or even better to CO_2. Advanced oxidation processes include the production of highly oxidizing species, like hydroxyl radicals ($^{\bullet}OH$) capable of rapidly reacting with a variety of organics. Advanced oxidation processes usually improve their efficiency by utilizing an appropriate catalyst or by the application of some kind of energy, like visible light or ultraviolet (Machulek et al. 2013, Rueda-Marquez et al. 2020).

TABLE 1 List of advantages and disadvantages of different waste-water treatment techniques.

S. No.	Technique	Main Characteristics	Advantages	Disadvantages
1	Chemical precipitation	Applicable for numerous pollutants and separation of chemical is must from the final products	• Technologically simple • Integrated physico-chemical process • Efficient and cost-effective • Its performance is better even under high concentration of pollutants • Effective for the elimination of fluorides and metals • Substantial lessening in chemical oxygen demand	• A high amount of chemicals are required • Required to monitor the value of pH of the effluent regularly • Inefficient for the elimination of metal ions to a very less concentration • Oxidation becomes necessary if existing metals within the wastewater are in the complex form • Produces high amount of sludge, then its handling and disposal problems
2	Coagulation/ flocculation	Neutralization of suspended particles	• Technologically simple process simplicity • Integrated physicochemical process • Extensive variety of chemicals are accessible, commercially • Cost-effective • Very effective for colloidal particles and sewage sludge too • Substantial lessening in biochemical and chemical oxygen demand • Better ability for bacterial inactivation ability • Effective and quick elimination of contaminants which are insoluble in water	• Involvement of such chemicals which are non-reusable • Required to monitor the value of pH of the effluent regularly • Produces high amount of sludge, then its handling and disposal problems
3	Flotation/ Froth flotation	Separation process	• Integrated physicochemical process • Very effective for the elimination of tiny and low-density particles • Metal selective • Short retention time • In the pulp and paper industry, this method is utilized as an effective tertiary treatment	• Expensive • Requirement of chemicals to regulator the relative hydrophobicity between particles and to retain appropriate froth characteristics • Selectivity is pH-dependent
4	Chemical oxidation	Utilization of oxidants	• Integrated physicochemical process • Quick, technologically simple, and effective • No storage-related hazards	• Pre-treatment is a crucial type of oxidant that greatly influence the efficiency • Utilization of chemicals • Short half-life

Contd.

			• Better reduction of odor and color • Very effective for the treatment of sulfide and cyanide • Intensify the biodegradability of the products • No production of sludge	• Production of intermediates • No reduction in chemical oxygen demand • Discharge of aromatic amines and volatile compounds • Production of sludge
5	Biological methods	Utilization of biological cultures	• Simple, cost-effective, and accepted • Huge number of species are utilized in pure (white-rot fungus) and mixed cultures (consortiums) • Pure culture produces an extensive range of extracellular enzymes having great biodegradability ability • Effectively removes NH^{4+}, NH_3, iron, and biodegradable organic matter • Great efficiency for the elimination of suspended solids and biochemical oxygen demand • Conclusive role of microbiological processes in the upcoming technologies utilized for the elimination of developing impurities from water	• It is mandatory to construct favorable conditions • Requirement, and proper maintenance of the microorganisms • Process is slow • Possiblity of sludge foaming and bulking • Production of biological sludge • Composition of mixed cultures possibly changed during the process of decomposition • Difficulty in the microbiological mechanisms • Necessary to have a good knowledge of the enzymatic processes leading to the degradation of the pollutants
6	Adsorption/ filtration	Nondestructive process use of a solid material	• Technically simple and acceptable to several treatment formats • Numerous pollutants can be eliminated • Very efficient process • Treated effluent is of exceptional quality • Comprehensive removal of pollutants however probably selective dependent on the type of adsorbent • Highly effective treatment when attached with coagulation to eliminate color, chemical oxygen, and chemical oxygen demand	• It is an expensive treatment • Non-selective and non-destructive processes • Performance is dependent on the type of absorbent • Numerous requirement of adsorbents • For the improvement of adsorption ability of the absorbent chemical derivatization is required • Quick saturation and blockage of reactors • Regeneration is required but it is an expensive process • Not applied in industries like textile and pulp and paper due to cost ineffectiveness

Contd.

7	Ion exchange	Non-destructive process	• Technically simple • Easy to control and low maintenance • Easy to utilize with other techniques • Regeneration ability is high • Quick and effective process • Treated effluent is of high quality • Concentrates mostly all kinds of pollutants • For certain metals, it can be selective • Effective technology to recover ceratin valued metals	• Expensive • Regeneration process is time-consuming • Quick saturation and blockage of reactors • Saturation of the cationic exchanger before the anionic resin (precipitation of metals and blocking of the reactor) • Beads effortlessly contaminated by organic matters and particulates • Requirement of physicochemical pre-treatment for example carbon adsorption or sand filtration • Matrix damages with the passage of time and with the production of certain waste materials • Effluent is pH-dependent • Not efficient for a few of the targeted pollutants • Removal of resin is required
8	Incineration Thermal oxidation	Destruction by combustion	• Technically simple • Beneficial for concentrated sludges • Extremely effective • Removes usually all kind of organic pollutants • Generation of energy	• Relatively expensive • Storage and transportation of the effluents is required • High running costs • Generation of secondary pollutants • Local population all the time opposed the occurrence of incinerating plant in the vicinity
9	Electro-coagulation	Electrolysis	• Effective technology to recover the valued metals like silver and gold • Adaptation to various concentration of pollutants and different fowling rates • Better biodegradability Highly efficient and quick separation of organic matter compared to traditional coagulation • Controlling pH is not essential • *In situ* production of coagulants	• Relatively expensive • Addition of chemicals is required like flocculants, coagulants, and salts • Passivation of anode and deposition of sludge onto the electrodes prohibits the electrolytic process in continuous operation • Involves the post-treatment for the removal of high concentrations of iron and aluminum ions • Separation efficiency is dependent on the size of bubbles

Contd.

			• Economically practicable and efficient for the removal of dissolved metals, suspended solids, dyes and tannins • Effective removal of SS, grease, oils, metals and colors • Extensively utilized in miming industries • Very efficient in treating drinking water supplies for small and medium-scale communities	• Production of sludge
10	Membrane filtration	Non-destructive separation	• Extensive variety of commercial membranes are available in market • Huge number of module configurations and applications • Requirement of small spaces, effective and quick, even at high loads of pollutants • Generates a high-quality-treated effluent • No requirement of chemicals • Production of solid waste is low • Removes all kinds of mineral derivatives, salts, and dyes • Effective removal of volatile and non-volatile organics, phenols, suspended micro-organisms, solids, zinc, cyanide, and dissolved organic matter • An extensive variety of applications such as multivalent ions, separation of polymers, salts from polymer solutions, clarification or sterile filtration, desalination and production of pure water • Well-known separation mechanisms are available: size exclusion (Microfiltration, ultrafiltration, and nanofiltration), solubility/diffusivity, charge	• Expensive for small and medium industries • High requirement of energy • Significant differences in the designing of membrane filtration systems • High operation and maintenance on costs • Quick clogging of the membrane • Low output • Flow rate is low • Ineffective at low solute feed concentrations • Selection of membrane is dependent on the particular application and removal of concentrate

Contd.

11	Liquid–liquid extraction	Separation technology	• Well established separation technology for recycling of wastewater • Mainly utilized where the concentration of the pollutants is high, i.e. large-scale operations • Easy to perform • Simple monitoring and control of the process • Economically feasible when both waste-water flow rates and solute concentrations are high • Low operating costs • Recyclability of extractants • Selectivity of the exchangers for effective removal of metal • Effective for the separation of phenol	• Expensive • Utilizes large volumes of organic extractants • Hydrodynamic restrictions • Poor quality of effluent due to entrainment of phases • Cross-contamination of aqueous stream • Poor separation due to emulsification of phase • Fire risk due to the emissions of volatile organic compounds • Utilization of toxic solvents • Not utilizes when at low concentration of solute feed
12	Advanced oxidation processes	Destructive technique	• *In situ* generation of reactive radicals • Negligible consumption of chemicals • Complete degradation of pollutants • No sludge production • Quick elimination of pollutants • Effective for intractable molecules • Elimination of total and chemical oxygen demand • Appropriate technology for the effluent which is too concentrated for biological treatment and too diluted for the incineration • Degradation of phenol in water solution • Insoluble organic matter is transformed into simpler soluble compounds without emissions of hazardous substances	• Utilizes at laboratory scale • Practically not economic for small and medium industries • Technical limitations • Productiontion of by-products • Low output • Requirement of energy-intensive and high-pressure conditions • pH dependency • Sometimes total degradation of pollutants not achievable

5. NANOMATERIALS AS PHOTOCATALYST

According to the literature, the utilization of a photocatalyst for water purification is done through an advanced oxidation process (Deng et al. 2012, Chen and Lu 2007, Lin et al. 2020, Mashuri et al. 2020). It possesses several advantages, such as it does not require non-renewable energy, it is low-cost, rapid, and completely degrades different types of pollutants present in wastewater. Most importantly, the process of photocatalysis is able to overcome most of the drawbacks of several other wastewater treatment techniques. For example, techniques like liquid-liquid extraction, membrane filtration, electro-coagulation, and ion exchange relatively require high cost while the process of photocatalysis requires very low cost for their operation. Chemical precipitation techniques require a high amount of chemicals for their operations, while in photocatalysis there is no or very little use of chemicals. There is the rapid degradation of pollutants in comparison to biological treatment techniques since biological methods are very slow. Generally, photocatalysis is classified into two major classes: (1) homogeneous catalysis and (2) heterogeneous catalysis. In the process of homogeneous catalysis, the reactants, as well as the photocatalyst, exist in the same phase while in the process of heterogeneous catalysis, the catalyst exists in a different phase from the reactants. Heterogeneous catalysis comprises a huge variety of reactions like hydrogen transfer, dehydrogenation, partial or total oxidation, and metal deposition (Kumar and Pandey 2017a). The area of heterogeneous photocatalysis has extended rapidly within the last four decades and has gone through numerous developments covering an extensive range of environmental applications like water remediation, solar water splitting, and environmental clean-up of oil spills and other pollutants (Ibhadon and Fitzpatrick 2013, Pawar et al. 2018, Mohamed et al. 2020). The photocatalysis process may be termed as a photoreaction accelerated in the presence of a catalyst, which is called a photocatalyst (Ibhadon and Fitzpatrick 2013, Mills and Hunte 1997). A photocatalyst is a kind of material that helps to quicken and augment the light-induced reaction without being spent in the process (Lee and Gouma 2012). Photocatalysts have the capability to transform light into chemical energy over a series of surface reactions and electronic processes (Hernandez-Alonso et al. 2009). An ideal photocatalyst should be non-toxic, inexpensive, highly photoactive, and easily available (Kumar et al. 2014). Generally, photoreactions are triggered by the absorption of a photon having a sufficient amount of energy (equal to or higher than the band-gap of the catalysts) (Carp et al. 2004, Kumar et al. 2014). Most of the photocatalysts are semiconductors and their high stability and mobility of charge carriers in photocatalysts enables their transport to the surface where they can interact with the adsorbed molecules of pollutants (Carp et al. 2004).

5.1 Nanoscale Semiconductors

Semiconductors are those materials whose valence and conduction band are separated by band-gap energy (Akpan and Hameed 2009). The separation between the valence band and conduction band is called band-gap energy (E_{gap}) (Tiwari et al. 2019). The foremost condition for a photocatalyst to be effective is the redox potential of charged electrons and holes should lie in the band-gap range of the photocatalyst. The energy level at the top of the valence band defines the oxidizing capability of photogenerated holes whereas the energy level at the bottom of the conduction band defines the reducing capability of photoinduced electrons (Carp et al. 2004).

Several properties of nano-crystalline semiconductor particles are different from those of bulk materials. Surface adsorption, as well as photocatalytic reactions, can be enhanced by nano-sized semiconductors because of reduced size. Reduced size of catalysts means availability of high specific surface area which leads to enhanced adsorption of pollutants molecules on the surface of the photocatalyst ultimately enhanced the rate of degradation of pollutants. Upon irradiation, a semiconductor absorbs photons having energy equal to or greater than its band-gap which leads to the separation of charge due to the promotion of an electron from the valence band to the conduction band hence generating a hole in the valence band (Gaya and Abdullah 2008, Kumar et al. 2014). The excited semiconductor nanoparticles go through charge separation and oxidize the organic pollutants at their surface (Hoffmann et al. 1995, Jain and Vaya 2017). It is reported that the use of semiconductor nanomaterials for wastewater treatment possesses several advantages over conventional technologies since nano-sized semiconductors completely oxidize most of the organic pollutants from the wastewater (Stylidi et al. 2003). In order to photodegrade, the pollutants present in wastewater are different semiconductors, like TiO_2, ZnO, Fe_2O_3, CdS, SnO_2, SiC, and ZnS, which are utilized in the form of nanoparticles or nanocomposites (Kumar et al. 2014, Jain and Vaya 2017, Fernandes et al. 2020, Varma et al. 2020, Lin et al. 2020).

Apart from possessing appropriate band-gap energy, an ideal semiconductor should possess properties like (a) excellent photo-activity, (b) biological and chemical non-reactivity, (c) stability toward photo-corrosion, (d) ability to utilize a broad solar spectrum, (e) inexpensive, and (f) non-toxic (Bhatkhande et al. 2001, Fujishima and Honda 1972). The process of electrons and hole recombination in which electron and hole lose their redox ability by generating heat energy is known as photo-corrosion (Pawar et al. 2018). Reported semiconductors like GaAs, PbS, and CdS are not satisfactorily stable in aqueous media as these semiconductors freely undergo photo-corrosion; additionally, these are toxic in nature (Mills and Hunte 1997). Since ZnO is unstable so it gets dissolved in aqueous media very easily (Bahnemann et al. 1987) and yields $Zn(OH)_2$ onto the surface of ZnO particle that deactivates the catalyst over time. Semiconductors, like WO_3, SnO_2, and Fe_2O_3, have a conduction band edge at an energy level under the reversible hydrogen potential. Therefore, the type of systems that utilize this kind of materials involves the use of an external electrical bias to complete the water-splitting reaction and to attain hydrogen evolution at the cathode (Sivula et al. 2009).

Numerous unique photocatalytic heterogeneous reactions have been described in literature at the interface of irradiated TiO_2 photocatalyst (Gupta and Tripathi 2011, Jain and Vaya 2017). Among all the reported semiconductors, TiO_2 has been found very close to being an ideal photocatalyst and set a standard due to its performance (Kumar et al. 2014, Wu et al. 2014, Huang et al. 2016, Andronic and Enesca 2020). TiO_2 has been part of intensive research due to its high photocatalytic activity under UV region and possibly under visible light too. There are numerous fields of study in which the photocatalytic activity of TiO_2 nanoparticles have been discussed, such as water remediation, photocatalytic water splitting, photocatalytic self-cleaning, and photovoltaics (Jain and Vaya 2017, Fujishima et al. 2008, Cho et al. 2004, Alrousan et al. 2012, Navntoft et al. 2007, Gamage and Zhang 2010).

6. TIO_2 NANOMATERIAL AS A PHOTOCATALYST

TiO_2 is a well-established photocatalyst due to its dielectric constant, outstanding incident photoelectric conversion efficiency, high oxidation efficiency, non-toxicity, long-time corrosion resistance, high photostability, chemical inertness, eco-friendly, high refractive index, extraordinary physical, optical, and electrical properties (Kumar et al. 2014, Dong et al. 2015, Huang et al. 2016, Tiwari et al. 2019). Additionally, TiO_2 has suitable band-gap energy of 3.20 eV, capable of absorbing photons of wavelength <350 nm (Kanan et al. 2019, Tiwari et al. 2019).

Synthesizes of nano-TiO_2 is done in several sizes and shapes like nanoparticles (Holland et al. 1999), nanotubes (Wang et al. 2002, Poudel et al. 2005), nanowires (Poudel et al. 2005), inverse opals (Richel and Johnson 2000), nano-ribbons (Yuan et al. 2002) and nano-sheet arrays (Li et al. 2013). Normally, nano-TiO_2 is synthesized by utilizing several titania precursors, like titanium tetra-iso-propoxide (Nagamine et al. 2008), tetrabutyltitanate (Yu et al. 2007), and titanium tetrachloride (Lee and Yang 2005).

Due to the high specific surface area, TiO_2 nanoparticles offer an enhanced rate of light absorption; the increased surface photo-induced carrier density in TiO_2 nanocatalysts leads to higher surface photoactivity and augments photocatalytic activity (Gupta and Tripathi 2011, Kanan et al. 2019, Jain and Jaya 2017). It is able to mineralize a wide range of pollutants, like pesticides, dyes, herbicides, and phenols under UV radiation (Lee et al. 2015). TiO_2 was discovered in 1795 and as a naturally occurring mineral (Gupta and Tripathi 2011, Tiwari et al. 2019). It usually exists in three forms: anatase (tetragonal), rutile (tetragonal) and brookite (orthorhombic) (Gianluca et al. 2008, Tiwari et al. 2019, Kanan et al. 2019). The band-gap energy of anatase is 3.21 eV, rutile is 3.0 eV, and brookite is 3.13 eV, suggesting that all forms of TiO_2 are active under UV radiations (Reyes-Coronado et al. 2018). All the forms of TiO_2 are composed of a Ti^{4+} in the center with adjoining oxygen atoms which act as a bridge and terminal ligands (Guo et al. 2016). The distinctive structure of all the TiO_2 complexes allows the surface vacancies on the Ti^{4+} to behave as active sites for photocatalytic degradation (Ye et al. 2017, Dong et al. 2015, Zada et al. 2018, Pan et al. 2014). It is reported that anatase and brookite are metastable phases while rutile is a stable mineral phase and at higher temperatures, anatase, and brookite transformed into rutile (Tiwari et al. 2018, Pawar et al. 2018). According to the literature, the activity of the rutile phase when it is utilized as a photocatalyst is usually very poor (Zhang et al. 2000). The anatase phase of TiO_2 is found to be a common applicant as photocatalyst due to its stable configuration which is suitable for photo-degradation of pollutants (Kanan et al. 2019). The anatase phase is transformed to the rutile phase only at temperatures exceeding 600°C, indicating that it is the most acceptable candidate for environmental photocatalytic applications due to its high thermal stability (Hu et al. 2017).

Akpan and Hameed (2009) reported that there are several factors due to which TiO_2 is used as photocatalyst at a commercial scale for waste-water treatment:

- Photocatalytic reactions do not undergo photolysis reactions which means there is no production of intermediates because organic pollutants are totally photodegraded to non-toxic substances like water, HCl, and CO_2
- Photocatalysis generally takes place at room temperature

TiO_2 is a low-cost photocatalyst and it can be supported on different types of substrates, like inorganic materials, fibers, glass, sand, activated carbon, and stainless steel permitting their continuous utilization.

6.1 Basic Principle of TiO_2 Photocatalyst

The mechanism which is involved in the heterogeneous photocatalysis utilizing TiO_2 semiconductor as a photocatalyst has been extensively described in the literature (Gaya and Abdullah 2008, Banerjee et al. 2014, Fujishima et al. 2000, Mohamed et al. 2020). Significantly five steps are involved in the heterogeneous photocatalysis, which occurs onto the surface of TiO_2 and these steps are (1) photoexcitation (when the light gets absorbed and the charge carriers are created), (2) diffusion (pollutant, moisture, and oxygen are diffused to the catalyst surface), (3) trapping (water molecules and hydroxyl groups available as electron donors, react with the migrated charge carriers producing highly reactive and strongly oxidizing hydroxyl radical), (4) recombination (excited-state conduction band electrons and valence band holes recombined and dissipate input energy in form of heat, react with electron acceptors and electron donors adsorbed on the surface of semiconductor), and (5) oxidation (holes react with the moisture that is present on the semiconductor surface and produces a hydroxyl radical) (Foo and Hameed 2010, Chong et al. 2010). In literature, photogeneration of radical species of TiO_2 photocatalyst under UV irradiation can be described as follows in Table 2 (Akpan and Hameed 2009, Fujishima et al. 2008):

TABLE 2 The reactions involved in the photocatalysis of organic pollutants (RH) using TiO_2 photocatalyst.

Reactions	Equation No.
Photoexcitation: $TiO_2 + hv(UV) \rightarrow TiO_2(e^-_{CB} + h^+_{VB})$	1
Trapping of charge – carrier e^-: $e^-_{CB} \rightarrow e^-_{TR}$	2
Trapping of charge – carrier h^+: $h^+_{VB} \rightarrow h^+_{TR}$	3
Recombination of electrons – holes: $e^-_{TR} + h^+_{VB}(h^+_{TR}) \rightarrow e^-_{CB} + heat$	4

Contd.

Oxidation of hydroxyls: $(OH^-)_{ads} + h^+ \rightarrow \bullet OH_{ads}$	5
Scavenging of photoexcited e^-: $(O_2)_{ads} + e^- \rightarrow O_2^-$	6
Protonation of superoxides: $O_2^- + \bullet OH \rightarrow HO_2^\bullet$	7
Co – scavenging of e^-: $HO_2^\bullet + e^- \rightarrow HO_2^-$	8
H_2O_2 formation: $HO_2^- + H^+ \rightarrow H_2O_2$	9
Photodegradation of $\bullet OH$: $RH + \bullet OH \rightarrow R^\bullet + H_2O$	10
Direct photoholes: $RH + h^+ \rightarrow R^\bullet \rightarrow$ Intermediate or final degradation products	11
$RH \xrightarrow{TiO_2/hv}$ intermediate $\rightarrow CO_2 + H_2O +$ inorganicions	12

The initial step of photocatalysis is the absorption of photons of wavelength which is equivalent to or greater than the band-gap energy of the photocatalyst. In the photocatalytic process of TiO_2, photons having energy equivalent or higher than the band-gap energy which is 3.2 eV get absorbed by the TiO_2 active sites, consequently electrons (e^-) and holes (h^+) are generated within the conduction band and in the valence band, respectively. The photo-generated holes and electrons greatly encourage the oxidation and reduction of pollutants adsorbed onto the surface of TiO_2, further promotes the oxidative photodegradation of pollutants present in wastewater via radical-induced photocatalytic reactions (Makwana et al. 2016, Pawar et al. 2018, De Souza and Corio 2013). This whole process is shown in Figure 2. Figure 2 illustrates the mechanism involved in the establishment of hole and electron pair when TiO_2 nanoparticle is illuminated with light of sufficient energy. The wavelength of light for such photon energy generally relates to $\lambda < 400$ nm. The excitation of electron leaves behind an empty space within the valence band, which generates pair of electrons and hole pair, Equation (1) to (3) are given in Table 2. The process of generation of electron-hole pair is called photoexcitation.

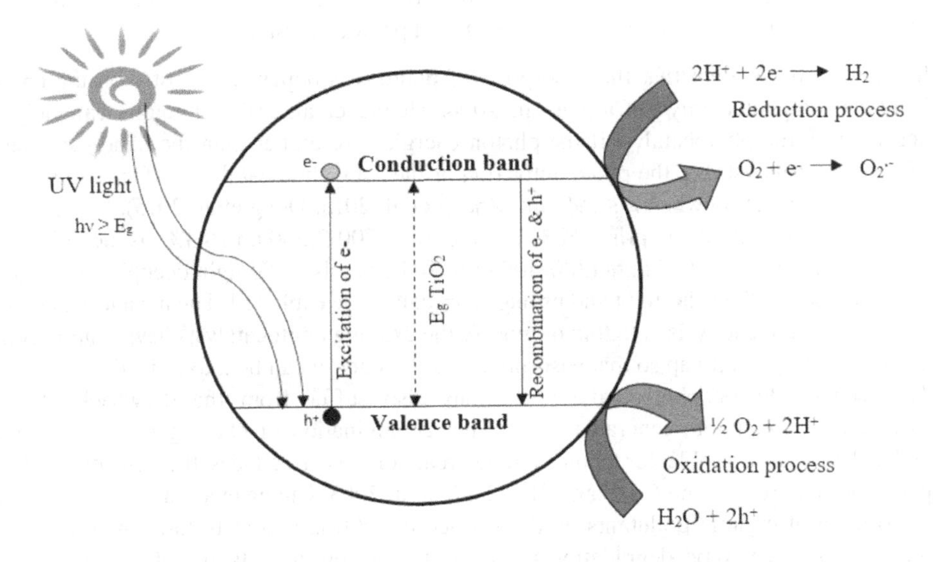

FIGURE 2 Principle of TiO_2 photocatalysis.

Equation (4) represents the surface trapped valence band electron, e^-_{TR} and conduction band hole, h^+_{TR}. Furube et al. (2015) reported that these surface trapped electrons, as well as holes, recombine gradually in nano-sized semiconductors than in the bulk phase. The significant implication of surface-trapped electrons and holes to the photooxidation of organic compounds was reported by Serpone et al. (1995). In the lack of electron scavengers (Equation 4), the photoexcited electrons recombine with the valence band holes within nanoseconds with instantaneous dissipation of heat energy. Therefore, the existence of electron scavengers is important for delaying the process of recombination and for the efficient operation of the photocatalyst. Photocatalytic reactions are generally carried out when targeted pollutants, air, water, and photocatalyst are present (Dong et al. 2015, Pawar et al. 2018). Adsorbed water present onto the surface of TiO_2 splits up into •OH and H^+ and this hydroxyl radical easily oxidizes the organic pollutants present in the wastewater during photo-degradation (Pawar et al. 2018). Normally, the holes react with the surface OH groups present on the surface of TiO_2 nanoparticles forming surface adsorbed hydroxyl radicals (•OH) Equation (5). Equation (6) illustrates that the presence of oxygen avoids the recombination of holes and electrons while permitting the creation of superoxide radical anion (O_2^-). This O_2^- further protonated to produce hydroperoxyl radical ($HO_2^•$) followed by H_2O_2, from Equation (7) to (9). It is reported that the $HO_2^•$ formed also exhibited the property of scavenging; therefore, the co-occurrence of these radical species can doubly extend the time of recombination of electrons and holes in the whole reactions during the process of photocatalysis. Photo-generated holes and the generated reactive species of oxygen; for example, •OH also participate in the photodegradation of organic pollutants, Equation (10) to (11) (Banerjee et al. 2014, Zhao and Yang 2003, Fujishima et al. 2000, Pawar et al. 2018). The heterogeneous photocatalysis reaction generally takes place on the photon-activated surface of the TiO_2 photocatalyst. So, it is crucial to understand all the steps included in the photodegradation of organic pollutants. With an increment in irradiation time, the liquid phase organic pollutants are photodegraded to their respective intermediates which get further degraded to inorganic ions, CO_2, and H_2O (see Equation 12).

6.2 Limitations in TiO_2-Based Photocatalytic Processes

According to literature, TiO_2 is one of the utmost effective photocatalysts as it utilizes solar energy and leads to the complete degradation of many of the organic pollutants present in the wastewater (Dong et al. 2015). However, few drawbacks of TiO_2 as a photocatalyst are:

- Since it is well-known that the absorption of photons is dependent mostly on the energy band-gap of photocatalyst (Carp et al. 2004, Huang et al. 2016), photons can only be absorbed by the photocatalyst if the photon energies are higher than the band-gap energy of the photocatalyst. So, the photo-activation of the TiO_2 photocatalyst surface can be done under UV irradiation when $\lambda \leq 390$ nm (Zheng et al. 2013, Dong et al. 2015). However, the spectrum of solar light comprises 52% infrared light (700-2,500 nm), 43% visible light (400-700 nm), and only 5% UV light (300-400 nm). To be precise, UV light occupies only a very slight fraction of the solar light and its huge part cannot be exploited. The most urgent task in the field of photocatalysis is to find or modify the existing photocatalysts having appropriate semiconducting band-gap so that adsorption of solar energy can be maximized.
- TiO_2 semiconductor photocatalytic reactions also suffer from the drawback of easy recombination of photo-generated holes. The recombination of photo-generated electrons and holes is unfavorable for photocatalytic reactions as it reduces the efficiency of the photocatalyst (Hoque and Guzman 2011, Guidi et al. 2003, Huang et al. 2016).
- Adsorption of organic pollutants on the surface of TiO_2 is relatively low which results in the slow photocatalytic degradation rates due to the low affinity of TiO_2 especially for hydrophobic organic pollutants (Dong et al. 2015).

- During the process of photocatalysis, nanoparticles of TiO_2 might experience aggregation due to the unstable nanoparticles, which may obstruct the incidence of light on the active centers and therefore decrease the photocatalytic activity of TiO_2 (Mallakpour and Nikkhoo 2014, Gao et al. 2011).

7. SYNTHESIS OF TIO_2-BASED NANOMATERIAL WITH IMPROVED PHOTO-ACTIVITY

To overcome the drawbacks of TiO_2 photocatalysis, a few countermeasures have been suggested in literature like (1) extending the range of wavelength for the photo-activation of TiO_2 photocatalyst under visible light and increasing the use of solar energy by modifying TiO_2 photocatalyst via doping (Bannat et al. 2009, Dong et al. 2011, Huang et al. 2016), (2) preparing a catalyst which comprises of small size particles having a well-defined crystal structure, and greater affinity for various organic pollutants by optimizing effecting parameters while synthesis (Lightcap et al. 2010, Makarova et al. 2000), and (3) designing and developing a second-generation TiO_2 catalyst having great separation capability and further regeneration and recovery efficiently (Nakayama and Hayashi 2007, Tang et al. 2013). So, the main focus of all the developments and adaptations is improvising the efficiency of the photocatalyst through mineralization of organic pollutants, improvement in reproducibility and stability, improvement in the reusing capabilities, and improvement in visible light absorption of TiO_2 (Dong et al. 2015). The subsequent sub-sections discuss various techniques used to synthesize nanomaterial with improved characteristics.

7.1 Doping

TiO_2 doping is very important in band-gap engineering to change the optical response of the photocatalysts (Gupta and Tripathi 2011). The primary objective of doping is to shrink the band-gap or introduce intra-band-gap states, which result in additional absorption of visible light leading to enhanced photocatalytic efficiency of TiO_2 (Crap et al. 2004, Huang et al. 2016, Akpan and Hamed 2009). To enhance the effectiveness of TiO_2 photocatalyst, TiO_2 can be doped with different metals and non-metals ions in it.

Generally, a doped ion introduces some extra levels of energy in the band structure, which can arrest the holes/electrons to separate the charge carriers and hence lets additional carriers be effectively drawn out to the surface of the photocatalyst. The chief motive of doping is to modify the electronic structure and band-gap of photocatalyst so that its optical properties can be optimized to harvest visible light (Umezawa and Ye 2012). Simultaneously, it is also essential to maintain the original structure of the crystal of the photocatalyst.

Degradation of pollutants presents in wastewater proceeds by oxidation either by reacting directly with generated holes or indirectly with free radicals of •OH (Mao and Weng 2012). Furthermore, the activity of the TiO_2 photocatalyst depends on nature as well as on the density of surface defected sites. Generally, the addition of dopants into TiO_2 crystals at titanium or oxygen sites can create mid-gap states (Huang et al. 2016). Additionally, interactions between photo-generated carriers and impurities present in TiO_2 are also able to alter the energy band-gap and electronic structure and further improve the performance of TiO_2 photocatalyst (Eslami et al. 2016). It is reported that excess doping can increase the crystal defects, thermal instability, carrier trapping, and carrier recombination centers (Martyanov et al. 2004, Naldoni et al. 2012). Chemically, TiO_2 doping corresponds to introducing defect sites of Ti^{3+} into the lattice. Oxidation of Ti^{3+} is speedy in comparison to the oxidation of Ti^{4+}. This difference in photoactivity is due to the alteration in diffusion length of minority carriers (Maruska and Ghosh 1979). According to Gautron et al. (1981), the scale of the potential drop through the space-charge layer must not be below 0.2 V to get the optimized separation of holes and electrons. The recombination of holes and electrons is affected by the concentration of the dopant by the Equation: $W = (2\ \varepsilon\varepsilon_o V_s/eN_d)$, where '$W$' is space-charge layer thickness, 'ε' is the static dielectric constant of the semiconductor, 'ε_o' is static dielectric constant in vacuum, 'V_s' is surface potential, 'N_d' is the number of dopant donor atoms, and 'e' is the charge of electrons (Fox and Duley 1995). As the concentration of dopant is increased, the space-charge region becomes thinner and electron-hole pairs in the region are effectively detached by the huge electric field before the process of recombination. With further increase in dopant concentration, the space-charge region becomes very thin and then the penetration depth of light into TiO_2 photocatalyst significantly goes beyond the thickness of the space-charge layer. Thus, the recombination rate of photo-generated holes and electrons in the photocatalyst rises as there is no driving force to separate them (Fox and Dulay 1995). So, the geometric structure and optimization of system-charge equilibria must be considered very carefully during the doping process (Huang et al. 2016).

7.1.1 Self-Doping

Self-doping controls the system-charge equilibria via the selection of raw material, synthesis method, and process control (Huang et al. 2016). It has been proved through several theoretical predictions and experimental results that Ti^{3+} self-doped TiO_2 can markedly enhance the associated intrinsic oxygen vacancies in TiO_2 crystals. It takes place because the surface chemistry of non-stoichiometric TiO_2 containing Ti^{3+} varies evidently from pure TiO_2. When compared to traditional impurity incorporation, Ti^{3+} self-doped TiO_2 is in itself efficient enough to lengthen the absorption of visible light (Cai et al. 2014a, Li et al. 2015b, Huo et al. 2014, Zheng et al. 2013).

In the fundamental process of self-doping, the electrons are trapped and inclined to turn Ti^{4+} cations to Ti^{3+} (Chen et al. 2005), the holes oxidize O^{2-} anions to form O^- trapped hole or to O_2 gas. Given below are the steps of charge transfer (Huang et al. 2016):

Reactions	Equation No.
$TiO_2 + hv \rightarrow e^-_{CB} + h^+_{VB}$	13
$e^-_{CB} + Ti^{4+} \rightarrow Ti^{3+}$ (trapped electron)	14
$h^+_{VB} + O^{2-} \rightarrow O^-$ (trapped hole)	15
$4h^+_{VB} + 2O^{2-} \rightarrow O_2$	16

Based on the above mechanism, numerous approaches are recognized, such as ionothermal method (Li et al. 2015a), hydrothermal method (Cai et al. 2014a, Wang et al. 2015, Qiu et al. 2015, Fu et al. 2014, Liu et al. 2013), solvo-thermal method (Xin et al. 2015, Sirisuk et al. 2008), solution-based reduction method (Tian et al. 2016, Mao et al. 2014), solution-based oxidative method (Liu et al. 2014, Grabstanowicz et al. 2013), evaporation-induced self-assembly (EISA) method, combustion method (Chen et al. 2005), vapor-fed aerosol flame synthesis (Huo et al. 2014), and metallic reduction method (Si et al. 2014, Wen et al. 2015). Among all the mentioned methods, Ti^{3+} self-doped TiO_2 is synthesized majorly by using the hydrothermal method. Wang et al. (2015) improved hydrothermal conditions to obtain Ti^{3+} self-doped TiO_2 nanoparticles. As per the results, the oxygen vacancies and contents of Ti^{3+} in TiO_2 crystals can practically be manipulated. To form uniform Ti^{3+} self-doped TiO_2 nanocrystals, solvo-thermal method which is based on oxidation was found to be an effective technique (Xin et al. 2015). The interface ion diffusion-redox reaction method was utilized to produce Ti^{3+} self-doped TiO_{2-x} nanoparticles by Liu et al. (2014). It resulted in better crystallinity and augmented visible light-driven photocatalytic oxidation of self-doped TiO_{2-x}.

7.1.1.1 *Mechanism of Enhanced Photocatalytic via Self-Doping Method*

At high concentration of oxygen vacancies, the Fermi level is considerably close to the tail of the conduction band. So, it can be concluded that the high oxygen vacancies lead to increased absorption of photon energy below the direct band-gap. The mid-gap positions below the conduction band turn broad at a higher concentration of oxygen vacancies shown in Figure 3, permitting photo-generated electrons to effortlessly move between the conduction and the valence band. The transfer of electron occurs from both oxygen vacancy level localized states and valence band to the tailed conduction band. Thus, the commencement of optical absorption of TiO_2 nanocrystals is depressed from 1,100 to 900 nm (He et al. 2012, Guo et al. 2016, Wang et al. 2014). In Figure 3, $h\nu$ (Ti^{4+}) and $h\nu$ (Ti^{3+}) corresponds to the energy required to excite electrons from the valence band to conduction band before and after self-doping, respectively.

Numerous studies are available, which proved that the local electrostatic balance is broken when host Ti^{4+} is reduced to Ti^{3+} and due to charge compensation, oxygen vacancies are introduced (Lu et al. 2001, Xiong et al. 2012). The central Ti^{3+} is expected to shift away from the oxygen vacancy since the effective charge of the oxygen vacancy is positive, establishing a special sub-level electric state. On the other hand, the shift of Ti^{3+} forces the O^{2-} ions to transport near the oxygen vacancy for the maintenance of electrostatic balance. Then the electrons can be photo-excited to the conduction band under visible light. In the meantime, the oxygen vacancies hinder the recombination of photo-generated holes and electrons. Moreover, Ti^{3+} ions behave as hole traps and inhibit the recombination process of carriers increasing the lifespan of the charges. The distortion of the outer orbitals of Ti^{3+} ions is responsible for the high light absorption in Ti^{3+} self-doped TiO_2 (Huang et al. 2016).

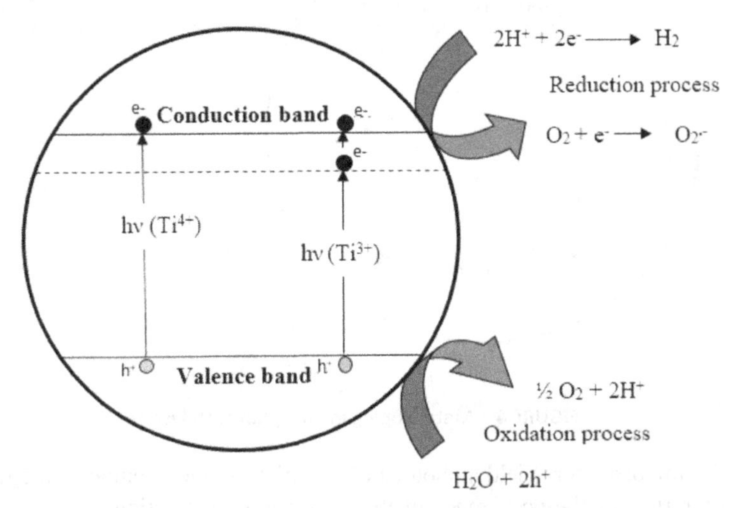

FIGURE 3 Self-doping process in TiO_2 photocatalyst.

Ti^{3+} self-doped TiO_2 shows improved photocatalytic performance and outstanding activities as a photocatalyst. Though, practically it is tough to incorporate Ti^{3+} into TiO_2 crystals since Ti^{3+} species are generally unstable which can simply be oxidized in the presence of oxygen. Thus, developing a phase-controlled method to synthesize stable Ti^{3+} self-doped TiO_2 photocatalysts is still a challenging task. Additionally, to have complete knowledge of the methods, Ti^{3+} generation and exploration of Ti^{3+} properties are very significant (Huang et al. 2016).

7.1.2 Doping with Metals

Substitution of metal ions such as transition metal doping (Fan et al. 2014, Chang and Liu 2014, Hahn et al. 2013), rare-earth metal doping (Ishii et al. 2013, Tobaldi et al. 2013, Lima et al. 2015), and other metal doping (Li et al. 2014, Klaysri et al. 2015, Zhao et al. 2011) can lead to an intra-band state which is adjacent to the edge of the conduction band, resulting in a redshift in the band-gap adsorption due to sub-band-gap energies. In solid materials, diffusion of metal ions is difficult under low temperature, leading to inhomogeneous dispersal of dopants and restricted depth near the sub-surface region. To get better homogenized metal-doped TiO_2 nanoparticles, sintering has to be directed at a high temperature. Moreover, metal dopants offer additional trapping sites for holes and electrons than in comparison to non-metal dopants.

Recently, metal-doped (such as Cu, Ni, Co, Cr, Mn, Ru, Fe, Au, Pt, and Ag) TiO_2 photocatalysts have been broadly explored to improve its photocatalytic performance under visible light (Arabatzis et al. 2003, Hu et al. 2006, Han et al. 2014), its details are given in Table 3. TiO_2 nanoparticles can simply be interstitially doped with different cations and form mixed oxides. The fundamental factors during the process of doping are the characteristics and concentration of the dopants and the applied thermal treatment. It is reported that the influence of metal-ion dopant on photocatalyst activity is generally a difficult problem (Dong et al. 2015). The photocatalyst activity of metal-doped TiO_2 under visible light is usually explained as a result of a new energy level created in the TiO_2 band-gap through the dispersal of metal nanoparticles in the matric of TiO_2. As presented in Figure 4, electrons are excited to the conduction band of TiO_2 from the defect state by means of photons having energy equivalent to hv_m. In Figure 4, hv and hv_m correspond to the energy required to excite electrons from valence band to conduction band before and after metal doping, respectively.

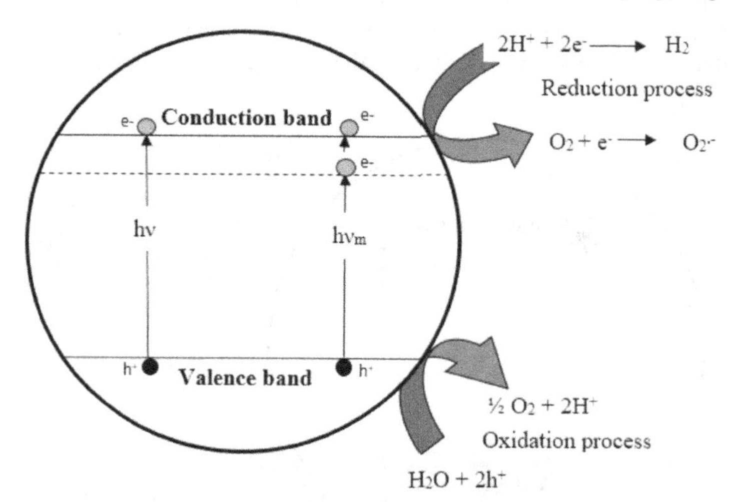

FIGURE 4 Metal doping in TiO_2 photocatalyst.

Metal-doped semiconductors yield a photocatalyst which is able to enhance the rate of trapping to recombination ratio. For the occurrence of the photocatalytic reactions, it is necessary that the trapped holes and electrons must be transported to the surface of the photocatalyst, which means, to permit the effective transfer of charge, any metal ions must be doped adjoining to the surface of the photocatalyst. The concentration of the dopant affects the activity of the metal-doped TiO_2 photocatalyst. Tong et al. (2008) and Ambrus et al. (2008) showed enhancement in the photocatalyst activity via improving Fe dopant concentration. Results showed that, as the concentration of the dopant reached an optimal level, it acts as electron-hole separation centers and improves the efficiency of the photocatalyst. However, if it goes beyond the optimal value, then it can behave as recombination centers for holes and electrons, which is unfavorable for the performance of the photocatalyst (Wu et al. 2009). Though, metal-doped semiconductors exhibited some shortcomings too like the thermal instability of doped TiO_2 which diminishes the reuse of the photocatalyst and creates the necessity for further expenses on ion-implantation facilities (Huang et al. 2016, Zhang et al. 2010). Various types of metals, such as noble metals, transition metals, and rare-earth metals, have been studied by many scientists for doping of TiO_2 and these are summarized in Table 3.

7.1.3 Doping with Non-Metals

Doping of TiO_2 is done with various non-metal having high electro-negativity as well as ionization energy such as carbon (C) (El-Sheikh et al. 2014, Yu et al. 2011), sulfur (S) (Li et al. 2012, Tipayarom et al. 2011), nitrogen (N) (Bolokang et al. 2015, Zhang et al. 2013a, Lynch 2015), chlorine (Cl) (Ji et al. 2012, Xu et al. 2008), boron (B) (Stengl et al. 2010, Wang et al. 2012a, Patel et al. 2015), and fluorine (F) (Senna et al. 2012, Kafizas et al. 2014). A list of non-metal dopants is given in Table 3. It is an efficient methodology to enhance the performance of TiO_2 under visible light due to reduction in energy band-gap and shift of absorption edge (Huang et al. 2016). According to the literature, non-metals are more appropriate to increase the photocatalytic activity of TiO_2 under the visible light region in comparison to metals because their contamination states are close to the valence band (Dong et al. 2015). Recently, various studies have been dedicated to developing efficient non-metal doped TiO_2 photocatalyst to enhance the light absorbance range under visible light region (Likodimos et al. 2013, Moustakas et al. 2013). During doping of TiO_2, non-metal dopants affect the valence band via interaction with 2*p* electrons of *O*. The *p* states of non-metal dopants commonly form the impurity levels which lie beyond the valence band and improve the optical absorption range of TiO_2. In contrast, non-metal dopants can exist inside a surface as an isolated atom instead of clusters. Subsequently, the dispersal of dopant states is overhead the valence band, which has the potential to augment the performance of the photocatalyst under visible light (Huang et al. 2016).

Non-metal doped-TiO_2 photocatalyst improved the performance of TiO_2; however, it provides few drawbacks too. These are as follows:

(a) Non-metals doping into the TiO_2 lattice resulted in the creation of oxygen vacancies that further can behave as centers for recombination of photo-induced holes and electrons and hampers the efficiency of the non-metal doped TiO_2 photocatalyst efficiency under visible light (Dong et al. 2011, Ozaki et al. 2007, Dong et al. 2010);

(b) In order to maintain the long-term efficiency of the non-metal doped TiO_2 photocatalyst, its stability is a major concern. Kitano et al. (2006) found that stability of N-doped TiO_2 turns out to be poorer after the process of photo-electrolysis of water under visible because the concentration of N ion at the surface decreased. According to the literature, N-doped TiO_2 got deactivated after a time of 150 hours when it was utilized under visible light for the oxidation of 2-propanol in aqueous media. This happened due to the release of doped N atoms (Nosaka et al. 2005);

(c) Another critical parameter is the synthesis method which is very important for practical applications. Generally, non-metal doping constantly requires high-temperature treatment. Moreover, toxic, expensive, or unstable precursors are utilized, objectionable gaseous byproducts are usually formed, and somewhat tedious procedures are followed during the preparation process, which altogether makes the synthesis difficult and costly when executed at large-scale. Consequently, the development of new photocatalysts and optimizing existing photocatalysts with improved performance under visible light and having high physical and chemical stability are essential for their utilization at a large scale. Several non-metals like C, S, and F have been utilized as dopants by numerous researchers for the degradation of pollutants and these are summarized in Table 3.

7.2 Co-Doping Technique

Although the mono-doped metals and non-metal TiO_2 photocatalyst can visibly improve the performance of photocatalyst, it constantly forms recombination centers due to the partly filled impurity bands. Theoretically, it has been established that co-doping can passivate the impurity bands as well as reduce the creation of recombination centers via improving the solubility limit of dopants. Besides, all these, co-doping is also able to modulate the charge equilibrium. Thus, co-doping is very efficient to augment the performance of photocatalysts (Huang et al. 2016). It is reported that co-doping of TiO_2 may be utilized as an efficient way to enhance the charge separation (Gupta and Tripathi 2011). Co-doped TiO_2 nanomaterials have been confirmed to show better performance as photocatalyst than non-doped or single-ion doped TiO_2, particularly under visible light regions (Dong et al. 2015). For example, mono-crystalline TiO_2 co-doped using Fe^{3+} and Eu^{3+} showed a synergistic effect. While comparing results, it was found that the enhancement in photocatalytic degradation of chloroform was around five times in comparison to pure TiO_2. Also, degradation of chloroform was around six and two times when co-doped TiO_2 was compared with Er^{3+} and Fe^{3+} doped TiO_2, respectively. Eu^{3+} acts as an electron trap while Fe^{3+} as a hole trap, speeding up the cathodic and anodic processes, respectively, through interfacial charge transfer enhancement (Yang et al. 2002). Vasiliu et al. (2009) reported that there is an improved performance of photocatalyst and redshift in the absorption spectrum under visible light for the catalytic oxidation reactions and degradation of phenol and styrene, respectively, when TiO_2 was doped with Eu and Fe. Several types of co-doping techniques available in literature including non-metal-non-metal, metal-metal, and non-metal-metal are given in Table 3.

Generally, co-doping enhances the activity of photocatalyst. However, there are some issues with the process of co-doping. Co-doping may destabilize some crystals and cause unwanted defects, introduce unfavorable mid-gap bands, and sometimes deactivate required defects. Thus, there is a compromise between the enhanced activity of photocatalyst and the disadvantages of the co-doping technique (Zhang et al. 2016).

TABLE 3 Doping and Co-doping of TiO_2 for the photodegradation of pollutants.

Doping/ Co-Doping	Type of Material	Examples	Results	Reference
Doping	**Noble Metals**	Ag and Pt	Significant improvement in the photocatalytic degradation of oxalic acid	Iliev et al. 2006
		Pt, Ag, Au, Pd, Ni, Rh, and Cu	Enhanced in the photocatalysis performance	Rupa et al. 2009, Adachi et al. 1994, Wu et al. 2004
		Ag, Pt, and Au	Efficiently delayed the electrons and holes recombination	Tran et al. 2006, Kowalska et al. 2008, Subramanian et al. 2001, Li et al. 2008, Bannat et al. 2009
	Transition	Fe, Mo, Ru, Os, Re, and V	Significant enhancement in the photochemical reactivity of TiO_2	Choi et al. 1994
		Ruthenium	Enhanced photocatalytic activity (>80%) for the degradation of methylene blue	Khan et al. 2009
		La, Nd, Sm, Eu, Gd, and Yb	Significant improvement in the photocatalytic activity of TiO_2	El-Bahy et al. 2009

Contd.

		Gd	Significant enhancement in the degradation rate of Reactive Brilliant Red even under visible light	Wang et al. 2010b
		Ce	Considerable improvement in the degradation rate of dye X-3B	Wang et al. 2010a
		Fe	Enhanced photocatalytic activity of TiO_2 even at a low dopant concentration Improvement in the degradation rate of acetic acid and formaldehyde Considerable shift in absorption toward visible region	Xin et al. 2007, Li et al. 2009, Vijayan et al. 2009, Manu and Khadar 2015, Yan et al. 2015
		Cr	Reduction in the band-gap of TiO_2	Li et al. 2013, Herrmann 2012, Diaz-Uribe et al. 2014
		Nb	Significant enhancement in the photocatalytic performance of TiO_2	Archana et al. 2010, Archana et al. 2011, Yang et al. 2012, Wang et al. 2014
		Nb	Remarkable improvement in the photocatalytic activity with 97.3% degradation of methylene blue	Joshi et al. 2013
		W	Remarkably inhibited the rate of recombination of electron-hole pairs	Liu et al. 2012, Mayoufi et al. 2014
		Cu	Enhancement in the photocatalytic activity of TiO_2	Sajjad et al. 2013, Huang et al. 2016, Yang et al. 2015, Zhao et al. 2015
		Co	Enhanced photocatalytic activity of TiO_2	Cai et al. 2014b
		Ta	Significant modifications on the morphological, structural, surface, electronic and optical properties of TiO_2	Sengele et al. 2016
		Zn	Significant enhancement in the photocatalytic activity of TiO_2	Saad et al. 2014
	Rare-earth metals	La	Significant enhancement in the degradation rate of methylene blue	Du et al. 2013
		Ce	Reduction in the band-gap and enhancement in the photocatalytic efficiency of TiO_2	Maddila et al. 2016

Contd.

		Ce	Reduction in the rate of recombination of electron-hole pairs	Huang et al. 2016
		Er	Improvement in the photocatalyst activity of TiO_2	Obregon and Colon 2012
		Gd	Enhancement in the photocatalytic activity of TiO_2	Choi et al. 2014
		Y	Enhancement in the photocatalytic activity of TiO_2 for the degradation of methyl orange	Wu et al. 2013
		N	Under visible light, it is one of the most efficient dopants for TiO_2 photocatalyst Prohibit the process of recombination of electrons and holes	Zaleska 2008, Asahi et al. 2001, Lin et al. 2005, Valentin et al. 2005, Gao et al. 2009, Emeline et al. 2008, Yates et al. 2006, Gai et al. 2012, Irie et al. 2003
	Non-metals	C	Reduction in the band-gap of TiO_2	Wu et al. 2013, Lin et al. 2013b, Etacheri et al. 2013, Cuomo et al. 2015, Khan et al. 2002, Shi et al. 2015, Zhang et al. 2015a, Shao et al. 2015, Zhang et al. 2013b, Yang et al. 2015, Sakthivel and Kisch 2003, Chen et al. 2007, Valentin et al. 2005, Lettmann et al. 2001, Sakthivel and Kisch 2003, Kang et al. 2008, Hahn et al. 2009, Janus et al. 2006, Xiao and Ouyang 2009, Shi et al. 2015
		S	Reduction in the band-gap of TiO_2 Enhanced photocatalytic performance	Goswami and Ganguli 2013, Ramacharyulu et al. 2015, Lin et al. 2014, Sharotri and Sud 2015
		F	Shrinkage in the band-gap Enhancement in the photocatalyst performance of TiO_2 Reduction in the recombination of photo-generated electrons and holes	Fang et al. 2014, Lozano et al. 2014, Rahimi et al. 2015, Dozzi et al. 2013
	Halogen	Cl	Enhancement in the rate of photo-degradation of butyl benzyl phthalate	Wang et al. 2012b

Contd.

		I	Shrinkage in the band-gap Extension in the intrinsic absorption edge under a visible light region	Lin et al. 2015, Siuzdak et al. 2015, He et al. 2010, Liu et al. 2011
		F	Augmentation in the light-harvesting under UV-vis range Shrinkage in the band-gap Enhancement in the photocatalyst performance of TiO_2 Reduction in the recombination process of photo-generated electrons and holes	Pan et al. 2011, Fang et al. 2014, Lozano et al. 2014, Rahimi et al. 2015, Dozzi et al. 2013
Co-doping	**Non-metal-Non-metal**	N, S	Improved photocatalytic performance of TiO_2 due to strong synergistic interaction between N and S	Chung et al. 2015, Pany et al. 2014, Xiang et al. 2011
		N, B	Efficiently reduced the band-gap and inhibit the transformation of anatase TiO_2 to the rutile phase	Czoska et al. 2011, Wang et al. 2015, Zhang et al. 2014a
		N, C	Resulted in a shift of absorption edge to lower energy by encouraging new band levels Formed a large amount of single electron-trapped oxygen vacancies	Liu et al. 2015
		N, H	Narrowing band-gap	Wei et al. 2015
		C, B	Enhancement in the photocatalytic activity of TiO_2 Reduction in the band-gap	Yu et al. 2013, Lin et al. 2013a
		C, F	Shrinkage in the band-gap Inhibited the recombination of photo-induced irradiation Improvement in the photocatalytic activity of TiO_2	Deng et al. 2015
		N, C, F	Improvement in the photocatalytic performance of TiO_2	Ramanathan and Bansal 2015
		Ni-Ti	Shrinkage in the band-gap Reduction in the rate of recombination of electrons and holes	Zhang et al. 2015b

Contd.

	Metal-Metal	Fe-Ti	Reduction in the band-gap Significant improvement in the photocatalytic performance under visible light	Chen et al. 2014
		Ag, W	Enhanced photocatalytic performance	Khan et al. 2015
		Zn, Mn	Reduction in the energy band-gap Improvement of light absorbance under UV region	Benjwal and Kar 2015
		Cu, V	Improved photocatalytic performance	Christoforidis and Fernández-García 2016
		Fe, Ce	Enhanced the light adsorption by trapping carriers Prohibited the recombination of photo-excited electrons and holes	Jaimy et al. 2012
		N, La	Enhanced photocatalytic activity of TiO_2	Sun et al. 2012, Yu et al. 2015
	Non-metal-metal	N, Ti	Reduction in the band-gap Improved the adsorption capacity and increased the rate of carrier transport	Li et al. 2015b
		N, Zn	Enhanced photocatalytic activity of TiO_2	Wang et al. 2011
		N, W	Condensed the band-gap	Lai and Wu 2015, Mishra et al. 2011
		N, Cu	Improvement in the photocatalytic performance of TiO_2	Wang et al. 2014
		N, Fe	Significant improvement in the photocatalytic activity of TiO_2 under visible light	Zhang et al. 2015c
		N, Ni	Reduction in the energy band-gap	Liu et al. 2015
		B, Co	Improved photocatalytic activity of TiO_2	Jaiswal et al. 2016
		Cu, Ce, B	Enhanced photocatalytic activity of TiO_2 Prohibited the recombination of electrons and holes and improvement in the concentration of photo-generated carriers	Li et al. 2015c
		Ag, V	Improvement in the photocatalytic activity of TiO_2	Yang et al. 2010

7.3 Coupled/Composite TiO_2

According to the literature, it is possible to formulate such colloidal structures where illumination of one semiconductor yields a response at the interface of another one and these are called coupled semiconductors (Mills and Hunte 1997). These coupled semiconductors display enhanced photocatalyst performance in liquid and gas phase reactions via enhancing the charge separation and the photo-excitation range (Paola et al. 2003). There are several factors like surface texture, particle size, and geometry of particles that are substantial in inter-particle electron transfer. For effective charge separation, optimized thickness and suitable location of individual semiconductors are very crucial.

Numerous researchers are concerned with coupling various semiconductors with photocatalyst TiO_2, (TiO_2-MoO_3, TiO_2-CdS, TiO_2-WO_3, TiO_2-Fe_2O_3, TiO_2-SnO_2, and Bi_2S_3-TiO_2) (Vogel et al. 1994, Bessekhouad et al. 2004, Grandcolas et al. 2008, Wang et al. 2009). Coupled CdS-TiO_2 colloidal particles have been extensively studied in the literature. Figure 5 shows that it is possible to illuminate CdS by the light of lesser energy in comparison to that required to electronically excite TiO_2 particles. Thus, the photo-generated electron is inserted into TiO_2 from CdS whereas the hole stays within CdS only (Mills and Hunte 1997). The electron transfer from CdS to TiO_2 raises the charge separation and the effectiveness of the photocatalytic process. The separated holes and electrons are free for electron transfer with adsorbates on the surface of the catalyst. Methyl viologen was almost entirely reduced under visible light when CdS-TiO_2 photocatalyst was used (Gopidas et al. 1990). In Figure 5, *hv* (CdS) corresponds to the energy required to excite electrons from the valence band to the conduction band for CdS.

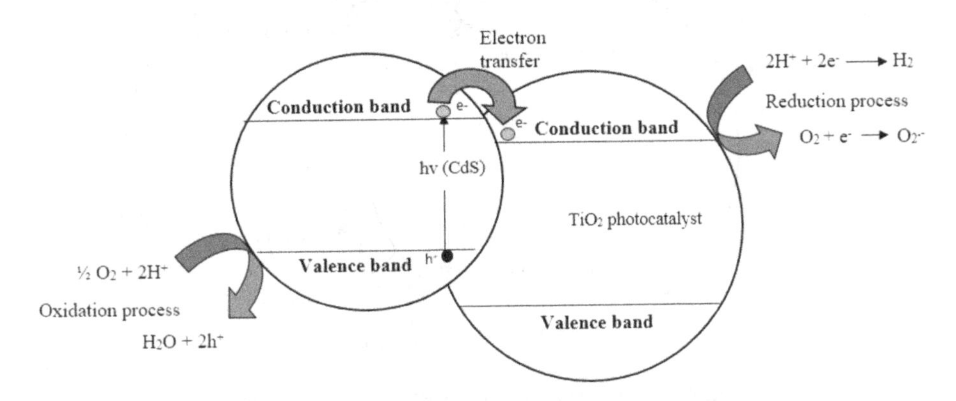

FIGURE 5 Mechanism of coupled CdS-TiO_2 photocatalyst.

For the photo-sensitization of TiO_2 nanocrystals, it is reported that Bi_2S_3 nanoparticles having an energy band-gap of 1.28 eV is a suitable material. Since the conduction band of Bi_2S_3 is less anodic than to conduction band of TiO_2 while the valence band is more cathodic than the valence band of TiO_2 (Nayak et al. 1983), thus the insertion of electron takes place from the excited state of Bi_2S_3 to TiO_2. A junction of Bi_2S_3-TiO_2 was synthesized by precipitation of different Bi_2S_3 concentrations onto the surface of TiO_2 and results showed that a huge portion of visible light up to 800 nm was absorbed when the junction contained 10 wt% Bi_2S_3 (Bessekhouad et al. 2004).

Another widely used coupled semiconductor is WO_3/TiO_2. The upmost edge of the valence band and the lowest edge of the conduction band of WO_3 are lower in comparison to TiO_2. The photo-sensitization of WO_3 is possible under visible light irradiation and the photo-generated holes can be transferred from WO_3 to TiO_2 (Gupta and Tripathi 2011). It is reported that the application of WO_3 onto the surface of TiO_2 enhanced the degradation of 1, 4-dichlorobenzene by about 5.9 times than pure TiO_2 (Song et al. 2001). It was concluded, electrons that are present in the conduction band of TiO_2 can easily be accepted by WO_3, as the standard reduction potential between W(VI) and W(V) is only -0.03 V. The electrons which are present in WO_3 would be transported to the molecules of oxygen and get adsorbed on the surface of TiO_2. Porogen template supported sol-gel production of coupled rutile-anatase TiO_2 having a large surface area was described by Grandcolas et al. (2008). Porogen was utilized as WO_3 support, stretched the absorption under the visible region and enhanced the performance of the photocatalyst. The mentioned system was utilized for the degradation of diethyl sulfide, production of hydrogen, and CO oxidation.

SnO_2/TiO_2 is another coupled semiconductor famous among researchers. For the SnO_2/TiO_2 coupled semiconductor, TiO_2 plays the part of a photosensitizer for SnO_2. Pure SnO_2 displays very less photocatalytic activity in comparison to TiO_2 as the band-gap of SnO_2 that is 3.5-3.8 eV is inadequate to initialize the photocatalytic reactions, even after ultraviolet irradiation. The electron affinity and work function of TiO_2 are both are about 4.2 eV through the electron affinity of SnO_2 is around 0.5 eV greater as compared to TiO_2, and the work function is about 4.4 eV. The Fermi energy level of SnO_2 is smaller than TiO_2. Since the work function of SnO_2 is large, so the transfer of electrons arises from the conduction band of TiO_2 to the conduction band of SnO_2 while hole transfer happens from the valence band of SnO_2 to the valence band of TiO_2 as given in Figure 6. In Figure 6, hv (SnO_2) and hv (TiO_2) corresponds to the energy required to excite electrons from valence band to conduction band for SnO_2 and TiO_2 catalyst. The photocatalyst degradation rate of various textile azo dyes was augmented by a factor of 10 by utilizing a SnO_2/TiO_2 composite system as a result of improved charge separation as reported by Vinodgopal et al. (1996).

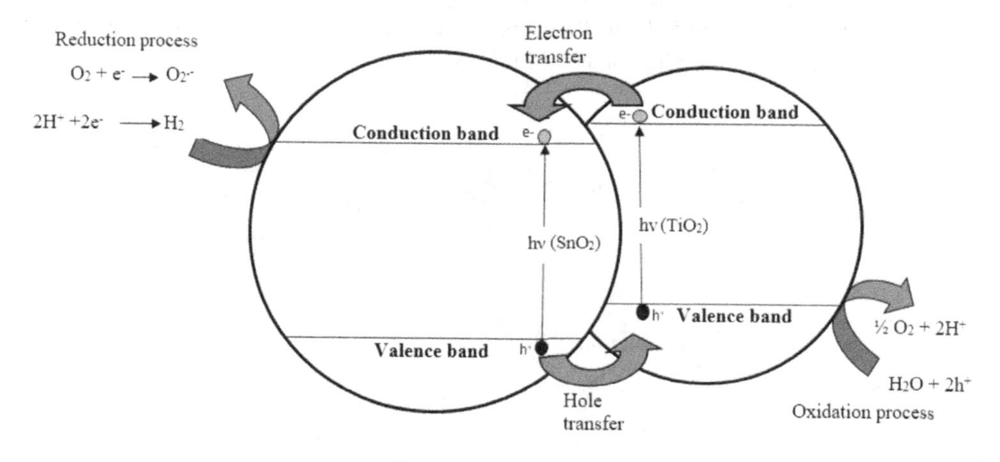

FIGURE 6 Mechanism of coupled SnO_2-TiO_2 photocatalyst.

8. FACTORS INFLUENCING THE PHOTOCATALYTIC DEGRADATION

There are numerous factors like the intensity of light, photocatalyst nature, structure, surface area and concentration, the concentration of dopant, initial concentration of the pollutants, temperature, pH of the sample (wastewater), and type of substrate which may govern the degradation of pollutants that exist in wastewater (Akpan and Hameed 2009, Kumar and Pandey 2017a, Reza et al. 2017, Chen et al. 2020).

8.1 Influence of Wavelength and Intensity of Light

It is reported that TiO_2 has an extensive energy band-gap, restricts the absorption of light under the UV region (Khan et al. 2002, Kuang et al. 2009, Atta et al. 2011). The photocatalytic degradation of pollutants present in aqueous media is usually affected by intensity, wavelength, and time of irradiation under UV light when TiO_2 photocatalyst was used in the reactor (Konstantinou and Albanis 2004, Gaya et al. 2008, Pawar et al. 2018). According to Chen et al. (2020), the intensity of light affected the rate of electron transfer from the valence band to the conduction band, and hence TiO_2 photocatalytic reaction rate was also influenced. Higher intensity generally leads to a high rate of photocatalysis. According to Asahi et al. (2001), the rate of photodegradation is directly dependent on the intensity of light at low intensities (25 mW/cm^2), whereas when intensity is intermediate (0-20 mW/cm^2), the rate of photodegradation is dependent on the square root of intensity of light. However, the rate of photocatalysis was found to be independent of the intensity of light is beyond an optimum level (Blake et al. 1991) due to the inadequate amount of active sites on the surface of photocatalyst (Cheng et al. 2018, Asahi et al. 2001). Nasirian and Mehrvar (2016) reported that at low intensity of light, the rate of photocatalysis follows a first-order order reaction, and at higher intensity of light, it follows a half-order. So, it can be said that adequate distribution of energy of photons provides a high photocatalysis rate. Maki et al. (2019) utilized Fe-N-Ce doped TiO_2 to degrade dye and found that as the LED current was increased from 0.22 to 0.68 amps, enhancement in the rate of photodegradation increased from 10% to 98%. Similarly, the photocatalyst performance rapidly increased to 56% when the intensity of light was increased from 0-3.0 mW/cm^2 when TiO_2 photocatalyst was utilized for the degradation of iso-propanol.

According to many reports, artificial UV irradiations are more reproducible than the original sunlight; additionally, it is more efficient to degrade textile dyes. As solar energy is non-hazardous, exists in abundance so it is understood that solar energy can emerge as a cost-effective alternative source of light (Muruganandham and Swaminathan 2006, Neppolian et al. (2002a). Generally, solar energy-based degradation reactions are done by directly exploiting solar irradiation (Gonc¸alves et al. 2005) or using parabolic collectors (Malato et al. 1998).

According to the results of Neppolian et al. (2002a), when TiO_2 photocatalyst was used for the degradation of dyes Reactive Red 2, Reactive Yellow 17, and Reactive Blue 4, UV irradiation was found to be more efficient in comparison to solar irradiation and the percentage photodegradation was found to increase with increment in the solar light intensity. The energy of UV irradiation was larger than the energy band-gap of the photocatalyst. Therefore, recombination of holes and electrons is avoided on a large scale under UV irradiation. However, only 5% of total sunlight radiations have optimum energy matching the energy band-gap for the excitation of electrons (Fatin et al. 2012). Therefore, the percentage photodegradation of textile dyes is found to be low under solar radiation.

Kundu et al. (2005) observed that photodegradation of 2,4-D was found to increase when the intensity of light was varied from 100 to 600 Lux. Hung and Yuan (2000) reported that as the intensity of light was increased from 215 to 586 l W/cm^2, the photodegradation reaction rates for Orange G were also increased. The effect of the intensity of light on the decolorization of Rhodamine B was studied by Chanathaworn et al. (2012). Results favored the increased decolorization of dye when the intensity of light was varied from 0 to 114 W/m^2. Three different intensities of light 1.24 mW/cm^2, 2.04 mW/m^2, and 3.15 mW/m^2 were utilized by Liu et al. (2006) for the decolorization of Acid Yellow 17, and it was found that maximum decolorization of the given dye was obtained at the maximum light intensity. Similar trends were also reported by other authors (So et al. 2002, Sakthivel et al. 2003).

In addition to the intensity of light, irradiation time also affects the rate of photodegradation reactions significantly. According to Kumar and Pandey (2017a), the rate of photodegradation was found to increase with increment in irradiation time and the photodegradation for methyl orange with Co and La modified titania was found to be highest at 3 hours of irradiation under visible light.

Irradiation wavelength also affects the degradation efficiency of the photocatalyst (Nguyen et al. 2014). The results were observed in the case of photoreduction of CO_2, 4-chlorophenol, and several organic pollutants with TiO_2 photocatalyst. As per observations, irradiation of a shorter wavelength promotes the generation of electrons and holes and thus improves the efficiency of the photocatalyst.

8.2 Nature, Form and Surface Area of the Photocatalyst

Several factors like nature, form, and structure of TiO_2 photocatalysts affected the degradation of pollutants present in wastewater. It is reported that among all the given forms of TiO_2 namely, Degussa P-25, nanotube, nanoribbons, multi-walled carbon nanotube TiO_2, and nano-flower like rutile TiO_2, the Degussa P-25 TiO_2 exhibited better performance than all other forms. Saquib and Muneer (2003) reported that the rate of photodegradation of Ethidium Bromide, Acridine Orange, and Triphenylmethane dye was found to be maximum with Degussa P-25 TiO_2 in comparison to Anatase and Rutile. Degussa P-25 showed enhanced performance due to the presence of 18% rutile and 73% anatase which shows a synergistic effect during photocatalytic activity. As the Anatase form of TiO_2 is composed of a large indirect band-gap due to which it displays a longer lifetime of holes and electrons compared to rutile TiO_2. Meanwhile, Rutile TiO_2 is composed of a smaller direct band-gap and hence it is able to take up photons with lower energy to generate pairs of holes and electrons and is further used by anatase TiO_2 (Faisal et al. 2007). According to Zhang et al. (2014b) and Bagbi et al. (2017), anatase TiO_2 possesses a low rate of recombination of holes and electrons and allows quicker migration of holes and electrons from the interior to the surface of the anatase TiO_2.

The structure of photocatalyst is also very significant during the measurement of its efficiency. According to the study of Khunphonoi and Grisdanurak (2016), the rate of photodecomposition of *p*-cresol was higher when TiO_2 nanorods were utilized as compared to P25 TiO_2 photocatalyst. The rate of photodegradation of phenolic pollutants was found to be higher with facet tailored TiO_2 compared to P25 TiO_2 (Liu et al. 2019). Kaur et al. (2007) evaluated the rate of photodegradation of Eriochrome Black T by utilizing different types of TiO_2. Nanoparticles of Titania P25-TiO_2 photocatalyst were found to be superior as it showed 86-94% degradation of Eriochrome Black T due to the presence of higher surface area in comparison to titania nanotubes and nanorods.

The surface morphology like the size of agglomerates or particles is a significant parameter that must be considered during the process of photocatalysis as these are directly associated with the surface coverage of the photocatalyst and organic compounds (Kormann et al. 1988, Watson et al. 2003, Chen et al. 2004, Xiaohong et al. 2003, Ding et al. 2005). The number of a photon striking the photocatalyst controls the rate of reaction which signifies that the reaction takes place only in the absorbed phase of the photocatalyst (Zhu et al. 2006). To increase the surface area of the photocatalyst, modification of particle size has been observed as a cost-efficient technique. According to the studies of Hu et al. (2017) and Zhou et al. (2018), there are significant differences in photochemical reactions and photodegradation activities between bulk and nanoscale TiO_2 systems. Liu et al. (2008) reported that as the particle size decreases, there is an increment in the internal recombination of holes and electrons as well as an increment in surface active sites. However, decreasing the size of the TiO_2 photocatalyst particles also increased the surface recombination rate of holes and electrons and lowers the use of photons. Consequently, an optimized size of the particles may exist where the equilibrium between the given conflicting effects is reached (Chen et al. 2020). In this regard, several studies were conducted to synthesize different forms of TiO_2 to achieving desirable characteristics of the TiO_2 photocatalyst. Kumar et al. (2015) utilized Titania and Cu-TiO_2 nanocomposites for the photodegradation of methyl red dye. The maximum rate of photodegradation was found in the case of Cu-TiO_2 due to the difference in the surface characteristics of Cu-TiO_2 from titania. Cu-TiO_2 had a zig-zag and rough surface that assisted in the adsorption of organic molecules on the surface of the Cu-TiO_2 photocatalyst. Reza et al. (2019) reported that the activity of the photocatalyst differs with the differences in the BET surface, lattice mismatch, and impurities on the surface of the photocatalyst. These affect the adsorption behavior of a pollutant on TiO_2 and the time-spam and rate of recombination rate of holes and electrons. According to Kogo et al. (1980), photocatalytic reaction efficiency can be influenced by the number of absorbed photons by photocatalysts surface.

Mostly used photocatalyst is the Degussa P-25 TiO_2 that is mainly in anatase form and has 50 m^2/g BET-surface area at 20-30 nm mean particle size (Bickley et al. 1991, Rachel et al. 2002). Lindner et al. (1997) reported that UV100 comprises 100% anatase having a 250 m^2/g specific BET-surface area and 5 nm primary particle size. PC500 photocatalyst has a 250 m^2/g BET-surface area with 100% anatase and 5-10 nm primary particle size (Rachel et al. 2002). TiO_2-Tytanpol A11 is totally in anatase form and has 37.3 nm crystalline size, 11.4 m^2/g BET specific surface area, a mean pore diameter 7.7 nm and an energy band-gap 3.31 eV (Zielin´ska et al. 2003). Three different photocatalysts P25, UV100, and PC500 were examined by Saquib and Muneer (2003) to analyze the photodegradation kinetics of gentian violet. From the results, it was found that photodegradation and mineralization of dye continue very quickly when P25 photocatalyst was utilized as compared to others. In P-25, a high concentration of holes and electrons are accessible to initiate photocatalytic reaction due to the slow recombination of holes and electrons. Qamar et al. (2005) also studied the photocatalytic performance of P25, UV100, and PC500 for the photodegradation of amido black 10B and chromotrope 2B. Results showed that photodegradation of both the dyes continues much quickly with Degussa P25. Giwa et al. (2012) analyzed the photodegradation of Reactive Violet 1 and Reactive Yellow 81 by utilizing TiO_2-P25 and TiO_2 anatase photocatalyst and results showed that Degussa P25 TiO_2 is more efficient compared to TiO_2. The high photocatalyst efficiency of TiO_2-P25 has been credited mainly to two factors: large surface area and slow recombination of holes and electrons (Muruganandham et al. 2006). According to the study of Zielin´ska et al. (2003), TiO_2-Degussa P25 was found to be more efficient compared to TiO_2-Tytanpol A11 for the degradation of Acid Blue 7 (AB7), Reactive Red 198 (RB198), and Acid Black 1.

Few studies are available in literature in which combustion synthesized TiO_2 was found to be a much better photocatalyst than P-25 TiO_2. According to the study of Sivalingam et al. (2003), methylene blue of initial concentration 200 ppm was completely degraded by combustion synthesized TiO_2 of loading 1 kg/m^3 under UV exposure in 65 minutes, while P-25 TiO_2 showed lessening in concentration only up to 100 ppm under similar conditions. Nosaka and Fox (1988) studied the degradation of methylene blue at an initial concentration of 100 ppm and 1 kg/m^3 photocatalyst loading under solar radiation. Complete photodegradation of methylene blue was obtained in 3 hours and 40 minutes with combustion synthesized TiO_2, while 5 hours was needed when P-25 photocatalyst was utilized under identical conditions. This specifies that the performance of the combustion synthesized TiO_2 was far better compared to photocatalyst Degussa P-25 for both solar radiations and UV irradiations. The enhanced performance of combustion synthesized TiO_2 could be accredited to the pure anatase crystal structure, high crystallinity, and high surface area.

8.3 Effect of Photocatalyst Concentration

Another important parameter that can influence the performance of TiO_2 photocatalyst is its concentration (Reza et al. 2017, Kumar and Pandey 2017a, Chen et al. 2020). It is deliberated that the photocatalyst activity can be augmented as the concentration of the photocatalyst increased (Krýsa et al. 2004, Kumar and Pandey 2017a) which is the feature of heterogeneous photocatalysis. According to the literature, an increased amount of photocatalyst intensifies the number of active sites on the surface of photocatalyst, hence initiating increment in the formation of $^{\bullet}OH$ radicals that participate in actual photodegradation of the dye solution. However, after a certain limit of concentration of the photocatalyst, the solution of dye turns turbid, so it restricts the UV radiation and percentage degradation starts declining (Coleman et al. 2007, Chun et al. 2000). Therefore, the optimum level of photocatalyst concentration has to be determined.

According to Konstantinou and Albanis (2004), in any of the reactors, initial rates were found to be directly proportional to the concentration of the photocatalyst when TiO_2-assisted photocatalyst was utilized for the photodegradation of azo dyes in an aqueous solution. Further, it was analyzed that there is a maximum limit of the concentration of the photocatalyst which may be utilized to degrade pollutants in wastewater and above which there will be a reduction in photocatalysis rate. These results were also supported by several other researchers (Chakrabarti and Dutta 2004, Huang et al. 2008, Sun et al. 2008, Liu et al. 2006).

Kumar and Pandey (2017a) studied the photodegradation of methyl orange dye at different concentrations of photocatalyst and found that the photodegradation of methyl orange dye increased with an increase in the concentration of photocatalyst. A similar trend was also reported by Kumar and Pandey (2017b) for the photodegradation of Eriochrome black T (EBT). Kumar and Pandey (2017c) observed enhancement in the rate of photodegradation of acetic acid with a rise in titania and Co, Ni modified titania concentration. To observe the influence of photocatalyst concentration on photodecomposition of gentian violet, Saquib and Muneer (2003) varied the concentration of P-25 TiO_2 from 0.5 to 5 g/l and showed an enhancement in the rate of photodegradation with an increase in the concentration of the photocatalyst. Qamar et al. (2005) studied the photodegradation rate of amido black 10B and chromotrope 2B were found to improve with a rise in the concentration of the photocatalyst. Neppolian et al. (2002b) studied the rate of photodegradation of Reactive Yellow 17, Reactive Blue 4, and Reactive Red 2 by changing the concentration of TiO_2 photocatalyst from 100 to 600 mg per 100 ml of dye solution. Results showed that there was a quick increase in the percentage photodegradation of all the dyes as the concentration of TiO_2 was increased from 100 to 300 mg/100 ml. However, it was found that the percentage photodegradation was enhanced slightly in the 300-500 mg concentration range of TiO_2 in the case of dyes Reactive Blue 4 and Reactive Red 2; however, percentage photodegradation was constant in the case of Reactive Yellow 17 (RY17) signifying that there is an upper level for the photocatalyst effectiveness. Additional increment in TiO_2 concentration from 500 to 600 mg for the degradation of Reactive Yellow 17 showed a reduction in the percentage photodegradation. Hung and Yuan (2000) found that the rate of photodegradation of Orange G was proportionate to the concentration of photocatalyst TiO_2 when varied in the range of 300 to 2,000 mg/l. Chakrabarti and Dutta (2004) stated the main reason behind this trend is that as the concentration of photocatalyst is increased, there is an increment in the number of active sites on the surface of the photocatalyst, which further increase the number of superoxide and hydroxyl radicals. Again, when the concentration of photocatalyst rises above the optimum value, the rate of photodegradation starts declining because of the interference of light via suspension. Huang et al. (2008) reported that the rise in the concentration of photocatalyst beyond the optimal value resulted in the accumulation of particles of the photocatalyst, therefore that part of the surface of photocatalyst becomes unavailable for the absorptions of photons and reduces the rate of photodegradation.

Increased concentration of photocatalyst enhances the number of photons absorbed on the surface of the photocatalyst, subsequently increases the rate of photodegradation (Muruganandham and Swaminathan 2006, Herrmann 1995). However, an increased amount of photocatalyst also intensified the opacity of the solution and led to a reduction in the penetration of photon flux within the reactor, thus decreasing the rate of photocatalytic degradation (Gogate and Pandit 2004, Kamble et al. 2003). Kaneco et al. (2004) also observed a loss in surface area by agglomeration of the photocatalyst at high concentrations.

8.4 Effect of Dopant Concentration

Effect of concentration of dopant on the photocatalyst activity of the photocatalyst has been investigated by numerous researchers (Wei et al. 2007, Ding et al. 2008, Rengaraj et al. 2006, Xu et al. 2002). It was observed that photocatalytic degradation of pollutants increases with an increase in the concentration of dopants. The main role of a dopant is to trap the electrons effectively and initiate the reaction with the molecules of water and form oxidative radicals for the degradation of pollutants. However, if the dopant is in excess amount, it inhibits the photocatalyst to capture the photons confining the creation of holes and electron pairs and reducing the efficiency of the photocatalyst (Gaidau et al. 2016, Arabatzis et al. 2003). Chun-Te Lin et al. (2018) reported that the rate of photodegradation increased with an increase in the concentration of dopant (0.07 to 0.21

mol%); however, when concentration was further increased from 0.28 and 0.35 mol%, the rate of photodegradation dropped.

According to the study of Zhiyong et al. (2007), discoloration of Orange II was improved if TiO_2 was doped with different mol% of Zn via utilizing $ZnSO_4$. Based on results, the optimum dopamine level was found to be 2 mol% Zn because, at 4 mol%, the adverse effect of Zn was seen on the surface area of TiO_2 inhibiting the absorption of reactants.

Li et al. (2007) explained the reason behind the enhancement in photoactivity of the photocatalyst with an increase in the concentration of the dopant as:

(1) In the case of alkaline earth metals, a space charge layer was formed on the surface of the photocatalyst TiO_2 due to the deposition of metal-oxide, which can help to separate the pairs of photoinduced holes and electrons. With the increase in the concentration of dopant, there is also an increment in the surface barrier, and the pairs of holes and electrons inside the region are then efficiently separated by the large electric field.
(2) As there is a difference in the electron negativity between Titania (Ti) and Metal (M), Ti–O–M formed through M^{2+} entering into the shallow surface of TiO_2 may encourage the transfer of charges and subsequently, there is a rise in the activity of the photocatalyst.
(3) Doping of alkaline earth metal ions can cause a deformation in the lattice, which further produces defects within the crystal. The presence of defects can obstruct the recombination of holes and electrons and ultimately, augment the photocatalytic activity.

Surplus amount or beyond the optimum value of the dopant lessens the specific area of photocatalyst, and there is a hindrance in the adsorption of reactant and ultimately constrains the photocatalytic activity (Liu et al. 2005). Excess amount of dopant significantly refines the TiO_2 from UV light which prevent the interfacial transfer of holes and electrons, resulting in a smaller photocatalytic activity (Li et al. 2007).

8.5 Concentration and Nature of Pollutants

The rate of photocatalytic degradation of pollutants is dependent on its nature, concentration, and other existing compounds present in wastewater. Studies are available in the literature that suggested that the rate of photocatalysis of TiO_2 was found to be dependent on the concentration of pollutants present in water (Chen and Mao 2007). During the process of photocatalysis, only the quantity of pollutant which gets adsorbed on the surface of the photocatalyst contributes to the photodegradation of the pollutant, but not the whole quantity which is present in the solution. So, the initial concentration of the pollutant affects its adsorption on the surface of the photocatalyst. In general, it can be said that the rate of photodegradation decreases with an increase in the number of pollutants while keeping the amount of catalyst constant (Reza et al. 2017).

According to Avasarala et al. (2010), as the initial concentration of pollutants increases, more molecules of pollutants will be available for excitation and energy transfer. The dependency is possibly associated with the development of numerous monolayers of adsorbed pollutants on the surface of TiO_2, which generally happens at a high concentration of pollutants until the optimized level is reached (Kiriakidou et al. 1999). On contrary, there is a reduction in the rate of photodegradation as the concentration of pollutants increases which may happen due to several reasons:

(i) With an increase in the initial concentration of pollutants, its molecules get adsorbed on the surface of the photocatalyst, and then a substantial amount of UV light gets absorbed by the molecules of pollutants and not the particles of TiO_2. So, the penetration of light on the surface of photocatalyst reduces (Kiriakidou et al. 1999, Daneshvar et al. 2003, Augugliaro et al. 2002, Saggioro et al. 2011, Liu et al. 2006, Segne et al. 2011, Kumar and Pandey 2017a) which decreases the photonic efficiency and deactivates the photocatalyst (Sobczynski and Dobosz 2001, Saggioro et al. 2011, Davis et al. 1994).

(ii) The formation of $^{\bullet}OH$ radicals reduces since the available active sites of the photocatalyst are engaged by the molecules of the pollutants (Grzechulska and Morawski 2002, Daneshvar et al. 2003, Kumar and Pandey 2017a).

(iii) The adsorbed molecules of pollutants on the surface of the photocatalyst inhibit the reaction of adsorbed molecules with $^{\bullet}OH$ radicals and photo-induced positive holes, as there is no direct contact of photocatalyst with them (Grzechulska and Morawski 2002, Daneshvar et al. 2003, Poulios and Aetopoulou 1999, Kiriakidou et al. 1999, Kumar and Pandey 2017a).

Neppolian et al. (2002b) studied the effect of initial concentration of pollutants on the rate of photodegradation of Reactive Yellow 17, Reactive Red 2, and Reactive Blue 4 when their initial concentration was varied from $8*10^{-4}$ to $1.2*10^{-3}$ M, $4.16*10^{-4}$ to $1.25*10^{-3}$ M and $1*10^{-4}$ to $5*10^{-4}$ M, respectively, keeping the optimized concentration of the photocatalyst. Results showed the rate of photodegradation reduces with an increase in the initial concentration of dye. These results were in agreement with the results obtained by several others researchers (Zhang et al. 2002, Saggioro et al. 2011, Giwa et al. 2012). So, it can be concluded that with an increase in the initial concentration of pollutants, the requirement of the surface of photocatalyst also rises. Since, the amount of photocatalyst, time of irradiation, and the $^{\bullet}OH$ radical that is a primary oxidant formed on the surface of the photocatalyst TiO_2 is also constant, therefore the relative number of free radicals targeting the pollutants reduces with the rise in the concentration of pollutants (Mengyue et al. 1995).

Moreover, the nature of the targeted pollutant affects the activity photocatalyst significantly. For instance, 4-chlorophenol generally takes a long duration of irradiation due to its conversion to intermediates in comparison to oxalic acid, which is converted directly to water and carbon dioxide (Tian et al. 2009). Besides, if the targeted pollutant adheres efficiently on the surface of the photocatalyst, then the process of photocatalysis would be more efficient in eliminating such kinds of pollutants from the given solution (Cheng et al. 2013).

8.6 Effect of Reaction Temperature

It is reported that as the temperature of the reaction rises, there is an improvement in the activity of the photocatalyst but if the reaction temperature is greater than 80°C then it leads to the recombination of electrons and holes further decreasing the adsorption of organic compounds on the surface of TiO_2 (Hashimoto et al. 2005). A reaction temperature less than 80°C is favorable for the adsorption, whereas when the reaction temperature approaches 0°C, it results in a rise in the activation energy of the photocatalyst (Peral et al. 1997). Consequently, temperature between 20-80°C has been found as the preferred range of temperature for the effective photodegradation of organic pollutants (Mamba et al. 2014).

According to Pawar et al. (2018), the efficiency of the photocatalyst decreases with a rise in reaction temperature which is due to the enhanced recombination of electrons and hole pairs and boosts the desorption of adsorbed species from the external surface of the photocatalyst.

Mozia et al. (2009) examined that the rate of photodegradation for Acid Red 18 was enhanced by 11 and 13% at 0.3 and 0.5 g/dm^3 of TiO_2 when solution temperature was varied from 313 to 323 K, respectively. Further increasing the temperature up to 333 K resulted in an additional 10% enhancement in the rate of photodegradation for both the concentration of photocatalyst. Reducing the temperature of the solution is favorable for the adsorption of the reactants, which is a spontaneous exothermic phenomenon. Additionally, depressing the temperature of the solution is also favorable for the adsorption of the final reaction products; whose desorption prevents the reaction. Mehrotra et al. (2005) reported exothermic adsorption of the reactants becomes unfavorable and tends to restrict the reaction when the temperature is more toward the boiling point of water. Soares et al. (2007) also studied the influence of temperature on the photodegradation reaction. Based on the results, the optimum temperature range was found between 40-50°C. At lower temperatures, desorption of the products confines the reaction as desorption is slower than the adsorption of the reactants, on the other hand when the temperature is high, the regulating stage turns out to be the adsorption of the pollutant on the surface of the photocatalyst TiO_2.

Kumar and Pandey (2017a) and Kumar et al. (2015) presented the influence of temperature on the rate of photodegradation of methyl orange and methyl red with TiO_2 and La and Co modified TiO_2. It was observed that when the temperature of the reaction was high then the rate of photodegradation was observed to be highest because the interaction between the molecules of the dye and the surface of the photocatalyst increases which further raises the rate of adsorption and hence enhanced the rate of photodegradation. However, the rate of photodegradation was found to be low at lower temperatures.

8.7 Substrate/Supported Types

When TiO_2 photocatalyst was used directly for the photodegradation of pollutants then it possesses several drawbacks like high cost of separation/recovery and low light utilization. So, according to the literature, selecting a substrate is an optimized way to solve the given problems. The photocatalytic influences can be considerably enhanced by integrating TiO_2 with different carbon-based nanomaterials like reduced graphene oxide, graphene oxide, graphene, fullerene, carbon dots, single and multi-walled carbon nanotubes (Suarez-Iglesias et al. 2017, Irie et al. 2003) inside the matrices of the semiconductor.

Graphene has received the attention of numerous researchers among all the listed carbon-based nanomaterials, (Akhavan et al. 2011, Novoselov et al. 2004, Zhang et al. 2010) due to its high specific surface area, high thermal and electrical conductivity, and high Young's modulus (Zhu et al. 2010). Its outstanding electron conductivity permits the flow of electrons from the semiconductor to its surface, preventing the recombination of holes and electrons (Hsu et al. 2013). Zhang et al. (2011) proved that the chemical bonds formed between graphene sheets and TiO_2 to some extent reduce the band-gap resulting in rapid migration of electrons (Zhang et al. 2011). The foam-like macro-structures of graphene increase the number of reactive sites for the process of photocatalysis, resulting in the augmentation of photocatalytic activities. Additionally, two and three-dimensional structures created by the hexagonal rings of sp^2 hybridized carbon atoms, support electron transport and charge separation properties (Atchudan et al. 2016). According to the study of Wang et al. (2012), the photodegradation of methylene blue was enhanced up to 82.5% when graphene-TiO_2 was utilized as photocatalyst due to the presence of highly reactive sites while photodegradation enhanced only up to 35.5% when in case of pure TiO_2 under visible light. It is also confirmed by several researchers that a graphene-based TiO_2 heterojunction provides high efficiency of the photocatalyst as compared to pure TiO_2. Graphene-TiO_2 is found to be an efficient photocatalyst for the degradation of anionic or cationic dyes. Nanocomposites that are produced by utilizing the hydrothermal method exhibited improved adsorption of methylene blue due to the negatively charged surface of the graphene oxide matrix (Zhang et al. 2010). Chen et al. (2020) observed that while utilizing graphene-based substrate, the 90% and 100% degradation of organic pollutants was achieved under UV and visible light, respectively. The close contact between graphene and TiO_2 permits the utilization of their synergistic effect and improves its adsorption capacity (Stengl et al. 2011). Chong et al. (2013) reported the photodegradation of methylene blue by 10% when TiO_2 was utilized for 120 minutes, whereas photodegradation was enhanced up to 33% with the usage of CuS-graphene oxide/TiO_2. Fernandez-Ibanez et al. (2015) reported that the required amount of solar UVA amount was 18.4 kJ/m^2 to deactivate 4 M CFU/mL of bacteria in the case of pure TiO_2, while in the case of graphene-TiO_2, the required amount of solar UVA was only 12.2 kJ/m^2 to deactivate the same bacteria. To attained optimal photodegradation, an adequate concentration of graphene is required to accomplish optimum photodegradation. The photodegradation of dyes was improved by increasing the amount of reduced graphene oxide <5%. The combination with >5% reduced graphene oxide offered considerably lesser activity of the photocatalyst. The reduction in the activity of the photocatalyst was possibly due to the fact that the surface of TiO_2 was shielded by reduced graphene oxide (Brindha and Sivakumar 2017). Due to tunable optical absorption and better performance in electron transfer for organic molecules, carbon dots were also deliberated as a potential metal-free substrate to support TiO_2 photocatalyst (Yu et al. 2016, Ge et al. 2016).

8.8 Effect of pH

The pH of the solution has a significant effect on the photodegradation of pollutants present in wastewater (Kumar and Pandey 2017c, Chen et al. 2020). However, the discussion on the influence of pH is a challenging task due to its numerous roles during the process of photodegradation (Konstantinou and Albanis 2004). The pH of the solution alters the interface of the electrical double layer of the solid electrolyte, thereby affecting the processes of sorption–desorption, and separation of photo-generated holes and electrons onto the surface of the photocatalyst. As TiO_2 is amphoteric in nature so either a negative or a positive charge can be developed on its surface (Guillard et al. 2003, Poulios et al. 2000, Senthilkumaar et al. 2006, Zielin´ska et al. 2003). Under alkaline or acidic conditions, the surface of the titania can be deprotonated or protonated, respectively, according to the following reactions (Davis et al. 1994).

Reactions	Equation No.
$TiOH + H^+ \rightarrow TiOH_2^+$	17
$TiOH + OH^- \rightarrow TiO^- + H_2O$	18

Thus, the surface of titania surface will remain negatively or positively charged in alkaline or acidic medium, respectively. Variation in pH can affect the adsorption of molecules of the given dye onto the surface of photocatalyst TiO_2 (Wang et al. 2008) by modifying the charge of the surface of photocatalyst TiO_2 which alters the potentials of photocatalytic reactions. It is reported that at lower pH, TiO_2 possesses higher oxidizing activity, but an excess amount of H^+ reduces the rate of reaction. TiO_2 acts as a strong Lewis acid due to the positively charged surface, while the anionic dye behaves as a strong Lewis base and can easily adsorb on the surface of the TiO_2 photocatalyst. This condition is favorable for the adsorption of dye under acidic conditions, while under alkaline conditions this process is unfavorable apparently due to the competitive adsorption by hydroxyl groups and the dye molecules in addition to the Coulombic repulsion of the negatively charged photocatalyst with the dye molecule (Mozia et al. 2009, Huang et al. 2008, Lu et al. 2017).

Chen et al. (2020) analyzed that the optimum pH for the photodegradation of organic pollutants is in the acidic range. Additionally, pH modifies the path for the photocatalytic degradation of organic pollutants. Li et al. (2008) reported that methyl orange was transformed to quinine under strongly acidic conditions and consequently quinine formed could be protected by zeolite. However, the absorption of methyl orange under basic or neutral conditions was only 3%. The photodegradation of Bromocresol Purple in a pH range of 4.5 to 8.0 was examined by Baran et al. (2008). A 6-fold enhancement in the adsorption of dye was evaluated when the wastewater solution was acidified from pH 8.0 to 4.5. The observed increment began with a change in the charge of the Bromocresol Purple. The immediate substantial enhancement in the rate of photodegradation of dye specifies that change in pH also affects the rate of decomposition of dye. Konstantinou and Albanis (2004) reported that the rate of photodegradation of azo dyes enhanced as there is a reduction in the pH value. It was observed that at a pH value of 6.8, molecules of dye are negatively charged under an alkaline medium and their adsorption is estimated to be affected by a rise in the density of TiO_2 groups on the surface of the photocatalyst. Therefore, due to Coulombic repulsion, dyes are barely adsorbed on the surface of the photocatalyst (Lachheb et al. 2002, Stylidi et al. 2003). A rapid photodegradation of Orange G was analyzed under more basic or acidic conditions, whereas photodegradation was very slow under neutral conditions (Hung and Yuan 2000). It was attributed to the fact that more number of OH^- exist in the solution, consequently lead to the creation of more •OH.

Bubacz et al. (2010) analyzed that with the rise in the value of pH, the rate of photodegradation of methylene blue was also increased. Ling et al. (2004) reported that at a higher value of pH, the electrostatic interactions between methylene blue cation and negative TiO^- leads to the strong adsorption with a corresponding higher rate of photodegradation. As methylene blue is a cationic dye, it is feasible that at higher pH, its adsorption is favored on a negatively charged surface. It is also found that the surface charge properties of TiO_2 change as the value of the pH change due to the amphoteric nature of TiO_2 (Guillard et al. 2003, Senthilkumaar et al. 2006, Zielin´ska et al. 2003).

According to the study of Akpan and Hamed (2009), it was observed that different types of pollutants or dyes behave in dissimilar ways during the process of photocatalysis. Some of the dyes are photodegraded at lower pH, whereas some of the dyes do so at a higher value of pH. This can be accredited to the nature of the dye which is to be photodegraded. Thus, it is essential to first analyze the nature of the dyes or pollutants and determine the exact value of pH for their complete photodegradation. Please summarize this section too.

9. CONCLUSIONS AND FUTURE RECOMMENDATIONS

The rapid growth in industrialization and its resulting by-products have adversely affected the environment by generating hazardous pollutants and poisonous gas fumes and smokes, which have been directly released into the environment like in water bodies and in air. Like development in industries, there is also a fast growth in the area of nanotechnology to treat wastewater. The treatment of pollutants present in wastewater is a great challenge and nanomaterials play a significant role in their remediation. The unique properties of nanomaterials could be applied to degrade numerous pollutants present in wastewater. Conventional technologies such as adsorption, biological treatment, ion exchange, and chemical oxidation have been applied for the treatment of numerous pollutants present in wastewater. Among all the technologies available, advanced oxidation techniques, i.e. photocatalysis possess several advantages such as it does not require energy, low-cost, rapid, and complete degradation of different types of pollutants present in wastewater. For the photodegradation of pollutants which exist in wastewater several semiconductors and transition metals, like TiO_2, ZnO, Fe_2O_3, CdS, SnO_2, SiC, and ZnS are utilized in the form of nanoparticles or nanocomposites (nanoscale semiconductors or metals). Surface adsorption and photocatalytic degradation of wastewater pollutants can be enhanced by nano-sized semiconductors because of the reduced size of the semiconductor which provides a high specific surface area. Among all the photocatalysts, TiO_2 is a well-established photocatalyst due to its high oxidation efficiency, excellent incident photoelectric conversion efficiency, non-toxicity, long-time corrosion resistance, high photostability, chemical inertness, and eco-friendly. Due to the presence of all these properties, TiO_2 is able to oxidize a wide range of pollutants, like dyes, phenols, and pesticides present in wastewater under UV radiations. So, the main focus of this chapter was more toward the removal of pollutants present in wastewater by utilizing TiO_2 nanoparticles as photocatalysts.

In spite of the utilization of TiO_2 at a larger scale and improvement in the past few decades, still there are some obstacles that hinder the commercialization of TiO_2 based photocatalytic technology. So, more research is required to overcome all the limitations of TiO_2 photocatalyst and further develop the use of TiO_2 nanoparticles for the photocatalytic degradation of the pollutants.

- Numerous studies related to the modification of the surface of TiO_2 to enhance its photocatalytic activity, but limited literature is available in which augmentation in the activity of the photocatalyst under visible light of magnetic TiO_2 nanoparticles. In addition to the physio-chemical stability of non-metal doped TiO_2 particles, there is no study available in which repetitive usage of non-metal doped TiO_2 particles has been reported. So, it is essential to propose a multi-functional TiO_2 photocatalyst that incorporates the features like high adsorption capacity, excellent photocatalytic activity under visible light, magnetic separability, and high stability.

- An exhaustive degradation of a few of the pollutants is still difficult to attain with the utilization of TiO_2 photocatalyst. So, more research should be carried out concerning the enhancement of photocatalyst activity of TiO_2. Moreover, it is also required to search the opportunities of combined usage of TiO_2 photocatalyst with the available technologies like biological method or adsorption to expand the scope of environmental remediation like wastewater treatment. Additionally, the intermediates or the ultimate products after photodegradation may or may not be non-toxic. Sometimes, the photodegraded products might cause more harm as compared to the original pollutant and become the reason for the production of secondary pollution and even causes a reduction in the rate of photodegradation. A very few reports are available in which the overall photocatalytic process of TiO_2 is reported particularly for the modified TiO_2 photocatalysts. So, it is necessary to quantify and characterize the toxicity, reactivity, and providence of all the modified photocatalysts.
- The photocatalyzed reactions of TiO_2 are non-selective in nature because these reactions are directed by a free radical mechanism, so the photodegradation rate of various wastewater pollutants is found to be close. While sometimes this absence of selectivity could be beneficial; however, a reduced selectivity suggests that the photocatalyst does not distinguish among the low or high toxic pollutants. Pollutants having low toxicity in nature can be easily photodegraded with the use of a biological method, whereas pollutants which are having high toxicity are non-biodegradable. Thus, it is required to improve such a system of photocatalyst which can selectively photodegrade the pollutants under visible light.
- The effect of pH on the photodegradation of various pollutants has been reported by several researchers. While studying the effect of pH on the photocatalytic degradation of pollutants, two things must be considered, that is (i) the industrial wastewater may not be neutral and (ii) the pH of waste-water can influence the surface-charge properties of the photocatalyst.
- Above all, there is an essential requirement of such a real system in which can perform the task of photodegradation of pollutants and can recover the photocatalyst particles from the treated water.

References

Adachi, K., Ohta, K. and Mizuno, T. 1994. Photocatalytic reduction of carbon dioxide to hydrocarbon using copper-loaded titanium dioxide. Sol. Energ. 53: 187-190.

Ademollo, N., Ferrara, F., Delise, M., Fabietti, F. and Funari, E. 2008. Nonylphenol and octylphenol in human breast milk. Environ. Int. 34: 984-987.

Ahammad, S.Z., David, W. and Dolfing, G.J. 2013. Wastewater Treatment: Biological. Encyclopedia of Environmental Management. Taylor & Francis. 2645-2656. doi: 10.1081/E-EEM-120046063.

Ahel, M., Giger, W. and Koch, M. 1994. Behaviour of alkylphenol polyethoxylate surfactants in the aquatic environment—i. Occurrence and transformation in sewage treatment. Water Res. 28: 1131-1142.

Ahmadimoghaddam, M., Mesdaghinia, A., Naddafi, K., Nasseri, S., Mahvi, A.H., Vaezi, F., et al. 2010. Degradation of 2,4-dinitrophenol by photo fenton process. Asian J. Chem. 22(2): 1009-1016.

Akerdi, A.G. and Bahrami, S.H. 2019. Application of heterogeneous nano-semiconductors for photocatalytic advanced oxidation of organic compounds: A review. J. Environ. Chem. Eng. 7: 103283.

Akhavan, O., Ghaderi, E. and Esfandiar, A. 2011. Wrapping bacteria by graphene nanosheets for isolation from environment, reactivation by sonication, and inactivation by near-infrared irradiation. J. Phys. Chem. B 115: 6279-6288.

Akpan, U.G. and Hameed, B.H. 2009. Parameters affecting the photocatalytic degradation of dyes using TiO_2 –based photocatalysts: A review. J. Hazard. Mater. 170: 520-529.

Akpor, O.B., Otohinoyi, D.A., Olaolu, T.D. and Aderiye, B.I. 2014. Pollutants in wastwater effluents: Impacts and remediation processes. IJERES 3: 050-059.

Alcock, R.E., Halsall, C.J., Harris, C.A., Johnston, A.E., Lead, W.A., Sanders, G., et al. 1994. Contamination of environmental samples prepared for PCB analysis. Environ. Sci. Technol. 28: 1838-1842.

Alrousan, D.M.A., Polo-Lopez, M.I., Dunlop, P.S.M., Fernández-Ibáñez, P. and Byrne, J.A. 2012. Solar photocatalytic disinfection of water with immobilised titanium dioxide in recirculating flow CPC reactors. Appl. Catal. B. 128: 126-134.

Alshammari, M.S. 2020. Assessment of sewage water treatment using grinded bauxite rock as a robust and low-cost adsorption. J. Chem. 2020: 1-5. doi:10.1155/2020/7201038.

Ambaye, T.G. and Hagos, K. 2020. Photocatalytic and biological oxidation treatment of real textile wastewater. Nanotechnol. Environ. Eng. 5: 28.

Ambrus, Z., Balazs, N., Alapi, T., Wittmann, G., Sipos, P., Dombi, A., et al. 2008. Synthesis, structure and photocatalytic properties of Fe(III)-doped TiO_2 prepared from $TiCl_3$. Appl. Catal. B. 81: 27-37.

Amin, M. 2013. Methods for preparation of nano-composites for outdoor insulation applications. Rev. Adv. Mater. Sci. 34: 173-184.

Andronic, L. and Enesca, A. 2020. Black TiO_2 synthesis by chemical reduction methods for photocatalysis applications. Front. Chem. 8: 565489. doi: 10.3389/fchem.2020.565489.

Anjaneyulu, Y., Sreedhara, C.N. and Samuel, S.R.D. 2005. Decolourization of industrial efuents: Available methods and emerging technologies: A review. Rev. Environ. Sci. Bio/Technol. 4: 245-273.

Anku, W.W., Mamo M.A. and Govender, P.P. 2017. Phenolic Compounds in Water: Sources, Reactivity, Toxicity and Treatment Methods. 419-443. http://dx.doi.org/10.5772/66927.

Arabatzis, I., Stergiopoulos, T., Bernard, M., Labou, D., Neophytides, S. and Falaras, P. 2003. Silver modified titanium dioxide thin films for efficient photodegradation of methyl orange. Appl. Catal. B. 42: 187-201.

Araújo, J., Delgado, F.I. and Paumgartten, F.J.R. 2016. Glyphosate and adverse pregnancy outcomes, a systematic review of observational studies. BMC Public Heath 16: 472-480.

Archana, P.S., Jose, R., Yusoff, M.M. and Ramakrishna, S. 2011. Near band-edge electron diffusion in electrospun Nb-doped anatase TiO_2 nanofibers probed by electrochemical impedance spectroscopy. Appl. Phys. Lett. 98: 152103.

Archana, P.S., Jose, R., Jin, T.M., Vijila, C., Yusoff, M.M. and Ramakrishnaw, S. 2010. Structural and electrical properties of Nb-doped anatase TiO_2 nanowires by electrospinning. J. Am. Ceram. Soc. 93: 4096-4102.

Asahi, R., Morikawa, T., Ohwaki, T., Aoki, K. and Taga, Y. 2001. Visible-light photocatalysis in nitrogen-doped titanium oxides. Science 293: 269-271.

Ashraf, M.A. 2017. Persistent organic pollutants (POPs): A global issue, a global challenge. Environ. Sci. Pollut. Res. 24: 4223-4227.

Atchudan, R., Edison, T.N.J.I., Perumal, S., Karthikeyan, D. and Lee, Y.R. 2016. Facile synthesis of zinc oxide nanoparticles decorated graphene oxide composite via simple solvothermal route and their photocatalytic activity on methylene blue degradation. J. Photochem. Photobiol. B, Biol. 162: 500-510.

Atta, N.F., Amin, H.M.A., Khalil, M.W. and Galal, A. 2011. Nanotube arrays as photoanodes for dye sensitized solar cells using metal phthalocyanine dyes. Int. J. Electrochem. Sci. 6: 3316-3332.

Augugliaro, V., Baiocchi, C., Prevot, A.B., Garcia-Lopez, E., Loddo, V., Malato, S., et al. 2002. Azo-dyes photocatalytic degradation in aqueous suspension of TiO_2 under solar irradiation. Chemosphere 49: 1223-1230.

Avasarala, B.K., Tirukkovalluri, S.R. and Bojja, S. 2010. Synthesis, characterization and photocatalytic activity of alkaline earth metal doped titania. Indian J. Chem. 49A: 1189-1196.

Baby, R., Saifullah, B. and Hussein, M.Z. 2019. Carbon nanomaterials for the treatment of heavy metal-contaminated water and environmental remediation. Nanoscale Res. Lett. 14: 341.

Bagbi, Y., Sarswat, A., Mohan, D., Pandey, A. and Solanki, P.R. 2017. Lead and chromium adsorption from water using L-cysteine functionalized magnetite (Fe_3O_4) nanoparticles. Sci. Rep. 7: 7672.

Bahnemann, D.W., Kormann, C. and Hoffmann, M.R. 1987. Preparation and characterization of quantum size zinc oxide: A detailed spectroscopic study. J. Phys. Chem. 91: 3789-3798.

Banerjee, S., Pillai, S.C., Falaras, P., O'Shea, K.E., Byrne, J.A. and Dionysion, D.D. 2014. New insights into the mechanism of visible light photocatalysis. J. Phys. Chem. Lett. 5(15): 2543-2554.

Bannat, I., Wessels, K., Oekermann, T., Rathousky, J., Bahnemann, D. and Wark, M. 2009. Improving the photocatalytic performance of mesoporous titania films by modification with gold nanostructures. Chem. Mater. 21: 1645-1653.

Barakat, M.A. 2011. New trends in removing heavy metals from industrial wastewater. Arab. J. Chem. 4: 361-377.

Baran, W., Makowski, A. and Wardas, W. 2008. The effect of UV radiation absorption of cationic and anionic dye solutions on their photocatalytic degradation in the presence TiO_2. Dyes Pigm. 76: 226-230.

Beard, J. 2006. DDT and human health. Sci. Total Environ. 355: 78-89.

Bellona, C., Drewes, J.E., Xu, P. and Amy, G. 2004. Factors affecting the rejection of organic solutes during NF/RO treatment: A literature review. Water Res. 38: 2795-2809.

Benjwal, P. and Kar, K.K. 2015. Removal of methylene blue from wastewater under a low power irradiation source by Zn, Mn co-doped TiO_2 photocatalysts. RSC Adv. 5: 98166-98176.

Bessekhouad, Y., Robert, D. and Weber, J.V. 2004. Bi_2S_3/TiO_2 and CdS/TiO_2 heterojunctions as an available configuration for photocatalytic degradation of organic pollutant. J. Photochem. Photobiol. A Chem. 163: 569-580.

Bhatkhande, D.S., Pangarkar, V.G. and Beenackers, A.A.C.M. 2001. Photocatalytic degradation for environmental applications: A review. J. Chem. Technol. Biotechnol. 77: 102-116.

Bibi, S., Khan, R.L. and Nazir, R. 2016. Heavy metals in drinking water of Lakki Marwat District, KPK, Pakistan. World Appl. Sci. J. 34: 15-19.

Bickley, R.I., Carreno, T.G., Lees, J.S., Palmisano, L. and Tilley, R.J.D. 1991. A structural investigation of titanium dioxide photocatalysts. J. Solid State Chem. 92: 178-190.

Blake, D.M., Webb, J., Turchi, C. and Magrini, K. 1991. Kinetic and mechanistic overview of TiO_2-photocatalyzed oxidation reactions in aqueous solution. Sol. Energy Mater. 24: 584-593.

Bolokang, A.S., Motaung, D.E., Arendse, C.J. and Muller, T.F.G. 2015. Morphology and structural development of reduced anatase-TiO_2 by pure Ti powder upon annealing and nitridation: Synthesis of TiO_x and TiO_xN_y powders. Mater. Charact. 100: 41-49.

Boulbar, E.L., Millon, E., Boulmer-Leborgne, C., Cachoncinlle, C., Hakim, B. and Ntsoenzok, E. 2014. Optical properties of rare earth-doped TiO_2 anatase and rutile thin films grown by pulsed-laser deposition. Thin Solid Films 553: 13-16.

Bouras, P., Stathatos, E. and Lianos, P. 2007. Pure versus metal-ion-doped nanocrystalline titania for photocatalysis. Appl. Catal. B: Environ. 73: 51-59.

Brigden, K., Hetherington, S., Wang, M., Santillo, D. and Johnston, P. 2013. Hazardous Chemicals in Branded Textile Products on Sale in 25 Countries/Regions during. Greenpeace: Amsterdam, The Netherlands.

Brindha, A. and Sivakumar, T. 2017. Visible active N, S co-doped TiO_2/graphene photocatalysts for the degradation of hazardous dyes. J. Photochem. Photobiol. A. 340: 146-156.

Bubacz, K., Choina, J., Dolat, D. and Morawski, A.W. 2010. Methylene blue and phenol photocatalytic degradation on nanoparticles of anatase TiO_2. Pol. J. Environ. Stud. 19: 685-691.

Cai, J., Huang, Z., Lv, K., Sun, J. and Deng, K. 2014a. Ti powder-assisted synthesis of Ti^{3+} self-doped TiO_2 nanosheets with enhanced visible-light photoactivity. RSC Adv. 4: 19588-19593.

Cai, L., Cho, I.S., Logar, M., Mehta, A., He, J., Lee, C.H., et al. 2014b. Sol-flame synthesis of cobalt-doped TiO_2 nanowires with enhanced electrocatalytic activity for oxygen evolution reaction. Phys. Chem. Chem. Phys. 16: 12299-12306.

Carmen, Z. and Daniela, S. 2012. Textile organic dyes—characteristics, polluting efects and separation/ elimination procedures from industrial efuents: A critical overview. *In*: Tomasz Puzyn and Aleksandra Mostrag [ed.]. Organic Pollutants Ten Years after the Stockholm Convention—Environmental and Analytical Update. 55-86.

Carp, O., Huisman, C.L. and Reller, A. 2004. Photoinduced reactivity of titanium dioxide. Prog. in Solid State Chem. 32: 33-117.

Chakrabarti, S. and Dutta, B.K. 2004. Photocatalytic degradation of model textile dyes in wastewater using ZnO as semiconductor catalyst. J. Hazard. Mater. B 112: 269-278.

Chambers, H.W., Boone, J.S., Carr, R.L. and Chambers, J.E. 2001. Chemistry of Organophosphorus Insecticides. Handbook of Pesticide Toxicology, 2nd edition, San Diego.

Chanathaworn, J., Bunyakan, C., Wiyaratn, W. and Chungsiriporn, J. 2012. Photocatalytic decolorization of basic dye by TiO_2 nanoparticle in photoreactor. Songklanakarin J. Sci. Technol. 34: 203-210.

Chang, S.M. and Liu, W.S. 2014. The roles of surface-doped metal ions (V, Mn, Fe, Cu, Ce, and W) in the interfacial behavior of TiO_2 photocatalysts. Appl. Catal. B Environ. 156-157: 466-475.

Chen, B., Haring, A.J., Beach, J.A., Li, M., Doucette, G.S., Morris, A.J., et al. 2014. Visible light induced photocatalytic activity of Fe^{3+}/Ti^{3+} co-doped TiO_2 nano-structures. RSC Adv. 4: 18033-18037.

Chen, C.C. and Lu, C.S. 2007. Photocatalytic degradation of Basic Violet 4: Degradation efficiency, product distribution, and mechanisms. J. Phys. Chem. C 111: 13922-13932.

Chen, D., Kannan, K., Tan, H., Zheng, Z., Feng, Y.L., Wu, Y., et al. 2016. Bisphenol analogues other than BPA: Environmental occurrence, human exposure, and toxicity: A review. Environ. Sci. Technol. 50: 5438-5453.

Chen, D., Cheng, Y., Zhou, N., Chen, P., Wang, Y., Li, K., et al. 2020. Photocatalytic degradation of organic pollutants using TiO_2-based photocatalysts: A review. J. Clean. Prod. 268: 121725.

Chen, D., Jiang, Z., Geng, J., Wang, Q. and Yang, D. 2007. Carbon and nitrogen co-doped TiO_2 with enhanced visible-light photocatalytic activity. Ind. Eng. Chem. Res. 46: 2741-2746.

Chen, X. and Mao, S.S. 2007. Titanium dioxide nanomaterials: Synthesis, properties, modifications, and applications. Chem. Rev. 107: 2891-2959.

Chen, X.Q., Liu, H.B. and Gu, G.B. 2005. Preparation of nanometer crystalline TiO_2 with high photocatalytic activity by pyrolysis of titanyl organic compounds and photocatalytic mechanism. Mater. Chem. Phys. 91: 317-324.

Chen, Y., Wang, K. and Lou, L. 2004. Photodegradation of dye pollutants on silica gel supported TiO_2 particles under visible light irradiation. J. Photochem. Photobiol. A 163: 281-287.

Cheng, H.Y., Chang, K.C., Lin, K.L. and Ma, C.M. 2018. Study on isopropanol degradation by UV/TiO_2 nanotube. AIP Conference Proceedings 1946: 020006: doi: 10.1063/1.5030310.

Cheng, X., Liu, H., Chen, Q., Li, J. and Wang, P. 2013. Construction of N, S codoped TiO_2 NCs decorated TiO_2 nano–tube array photoelectrode and its enhanced visible light photocatalytic mechanism. Electrochimica Acta. 103: 134-142.

Cho, M., Chung, H., Choi, W. and Yoon, J. 2004. Linear correlation between inactivation of E. coli and OH radical concentration in TiO_2 photocatalytic disinfection. Water Res. 38: 1069-1077.

Choi, J., Sudhagar, P., Lakshmipathiraj, P., Lee, J.W., Devadoss, A., Lee, S., et al. 2014. Three-dimensional Gd-doped TiO_2 fibrous photoelectrodes for efficient visible light-driven photocatalytic performance. RSC Adv. 4: 11750-11757.

Choi, S.D., Baek, S.Y. and Chang, Y.S. 2008. Atmospheric levels and distribution of dioxin-like polychlorinated biphenyls (PCBs) and polybrominated diphenyl ethers in the vicinity of an iron and steel making plant. Atmos. Environ. 42: 2479-2488.

Choi, Y., Termin, A. and Hoffmann, M.R. 1994. The role of metal ion dopants in quantum-sized TiO_2: Correlation between photoreactivity and charge carrier recombination dynamics. J. Phys. Chem. 98: 13669-13679.

Chong, M.N., Jin, B., Chow, C.W.K. and Saint, C. 2010. Recent developments in photocatalytic water treatment technology: A review. Water Res. 44: 2997-3027.

Chong Yeon, P., Ghosh, T., ZeDa, M., Kefayat, U., Vikram, N. and WonChun, O. 2013. Preparation of CuS-graphene oxide/TiO_2 composites designed for high photonic effect and photocatalytic activity under visible light. Chinese J. Catal. 34: 711-717.

Christoforidis, K.C. and Fernández-García, M. 2016. Photoactivity and charge trapping sites in copper and vanadium doped anatase TiO_2 nano-materials. Catal. Sci. Technol. 6: 1094-1105.

Chun, H., Yizhong, W. and Hongxiao, T. 2000. Destruction of phenol aqueous solution by photocatalysis or direct photolysis. Chemosphere 41: 1205-1209.

Chun-Te Lin, J., Sopajaree, K., Jitjanesuwan, T. and Lu, M.C. 2018. Application of visible light on copper-doped titanium dioxide catalyzing degradation of chlorophenols. Sep. Purif. Technol. 191: 233-243.

Chung, J., Chung, J.W. and Kwak, S.Y. 2015. Adsorption-assisted photocatalytic activity of nitrogen and sulfur codoped TiO_2 under visible light irradiation. Phys. Chem. Chem. Phys. 17: 17279-17287.

Coleman, H.M., Vimonses, V., Leslie, G. and Amal. R. 2007. Degradation of 1,4–dioxane in water using TiO_2 based photocatalytic and H_2O_2/UV processes. J. Hazard Mater. 146: 496-501.

Connell, D.W., Wu, R.S.S., Richardson, B.J. and Lam, P.K.S. 2006. Chemistry of organic pollutants, including agrochemicals. Environ. Ecol. Chem. 3: 1-10.

Cortes, D.R. and Hites, R.A. 2000. Detection of statistically significant trends in atmospheric concentrations of semivolatile compounds. Environ. Sci. Technol. 34: 2826-2829.

Cosnefroy, A., Brion, F., Guillet, B., Laville, N., Porcher, J.M., Balaguer, P., et al. 2009. A stable fish reporter cell line to study estrogen receptor transactivation by environmental (xeno) estrogens. Toxicol. Vitro 23: 1450-1454.

Costa, A.S., Romão, L.P.C., Araújo, B.R., Lucas, S.C.O., Maciel, S.T.A., Wisniewski, A. et al. 2012. Environmental strategies to remove volatile aromatic fractions (BTEX) from petroleum industry wastewater using biomass. Bioresour. Technol. 105: 31-39.

Costner, P. 1998. Intersessional Group Intergovernamenatl Forum on Chemical Safety. Yokohama, Japan.

Crini, G., Lichtfouse, E., Wilson, L.D. and Morin-Crini, N. 2019. Conventional and non-conventional adsorbents for wastewater treatment. Environ. Chem. Lett. 17: 195-213.

Crini, G. and Badot, P.M. 2010. Sorption Processes and Pollution. PUFC, Besançon.

Crini, G. and Lichtfouse, E. 2019. Advantages and disadvantages of techniques used for wastewater treatment. Environ. Chem. Lett. 17: 145-155.

Cuomo, F., Venditti, F., Ceglie, A., Leonardis, A.D., Macciola, V. and Lopez, F. 2015. Cleaning of olive mill wastewaters by visible light activated carbon doped titanium dioxide. RSC Adv. 5: 85586-85591.

Czoska, A.M., Livraghi, S., Paganini, M.C., Giamello, E., Di Valentin, C. and Pacchioni, G. 2011. The nitrogen-boron paramagnetic center in visible light sensitized N-B co-doped TiO_2: Experimental and theoretical characterization. Phys. Chem. Chem. Phys. 13: 136-143.

Dai, Q., Min, X. and Weng, M. 2016. A review of polychlorinated biphenyls (PCBs) pollution in indoor air environment. J. Air Waste Manag. Assoc. 66: 941-950.

Daneshvar, N., Salari, D. and Khataee, A.R. 2003. Photocatalytic degradation of azo dye acid red 14 in water: Investigation of the effect of operational parameters. J. Photochem. Photobiol. A 157: 111-116.

Das, S., Kuppanan, N., Channashettar, V.A. and Lal, B. 2018. Remediation of oily sludge- and oil-contaminated soil from petroleum industry: Recent developments and future prospects. pp. 165-177. *In*: Adhya, T.K. et al. [eds.]. Advances in Soil Microbiology: Recent Trends and Future Prospects. Springer.

Davis, R.J., Gainer, J.L., O'Neal, G. and Wu. I.W. 1994. Photocatalytic decolorization of wastewater dyes. Water Environ. Res. 66: 50-53.

De-Coster, S. and Larebeke, V.N. 2012. Endocrine-disrupting chemicals: Associated disorders and mechanisms of action. J. Environ. Public Health 2012: 1-52. doi: 10.1155/2012/713696.

Dean, J.G., Bosqui, F.L. and Lanouette, K.H. 1972. Removing heavy metals from waste water. Environ. Sci. Technol. 6: 518-522.

Demirbas, A. 2009. Agricultural based activated carbons for the removal of dyes from aqueous solutions: A review. J. Hazard. Mater. 167: 1-9.

Deng, F., Li, Y., Luo, X., Yang, L. and Tu, X. 2012. Preparation of conductive polypyrrole/TiO_2 nanocomposite via surface molecular imprinting technique and its photocatalytic activity under simulated solar light irradiation. Colloids Surf. A. Physicochem Eng. Asp. 395: 183-189.

Deng, Q., Liu, Y., Mu, K., Zeng, Y., Yang, G., Shen, F., et al. 2015. Preparation and characterization of F-modified C-TiO_2 and its photocatalytic properties. Phys. Status Solidi A 21: 691-697.

De Souza, M.L. and Corio, P. 2013. Effect of silver nanoparticles on TiO_2-mediated photodegradation of Alizarin Red. S. Appl. Catal. B. 136-137: 325-333.

Di, P. and Chang, D.P.Y. 2011. Investigation of polychlorinated biphenyl removal from contaminated soil using microwave-generated steam. J. Air Waste Manage. Assoc. 51: 482-488.

Diaz-Uribe, C., Vallejo, W. and Ramos, W. 2014. Methylene blue photocatalytic mineralization under visible irradiation on TiO_2 thin films doped with chromium. Appl. Surf. Sci. 319: 121-127.

Dindar, E., Şağban, F.O.T. and Başkaya, H.S. 2013. Bioremediation of petroleum contaminated soil. J. Biol. Environ. Sci. 7: 39.

Ding, H., Sun, H. and Shan, Y. 2005. Preparation and characterization of mesoporous SBA-15 supported dye-sensitized TiO_2 photocatalyst. J. photochem. Photobiol. A 169: 101-107.

Ding, X., An, T., Li, G., Zhang, S., Chen, J., Yuan, J., et al. 2008. Preparation and characterization of hydrophobic TiO_2 pillared clay: The effect of acid hydrolysis catalyst and doped Pt amount on photocatalytic activity. J. Colloid. Interface Sci. 320: 501-507.

Dong, F., Wang, H., Wu, Z. and Qiu, J. 2010. Marked enhancement of photocatalytic activity and photochemical stability of N–doped TiO_2 nanocrystals by Fe^{3+}/Fe^{2+} surface modification. J. Colloid Interface Sci. 343: 200-208.

Dong, F., Guo, S., Wang, H., Li, X. and Wu, Z. 2011. Enhancement of the visible light photocatalytic activity of C-doped TiO_2 nanomaterials prepared by a green synthetic approach. J. Phys. Chem. 115: 13285-13292.

Dong, H., Zeng, G., Tang, L., Fan, C., Zhang, C., He, X., et al. 2015. An overview on limitations of TiO_2-based particles for photocatalytic degradation of organic pollutants and the corresponding countermeasures. Water Res. 79: 128-146.

Dozzi, M.V., D'Andrea, C., Ohtani, B., Valentini, G. and Selli, E. 2013. Fluorine-doped TiO_2 materials: Photocatalytic activity vs time-resolved photoluminescence. J. Phys. Chem. C 117: 25586-25595.

Du, J., Li, B., Huang, J., Zhang, W., Peng, H. and Zou, J. 2013. Hydrophilic and photocatalytic performances of lanthanum doped titanium dioxide thin films. J. Rare Earth 31: 992-996.

El-Bahy, Z.M., Ismail, A.A. and Mohamed, R.M. 2009. Enhancement of titania by doping rare earth for photodegradation of organic dye (direct blue). J. Hazard. Mater. 166: 138-143.

El-Sheikh, S.M., Zhang, G., El-Hosainy, H.M., Ismail, A.A., O'Shea, K.E., Falaras P., et al. 2014. High performance sulfur, nitrogen and carbon doped mesoporous anatase-brookite TiO_2 photocatalyst for the removal of microcystin-LR under visible light irradiation. J. Hazard. Mater. 280: 723-733.

Elshafei, G.M.S., Yehia, F.Z., Dimitry, O.I.H., Badawi, A.M. and Eshaq, G. 2014. Ultrasonic assisted-Fenton-like degradation of nitrobenzene at neutral pH using nanosized oxides of Fe and Cu. Ultrason Sonochem 21: 1358-1365.

Emeline, E.V., Kuznetsov, V.N., Rybchuk, V.K. and Serpone, N. 2008. Visible-light-active titania photocatalysts: The case of N-doped TiO_2-properties and some fundamental issues. Int. J. Photoenergy 2008: 258394.

Ersoy, B. and Çelik, M.S. 2004. Uptake of aniline and nitrobenzene from aqueous solution by organo-zeolite. Environ. Technol. 25: 341-348.

Eslami, A., Amini, M.M., Yazdanbakhshc, A.R., Mohseni-Bandpei, A., Safari, A.A. and Asadi, A. 2016. N, S codoped TiO_2 nanoparticles and nanosheets in simulated solar light for photocatalytic degradation of nonsteroidal anti-inflammatory drugs in water: A comparative study. J. Chem. Technol. Biot. 91(10): 2693-2704.

Etacheri, V., Michlits, G., Seery, M.K., Hinder, S.J. and Pillai, S.C. 2013. A highly efficient $TiO_{2-x}C_x$ nano-heterojunction photocatalyst for visible light induced antibacterial applications. ACS Appl. Mater. Inter. 5: 1663-1672.

European Parliament and Council. 2008. Directive 2008/105/EC of the European Parliament and of the Council of 16 December 2008 on Environmental Quality Standards in the Field of Water Policy, Amending and Subsequently Repealing Council Directives 82/176/EEC, 83/513/EEC, 84/156/EEC, 84/491/EEC, 86/280/EEC and Amending Directive 2000/60/EC of the European Parliament and of the Council; 2008/105/EC; European Parliament and Council: Strasbourg, France.

Ezhilkumar, P., Selvakumar, K.V., Jenani, R., Devi, N.K., Selvarani, M. and Sivakumar, V.M. 2016. Studies on removal of formaldehyde from industrial wastewater by photocatalytic method. JCPS 9: 259-264.

Faisal, M., Tariq, M.A. and Muneer, M. 2007. Photocatalysed degradation of two selected dyes in UV irradiated aqueous suspensions of titania. J. Dyes Pigm. 72: 233-239.

Fallgren, P.H and Jin, S. 2008. Biodegradation of petroleum compounds in soil by a solid-phase circulating bioreactor with poultry manure amendments. J. Environ. Sci. Health A. 43: 125-131.

Fan, W.Q., Bai, H.Y., Zhang, G.H., Yan, Y.S., Liu, C.B. and Shi, W.D. 2014. Titanium dioxide macroporous materials doped with iron: Synthesis and photo-catalytic properties. Cryst. Eng. Comm. 16: 116-122.

Fang, W.Q., Wang, X.L., Zhang, H., Jia, Y., Huo, Z., Li, Z., et al. 2014. Manipulating solar absorption and electron transport properties of rutile TiO_2 photocatalysts via highly *n*-type F-doping. J. Mater. Chem. A 2: 3513-3520.

Fatin, S.O., Lim, H.N., Tan, W.T. and Huang, N.M. 2012. Comparison of photocatalytic activity and cyclic voltammetry of zinc oxide and titanium dioxide nanoparticles toward degradation of methylene blue. Int J. Electrochem. Sci. 7: 9074-9084.

Fernandes, A., Makoś, P., Wang, Z. and Boczkaj, G. 2020. Synergistic effect of TiO_2 photocatalytic advanced oxidation processes in the treatment of refinery effluents. Chem. Eng. J. 391-405.

Fernandez-Ibanez, P., Polo-López, M., Malato, S., Wadhwa, S., Hamilton, J.W.J., Dunlop, P.S.M., et al. 2015. Solar photocatalytic disinfection of water using titanium dioxide graphene composites. Chem. Eng. J. 261: 36-44.

Fitzgerald, E.F., Belanger, E.E., Gomez, M.I., Hwang, S.A., Jansing, R.L. and Hicks, H.E. 2007. Environmental exposures to polychlorinated biphenyls (PCBs) among older residents of upper Hudson River communities. Environ. Res. 104: 352-360.

Foo, K.Y. and Hameed, B.H. 2010. Decontamination of textile wastewater via TiO_2/activated carbon composite materials. Adv. Colloid. Interface Sci. 159: 130-143.

Forte, M., Di Lorenzo, M., Carrizzo, A., Valiante, S., Vecchione, C., Laforgia, V., et al. 2016. Nonylphenol effects on human prostate non tumorigenic cells. Toxicology 357-358: 21-32.

Fox, M.A. and Dulay, M.T. 1995. Heterogeneous photocatalysis. Chem. Rev. 93: 341-357.

Fu, R., Gao, S., Xu, H., Wang, Q., Wang, Z., Huang, B., et al. 2014. Fabrication of Ti^{3+} self-doped TiO_2(A) nanoparticle/TiO_2(R) nanorod heterojunctions with enhanced visible-light-driven photocatalytic properties. RSC Adv. 4: 37061-37069.

Fujishima, A. and Honda, K. 1972. Electrochemical photolysis of water at a semiconductor electrode. Nature 238: 37-38.

Fujishima, A., Rao, T.N. and Tryk, D. 2000. Titanium dioxide photocatalysis. J. Photochem. Photobiol. C: Photochem. Rev. 1: 1-21.

Fujishima, A., Zhang, X. and Tryk, D.A. 2008. TiO_2 photocatalysis and related surface phenomena. Surf. Sci. Rep. 63(12): 515-582.

Furube, A., Asahi, T., Masuhara, H., Yamashita, H. and Anpo, M. 2015. Direct observation of a picosecond charge separation process in photoexcited platinum-loaded TiO_2 particles by femtosecond diffuse reflectance spectroscopy. Chem. Phys. Lett. 336(2001): 424-430.

Gad-Allah, T.A., Kato, S., Satokawa, S. and Kojima, T. 2007. Role of core diameter and silica content in photocatalytic activity of $TiO_2/SiO_2/Fe_3O_4$ composite. Solid State Sci. 9(2007): 737-743.

Gai, L., Duan, X., Jiang, H., Mei, Q., Zhou, G., Tian, Y., et al. 2012. One-pot synthesis of nitrogen-doped TiO_2 nanorods with anatase/brookite structures and enhanced photo-catalytic activity. Cryst. Eng. Comm. 14: 7662-7671.

Gaidau, C., Petica, A., Ignat, M., Iordache, O., Ditu, L.M. and Ionescu, M. 2016. Enhanced photocatalysts based on Ag-TiO_2 and Ag-N-TiO_2 nanoparticles for multifunctional leather surface coating. Open Chem. J. 14: 383-392.

Gamage, J. and Zhang, Z. 2010. Applications of photocatalytic disinfection. Int. J. Photoenergy 2010: 11.

Gao, B., Yap, P.S., Limb, T.M. and Lim, T.T. 2011. Adsorption-photocatalytic degradation of Acid Red 88 by supported TiO_2: Effect of activated carbon support and aqueous anions. Chem. Eng. J. 171: 1098-1107.

Gao, H., Zhou, J., Dai, D. and Qu, Y. 2009. Photocatalytic activity and electronic structure analysis of N-doped anatase TiO_2: A combined experimental and theoretical study. Chem. Eng. Technol. 32: 867-872.

Gautron, J., Lemasson, P. and Marucco, J.M. 1981. Correlation between the non-stoichiometry of titanium dioxide and its photoelectrochemical behavior. Faraday Discuss. Chem. Soc. 70: 81-91.

Gaya, U.I. and Abdullah, A.H. 2008. Heterogeneous photocatalytic degradation of organic contaminants over titanium dioxide: A review of fundamentals, progress and problems. JPPC 9: 1-12.

Ge, J., Lan, M., Liu, W., Jia, Q., Guo, L., Zhou, B., et al. 2016. Graphene quantum dots as efficient, metal-free, visible-light-active photocatalysts. Sci. China. Mater. 59: 12-19.

Gennadiev, A., Yu, N., Pikovskii, I., Tsibart, A.S. and Smirnova, M.A. 2015. Hydrocarbons in soils: Origin, composition, and behavior (review). Eur. Soil Sci. 48: 1076-1089.

Ghaedi, M., Hassanzadeh, A. and Nasiri Kokhdan, S. 2011. Multiwalled carbon nanotubes as adsorbents for the kinetic and equilibrium study of the removal of Alizarin Red S and Morin. J. Chem. Eng. 56: 2511-2520.

Gianluca, L.P., Bono, A., Krishnaiah, D. and Collin, J.G. 2008. Preparation of titanium dioxide photocatalyst loaded onto activated carbon support using chemical vapor decomposition: A review paper. J. Hazard. Mater. 157: 209-219.

Gita, S., Hussan, A. and Choudhury, T.G. 2017. Impact of textile dyes waste on aquatic environments and its treatment. Environ. Ecol. 35: 2349-2353.

Giwa, A., Nkeonye, P.O., Bello, K.A. and Kolawole, E.G. 2012. Solar photocatalytic degradation of reactive yellow 81 and reactive violet 1 in aqueous solution containing semiconductor oxides. Int. J. Appl. Sci. Technol. 2: 90-105.

Gogate, P.R. and Pandit, A.B. 2004. A review of imperative technologies for wastewater treatment I: Oxidation technologies at ambient conditions. Adv. Environ. Res. 8: 501-551.

Golding, J., Emmett, P., Iles-Caven, Y., Steer, C. and Lingam, R. 2014. A review of environmental contributions to childhood motor skills. J. Child Neurol. 29: 1531-1547.

Goncharov, A., Haase, R.F., Santiago-Rivera, A., Morse, G., Mccaffrey, R.J. and Rej, R. 2008. High serum PCBs are associated with elevation of serum lipids and cardiovascular disease in a Native American population. Environ. Res. 106: 226-239.

Gonçalves, M.S.T., Pinto, E.M.S., Nkeonye, P. and Oliveira-Campos, A.M.F. 2005. Degradation of C.I. reactive orange 4 and its simulated dyebath wastewater by heterogeneous photocatalysis. Dyes Pigm. 64: 135-139.

Gopidas, K.R., Bohorquez, M. and Kamat, P.V. 1990. Photophysical and photochemical aspects of coupled semiconductors: Charge-transfer processes in colloidal cadmium sulfide-titania and cadmium sulfide-silver(I) iodide systems. J. Phys. Chem. 94: 6435-6440.

Gordon, P.F. and Gregory, P. 1983. Organic Chemistry in Colour. Springer, Berlin.

Goswami, P. and Ganguli, J.N. 2013. A novel synthetic approach for the preparation of sulfated titania with enhanced photocatalytic activity. RSC Adv. 3: 8878-8888.

Grabstanowicz, L.R., Gao, S., Li, T., Richard, M., Rajh, T., Liu, D.J., et al. 2013. Facile oxidative conversion of TiH_2 to high-concentration Ti^{3+}-self-doped rutile TiO_2 with visible light photoactivity. Inorg. Chem. 52(7): 3884-3890.

Grandcolas, M., Du, K.L. and Louvet, F.B.A. 2008. Porogen template assisted TiO_2 rutile coupled nanomaterials for improved visible and solar light photocatalytic applications. Catal. Lett. 123: 65-71.

Gregoraszczuk, E.L. and Ptak, A. 2013. Review article endocrine-disrupting chemicals: Some actions of POPs on female reproduction. Int. J. Endocrin. 2013: 1-9.

Grzechulska, J. and Morawski, A.W. 2002. Photocatalytic decomposition of azo-dye acid black 1 in water over modified titanium dioxide. Appl. Catal. B 36: 45-51.

Guidi, V., Carotta, M.C., Ferroni, M., Martinelli, G. and Sacerdoti, M. 2013. Effect of dopants on grain coalescence and oxygen mobility in nanostructured titania anatase and rutile. J. Phys. Chem. B 107: 120-124.

Guillard, C., Lachheb, H., Houas, A., Ksibi, M., Elaloui, E. and Herrmann, J.M. 2003. Influence of chemical structure of dyes, of pH and of inorganic salts on their photocatalytic degradation by TiO_2 comparison of the efficiency of powder and supported TiO_2. J. Photochem. Photobiol. A 158: 27-36.

Guo, Q., Zhou, C., Ren, M.Z., Fan, H. and Yang, X. 2016. Elementary Photocatalytic Chemistry on TiO_2 Surfaces. Chem. Soc. Rev. 45: 3701-3730.

Gupta, S.M. and Tripathi, M. 2011. A review of TiO_2 nanoparticles. Chin. Sci. Bull. 56: 1639-1657.

Hahn, R., Schmidt-Stein, F., Salonen, J., Thiemann, S., Song, Y.Y., Kunze, J., et al. 2009. Semimetallic TiO_2 nanotubes. Angew. Chem. Int. Ed. 48: 7236-7239.

Hahn, R., Stark, M., Killian, M.S. and Schmuki, P. 2013. Photocatalytic properties of *in situ* doped TiO_2-nanotubes grown by rapid breakdown anodization. Catal. Sci. Technol. 3: 1765-1770.

Han, C., Likodimos, V., Khan, J.A., Nadagouda, M.N., Anderson, J., Falaras, P., et al. 2014. UV–visible light-activated Ag-decorated, monodisperse TiO_2 aggregates for treatment of the pharmaceutical oxytetracycline. Environ. Sci. Pollut. Res. 21(20): 11781-11793.

Hasanbeigi, A. and Price, L. 2015. A technical review of emerging technologies for energy and water efficiency and pollution reduction in the textile industry. J. Clean. Prod. 95: 30-44.

Haseena, M., Malik, M.F., Javed, A., Arshad, S., Asif, N., Zulfiqar, S., et al. 2017. Water pollution and human health. Environ. Risk. Assess. Remediat. 1(3): 16-19.

Hashimoto, K., Irie, H. and Fujishima, A. 2005. TiO_2 photocatalysis: A historical overview and future prospects. Jap. J. Appl. Phys. 44: 8269-8285.

Haseena, M., Malik, M.F., Javed, A., Arshad, S., Asif, N. Zulfiqar, S., et al. 2017. Health and environmental impacts of dyes: Mini review. Am. J. Environ. Sci. Eng. 1: 64-67.

Hassan, M.M. and Carr, C.M. 2018. A critical review on recent advancements of the removal of reactive dyes from dyehouse effluent by ion-exchange adsorbents. Chemosphere 209: 201-219.

He, Z., Xie, L., Tu, J., Song, S. and Liu, W. 2010. Visible light-induced degradation of phenol over iodine-doped titanium dioxide modified with platinum: Role of platinum and the reaction mechanism. J. Phys. Chem. C 114: 526-532.

He, Z., Que, W., Xie, H., Chen, J., Yuan, Y. and Sun, P. 2012. Facile synthesis of self-sensitized TiO_2 photocatalysts and their higher photocatalytic activity. J. Am. Ceram. Soc. 95: 3941-3946.

Hena, S. 2010. Removal of chromium hexavalent ion from aqueous solutions using biopolymer chitosan coated with poly 3-methyl thiophene polymer. J. Hazard. Mater. 181: 474-479.

Hentati, O., Lachhab, R., Ayadi, M. and Ksibi, M. 2013. Toxicity assessment for petroleumcontaminated soil using terrestrial invertebrates and plant bioassays. Environ. Monit. Assess. 185: 2989-2998.

Herrmann, J.M. 1995. Heterogeneous photocatalysis: An emerging discipline involving multiphase systems. Catal Today 24: 157-164.

Herrmann, J.M. 2012. Detrimental cationic doping of titania in photocatalysis: Why chromium Cr^{3+}-doping is a catastrophe for photocatalysis, both under UV- and visible irradiations. N.J. Chem. 36: 883-890.

Hernandez-Alonso, M.D., Fresno, F., Suarez, S. and Coronado, J.M. 2009. Development of alternative photocatalysts to TiO_2: Challenges and opportunities. Energy Environ. Sci. 2: 1231-1257.

Ho, H. and Watanabe, T. 2017. Distribution and removal of nonylphenol thoxylates and nonylphenol from textile wastewater—a comparison of a cotton and a synthetic fiber factory in Vietnam. Water 9: 386.

Hoffmann, M.R., Martin, S.T., Choi, W. and Bahenemann, D.W. 1995. Environmental application of semiconductor photocatalysis. Chem. Rev. 95: 69-96.

Holland, B.T., Blanford, C.F., Do, T. and Stein, A. 1999. Synthesis of highly ordered, three-dimensional, macroporous structures of amorphous or crystalline inorganic oxides, phosphates, and hybrid composites. Chem. of Mater. 11: 795.

Hoppin, J.A., Umbach, D.A., London, S.J., Lynch, C.F., Alavanja, M.C. and Sandler, D.P. 2006. Pesticides and adult respiratory outcomes in the agricultural health study. Ann. N.Y. Acad. Sci. 1076: 343-354.

Hoque, M. and Guzman, M. 2011. Photocatalytic activity: Experimental features to report in heterogeneous photocatalysis. Materials 11: 1990-2000.

Hsu, H.C., Shown, I., Wei, H.Y., Chang, Y.C., Du, H.Y., Lin, Y.G., et al. 2013. Graphene oxide as a promising photocatalyst for CO_2 to methanol conversion. J. Nanoscale 5: 262-268.

Hu, C., Lan, Y., Qu, J., Hu, X. and Wang, A. 2006. Ag/AgBr/TiO_2 visible light photocatalyst for destruction of azodyes and bacteria. J. Phys. Chem. B 110: 4066-4072.

Hu, X., Hu, X., Tang, C., Wen, S., Wu, X., Long, J., et al. 2017. Mechanisms underlying degradation pathways of microcystin-LR with doped TiO_2 photocatalysis. Chem. Eng. J. 330: 355-371.

Huang, F., Yan, A. and Zhao, H. 2016. Influences of doping on photocatalytic properties of TiO_2 photocatalyst. Semiconductor Photocatalysis: Materials, Mechanisms and Applications. doi: 10.5772/63234.

Huang, M., Xu, C., Wu, Z., Huang, Y., Lin, J. and Wu, J. 2008. Photocatalytic discolorization of methyl orange solution by Pt modified TiO_2 loaded on natural zeolite. J. Dyes Pigm. 77: 327-334.

Hung, C.H. and Yuan, C. 2000. Reduction of Azo-dye via TiO_2–photocatalysis. J. Chin. Inst. Environ. Eng. 10: 209-216.

Hung, M.H., Ling, F.H. and Hoffmann, M.R. 2000. Kinetics and mechanism of the enhanced reductive degradation of nitrobenzene by elemental iron in the presence of ultrasound. Environ. Sci. Technol. 34: 1758-1763.

Huo, J., Hu, Y., Jiang, H. and Li, C. 2014. *In situ* surface hydrogenation synthesis of Ti^{3+} self-doped TiO_2 with enhanced visible light photoactivity. Nanoscale 6: 9078-9084.

Ibhadon, A.O. and Fitzpatrick, P. 2013. Heterogeneous photocatalysis: Recent advances and applications. Catalysts 3: 189-218.

Igbedioh, S.O. 1991. Effects of agricultural pesticides on humans, animals and higher plants in developing countries. Arch. Environ. Health 46: 218-226.

Iliev, V., Tomova, D., Bilyarska, L., Eliyas, A. and Petrov, L. 2006. Photocatalytic properties of TiO_2 modified with platinum and silver nanoparticles in the degradation of oxalic acid in aqueous solution. Appl. Catal. B. 63: 266-271.

Irie, H., Watanabe, Y. and Hashimoto, K. 2003. Carbon-doped anatase TiO_2 powders as a visible-light sensitive photocatalyst. Chem. Lett. 32: 772-773.

Irie, H., Watanabe, Y. and Hashimoto. K. 2003. Nitrogen-concentration dependence on photocatalytic activity of $TiO_{2-x}N_x$ powders. J. Phys. Chem. B 107: 5483-5486.

Ishii, M., Towlson, B., Harako, S., Zhao, X.W., Komuro, S. and Hamilton, B. 2013. Roles of electrons and holes in the luminescence of rare-earth-doped semiconductors. Electr. Commun. Jpn. 96: 1-7.

Jacob, J. 2013. A Review of the Accumulation and Distribution of Persistent Organic Pollutants in the Environment. IJBBB. 3(6): 657-661.

Jaimy, K.B., Safeena, V.P., Ghosh, S., Hebalkar, N.Y. and Warrier, K.G.K. 2012. Photocatalytic activity enhancement in doped titanium dioxide by crystal defects. Dalton Trans. 41: 4824-4832.

Jain, A. and Dipti, V. 2017. Photocatalytic activity of TiO_2 nanomaterial. J. Chil. Chem. Soc. 62(4): 3683-3690.

Jaiswal, R., Patel, N., Dashora, A., Fernandes, R., Yadav, M., Edla, R., et al. 2016. Efficient Co-B-codoped TiO_2 photocatalyst for degradation of organic water pollutant under visible light. Appl. Catal. B Environ. 183: 242-253.

Janus, M., Inagaki, M., Tryba, B., Toyod, M. and Morawski, A.W. 2006. Carbon-modified TiO_2 photocatalyst by ethanol carbonization. Appl. Catal. B. 63: 272-276.

Jesus, J.M., Danko, A.S., Fiúza, A. and Borges, M.T. 2015. Phytoremediation of saltaffected soils: A review of processes, applicability, and the impact of climate change. Environ. Sci. Pollut. Res. 22: 6511-6525.

Ji, P.L., Kong, X.Z., Wang, J.G. and Zhu, X.L. 2012. Characterization and photocatalytic properties of silver and silver chloride doped TiO_2 hollow nanoparticles. Chin. Chem. Lett. 23: 1399-1402.

Jiang, B.C., Lu, Z.Y., Liu, F.Q., Li, A.M., Dai, J.J., Xu, L., et al. 2011. Inhibiting 1, 3-dinitrobenzene formation in Fenton oxidation of nitrobenzene through a controllable reductive pretreatment with zero-valent iron. Chem. Eng. J. 174: 258-265.

Jing, L., Zhou, W., Tian, G. and Fu, H. 2013. Surface tuning for oxide-based nanomaterials as efficient photocatalysts. Chem. Soc. Rev. 42: 9509-9549.

Jing, R., Fusi, S. and Kjellerup, B.V. 2018. Remediation of polychlorinated biphenyls (PCBs) in contaminated soils and sediment: State of knowledge and perspectives. Environ. Sci. 6: 1-17.

Jones, K.C. and De Voogt, P. 1999. Persistent organic pollutants (POPs): State of the science. Environ. Pollut. 100: 209-221.

Joshi, B.N., Yoon, H., van Hest, M.F.A.M. and Yoon, S.S. 2013. Niobium-doped titania photo-catalyst film prepared via a nonaqueous sol-gel method. J. Am. Ceram. Soc. 96: 2623-2627.

Joshi, M.M., Labhsetwar, N.K., Mangrulkar, P.A., Tijare, S.N., Kamble, S.P. and Rayalu, S.S. 2009. Visible light induced photoreduction of methyl orange by N-doped mesoporous titania. App. Catal. A: General 357: 26-33.

Ju, K.S. and Parales, R.E. 2010. Nitroaromatic compounds, from synthesis to biodegradation microbiol. Mol. Biol. R. 74: 250-272.

Juneja, T. and Chauhdary, A. 2013. Assessment of water quality and its effect on the health of residents of Jhunjhunu district, Rajasthan: A cross sectional study. JPHE 5: 186-191.

Kafilzadeh, F. 2015. Assessment of organochlorine pesticide residues in water, sediments and fish from Lake Tashk, Iran. Arch. Life Sci. 9: 107-111.

Kafizas, A., Noor, N., Carmichael, P., Scanlon, D.O., Carmalt, C.J. and Parkin, I.P. 2014. Combinatorial atmospheric pressure chemical vapor deposition of F:TiO_2: The relationship between photocatalysis and transparent conducting oxide properties. Adv. Funct. Mater. 24: 1758-1771.

Kalia, S., Kango, S., Kumar, A., Haldorai, Y., Kumari, B. and Kumar, R. 2014. Magnetic polymer nanocomposites for environmental and biomedical applications. Colloid Polym. Sci. 292: 2025-2052.

Kamble, S.P., Sawant, S.B. and Pangarkar, V.G. 2003. Batch and continuous photocatalytic degradation of benzenesulfonic acid using concentrated solar radiation. Ind. Eng. Chem. Res. 42: 6705-6713.

Kanan, S., Moyet, M.A., Arthur, R.B. and Patterson, H.H. 2019. Recent advances on TiO_2-based photocatalysts toward the degradation of pesticides and major organic pollutants from water bodies. Catal. Rev. 1-65.

Kaneco, S., Rahman, M.A., Suzuki, T., Katsumata, H. and Ohta, K. 2004. Optimization of solar photocatalytic degradation conditions of bisphenol A in water using titanium dioxide. J. Photochem. Photobiol. A 163: 419-424.

Kang, In-C., Zhang, Q., Yin, S., Sato, T. and Satio, F. 2008. Improvement in photocatalytic activity of TiO_2 under visible irradiation through addition of N-TiO_2. Environ. Sci. Technol. 42: 3622-3626.

Kant, R. 2012. Textile dyeing industry an environmental hazard. J. Nat. Sci. 4: 22-26.

Karpińska, J. and U. Kotowaska. 2019. Removal of organic pollution in the water environment. Water 11(10). 1-7.

Kaur, S. and Singh, V. 2007. Visible light induced sonophotocatalytic degradation of reactive red dye using dye sensitized TiO_2. Ultrason. Sonochem. 14: 531-537.

Khan, M., Gul, S.R., Li, J. and Cao, W. 2015. Tungsten concentration influence on the structural, electronic, optical and photocatalytic properties of tungsten-silver codoped titanium dioxide. Ceram. Int. 41: 6051-6054.

Khan, M.A., Han, D.H. and Yang, O.B. 2009. Enhanced photoresponse towards visible light in Ru doped titania nanotube. Appl. Surf. Sci. 255: 3687-3690.

Khan, N., Hussain, S.T., Saboor, A., Jamila, N. and Kim, K.S. 2013. Physiochemical investigation of the drinking water sources from Mardan, Khyber Pakhtunkhwa, Pakistan. IJPS 8: 1661-1671.

Khan, S.U., Al-Shahry, M. and Ingler, W.B. 2002. Efficient photochemical water splitting by a chemically modified n-TiO_2. Science 297: 2243-2245.

Khunphonoi, R. and Grisdanurak, N. 2016. Mechanism pathway and kinetics of p-cresol photocatalytic degradation over titania nanorods under UV–visible irradiation. Chem. Eng. J. 296: 420-427.

Kim, D.G., Choi, K.I. and Lee, D.H. 2011. Gas-particle partitioning and behavior of dioxin-like PCBs in the urban atmosphere of Gyeonggi-do, South Korea. Atmos. Res. 101: 386-395.

Kim, S.H., Nam, K.H., Hwang, K.A. and Choi, K.C. 2016. Influence of hexabromocyclododecane and 4-nonylphenol on the regulation of cell growth, apoptosis and migration in prostatic cancer cells. Toxicol. Vitro 32: 240-247.

Kimbell-Dunn, M.R., Fishwick, R.D., Bradshaw, L., Erkinjuntti-Pekkanen, R. and Pearce, N. 2001. Work related symptoms in New Zealand Farmers. Am. J. Ind. Med. 39: 292-300.

Kiriakidou, F., Kondarides, D.I. and Verykios, X.E. 1999. The effect of operational parameters and TiO_2-doping on the photocatalytic degradation of azo-dyes. Catal. Today 54: 119-130.

Kitano, M., Funatsu, K., Matsuoka, M., Ueshima, M. and Anpo, M. 2006. Preparation of nitrogen-substituted TiO_2 thin film photocatalysts by the radio frequency magnetron sputtering deposition method and their photocatalytic reactivity under visible light irradiation. J. Phys. Chem. B 110: 25266-25272.

Klauck, C.R., Giacobbo, A., de Oliveira, E.D.L., da Silva, L.B. and Rodrigues, M.A.S. 2017. Evaluation of acute toxicity, cytotoxicity and genotoxicity of landfill leachate treated by biological lagoon and advanced oxidation processes. J. Environ. Chem. Eng. 5: 6188-6193.

Klausen, J., Ranke, J. and Schwarzenbach, R.P. 2001. Influence of solution composition and column aging on the reduction of nitroaromatic compounds by zero-valent iron. Chemosphere 44: 511-517.

Klaysri, R., Wichaidit, S., Tubchareon, T., Nokjan, S., Piticharoenphun, S. and Mekasuwandumrong O. 2015. Impact of calcination atmospheres on the physiochemical and photocatalytic properties of nanocrystalline TiO_2 and Si-doped TiO_2. Ceram. Int. 41: 11409-11417.

Kogevinas, M. 2001. Human health effects of dioxins: Cancer, reproductive and endocrine system effects. Hum. Reprod. Update 7: 331-339.

Kogo, K., Yoneyama, H. and Tamura, H. 1980. Photocatalytic oxidation of cyanide on platinized titanium dioxide. J. Phys. Chem. 84: 1705-1710.

Konstantinou, I.K. and Albanis, T.A. 2004. TiO_2-assisted photocatalytic degradation of azo dyes in aqueous solution: Kinetic and mechanistic investigations: A review. Appl. Catal. B 49: 1-14.

Kormann, C., Bahnemann, D.W. and Hoffman, M.R. 1988. Photocatalytic production of H_2O_2 and organic peroxides in aqueous suspensions of TiO_2, ZnO and desert sand. Environ. Sci. Technol. 22: 798-806.

Koumaki, E., Mamais, D. and Noutsopoulos, C. 2018. Assessment of the environmental fate of endocrine disrupting chemicals in rivers. Sci. Total Environ. 628-629: 947-958.

Kowalska, E., Remita, H., Colbeau-Justin, C., Hupka, J. and Belloni, J. 2008. Modification of titanium dioxide with platinum ions and clusters: Application in photocatalysis. J. Phys. Chem. C 112: 1124-1131.

Krýsa, J., Keppert, M., Jirkovský, J., Štengl, V. and Šubrt. J. 2004. The effect of thermal treatment on the properties of TiO_2 photocatalyst. Mater. Chem. Phys. 86: 333-339.

Kuang, S., Yang, L., Luo, S. and Cai, Q. 2009. Fabrication, characterization and photoelectrochemical properties of Fe_2O_3 modified TiO_2 nanotube arrays. Appl. Surf. Sci. 255: 7385-7388.

Kulkarni, S.J. and Kaware, D.J. 2013. Review on research for removal of phenol from waste-water. IJSRP 3: 1-4.

Kumar, A., Hitkari, G., Gautam, M., Singh, S. and Pandey. G. 2015. Synthesis, characterization and application of Cu–TiO_2 nanaocomposites in photodegradation of methyl red (MR). IARJSET 2: 50-55.

Kumar, A. and Pandey, G. 2017a. Photodegradation of methyl orange in aqueous solution by the visible light active Co:La:TiO_2 nanocomposite. Chem. Sci. J. 8: 164.

Kumar, A. and Pandey, G. 2017b. Photocatalytic degradation of Eriochrome Black-T by the Ni:TiO_2 nanocomposites. Desalin. Water Treat. 71: 406-419.

Kumar, A. and Pandey, G. 2017c. Photocatalytic activity of Co:TiO_2 nanocomposites and their application in photodegradation of acetic acid. Chem. Sci. Trans. 6: 385-392.

Kumar, S., Ahlawat, W., Bhanjana, G., Heydarifard, S., Nazhad, M.M. and Dilbaghi, N. 2014. Nanotechnology-based water treatment strategies. J. Nanosi. Nanotechnol. 14: 1838-1858.

Kundu, S., Pala, A. and Dikshitb, A.K. 2005. UV induced degradation of herbicide 2,4-D: kinetics, mechanism and effect of various conditions on the degradation. Sep. Purif. Technol. 44: 121-129.

Lachheb, H., Puzenat, E., Houas, A., Ksibi, M., Elaloui, E., Guillard, C. et al. 2002. Photocatalytic degradation of various types of dyes (Alizarin S, Crocein Orange G, Methyl Red, Congo Red, Methylene Blue) in water by UV-irradiated titania. Appl. Catal. B. 39: 75-90.

Lai, L.L. and Wu, J.M. 2015. A facile solution approach to W, N co-doped TiO_2 nanobelt thin films with high photocatalytic activity. J. Mater. Chem. A 3: 15863-15868.

Lallas, P.L. 2001. The stockholm convention on persistent organic pollutants. Am. J. Int. Law 95: 692-708.

Leadprathom, N., Parkpian, P., Satayavivad, J., Delaune, R.D. and Jugsujinda, A. 2009. Transport and deposition of organochlorine pesticides from farmland to estuary under tropical regime and their potential risk to aquatic biota. J. Environ. Sci. Health Part B 44: 249-261.

Lee, J.H. and Yang, Y.S. 2005. Effect of HCl concentration and reaction time on the change in the crystalline state of TiO_2 prepared from aqueous $TiCl_4$ solution by precipitation. J. Europ. Ceramic Soc. 25(16): 3573-3578.

Lee, H., Park, Y.K., Kim, S.J., Kim, B.H. and Jung, S.C. 2015. Titanium dioxide modification with cobalt oxide nanoparticles for photocatalysis. J. Ind. Eng. Chem. 32: 259-263.

Lee, J. and Gouma, P.I. 2012. Sol-gel processed oxide photocatalysts. Sol-Gel Processing for Conventional and Alternative Energy. Springer, Center for Nanomaterials and Sensor Development, State University of New York at Stony Brook. 217-237.

Lehtiniemi, M., Engström-Öst, J. and Viitasalo, M. 2005. Turbidity decreases anti-predator behaviour in pike larvae, Esox Lucius. Environ. Biol. Fishes. 73: 1-8.

Lettinga, G. 1996. Sustainable integrated biological wastewater treatment. Water Sci. Technol. 33: 85-98.

Lettmann, C., Hildenbrand, K., Kisch, H., Macyk, W. and Maier, W.F. 2001. Visible light photodegradation of 4-chlorophenol with a coke-containing titanium dioxide photocatalyst. Appl. Catal. B. 32: 215-227.

Li, F., Sun, S., Jiang, Y., Xia, M., Sun, M. and Xue, B. 2008. Photodegradation of an azo dye using immobilized nanoparticles of TiO_2 supported by natural porous mineral. J. Hazard. Mater. 152: 1037-1044.

Li, G., Lian, Z., Li, X., Weng, W., Zhang, D., Tian, F., et al. 2015a. Ionothermal synthesis of black Ti^{3+}-doped single-crystal TiO_2 as an active photocatalyst for pollutant degradation and H_2 generation. J. Mater. Chem. A. 3: 3748-3756.

Li, G., Li, J., Li, G. and Jiang, G. 2015b. N and Ti^{3+} co-doped 3D anatase TiO_2 superstructures composed of ultrathin nanosheets with enhanced visible light photocatalytic activity. J. Mater. Chem. A 3: 22073-22080.

Li, H., Xing, J., Xia, Z. and Chen. J. 2014. Preparation of extremely smooth and boron-fluorine co-doped TiO_2 nanotube arrays with enhanced photoelectrochemical and photocatalytic performance. Electrochim. Acta 139: 331-336.

Li, J.Q., Wang, D.F., Liu, H. and Zhu, Z.F. 2012. Multilayered Mo-doped TiO_2 nanofibers and enhanced photocatalytic activity. Mater. Manuf. Process. 27: 631-635.

Li, L., Shi, J., Li, G., Yuan, Y., Li, Y., Zhao, W., et al. 2013. One-pot pyrolytic synthesis of C-N-codoped mesoporous anatase TiO_2 and its highly efficient photo-degradation properties. N. J. Chem. 37: 451-457.

Li, R., Chen, W. and Wang, W. 2009. Magnetoswitchable controlled photocatalytic system using ferromagnetic Fe-doped titania nanorods photocatalysts with enhanced photoactivity. Sep. Purif. Technol. 66: 171-176.

Li, R., Dong, G. and Chen. G. 2015c. Synthesis, characterization and performance of ternary doped Cu-Ce-B/TiO_2 nanotubes on the photocatalytic removal of nitrogen oxides. N. J. Chem. 39: 6854-6863.

Li, X., Wang, D., Cheng, G., Luo, Q., An, J. and Wang, Y. 2008. Preparation of polyaniline-modified TiO_2 nanoparticles and their photocatalytic activity under visible light illumination. Appl. Catal. B. 81: 267-273.

Li, X., Guo, Z. and He, T. 2013. The doping mechanism of Cr into TiO_2 and its influence on the photocatalytic performance. Phys. Chem. Chem. Phys. 15: 20037-20045.

Li, Y., Peng, S., Jiang, F., Lu, G. and Li, S. 2007. Effect of doping TiO_2 with alkaline-earth metal ions on its photocatalytic activity. JSCS 72: 393-402.

Li, Z., Wnetrzak, R., Kwapinski, W. and Leahy, J.J. 2012. Synthesis and characterization of sulfated TiO_2 nanorods and ZrO_2/TiO_2 nanocomposites for the esterification of biobased organic acid. ACS Appl. Mater. Inter. 4: 4499-4505.

Li, Z. 2015. Synthesis of N doped and N, S co-doped 3D TiO_2 hollow spheres with enhanced photocatalytic efficiency under nature sunlight. Ceram. Int. 41: 10063-10069.

Liang, S.X., Zhang, H.K. and Lu, D. 2007. Determination of nitrobenzene in wastewater using a hanging mercury drop electrode. Environ. Monit. Assess. 129: 331-337.

Lightcap, I.V., Kosel, T.H. and Kamat, P.V. 2010. Anchoring semiconductor and metal nanoparticles on a two-dimensional catalyst mat. Storing and shuttling electrons with reduced graphene oxide. Nano Lett. 10: 577-583.

Likodimos, V., Han, C., Pelaez, M., Kontos, A.G., Liu, G., Zhu, D. et al. 2013. Anion-Doped TiO_2 Nanocatalysts for Water Purification under Visible Light 52(39): 13957-13964.

Lima, J.F., Harunsani, M.H., Martin, D.J., Kong, D., Dunne, P.W., Gianolio, D., et al. 2015. Control of chemical state of cerium in doped anatase TiO_2 by solvothermal synthesis and its application in photocatalytic water reduction. J. Mater. Chem. A 3: 9890-9898.

Lin, H., Deng, W., Zhou, T., Ning, S., Long, J. and Wang, X. 2015. Iodine-modified nanocrystalline titania for photocatalytic antibacterial application under visible light illumination. Appl. Catal. B Environ. 176: 36-43.

Lin, L., Jiang, W., Chen, L., Xu, P. and Wang. H. 2020. Treatment of produced water with photocatalysis: Recent advances, affecting factors and future research prospects. Catalysts 10: 924.

Lin, Y., Jiang, Z., Zhu, C., Hu, X., Zhang, X., Zhu, H., et al. 2013a. C/B codoping effect on band gap narrowing and optical performance of TiO_2 photocatalyst: A spin-polarized DFT study. J. Mater. Chem. A 1: 4516-4524.

Lin, Y.H., Chou, S.H. and Chu, H. 2014. A kinetic study for the degradation of 1,2-dichloroethane by S-doped TiO_2 under visible light. J. Nanopart. Res. 16: 1016-1030.

Lin, Y.T., Weng, C.H., Lin, Y.H., Shiesh, C.C. and Chen, F.Y. 2013b. Effect of C content and calcination temperature on the photocatalytic activity of C-doped TiO_2 catalyst. Sep. Purif. Technol. 116: 114-123.

Lin, Z., Orlov, A., Lambert, R.M. and Payne, M.C. 2005. New insights into the origin of visible light photocatalytic activity of nitrogen-doped and oxygen-deficient anatase TiO_2. J. Phys. Chem. B 109: 20948-20952.

Lindner, M., Bahnemann, D.W., Hirthe, B. and Griebler, W.D. 1997. Solar water detoxification: Novel TiO_2 powders as highly active photocatalysts. J. Sol. Energy. Eng. 119: 120-125.

Ling, C.H., Mohamed, A.R. and Bhatia, S. 2004. Perfomance of photocatalytic reactors using immobilized TiO_2 film for degradation of phenol and methylene blue dye present in water stream. Chemosphere 57: 547-554.

Liu, C., Min, Y., Zhang, A.Y., Si, Y., Chen, J.J. and Yu, H.Q. 2019. Electrochemical treatment of phenol-containing wastewater by facet-tailored TiO_2: Efficiency, characteristics and mechanisms. Water Res. 165: 114980.

Liu, C.C., Hsieh, Y.H., Lai, P.F., Li, C.H. and Kao, C.L. 2006. Photodegradation treatment of azo dye wastewater by UV/TiO_2 process. Dyes Pigm. 68: 191-195.

Liu, G., Zhang, X., Xu, Y., Niu, X., Zheng, L. and Ding, X. 2005. The preparation of Zn^{2+}-doped TiO_2 nanoparticles by sol-gel and solid phase reaction methods respectively and their photocatalytic activities. Chemosphere 59: 1367-1371.

Liu, G., Sun, C., Wang, L., Smith, S.C., Lu, G.Q. and Cheng, H.M. 2011. Bandgap narrowing of titanium oxide nanosheets: Homogeneous doping of molecular iodine for improved photoreactivity. J. Mater. Chem. 21: 14672-14679.

Liu, J., Zhang, Q., Yang, J., Ma, H., Tade, M.O., Wang, S., et al. 2014. Facile synthesis of carbon-doped mesoporous anatase TiO_2 for the enhanced visible-light driven photocatalysis. Chem. Commun. 50: 13971-13974.

Liu, Q., Ding, D., Ning, C. and Wang, X. 2015. Reduced N/Ni-doped TiO_2 nanotubes photoanodes for photoelectrochemical water splitting. RSC Adv. 5: 95478-95487.

Liu, S., Yang, J.H. and Choy, J.H. 2006. Microporous SiO_2-TiO_2 nanosols pillared montmorillonite for photocatalytic decomposition of methyl orange. J. Photochem. Photobiol. A: Chem. 179: 75-80.

Liu, S., Guo, E. and Yin, L. 2012. Tailored visible-light driven anatase TiO_2 photocatalysts based on controllable metal ion doping and ordered mesoporous structure. J. Mater. Chem. 22: 5031-5041.

Liu, T. and Zhang, H. 2013. Novel Fe-doped anatase TiO_2 nanosheet hierarchical spheres with 94% {001} facets for efficient visible light photodegradation of organic dye. RSC Adv. 3: 16255-16258.

Liu, X., Gao, S., Xu, H., Lou, Z., Wang, W., Huang, B., et al. 2013. Green synthetic approach for Ti^{3+} self-doped TiO_{2-x} nanoparticles with efficient visible light photocatalytic activity. Nanoscale 5: 1870-1875.

Liu, X., Xu, H., Grabstanowicz, L.R., Gao, S., Lou, Z., Wang, W., et al. 2014. Ti^{3+} self-doped TiO_{2-x} anatase nanoparticles via oxidation of TiH_2 in H_2O_2. Catal. Today 225: 80-89.

Liu, X., Chen, Y., Cao, C., Xu, J., Qian, Q., Luo, Y., et al. 2015. Electrospun nitrogen and carbon co-doped porous TiO_2 nanofibers with high visible light photocatalytic activity. N. J. Chem. 39: 6944-6950.

Liu, Z., Li, Z., Zhong, H., Zeng, G., Liang, Y., Chen, M., et al. 2017. Recent advances in the environmental applications of biosurfactant saponins: A review. J. Environ. Chem. Eng. 5: 6030-6038.

London, L. and Myers, J. 1995. General patterns of agricultural chemical usage in the southern regions of South Africa. South Afr. J. Sci. 91: 508.

Lozano, D., Hernandez-Lopez, J.M., Esbrit, P., Arenas, M.A., Gomez-Barrena, E., de Damborenea, J. et al. 2014. Influence of the nanostructure of F-doped TiO_2 films on osteoblast growth and function. Soc. Biomater. 103(6): 1985-1990.

Lu, D., Yang, M., Fang, P., Li, C. and Jiang, L. 2017. Enhanced photocatalytic degradation of aqueous phenol and Cr(VI) over visible-light-driven Tb_xO_y loaded TiO_2-oriented nanosheets. Appl. Surf. Sci. 399: 167-184.

Lu, T.C., Wu, S.Y., Lin, L.B. and Zheng, W.C. 2001. Defects in the reduced rutile single crystal. Physica. B 304: 147-151.

Lv, T., Wu, S., Hong, H., Chen, L. and Dong, R. 2013. Dynamics of nitrobenzene degradation and interactions with nitrogen transformations in laboratory-scale constructed wetlands. Bioresour. Technol. 133: 529-536.

Lynch, J., Giannini, C., Cooper, J.K., Loiudice, A., Sharp, I.D. and Buonsanti, R. 2015. Substitutional or interstitial site-selective nitrogen doping in TiO_2 nanostructures. J. Phys. Chem. C 119: 7443-7452.

Machulek, J.A., Oliveira, S.C., Osugi, M.E., Ferreira, V.S., Quina, F.H., Dantas, R.F., et al. 2013. Application of different advanced oxidation processes for the degradation of organic pollutants. pp. 238. *In*: Nageeb Rashed, M. [ed.]. Organic Pollutants – Monitoring, Risk and Treatment. InTech. London.

Maddila, S., Oseghe, E.O. and Jonnalagadda, S.B. 2016. Photocatalyzed ozonation by Ce doped TiO_2 catalyst degradation of pesticide Dicamba in water. J. Chem. Technol. Biotechnol. 91: 385-393.

Makarova, O.V., Rajh, T. and Thurnauer, M.C. 2000. Surface modification of TiO_2 nanoparticles for photochemical reduction of nitrobenzene. Environ. Sci. Technol. 34: 4797-4803.

Maki, L.K., Maleki, A., Rezaee, R., Daraei, H. and Yetilmezsoy. K. 2019. LED-activated immobilized Fe-Ce-N tri-doped TiO_2 nanocatalyst on glass bed for photocatalytic degradation organic dye from aqueous solutions. Environ. Technol. Innov. 15: 100411.

Makwana, N., Tighe, C.J., Gruar, R.I., Mcmillan, P. and Darr. J. 2016. Pilot plant scale continuous hydrothemal synthesis of nano-titania: Effect of size on photocatalytic activity. Mater. Sci. Semicond. Process. 42(1): 131-137.

Malato, S., Blanco, J., Richter, C., Braun, B. and Maldonado, M.I. 1998. Enhancement of the rate of solar photocatalytic mineralization of organic pollutants by inorganic oxidizing species. Appl. Catal. B 17: 347-356.

Mallakpour, S. and E. Nikkhoo. 2014. Surface modification of nano-TiO_2 with trimellitylimido-amino acid-based diacids for preventing aggregation of nanoparticles. Adv. Powder Technol. 25: 348-353.

Mamba, G., Mamo, M.A., Mbianda, X. and Mishra, A.K. 2014. Nd, N, S–TiO_2 decorated on reduced graphene oxide for a visible light active photocatalyst for dye degradation: Comparison to its MWCNT/Nd,N,S–TiO_2 analogue. Ind. Eng. Chem. Res. 53: 14329-14338.

Manu, S. and Khadar, M.A. 2015. Non-uniform distribution of dopant iron ions in TiO_2 nano-crystals probed by X-ray diffraction, Raman scattering, photoluminescence and photocatalysis. J. Mater. Chem. C 3: 1846-1853.

Mao, C., Zuo, F., Hou, Y., Bu, X. and Feng, P. 2014. *In situ* preparation of a Ti^{3+} self-doped TiO_2 film with enhanced activity as photoanode by N_2H_4 reduction. Angew. Chem. Int. Ed. 53: 10485-10489.

Mao, C.C. and Weng, H.S. 2012. Effect of heat treatment on photocatalytic activity of titania incorporated with carbon black for degradation of methyl orange. Environ. Prog. Sustain. Energ. 31(2): 306-317.

Mao, Z., Zheng, Y.L. and Zhang, Y.Q. 2010. Behavioral impairment and oxidative damage induced by chronic application of nonylphenol. Int. J. Mol. Sci. 12: 114-127.

Mari, M., Schuhmacher, M., Feliubadal6, J. and Domingo, J.L. 2008. Air concentrations of PCDD/Fs, PCBs and PCNs using active and passive air samplers. Chemosphere 70: 1637-1643.

Marques-Pinto, A. and Carvalho, D. 2013. Human infertility: Are endocrine disruptors to blame. Endocr. Connect. 2: 15-19.

Martyanov, I.N., Uma, S., Rodrigues, S. and Klabunde, K.J. 2004. Structural defects cause TiO_2-based photocatalysts to be active in visible light. Chem. Commun. 21: 2476-2477.

Maruska, H.P. and Ghosh, A.K. 1979. Transition-metal dopants for extending the response of titanate photoelectrolysis anodes. Sol. Energy Mater. 1: 237-247.

Mashuri, S.I.S., Ibrahim, M.L., Kasim, M.F., Mastuli, M.S., Rashid, U., Abdullah, A.H., et al. 2020. Photocatalysis for organic wastewater treatment: From the basis to current challenges for society. Catalysts 10: 1260.

Mayoufi, A., Nsib, M.F. and Houas, A. 2014. Doping level effect on visible-light irradiation W-doped TiO_2-anatase photocatalysts for Congo red photodegradation. C. R. Chimie 17: 818-823.

Mehndiratta, P.A.J., Srivastava, S. and Gupta, N. 2013. Environmental pollution and nanotechnology. Environ. Pollut. 2: 10.

Mehrotra, K., Yablonsky, G.S. and Ray, A.K. 2005. Macro kinetic studies for photocatalytic degradation of benzoic acid in immobilized systems. Chemosphere 60: 1427-1436.

Melcer, H. and Klecka, G. 2011. Treatment of wastewaters containing bisphenol A: State of the science review. Water Environ. Res. 83: 605-666.

Mengyue, Z., Shifu, C. and Yaowu, T. 1995. Photocatalytic degradation of organophosphorus pesticides using thin films of TiO_2. J. Chem. Technol. Biotechnol. 64: 339-344.

Mills, A. and Hunte, A.J. 1997. An overview of semiconductor photocatalysis. J. Photochem. Photobiol. A: Chem. 108: 1-35.

Mishra, A.K. and Kumar, G.S. 2015. Weathering of oil spill: Modeling and analysis. Aquatic Procedia 4: 435-442.

Mishra, T., Mahato, M., Aman, N., Patel, J.N. and Sahu, R.K. 2011. A mesoporous WN co-doped titania nanomaterial with enhanced photocatalytic aqueous nitrate removal activity under visible light. Catal. Sci. Technol. 1: 609-615.

Mohamed, A., Yousef, S., Nasser, W.S., Osman, T.A., Knebel, A., Sanchez, E.P.V., et al. 2020. Rapid photocatalytic degradation of phenol from water using composite nanofbers under UV. Environ. Sci. Eur. 32: 160.

Mohan, D. and Pittman, C.U. 2007. Arsenic removal from waste/wastewater using adsorbents: A critical review. J. Hazard Mater. 142: 1-53.

Mohapatra, D.P., Brar, S.K., Tyagi, R.D. and Surampalli, R.Y. 2011. Occurrence of bisphenol A in wastewater and wastewater sludge of sludge of CUQ treatment plant. J. Xenobiot. 1: 9-16.

Moreira, C.G., Moreira, M.H., Silva, V.M.O.C., Santos, H.G., Bila, D.M. and Fonseca, F.V. 2019. Treatment of bisphenol A (BPA) in water using UV/H_2O_2 and reverse osmosis (RO) membranes: Assessment of estrogenic activity and membrane adsorption. Water Sci. Technol. 80(11): 2169-2178.

Moussavi, G. and Mahmoudi, M. 2009. Removal of azo and anthraquinone reactive dyes from industrial wastewaters using MgO nanoparticles. J. Hazard. Mater. 168: 806-812.

Moustakas, N.G., Kontos, A.G., Likodimos, V., Katsaros, F., Boukos, N., Tsoutsou, D., et al. 2013. Inorganic–organic core–shell titania nanoparticles for efficient visible light activated photocatalysis. Appl. Catal. B. 130-131: 14-24.

Mozia, S., Morawski, A.W., Toyoda, M. and Inagaki, M. 2009. Application of anatase-phase TiO_2 for decomposition of azo dye in a photocatalytic membrane reactor. Desalination 241: 97-105.

Mu, Y., Yu, H.Q., Zheng, J.C., Zhang, S.J. and Sheng, G.P. 2004. Reductive degradation of nitrobenzene in aqueous solution by zero-valent iron. Chemosphere 54: 789-794.

Mukerjee, D. 1998. Assessment of Risk from multimedia exposures of children to environmental chemicals. J. Air Waste Manage. 48: 483-501.

Muruganandham, M., Shobana, N. and Swaminathan, M. 2006. Optimization of solar photocatalytic degradation conditions of reactive yellow 14 azo dye in aqueous TiO_2. J. Mol. Catal. A: Chem. 246: 154-161.

Muruganandham, M. and Swaminathan, M. 2006. Photocatalytic decolourisation and degradation of reactive orange 4 by TiO_2-UV process. Dyes Pigm. 68: 133-142.

Nagamine, S., Sugioka, A., Iwamoto, H. and Konishi, Y. 2008. Formation of TiO_2 hollow microparticles by spraying water droplets into an organic solution of titanium tetraisopropoxide (TTIP) – effects of TTIP concentration and TTIP – protecting additives. Powder Technol. 186(2): 168-175.

Nakayama, N. and Hayashi, T. 2007. Preparation and characterization of poly(L-lactic acid)/TiO_2 nanoparticle nanocomposite films with high transparency and efficient photodegradability. Polym. Degrad. Stab. 92(2007): 1255-1264.

Naldoni, A., Allieta, M., Santangelo, S., Marelli, M., Fabbri, F., Cappelli S., et al. 2012. Effect of nature and location of defects on bandgap narrowing in black TiO_2 nanoparticles. J. Am. Chem. Soc. 134: 7600-7603.

Nasirian, M. and Mehrvar, M. 2016. Modification of TiO_2 to enhance photocatalytic degradation of organics in aqueous solutions. J. Environ. Chem. Eng. 4: 4072-4082.

Navntoft, C., Araujo, P., Litter, M.I., Apella, M.C., Fernandez, D., Puchulu, M.E., et al. 2007. Field tests of the solar water detoxification SOLWATER reactor in Los Pereyra, Tucuman, Argentina. J. Sol. Energy Eng. 129: 127-134.

Nayak, B.B., Acharya, H.N. and Mitra, G.B. 1983. Structural characterization of $Bi_{2-x}Sb_xS_3$ films prepared by the dip-dry method. Thin Solid Film 105: 17-24.

Neppolian, B., Choi, H.S., Sakthivel, S., Arabindoo, B. and Murugesan, V. 2002a. Solar light induced and TiO_2 assisted degradation of textile dye reactive blue 4. Chemosphere 46: 1173-1181.

Neppolian, B., Choi, H.C., Sakthivel, S., Arabindoo, B. and Murugesan, V. 2002b. Solar/UV-induced photocatalytic degradation of three commercial textile dyes. J. Hazard Mater. B 89: 303-317

Nguyen, V.H., Shawn, D.L., Wu, J.C.S. and Bai, H. 2014. Artificial sunlight and ultraviolet light induced photo-epoxidation of propylene over V-Ti/MCM-41 photocatalyst. J. Nanotechnol. 5: 566-576.

Nosaka, Y. and Fox, M.A. 1988. Kinetics for electron transfer from laserpulse irradiated colloidal semiconductors to adsorbed methylviologen: Dependence of the quantum yield on incident pulse width. J. Phys. Chem. 92: 1893-1897.

Nosaka, Y., Matsushita, M., Nishino, J. and Nosaka, A.Y. 2005. Nitrogen-doped titanium dioxide photocatalysts for visible response prepared by using organic compounds. Sci. Technol. Adv. Mater. 6: 143-148.

Novoselov, K.S., Geim, A.K., Morozov, S.V., Jiang, D., Zhang, Y., Dubonos, S.V., et al. 2004. Electric field effect in atomically thin carbon films. J. Science 306: 811 666-669.

Obregon, S. and Colon, G. 2012. Evidence of upconversion luminescence contribution to the improved photoactivity of erbium doped TiO_2 systems. Chem. Commun. 48: 7865-7867.

Olatunji, O.S. 2019. Evaluation of selected polychlorinated biphenyls (PCBs) congeners and dichlorodiphenyltrichloroethane (DDT) in fresh root and leafy vegetables using GC-MS. Sci. Rep. 9: 538.

Ozaki, H., Iwamoto, S. and Inoue, M. 2007. Marked promotive effect of iron on visible-light-induced photocatalytic activities of nitrogen- and silicon-codoped titanias. J. Phys. Chem. C 111: 17061-17066.

Pan, H., Zhang, Y.W., Shenoy, V.B. and Gao, H. 2011. Effects of H-, N-, and (H, N)-doping on the photocatalytic activity of TiO_2. J. Phys. Chem. C 11: 12224-12231.

Pan, J. and Guan, B. 2010. Adsorption of nitrobenzene from aqueous solution on activated sludge modifed by cetyltrimethylammonium bromide. J. Hazard. Mater. 183: 341-346.

Pan, J.H., Cai, Z., Yu, Y. and Zhao, X.S. 2011. Controllable synthesis of mesoporous F-TiO_2 spheres for effective photocatalysis. J. Mater. Chem. 21: 11430-11438.

Pan, Z., Stemmler, E.A., Cho, H.J., Fan, W., LeBlanc, L.A., Patterson, H.H., et al. 2014. Photocatalytic degradation of 17α-ethinylestradiol (EE_2) in the presence of TiO_2-doped zeolite. J. Hazard. Mater. 279: 17-25.

Pany, S. and Parida, K.M. 2014. Sulfate-anchored hierarchical meso-macroporous N-doped TiO_2: A novel photocatalyst for visible light H_2 evolution. ACS Sustain. Chem. Eng. 2: 1429-1438.

Paola, A.D., Augugliaro, V., Palmisano, L., Pantaleo, G., Savinov, E. 2003. Heterogeneous photocatalytic degradation of nitrophenols. J. Photochem. Photobiol. A: Chem. 155: 207-218.

Patel, N., Dashora, A., Jaiswal, R., Fernandes, R., Yadav, M., Kothari, D.C., et al. 2015. Experimental and theoretical investigations on the activity and stability of substitutional and interstitial boron in TiO_2 photocatalyst. J. Phys. Chem. C 119: 18581-18590.

Pawar, M., Sendoğdular, S.T. and Gouma, P. 2018. A brief overview of TiO_2 photocatalyst for organic dye remediation: Case study of reaction mechanisms involved in Ce-TiO_2 photocatalysts system. J. Nanomater. 2018: 1-13.

Pawari, M.J. and Gawande, S. 2015. Ground water pollution & its consequence. IJERGS. 3: 773-776.

Peng, N., Hu, D., Zeng, J., Li, Y., Liang, L. and Chang, C. 2016. Superabsorbent cellulose–clay nanocomposite hydrogels for highly efficient removal of dye in water. ACS Sustain. Chem. Eng. 4: 7217-7224.

Peral, J., Domenech, X. and Ollis, D.F. 1997. Heterogeneous photocatalysis for purification, decontamination and deodorization of air. J. Chem. Technol. Biotechnol. 70: 117-140.

Pereira, L. and Alves, M. 2012. Dyes-environmental impact and remediation. pp. 111-162. *In*: Malik, A., Grohmann, E. [eds.]. Environmental Protection Strategies for Sustainable 111 Development, Strategies for Sustainability. Springer, Germany.

Pope, C.N. 1999. Organophosphorus pesticides: Do they all have the same mechanism of toxicity. J. Toxicol. Env. Heal. B. 2: 161-181.

Poudel, B., Wang, W.Z., Dames, C., Huang, J.Y., Kunwar, S., Wang, D., et al. 2005. Formation of crystallized titania nanotubes and their transformation into nanowires. Nanotechnology 16(9): 1935-1940.

Poulios, I., Avranas, A., Rekliti, E. and Zouboulis, A. 2000. Photocatalytic oxidation of Auramine O in the presence of semiconducting oxides. J. Chem. Technol. Biotechnol. 75: 205-212.

Poulios, I. and Aetopoulou, I. 1999. Photocatalytic degradation of the textile dye reactive orange 16 in the presence of TiO_2 suspensions. Environ. Technol. 20: 479-487.

Priyadharshini, U.K., Latha, R., Kavitha, U. and Nirmala, N. 2017. Effects of organophosphorus pesticides on cardiorespiratory parameters among the farmers. J. Clin. Diagn. Res. 11: CC01-CC04.

Qamar, M., Saquib, M. and Muneer, M. 2005. Photocatalytic degradation of two selected dye derivatives, chromotrope 2B and amido black 10B, in aqueous suspensions of titanium dioxide. Dyes Pigm. 65: 1-9.

Qiu, B., Zhou, Y., Ma, Y., Yang, X., Sheng, W., Xing, M., et al. 2015. Facile synthesis of the Ti^{3+} self-doped TiO_2-graphene nanosheet composites with enhanced photocatalysis. Sci. Rep. 5: 8591.

Rachel, A., Sarakha, M., Subrahmanyam, M. and Boule, P. 2002. Comparison of several titanium dioxides for the photocatalytic degradation of benzenesulfonic acids. Appl. Catal. B 37: 293-300.

Rahimi, R., Saadati, S. and Fard, E.H. 2015. Fluorine-doped TiO_2 nanoparticles sensitized by tetra(4-carboxyphenyl)porphyrin and zinc tetra(4-carboxyphenyl)porphyrin: Preparation, characterization, and evaluation of photocatalytic activity. Environ. Prog. Sustain. Energ. 34: 1341-1348.

Ramacharyulu, P.V.R.K., Nimbalkar, D.B., Kumar, J.P., Prasad, G.K. and Ke, S.C. 2015. N-doped, S-doped TiO_2 nanocatalysts: Synthesis, characterization and photocatalytic activity in the presence of sunlight. RSC Adv. 5: 37096-37101.

Ramanathan, R. and Bansal, V. 2015. Ionic liquid mediated synthesis of nitrogen, carbon and fluorine-codoped rutile TiO_2 nanorods for improved UV and visible light photocatalysis. RSC Adv. 5: 1424-1429.

Rani, M. and Shanker, U. 2017. Degradation of traditional and new emerging pesticides in water by nanomaterials: Recent trends and future recommendations. Int. J. Environ. Sci. Technol. 15: 1347-1380.

Rani, M., Shanker, U. and Jassal, V. 2017. Recent strategies for removal and degradation of persistent & toxic organochlorine pesticides using nanoparticles: A review. J. Environ. Manage. 190: 208-222.

Rathna, R. and Nakkeeran, E. 2018. Phenol degradation from industrial wastewater by engineered microbes. pp. 253-276. *In*: Sunita, J.V., Agarwal, A.K., Gnansounou, E. and Gurunathan, B. (eds.). Bioremediation: Applications for Environmental Protection and Management. Springer, Germany.

Rengaraj, S. and Li, X.Z. 2006. Photocatalytic degradation of bisphenol A as an endocrine disruptor in aqueous suspension using Ag-TiO_2 catalysts. Int. J. Environ. Pollut. 27: 20-37.

Reyes-Coronado, D., Gattorno, G.R., Pesqueira, M.E.E., Cab, C., de Coss, R. and Oskam, G. 2018. Phase-pure TiO_2 nanoparticles: Anatase, brookite and rutile. Nanotechnol. 19: 145605.

Reza, K.M., Kurny, A.S.W. and Gulshan, F. 2017. Parameters affecting the photocatalytic degradation of dyes using TiO_2: A review. Appl. Water Sci. 7: 1569-1578.

Ribeiro, E., Ladeira, C. and Viegas, S. 2017. Review EDCs mixtures: A stealthy hazard for human health. Toxics 5: 5-22.

Richel, N.P. and Johnson, D.W. 2000. McComb. Appl. Physics Lett. 76: 1816.

Rocha, O.R.S., Dantas, R.F., Duarte, M.M.M.B., Duarte, M.M.L. and da Silva. V.L. 2010. Oil sludge treatment by photocatalysis applying black and white light. Chem. Eng. J. 157: 80-85.

Rueda-Marquez, J.J., Levchuk, I., Ibanez, P.F. and Sillanpaa, M. 2020. A critical review on application of photocatalysis for toxicity reduction of real wastewaters. J. Clean. Prod. 258. 120694.

Rupa, A.V., Divakar, D. and Sivakumar, T. 2009. Titania and noble metals deposited titania catalysts in the photodegradation of tartrazine. Catal. Lett. 132: 259-267.

Saad, S.K.M., Umar, A.A., Nguyen, H.Q., Dee, C.F., Salleh, M.M. and Oyama, M. 2014. Porous (001)-faceted Zn-doped anatase TiO_2 nanowalls and their heterogeneous photocatalytic characterization. RSC Adv. 4: 57054-57063.

Sachdeva, S. and Kumar, A. 2009. Preparation of nanoporous composite carbon membrane for separation of rhodamine B dye. J. Membr. Sci. 329: 2-10.

Saggioro, E.M., Oliveira, A.S., Pavesi, T., Maia, C.G., Ferreira, L.F.V. and Moreira, J.C. 2011. Use of titanium dioxide photocatalysis on the remediation of model textile wastewaters containing azo dyes. Molecules 16: 10370-10386.

Sajjad, S., Leghari, S.A.K. and Zhang, J. 2013. Copper impregnated ionic liquid assisted meso-porous titania: Visible light photocatalyst. RSC Adv. 3: 12678-12687.

Sakthivel, S., Neppolian, B., Shankar, M.V., Arabindoo, B., Palanichamy, M. and Murugesan, V. 2003. Solar photocatalytic degradation of azo dye: Comparison of photocatalytic efficiency of ZnO and TiO_2. Sol. Energy Mater. Sol. Cells 77: 65-82.

Sakthivel, S. and Kisch, H. 2003. Daylight photocatalysis by carbon-modified titanium dioxide. Angew. Chem. Int. Ed. 42: 4908-4911.

Samir, S. and Ibrahim, M.S. 2008. Assessment of heavy metals pollution in water and sediments and their effect on oreochromisniloticus. In the Northern Delta Lakes, Egypt. ISTA. 8: 475-489.

Sammarco, P.W., Kolian, S.R., Warby, R.A.F., Bouldin, J.L., Subra, W.A. and Porter, S.A. 2016. Concentrations in human blood of petroleum hydrocarbons associated with the BP/deepwater horizon oil spill, Gulf of Mexico. Arch. Toxicol. 90: 829-837.

Saquib, M. and Muneer, M. 2003. TiO_2-mediated photocatalytic degradation of a triphenylmethane dye (gentian violet), in aqueous suspensions. Dyes Pigm. 56: 37-49.

Sarma, G.K., Gupta S.S. and Bhattacharyya, K.G. 2019. Nanomaterials as versatile adsorbents for heavy metal ions in water: A review. ESPR. 26: 6245-6278.

Schecter, A., Birnbaum, L., Ryan, J.J. and Constable, J.D. 2006. Dioxins: An overview. Environ. Res. 101: 419-428.

Schindler, D.W. 2006. Recent advances in the understanding and management of eutrophication. Limnol. Oceanogr. 51: 356-363.

Segne, T.A., Tirukkovalluri, S.R. and Challapalli, S. 2011. Studies on characterization and photocatalytic activities of visible light sensitive TiO_2 nano catalysts co-doped with magnesium and copper. Int. Res. J. Pure Appl. Chem. 1: 84-103.

Senanayake, N. and Karalliedde, L. 1987. Neurotoxic effects of organophosphate insecticides: An intermediate syndrome. N. Eng. J. Med. 316: 761-763.

Sengele, A., Robert, D., Keller, N., Keller, V., Herissan, A. and Colbeau-Justin, C. 2016. Ta-doped TiO_2 as photocatalyst for UV: A activated elimination of chemical warfare agent simulant. J. Catal. 334: 129-141.

Senna, M., Sepelak, V., Shi, J., Bauer, B., Feldhoff, A., Laporte, V., et al. 2012. Introduction of oxygen vacancies and fluorine into TiO_2 nanoparticles by co-milling with PTFE. J. Solid State Chem. 187: 51-57.

Senthilkumaar, S., Porkodi, K., Gomathi, R., Geetha Maheswari, A. and Manonmani, N. 2006. Sol-gel derived silver doped nanocrystalline titania catalysed photodegradation of methylene blue from aqueous solution. Dyes Pigm. 69: 22-30.

Serpone, N., Lawless, D., Khairutdinovand, R. and Pelizzetti, E. 1995. Subnanosecond relaxation dynamics in TiO_2 colloidal sols. Relevance to heterogeneous photocatalysis. Phys. Chem. 99: 16655-16661.

Shakoor, S. and Nasar, A. 2016. Removal of methylene blue dye from artificially contaminated water using citrus limetta peel waste as a very low cost adsorbent. J. Taiwan Inst. Chem. Eng. 66: 154-163.

Shanker, U., Rani, M. and Jassal, V. 2017. Degradation of hazardous organic dyes in water by nanomaterials. Environ. Chem. Lett. 15: 623-642.

Shao, P., Tian, J., Zhao, Z., Shi, W., Gao, S. and Cui, F. 2015. Amorphous TiO_2 doped with carbon for visible light photodegradation of rhodamine B and 4-chlorophenol. Appl. Surf. Sci. 324: 35-43.

Sharma, B., Dangi, A.K. and Shukla, P. 2018. Contemporary enzyme based technologies for bioremediation: A review. J. Environ. Manage. 210: 10-22.

Sharma, S.K. and Sanghi, R. 2012. Advances in Water Treatment and Pollution Prevention. Springer, Dordrecht.

Sharotri, N. and Sud, D. 2015. A greener approach to synthesize visible light responsive nano-porous S-doped TiO_2 with enhanced photocatalytic activity. N. J. Chem. 39: 2217-2223.

Shi, J.W., Liu, C., He, C., Li, J., Xie, C., Yang, S., et al. 2015. Carbon-doped titania nanoplates with exposed {001} facets: Facile synthesis, characterization and visible-light photocatalytic performance. RSC Adv. 5: 17667-17675.

Shukla, S.S., Dorris, K.L. and Chikkaveeraiah, B.V. 2009. Photocatalytic degradation of 2,4-dinitrophenol. J. Hazard. Mater. 164: 310-314.

Shuxin, D., Xiaoli, W. and Tiejun, W. 2004. Support vector machine for ultraviolet spectroscopic water quality analyzers. Chinese J. Anal. Chem. 32(9): 1227-1230.

Si, L., Huang, Z., Lv, K., Tang, D. and Yang, C. 2014. Facile preparation of Ti^{3+} self-doped TiO_2 nanosheets with dominant {001} facets using zinc powder as reductant. J. Alloy Compd. 601: 88-93.

Singh, A., Kumar, V. and Thakur, A.C. 2011. Paraoxonase-1 genetic polymorphisms and susceptibility to DNA damage in workers occupationally exposed to organophosphate pesticides. Toxicol. Appl. Pharmacol. 252: 130-137.

Sirisuk, A., Klansorn, E. and Praserthdam, P. 2008. Effects of reaction medium and crystallite size on Ti^{3+} surface defects in titanium dioxide nanoparticles prepared by solvothermal method. Catal. Commun. 9(9): 1810-1814.

Siuzdak, K., Szkoda, M., Sawczak, M., Lisowska-Oleksiak, A., Karczewski, J. and Ryl, J. 2015. Enhanced photoelectrochemical and photocatalytic performance of iodine-doped titania nanotube arrays. RSC Adv. 5: 50379-50391.

Sivakumar, P. and Palanisamy, P.N. 2008. Low-cost non-conventional activated carbon for the removal of reactive Red 4: Kinetic and isotherm studies. Rasayan J. Chem. 1: 871-883.

Sivalingam, G., Nagaveni, K., Hegde, M.S. and Madras, G. 2003. Photocatalytic degradation of various dyes by combustion synthesized nano anatase TiO_2. Appl. Catal. B 45: 23-38.

Sivula, K., Formal, F.L. and Grätzel, M. 2009. WO_3-Fe_2O_3 photoanodes for water splitting: A host scaffold, guest absorber approach. Chem. Mater. 21: 2862-2867.

So, C.M., Cheng, M.Y., Yu, J.C. and Wong, P.K. 2002. Degradation of azo dye procion red MX-5B by photocatalytic oxidation. Chemosphere 46: 905-912.

Soares, E.T., Lansarin, M.A. and Moro, C.C. 2007. A study of process variables for the photocatalytic degradation of Rhodamine B. Braz. J. Chem. Eng. 24: 29-36.

Sobczynski, A. and Dobosz, A. 2001. Water purification by photocatalysis on semiconductors. Pol. J. Environ. Stud. 10: 195-205.

Song, K.Y., Park, M.K. and Won, KY.T. 2001. Preparation of transparent particulate MoO_3/TiO_2 and WO_3/TiO_2 films and their photocatalytic properties. Chem. Mater. 13: 2349-2355.

Srinivasan, A. and Viraraghavan, T. 2010. Decolorization of dye wastewaters by biosorbents: A review. J. Environ. Manage. 91: 1915-1929.

Srivastava, M., Srivastava, A., Yadav, A. and Rawat. V. 2019. Source and Control of Hydrocarbon Pollution. Source and Control of Hydrocarbon Pollution. Hydrocarbon Pollution and its Effect on the Environment. doi:10.5772/intechopen.86487.

Štengl, V., Houšková, V., Bakardjieva, S. and Murafa, N. 2010. Photocatalytic activity of boron-modified titania under UV and visible-light illumination. ACS Appl. Mater. Inter. 2: 575-580.

Stengl, V., Popelkova, D. and Vlacil, P. 2011. TiO_2–graphene nanocomposite as high performace photocatalysts. J. Phys. Chem. C. 115: 25209-25218.

Strong, P.J. and J.E. Burgess. 2008. Treatment methods for wine-related and distillery wastewaters: A review. Bioremediat. J. 12(2): 70-87.

Stroud, J.L., Paton, G.I. and Sempel, K.T. 2009. Predicting the biodegradation of target hydrocarbons in the presence of mixed contaminants in soil. Chemosphere 74: 563-567.

Stylidi, M., Kondarides, D.I. and Verykios, X.E. 2003. Pathways of solar light-induced photocatalytic degradation of azo dyes in aqueous TiO_2 suspension. Appl. Catal. B: Environ. 40: 271-286.

Suarez-Iglesias, O., Collado, S., Oulego, P. and Diaz, M. 2017. Graphene-family nanomaterials in wastewater treatment plants. Chem. Eng. J. 313: 121-135.

Subramanian, V., Wolf, E. and Kama, P.V. 2001. Semiconductor-metal composite nanostructures. To what extent do metal nanoparticles improve the photocatalytic activity of TiO_2 Films? J. Phys. Chem. B 105: 11439-11446.

Subbaramaiah, V., Srivastava, V.C. and Mall, I.D. 2014. Catalytic oxidation of nitrobenzene by copper loaded activated carbon. Sep. Purif. Technol. 125: 284-290.

Sun, J., Qiao, L., Sun, S. and Wang, G. 2008. Photocatalytic degradation of Orange G on nitrogen-doped TiO_2 catalysts under visible light and sunlight irradiation. J. Hazard. Mater. 155: 312-319.

Sun, L., Zhao, X., Cheng, X., Sun, H., Li, Y., Li, P., et al. 2012. Synergistic effects in La/N codoped TiO_2 anatase (101) surface correlated with enhanced visible-light photocatalytic activity. Langmuir 28: 5882-5891.

Sun, X., Wang, C., Li, Y., Wang, W. and We, J. 2015. Treatment of phenolic wastewater by combined UF and NF/RO processes. Desalination 355: 68-74.

Tan, I.A.W., Ahmad, A.L. and Hameed, B.H. 2008. Adsorption of basic dye on high-surface-area activated carbon prepared from coconut husk: Equilibrium, kinetic and thermodynamic studies. J. Hazard. Mater. 154: 337-346.

Tang, Y., Zhang, G., Liu, C., Luo, S., Xu, X., Chen, L., et al. 2013. Magnetic TiO_2-graphene composite as a high-performance and recyclable platform for efficient photocatalytic removal of herbicides from water. J. Hazard. Mater. 252-253: 115-122.

Tchobanoglous, G., Burton, F.L., Stensel, H.D., Metcalf and Eddy. 2003. Fundamentals of biological treatment. *In*: Wastewater Engineering: Treatment and Resue, 6th Ed.; McGraw-Hill: New York.

Tian, G., Pan, K., Fu, H., Jing, L. and Zhou. W. 2009. Enhanced photocatalytic activity of S-doped TiO_2–ZrO_2 nanoparticles under visible–light irradiation. J. Hazard. Mater. 166: 939-944.

Tian, J., Hu, X., Yang, H., Zhou, Y., Cui, H. and Liu, H. 2016. High yield production of reduced TiO_2 with enhanced photocatalytic activity. Appl. Surf. Sci. 360: 768-743.

Tipayarom, D., Wantala, K. and Grisdanurak, N. 2011. Optimization of alachlor degradation on S-doped TiO_2 by sonophotocatalytic activity under visible light. Fresen. Environ. Bull. 20: 1425-1431.

Tiwari, A., Shukla, A., Choi, S.S. and Lee, S.M. 2018. Surface modified nanostructured-TiO_2 thin films for removal of congo red. Korean J. Chem. Eng. 35: 2133-2137.

Tiwari, A., Shukla, A., Tiwari, D., Choi, S.S., Shin, H.G. and Lee, S.M. 2019. Titanium dioxide nanomaterials and its derivatives in the remediation of water: Past, present and future. Appl. Chem. Eng. 30: 261-279.

Tobaldi, D.M., Pullar, R.C., Gualtieri, A.F., Seabra, M.P. and Labrincha, J.A. 2013. Sol-gel synthesis, characterisation and photocatalytic activity of pure, W-, Ag- and W/Ag co-doped TiO_2 nanopowders. Chem. Eng. J. 214: 364-375.

Tong, T., Zhang, J., Tian, B., Chen, F. and He, D. 2008. Preparation of Fe^{3+}-doped TiO_2 catalysts by controlled hydrolysis of titanium alkoxide and study on their photocatalytic activity for methyl orange degradation. J. Hazard. Mater. 155: 572-579.

Torosyan, G.H. and Simonyan, H.A. 2019. The detoxification nitrobenzene in waste water on zeolites. MOJ Biorg. Org. Chem. 3: 51-53.

Tran, H., Scott, J., Chiang, K. and Amal, R. 2006. Clarifying the role of silver deposits on titania for the photocatalytic mineralisation of organic compounds. J. Photochem. Photobiol. 183: 41-52.

Truskewycz, A., Gundry, T.D., Khudur, L.S., Kolobaric, A., Taha, M., Aburto-Medina, A., et al. 2019. Petroleum hydrocarbon contamination in terrestrial ecosystems—fate and microbial responses. Molecules 24: 3400.

Turabi, A., Danyal, A., Hassan, S., Ahmad, Y.S.M., Rashid, M.A. and Asif, A.H. 2007. Evaluation of suspected chronic pesticide poisoning among residents near agriculture fields. Biomedica. 23: 76-82.

Turner, M., Golovko, V.B., Vaughan, O.P.H., Abdulkin, P., A.B-Murcia Tikhov, M.S., et al. 2008. Selective oxidation with dioxygen by gold nanoparticle catalysts derived from 55-atom clusters. Nature 454: 981-983.

Ucankus, G., Deniz Uzunoglu, M.E. and Culha, M. 2018. Methods for preparation of nanocomposites in environmental remediation. New Polymer Nanocomposites for Environmental Remediation 1-28. https://doi.org/10.1016/B978-0-12-811033-1.00001-9.

Umezawa, N. and Ye, J. 2012. Role of complex defects in photocatalytic activities of nitrogen-doped anatase TiO_2. Phys. Chem. Chem. Phys. 14: 5924-5934.

Urban, J.D., Wikoff, D.S., Bunch, A.T., Harris, M.A. and Haws, L.C. 2014. A review of background dioxin concentrations in urban/suburban and rural soils across the United States: Implications for site assessments and the establishment of soil cleanup levels. Sci. Total Environ. 466: 586-597.

Valentin, C.D., Gianfranco, P. and Selloni, A. 2005. Theory of carbon doping of titanium dioxide. Chem. Mater. 17: 6656-6665.

Van den Berg, M., Birnbaum, L. and Bosveld, A.T.C. 1998. Toxic equivalency factors (TEFs) for PCBs, PCDDs, PCDFs for humans and wildlife. Environ. Health Perspect. 106: 775-792.

Van den Berg, M., Birnbaum, L.S., Denison, M., De Vito, M., Farland, W. and Feeley, M. 2006. The 2005 World Health Organization reevaluation of human and mammalian toxic equivalency factors for dioxins and dioxin-like compounds. Toxicol. Sci. 93: 223-241.

Van Gerven, T., Geysen, D. and Vandecasteele, C. 2004. Estimation of the contribution of a municipal waste incinerator to the overall emission and human intake of PCBs in Wilrijk, Flanders. Chemosphere 54: 1303-1308.

Varma, K.S., Tayade, R.J., Shah, K.J., Joshi, P.A., Shukla, A.D. and Gandhi, V.G. 2020. Photocatalytic degradation of pharmaceutical and pesticide compounds (PPCs) using doped TiO_2 nanomaterials: A review. Water-Energy Nexus 3: 46-61.

Varma, R. and Varma, D.R. 2005. The bhopal disaster of 1984. Bull. Sci. Technol. Soc. 23: 1-9.

Vasiliu, F., Diamandescu, L., Macovei, D., Teodorescu, C.M., Tarabasanu-Mihaila, D., Vlaicu, A.M., et al. 2009. Fe-and Eu-doped TiO_2 photocatalytical materials prepared by high energy ball milling. Top Catal. 52: 544-556.

Verma, S.P. and Sarkar, B. 2018. Simultaneous removal of Cd(II) and p-cresol from wastewater by micellar-enhanced ultrafiltration using rhamnolipid: Flux decline, adsorption kinetics and isotherm studies. J. Environ. Manage. 213: 217-235.

Vijayan, P., Mahendiran, C., Suresh, C. and Shanthi, K. 2009. Photocatalytic activity of iron doped nanocrystalline titania for the oxidative degradation of 2,4,6-trichlorophenol. Catal. Today 141: 220-224.

Villegas, L.G.C., Mashhadi, N., Chen, M. Mukherjee, D. Taylor, K.E. and Biswas, N. 2016. A short review of techniques for phenol removal from wastewater. Curr. Pollution. Rep. 2: 157-167.

Vinodgopal, K., Bedja, I. and Kamat, P.V. 1996. Nanostructured semiconductor films for photocatalysis. Photoelectrochemical behavior of SnO_2/TiO_2 composite systems and its role in photocatalytic degradation of a textile azo dye. Chem. Mater. 8: 2180-2187.

Vogel, R., Hoyer, P. and Weller, H. 1994. Quantum-sized PbS, CdS, Ag_2S, Sb_2S_3, and Bi_2S_3 particles as sensitizers for various nanoporous widebandgap semiconductors. J. Phys. Chem. 98: 3183-3188.

Vosges, M., Kah, O., Hinfray, N., Chadili, E., Page, Y.L., Combarnous, Y., et al. 2012. 17α-ethinylestradiol and nonylphenol affect the development of forebrain gnrh neurons through an estrogen receptors-dependent pathway. Reprod. Toxicol. 33: 198-204.
Wang, C., Shao, C. and Zhang, X. 2009. SnO_2 Nanostructures-TiO_2 nanofibers heterostructures: Controlled fabrication and high photocatalytic properties. Inorg. Chem. 48: 7261-7268.
Wang, C., Ao, Y., Wang, P., Hou, J., Qian, J. and Zhang, S. 2010a. Preparation, characterization, photocatalytic properties of titania hollow sphere doped with cerium. J. Hazard Mater. 178: 517-521.
Wang, C., Ao, Y., Wang, P., Huo, J. and Qian. J. 2010b. Photocatalytic performance of Gd ion modified titania porous hollow spheres under visible light. Mat. Lett. 64: 1003-1006.
Wang, C., Wang, X., Gong, P. and Yao, T. 2016. Residues, spatial distribution and risk assessment of DDTs and HCHs in agricultural soil and crops from the Tibetan plateau. Chemosphere 140: 358-365.
Wang, E., He, T., Zhao, L., Chen, Y. and Cao, Y. 2011. Improved visible light photocatalytic activity of titania doped with tin and nitrogen. J. Mater. Chem. 21: 144-150.
Wang, H., Liu, Z., Zhang, J., Huang, R., Yin, H., Dang, Z., et al. 2019. Insights into removal mechanisms of bisphenol A and its analogous in municipal wastewater treatment plants. Sci. Total Environ. 692: 107-116.
Wang, N., Li, J., Zhu, L., Dong, Y. and Tang, H. 2008. Highly photocatalytic activity of metallic hydroxide/titanium dioxide nanoparticles prepared via a modified wet precipitation process. J. Photochem. Photobiol. A 198: 282-287.
Wang, S., Yang, X., Wang, Y., Liu, L., Guo, Y. and Guo, H. 2014a. Morphology-controlled synthesis of Ti^{3+} self-doped yolk–shell structure titanium oxide with superior photocatalytic activity under visible light. J. Solid State Chem. 213: 98-103.
Wang, S., Zhang, X., Ma, D., Yu, Z., Wang, X. and Niu, Y. 2014b. Photocatalysis performance of Nb-doped TiO_2 film *in situ* growth prepared by a micro plasma method. Rare Metal Mater. Eng. 43: 1549-1552.
Wang, X., M. Blackford, K. Prince and R.A. Caruso. 2012a. Preparation of boron-doped porous titania networks containing gold nanoparticles with enhanced visible-light photocatalytic activity. ACS Appl. Mater. Inter. 4: 476-482.
Wang, X., Li, Y., Liu, X., Gao, S., Huang, B. and Dai, Y. 2015a. Preparation of Ti^{3+} self-doped TiO_2 nanoparticles and their visible light photocatalytic activity. Chin. J. Catal. 36: 389-399.
Wang, X., Wang, W., Wang, X., Zhang, J., Gu, Z., Zhou, L., et al. 2015b. Enhanced visible light photocatalytic activity of a floating photocatalyst based on B-N-codoped TiO_2 grafted on expanded perlite. RSC Adv. 5: 41385-41392.
Wang, X., Yan, Y., Hao, B. and Chen. G. 2014c. Biomimetic layer-by-layer deposition assisted synthesis of Cu, N co-doped TiO_2 nanosheets with enhanced visible light photocatalytic performance. Dalton Trans. 43: 14054-14060.
Wang, X.K., Wang, C., Jiang, W.Q., Guo, W.L. and Wang, J.G. 2012b. Sonochemical synthesis and characterization of Cl-doped TiO_2 and its application in the photodegradation of phthalate ester under visible light irradiation. Chem. Eng. J. 189-190: 288-294.
Wang, Y.Q., Hu, G.Q., Duan, X.F., Sun, H.L. and Xue, Q.K. 2002. Microstructure and formation mechanism of titanium dioxide nanotubes. Chem. Phys. Lett. 365: 427-431.
Watson, S.S., Beydoun, D., Scott, J.A. and Amal, R. 2003. The effect of preparation method on the photoactivity of crystalline titanium dioxide particles. Chem. Eng. J. 95: 213-220.
Wei, C.H., Tang, X.H., Liang, J.R. and Tan, S. 2007. Preparation, characterization and photocatalytic activity of boron- and cerium-codoped TiO_2. J. Environ. Sci. 19: 90-96.
Wei, S., Wu, R., Jian, J., Chen, F. and Sun, Y. 2015. Black and yellow anatase titania formed by (H, N)-doping: Strong visible-light absorption and enhanced visible-light photocatalysis. Dalton Trans. 44: 1534-1538.
Wen, M., Zhang, S., Dai, W., Li, G. and Zhang, D. 2015. *In situ* synthesis of Ti^{3+} self-doped meso-porous TiO_2 as a durable photocatalyst for environmental remediation. Chin. J. Catal. 36: 2095-2102.
Wen, Q., Chen, Z., Lian, J., Feng, Y. and Ren, N. 2012. Removal of nitrobenzene from aqueous solution by a novel lipoid adsorption material (LAM). J. Hazard Mater. 209: 226-232.
Wu, M.J., Bak, T., O'Doherty, P.J., Moffitt, M.C., Nowotny, J., Bailey, T.D., et al. 2014. Photocatalysis of titanium dioxide for water disinfection: Challenges and future perspectives. Int. J. Photochem. 2014: 1-9.
Wu, N.L. and Lee, M.S. 2004. Enhanced TiO_2 photocatalysis by Cu in hydrogen production from aqueous methanol solution. Inter. J. Hydro. Energ. 29: 1601-1605.

Wu, X., Yin, S., Dong, Q., Guo, C., Li, H., Kimura, T., et al. 2013b. Synthesis of high visible light active carbon doped TiO_2 photocatalyst by a facile calcination assisted solvothermal method. Appl. Catal. B. 142-143: 450-457.

Wu, Y., Liu, H., Zhang, J. and Chen, F. 2009. Enhanced photocatalytic activity of nitrogen-doped titania by deposited with gold. J. Phys. Chem. C 113: 14689-14695.

Wu, Y., Zhang, Q., Yin, X. and Cheng, H. 2013a. Template-free synthesis of mesoporous anatase yttrium-doped TiO_2 nanosheet-array films from waste tricolor fluorescent powder with high photocatalytic activity. RSC Adv. 3: 9670-9676.

Xiang, Q., Yu, J. and Jaroniec, M. 2011. Nitrogen and sulfur co-doped TiO_2 nanosheets with exposed {001} facets: Synthesis, characterization and visible-light photocatalytic activity. Phys. Chem. Chem. Phys. 13: 4853-4861.

Xiao, Q. and Ouyang, L. 2009. Photocatalytic activity and hydroxyl radical formation of carbon-doped TiO_2 nanocrystalline: Effect of calcination temperature. Chem. Eng. J. 148: 248-253.

Xiaohong, W., Zhaohua, J., Huiling, L., Shigang, X. and Xinguo, H. 2003. Photo-catalytic activity of titanium dioxide thin films prepared by micro-plasma oxidation method. Thin Solid Films 441: 130-134.

Xie, T., Zhaoqian, J., Hu, J., Yuan, P., Liu, Y. and Cao, S. 2018. Degradation of nitrobenzene-containing wastewater by a microbial-fuel-cellcoupled constructed wetland. Ecol. Eng. 112: 65-71.

Xin, B., Ren, Z., Wang, P., Liu, J., Jing, L. and Fu. H. 2007. Study on the mechanisms of photoinduced carriers separation and recombination for Fe^{3+}-TiO_2 photocatalysts. App. Surf. Sci. 253: 4390-4395.

Xin, X., Xu, T., Yin, J., Wang, L. and Wang, C. 2015. Management on the location and concentration of Ti^{3+} in anatase TiO_2 for defects-induced visible-light photocatalysis. Appl. Catal. B Environ. 176-177: 354-362.

Xing, Y., Lu, Y., Dawson, R.W., Shi, Y., Zhang, H., Wang, T., et al. 2005. A spatial temporal assessment of pollution from PCBs in China. Chemosphere 60: 731-39.

Xiong, L.B., Li, J.L., Yang, B. and Yu, Y. 2012. Ti^{3+} in the surface of titanium dioxide: Generation, properties and photocatalytic application. J. Nanomater. 2012: 1-13.

Xu, A.W., Gao, Y. and Liu, H.Q. 2002. The preparation, characterization, and their photocatalytic activities of rare-earth-doped TiO_2 nano-particles. J. Catal. 207: 151-157.

Xu, H., Zheng, Z., Zhang, L., Zhang, H. and Deng, F. 2008. Hierarchical chlorine-doped rutile TiO_2 spherical clusters of nanorods: Large-scale synthesis and high photocatalytic activity. J. Solid State Chem. 181: 2516-2522.

Xu, P., Xu, T., Lu, J., Gao, S., Hosmane, N.S., Huang, B., et al. 2010. Visible-light-driven photocatalytic S- and C-codoped meso/nanoporous TiO_2. Energ. Environ. Sci. 3: 1128-1134.

Xu, W.Y., Gao, T.Y. and Fan, J.H. 2005. Reduction of nitrobenzene by the catalyzed Fe-Cu process. J. Hazard. Mater. 123: 232-241.

Yan, C., Yi, W., Yuan, H., Wu, X. and Li, F. 2014. A highly photoactive S, Cu-codoped nano-TiO_2 photocatalyst: Synthesis and characterization for enhanced photocatalytic degradation of neutral red. Environ. Prog. Sustain. Energ. 33: 419-429.

Yan, J., Zhang, Y., Liu, S., Wu, G., Li, L. and Guan, N. 2015. Facile synthesis of an iron doped rutile TiO_2 photocatalyst for enhanced visible-light-driven water oxidation. J. Mater. Chem. A 3: 21434-21438.

Yang, B., Zuo, J., Li, P., Wang, K., Yu, X. and Zhang, M. 2016. Effective ultrasound electrochemical degradation of biological toxicity and refractory cephalosporin pharmaceutical wastewater. Chem. Eng. J. 287: 30-37.

Yang, J., Zhang, X., Wang, C., Sun, P., Wang, L., Xia, B., et al. 2012. Solar photocatalytic activities of porous Nb-doped TiO_2 microspheres prepared by ultrasonic spray pyrolysis. Solid State Sci. 14: 139-144.

Yang, M., Thompson, D.W. and Meye, G.J. 2002. Charge-transfer studies of iron cyano compounds bound to nanocrystalline TiO_2 surfaces. Inorg. Chem. 41: 1254-1262.

Yang, P., Luo, S., Liu, Y. and Jiao, W. 2018. Degradation of nitrobenzene wastewater in an acidic environment by Ti(IV)/H_2O_2/O_3 in a rotating packed bed. ESPR 25(25): 25060-25070.

Yang, X., Ma, F., Li, K., Guo, Y., Hu, J., Li, W., et al. 2010. Mixed phase titania nanocomposite codoped with metallic silver and vanadium oxide: New efficient photocatalyst for dye degradation. J. Hazard. Mater. 175: 429-438.

Yang, X.J., Wang, S., Sun, H.M., Wang, X.B. and Lian, J.S. 2015. Preparation and photocatalytic performance of Cu-doped TiO_2 nanoparticles. Trans. Nonferrous Met. Soc. China 25: 504-509.

Yang, Y., Ni, D., Yao, Y., Zhong, Y., Ma, Y. and Yao, J. 2015. High photocatalytic activity of carbon doped TiO_2 prepared by fast combustion of organic capping ligands. RSC Adv. 5: 93635-93643.

Yaseen, D.A. and Scholz, M. 2019. Textile dye wastewater characteristics and constituents of synthetic efuents: A critical review. Int. J. Environ. Sci. Technol. 16: 1193-1226.

Yates, H.M., Nolan, M.G., Sheel, D.W. and Pemble, M.E.J. 2006. The role of nitrogen doping on the development of visible light-induced photocatalytic activity in thin TiO_2 films grown on glass by chemical vapour deposition. J. Photochem. Photobiol. A 179: 213-223.

Ye, Y., Feng, Y., Bruning, H., Yntema, D. and Rijnaarts, H. 2017. Photocatalytic degradation of metoprolol by TiO_2 nanotube arrays and UV-Led: Effects of catalyst properties, operational parameters, commonly present water constituents, and photo-induced reactive species. Appl. Catal. B Environ. 220: 171-181.

Ying, G. and Fate, G. 2006. Behavior and effects of surfactants and their degradation products in the environment. Environ. Int. 32: 417-431.

Yu, G., Wang, B. Cheng. and M. Zhou. 2007. Effects of hydrothermal temperature and time on the photocatalytic activity and microstructures of bimodal mesoporous TiO_2 powders. Appl. Catal. B. 69(3-4): 171-180.

Yu, H., Zheng, X., Yin, Z., Tao, F., Fang, B. and Hou, K. 2007. Preparation of nitrogen-doped TiO_2 nanoparticle catalyst and its catalytic activity under visible light. Chin. J. Chem. Eng. 15: 802-807.

Yu, H., Shi, R., Zhao, Y., Waterhouse, G.I.N., Wu, L.Z., Tung, C.H., et al. 2016. Smart utilization of carbon dots in semiconductor photocatalysis. Adv. Mater. 28: 9454-9477.

Yu, J., Dai, G., Xiang, Q. and Jaroniec, M. 2011. Fabrication and enhanced visible-light photocatalytic activity of carbon self-doped TiO_2 sheets with exposed {001} facets. J. Mater. Chem. 21: 1049-1057.

Yu, J., Li, Q., Liu, S. and Jaroniec, L.M. 2013. Ionic-liquid-assisted synthesis of uniform fluorinated B/C-codoped TiO_2 nanocrystals and their enhanced visible-light photocatalytic activity. Chem. Eur. J. 19: 2433-2441.

Yu, L., Yang, X., He, J., He, Y. and Wang, D. 2015. One-step hydrothermal method to prepare nitrogen and lanthanum co-doped TiO_2 nanocrystals with exposed {001} facets and study on their photocatalytic activities in visible light. J. Alloy Compd. 637: 308-314.

Yuan, Z.Y., Colomer, J.F. and Su, B.L. 2002. Titanium oxide nanoribbons. Chem. Physics Lett. 363: 362.

Zada, A., Qu, Y., Ali, S., Sun, N., Lu, H., Yan, R. et al. 2018. Improved visible-light activities for degrading pollutants on TiO_2/g-C_3N_4 nanocomposites by decorating Spr Au nanoparticles and 2,4-dichlorophenol decomposition Path. J. Hazard. Mater. 342: 715-723.

Zaleska, A. 2008. Doped-TiO_2: A review. Recent Pat. Eng. 2: 157-164.

Zhang, H., Lv, X., Li, Y., Wang, Y. and Li, J. 2010. P25-graphene composite as a high performance photocatalyst. J. ACS Nano 4(1): 380-386.

Zhang, H., Z. Xing, Y. Zhang, Z. Li, X. Wu, C. Liu, et al. 2015b. Ni^{2+} and Ti^{3+} co-doped porous black anatase TiO_2 with unprecedented-high visible-light-driven photocatalytic degradation performance. RSC Adv. 5: 107150-107157.

Zhang, J., Fu, W., Xi, J., He, H., Zhao, S., Lu, H., et al. 2013a. N-doped rutile TiO_2 nano-rods show tunable photocatalytic selectivity. J. Alloy Compd. 575: 40-47.

Zhang, J., Tse, K., Wong, M., Zhang, Y. and Zhu, J. 2016. A brief review of co-doping. Front. Phys. 11: 117405.

Zhang, J., Wu, Y., Xing, M., Leghari, S. and Sajjad, S. 2010. Development of modified N doped TiO_2 photocatalyst with metals, nonmetals and metal oxides. Energ. Environ. Sci. 3: 715-726.

Zhang, J., Xi, J. and Ji, Z. 2012. (Mo, N)-codoped TiO_2 sheets with dominant {001} facets for enhancing visible-light photocatalytic activity. J. Mater. Chem. 22: 17700-17708.

Zhang, J., Zhou, P., Liu, J. and Yu, J. 2014b. New understanding of the difference of photocatalytic activity among anatase, rutile and brookite TiO_2. Phys. Chem. Chem. Phys. 16: 20382-20386.

Zhang, K., Wang, X., He, T., Guo, X. and Feng, Y. 2014a. Preparation and photocatalytic activity of B-N co-doped mesoporous TiO_2. Powder Technol. 253: 608-613.

Zhang, L., Tse, M.S., Tan, O.K., Wang, Y.X. and Han, M. 2013b. Facile fabrication and characterization of multi-type carbon-doped TiO_2 for visible light-activated photocatalytic mineralization of gaseous toluene. J. Mater. Chem. A 1: 4497-4507.

Zhang, Q., Gao, L. and Guo, J. 2000. Effects of calcination on the photocatalytic properties of nanosized TiO_2 powders prepared by $TiCl_4$ hydrolysis. Appl. Catal. B Environ. 26: 207-215.

Zhang, T., Oyama, T., Horikoshi, S., Hidaka, H., Zhao, J. and Serpone, N. 2002. Photocatalyzed N-demethylation and degradation of methylene blue in titania dispersions exposed to concentrated sunlight. Sol. Energy Mater. Sol. Cells 73: 287-303.

Zhang, X., Huang, Q., Deng, F., Huang, H., Wana, Q., Liu, M., et al. 2017. Mussel-inspired fabrication of functional materials and their environmental applications: Progress and prospects. Appl. Mater. Today 7: 222-238.

Zhang, Y., Zhao, Z., Chen, J., Cheng, L., Chang, J., Sheng, W., et al. 2015a. C-doped hollow TiO_2 spheres: *In situ* synthesis, controlled shell thickness, and superior visible-light photocatalytic activity. Appl. Catal. B Environ. 165: 715-722.

Zhang, Y., Zhu, W., Cui, X., Yao, W. and Duan, T. 2015c. One-step hydrothermal synthesis of iron and nitrogen co-doped TiO_2 nanotubes with enhanced visible-light photocatalytic activity. CrystEngComm. 17: 8368-8376.

Zhang, Y.P., Xu, J.J., Sun, Z.H., Li, C.Z. and Pan, C.X. 2011. Preparation of graphene and TiO_2 layer 998 by layer composite with highly photocatalytic efficiency. Pro. Nat. Sci. Mater. 21(6): 467-471.

Zhao, C., Shu, X., Zhu, D., Wei, S., Wing, Y.X., Tu, M.J., et al. 2015. High visible light photocatalytic property of Co^{2+} doped TiO_2 nanoparticles with mixed phases. Super-lattice Microst. 88: 32-42.

Zhao, J. and Yang, X. 2003. Photocatalytic oxidation for indoor air puri&cation: A literature review. Build. Environ. 38: 645-654.

Zhao, L., Ma, W., Ma, J., Wen, G. and Liu, Q. 2015. Relationship between acceleration of hydroxyl radical initiation and increase of multiple-ultrasonic feld amount in the process of ultrasound catalytic ozonation for degradation of nitrobenzene in aqueous solution. Ultrason. Sonochem. 22: 198-204.

Zhao, Y., Liu, J., Shi, L., Yuan, S., Fang, J., Wang, Z. and Zhang, M. 2011. Solvothermal preparation of Sn^{4+} doped anatase TiO_2 nanocrystals from peroxo-metal-complex and their photocatalytic activity. Appl. Catal. B: Environ. 103: 436-443.

Zhao, Y.F., Li, C., Lu, S., Yan, L.J., Gong, Y.Y., Niu, L.Y., et al. 2016. Effects of oxygen vacancy on 3d transition-metal doped anatase TiO_2: First principles calculations. Chem. Phys. Lett. 647: 36-47.

Zheng, C., Zhao, L., Zhou, X., Fu, Z. and Li, A. 2013. Treatment Technologies for Organic Wastewater. doi: 10.5772/52665.

Zheng, Z., Huang, B., Meng, X., Wang, J., Wang, S., Lou, Z., et al. 2013. Metallic zinc-assisted synthesis of Ti^{3+} selfdoped TiO_2 with tunable phase composition and visiblelight photocatalytic activity. Chem. Commun. 49: 868-870.

Zhiyong, Y., Bensimon, M., Sarria, V., Stolitchnov, I., Jardim, W., Laub, D., et al. 2007. $ZnSO_4$–TiO_2 doped catalyst with higher activity in photocatalytic processes. Appl. Catal. B: Environ. 76: 185-195.

Zhou, X., Shao, C., Li, X., Wang, X., Guo, X. and Liu, Y. 2018. Three dimensional hierarchical 1012 heterostructures of g-C3N4 nanosheets/TiO_2 nanofibers: Controllable growth via gas-solid reaction and enhanced photocatalytic activity under visible light. J. Hazard. Mater. 344: 113-122.

Zhu, J., Deng, Z., Chen, F., Zhang, J., Chen, H., Anpo, M., et al. 2006. Hydrothermal doping method for preparation of Cr^{3+}–TiO_2 photocatalysts with concentration gradient distribution of Cr^{3+}. Appl. Catal. B. 62: 329-335.

Zhu, Y., Murali, S., Cai, W., Li, X., Suk, J.W., Potts, J.R., et al. 2010. Graphene and graphene oxide: Synthesis, properties, and applications. Adv. Mater. 22(35): 3906-3924.

Zielińska, B., Grzechulska, J., Kaleńczuk, R.J. and Morawski, A.W. 2003. The pH influence on photocatalytic decomposition of organic dyes over A11 and P25 titanium dioxide. Appl. Catal. B 45: 293-300.

Graphene-Based Nano-Composite Material for Advanced Nuclear Reactor: A Potential Structural Material for Green Energy

Nisha Verma[1,*] and Soupitak Pal[2]

1. INTRODUCTION

The demand for C-free, safe, sustainable and reliable energy alternatives is increasing day by day with the increasing world population. Electricity has not only improved the quality of life and life expectancy, but the easy availability of electricity also has a direct impact on any countries economic prosperity and industrial growth. To date, fossil fuel (coal, natural gas and petroleum) serves as the largest stockholder in fulfilling the global energy need; at the same time, they are the largest contributor (≈66%) to the production of greenhouse gases and climate change. A recent report advocates that the baseline for global energy requirement will grow up to ≈ 25-60 TW from the current consumption of ≈14 TW around 2030 (Zinkle 2005). Tackling the problem of greenhouse gases, preservation of the clean environment, an alternative source of C-free energy is really essential. Nuclear energy has stood out as a promising alternative for clean energy resources without emission of greenhouse gases for current as well as future generation technologies. To date, 10% of the global energy need is supplied by nuclear technology. However, safety and reliability, as well as proliferation resistance of nuclear power, always raises questions in its way to expansion. In a typical fission reactor, a high-energy neutron collides with U^{238} isotopes at the reactor core to release heat through fission reaction, heats the water above ≈280°C to produce steam, drives the turbine to generate electricity. Apparently, this simple technology possesses severe safety and reliability challenges due to the production of the high-energy neutron, α, β particles and γ-ray irradiation.

2. FUTURE GEN-IV NUCLEAR REACTOR FOR GREEN ENERGY

There are 440 total nuclear reactors operating around the 30 countries worldwide and producing ≈13% of the total electricity need of the globe (Nuclear electricity generation worldwide from 1985 to 2020, © Statista 2021 'World Nuclear Power Reactors & Uranium Requirements' 2020). Most of the commercial reactors are generation-II and generation-III reactors, recently introduced to the market, and generation-III+ just started to commercialize. Although the safety and reliability of these reactors are well proven, they are still lacking in efficiency and prolong service life. The

[1] Materials Research Center, Malaviya Institute of Technology Jaipur, Rajasthan, Pin Code - 302017.

[2] Advanced Instrumentation for Nano-Analytics (AINA), Luxembourg Institute of Science and Technology (LIST), Materials Research and Technology Department, 41, Rue Du Brill, L-4422 Belvaux, Luxembourg.

[*] Corresponding author: nisha.mrc@mnit.ac.in.

current operating temperature for most of the functional reactors is confined within a maximum temperature of ≈350°C, which largely restricts the reactors to operate at their full efficiency (Murty and Charit 2008). Moreover, these reactors are licensed up to 30 years of service time. With an ever-increasing demand for the clean energy landscape, in 2000, the US Department of Energy launched a new program called Generation-IV Initiative in collaboration with the European Union and other countries to increase the utilization of nuclear energy by making further advances in nuclear energy systems design. In this initiative, prime focus has been given to higher operating temperature (above 600°C) with a service life of approximately 60 years or more along with safety, reliability, proliferation-resistance and profitability. Six different designs have been prioritized, and they are listed in Table 1 along with their coolant, neutron spectrum, core outlet temperatures, the pressure exerted and electrical efficiency.

TABLE 1 Different designs of Gen.-IV reactor systems (Corwin 2006, Zinkle and Was 2013).

System	Neutron Spectrum, Max. Dose (dpa)	Coolant	Outlet Temp. (°C)	Pressure (MPa)	Net Electrical Efficiency (%)
VHTR	Thermal, <20	Helium	850-1,000	7	50
SFR	Fast, 200	Sodium	500-550	0.1	>40
SCWR	Thermal, 30/fast, 70	Water	510-625	25	44
GFR	Fast, 80	Helium	850	7	45-48
LFR	Fast, 150	Lead	480-800	0.1	45
MSR	Thermal/fast, 200	Fluoride salts	700-1,000	0.1	44-50
PWR	Thermal, ~80	Water (single phase)	320	16	
BWR	Thermal, ~7	Water (two-phase)	288	7	

These harsh operating conditions put even more restrictions upon selecting, developing and qualifying the structural materials for Gen-IV nuclear reactors. Some of the desirable characteristics for selecting the new materials can be summarized as follows:

1. Excellent dimensional stability against thermal and irradiation creep and void swelling.
2. Outstanding room and high temperatures mechanical properties, such as strength, fracture, and fatigue.
3. Irradiation resistance under high neutron dose, up to 10-150 dpa with high resistance to helium embrittlement.
4. A high degree of chemical compatibility with the fuel and structural material along with sufficient resistance to stress-corrosion cracking (SCC) and irradiation-assisted stress-corrosion cracking (ISCC).
5. Finally, it has workability, weldability and processing cost.

All of these aforementioned criteria are related to fundamental high-temperature material degradation mechanisms, oxidation, phase instability, radiation-induced segregation and so forth.

2.1 Radiation Damage in Materials

Radiation damages are initiated by the exchange of energy between incoming energetic particles and atoms in solids (target atoms), leading to knocking off of the target atoms, generation of point defects in the form of vacancy-interstitial pairs and formation of point defect clusters. Schematic representation of all the leading defects generated during this process is shown in Figure 1. These non-equilibrium point defects become mobile at elevated temperatures and cause detrimental effect to the stability of microstructure. This is generally referred to as radiation damage as it leads to deterioration of alloys through embrittlement, void swelling and creep (Was 2007).

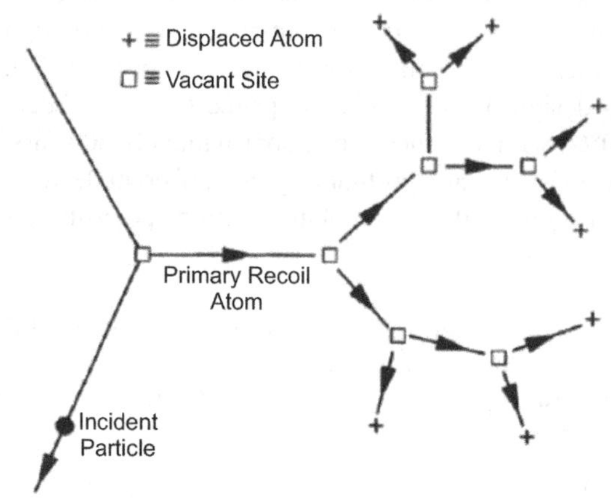

FIGURE 1 Schematic representation of cascade displacement created by a Primary Knock-on Atom (PKA) (Nastasi and Mayer 1996) (Starikov et al. 2011).

Many have explained the details of defect production during the radiation damage, and documented it well for metals and dilute alloys. The primary impact of the incoming energetic particle (i.e. neutron) is to recoil the atoms away from their primary lattice site, which happens in the following steps explained next. First, the energy transfer from incoming energetic particles to lattice atoms creates, Primary Knock-on Atom (PKA). If the incoming atoms transfer energy equivalent to threshold displacement energy to the PKA, it gets displaced from its original site and create a cascade of collisions with other lattice atoms. All the recoil atoms will continue this process till their energy is unable to displace further atoms. Moreover, the recoil atoms lose their energy during two separate processes, either by collision with other atoms or creating electronic transitions. Whereas, the energy loss during electronic excitation is not accounted for any defect production for metals and semiconductors. SRIM (Ziegler et al. 2010) is a well-known software used to calculate the number of defect pairs, i.e. vacancy and interstitial pairs (Frenkel pairs). However, a rough estimate can be obtained from the Kinchin-Pease formulation of defect pairs, which is reproduced below in Equation 1 (Pease RS 1955, Norgett and Robinson 1975).

$$N_{FP} = \frac{0.8E_D}{2E_d} \tag{1}$$

Here, 'E_D' is the damage energy, which is the net energy gained through recoil less the energy consumed in electronic excitation. 'E_d' is called displacement energy, which is equivalent to the average energy consumed during the creation of a vacancy-interstitial pair, which roughly corresponds to 25 eV. It is known that the recoil energy of a fission neutron in steel is ~ 50 eV, where the Kinchin-Pease equation estimate that this energy is capable of producing ~ 800 FP's (Frenkel pairs); here electron excitation energy loss has been neglected, which is indicative of the fact that a displacement cascade would create a substantial amount of defects pairs by displacing atoms from their respective locations. Whereas, the fact to be noted is that these defects are created within a special range of 5 nm and within a time frame of few picoseconds. Moreover, the overall damage can be quantified by accounting for the dose of irradiation particles. Conventionally, the actual dose is specified by displacement per atom (*dpa*), rather than specifying neutron/cm^2 or ion/cm^2, where *dpa* represents the ratio of the count of Frenkel pairs produced after irradiation by the total number of atoms present in irradiated volume. Presently, the research is based on designing structural materials for a nuclear reactor that can withstand a dose of ~ 200 dpa, implying that each atom within irradiated volume would become interstitial around 200 times during their service life. Many microstructure changes occur during the irradiation process, a few of them namely atomic mixing, radiation enhanced diffusion and irradiation-induced creep are discussed below.

2.1.1 Atomic Mixing

The impact of an incoming energetic particle on the lattice atoms is not only limited in creating defect pairs, i.e. Frankel pairs but it also causes events that result in atomic mixing by simply exchanging atoms from their original lattice sites. The main facilitator of this mixing is localized thermal spikes, where within these collision cascades high temperatures enable diffusion of atomic species. The local temperature within these cascades can get as high as the melting temperatures. Phase stability and nonequilibrium processes are the ones which mainly get affected by this intricate way of atomic mixing. Many reports exist which demonstrate the fact that through low-temperature irradiation, metastable phases can be realized, particularly for multilayer thin films (B. Tsaur 1981, J. Mayer et al. 1981, L. Hung et al. 1983, Hirvonen and Nastasi 1986, D. Williamson et al. 1994). Interestingly, the alloys that exhibit liquid phase immiscibility due to strong chemical interactions, i.e. large heat of mixing are shown to undergo during mixing, which also leads to the formation of clustering of solutes and precipitates within the cascade due to thermal spike. However, the large solute cluster would require several cascade overlaps as each thermal spike will relocate atoms only once or twice. Moreover, alloys with weak chemical interactions would homogenize (randomize) under irradiation by energetic ions. Mostly due to ballistic mixing (random mixing) that overcome the cascade mixing due to weak interaction in comparison to that involved in the thermal spike (Martin 1984). It holds utmost importance in the nuclear research community to evaluate the new materials for their phase stability and non-equilibrium phases for providing a solution to the standing challenges for future reactors. This understanding will help in designing new and unique engineering materials which remain stable under ion irradiation.

2.1.2 Radiation Enhanced Diffusion (RED)

High temperature along with irradiation can bring point defects in motion and make them active contributors to the diffusion process, which is as follows:

$$D_{RED} = f_i c_i D_i + f_v c_v D_v \qquad (2)$$

Here $c_{i(v)}$ is the interstitials (vacancies) concentration

$D_{i(v)}$ is the defect diffusivity

$f_{i(v)}$ is the appropriate correlation factor, for convenience, this factor is considered unity.

Here the concentration of point defects includes equilibrium point defects as well as defects produced through irradiation. The defect pairs produced during irradiation are usually supersaturated in nature. Almost the majority of point defects generated during irradiation get eliminated through self-recombination inside the cascade. However, a small fraction (ξ_{FM}) remains mobile within the lattice (Naundorf and Macht 1992, L. Wei et al. 1999). Eventually, these defects also annihilate, meeting sinks (matrix/precipitate interfaces, grain boundaries, free surfaces, and dislocations) (Siegel et al. 1980). Also, recombination can eliminate the defects. The recombining defects contribute to diffusion and annihilating defects lead to void swelling, creep and non-equilibrium segregation, which is more detrimental to the material response.

Two different regimes exist based on temperature range, high temperature is dominated by equilibrium defects and at low-temperature radiation-induced defects dominate. Vacancies and interstitials can freely migrate during high temperatures and get annihilated at sinks. Whereas, the defect concentration is not sufficient to promote recombination and thus this zone is termed as a 'sink-limited' regime. On the other hand, during low temperatures, the irradiation-induced defects migrate slowly and have enough time for accumulation and recombination. This regime is accordingly called a 'recombination-limited' regime. The separation of different regimes in space is based on defect migration enthalpies, the sink density and defect production rate. The separation of different regimes can also be pictured based on the temperature dependence of D_{RED}. Activation enthalpy for diffusion at high temperature is single-handedly governed by ΔH_{eq} and the self-diffusion activation energy in equilibrium. Subsequently, D_{RED} can be expressed as:

$$D_{RED} = 2 \cdot \left(\frac{\Omega}{4\pi r_{vs}} \right) K_0 \tag{3}$$

Here, D_{RED} is radiation diffusion coefficient, Ω is atomic volume, r_{vs} is vacancy sink interaction radius and K_0 is expressed in dpa-s^{-1} as it is the rate at which freely migrating defects generate. Here, the assumption is $\Delta H_{RED} \sim 0$ as migration distance for all defects to sink is the same. Similarly, at low-temperature, D_{RED} can be expressed as

$$D_{RED} = 2 \cdot \left(\frac{\Omega}{4\pi r_{iv}} \right)^{1/2} K_0^{1/2} D_v^{1/2} \tag{4}$$

Here, r_{iv} is vacancy interstitial interaction radius (typically ~ two times the lattice parameter) and D_v is vacancy diffusivity. The above equation is based on assumption that vacancies are the slowest migrating defect and $\Delta H_{RED} \sim \Delta H_{i,v}^{m}/2$ where $\Delta H_{i,v}^{m}$ is the enthalpy of interstitial or vacancy, whichever moves slowest.

2.1.3 Self-Organization Under Irradiation

Self-organization reaction under irradiation as first recognized by Nelson et al. for an alloy consisting of two phases, Ni-Ni_3Al (i.e. γ-γ') (Nelson et al. 1972). A steady-state was recognized which was associated with the arrangement of precipitates at a characteristic length scale. This steady-state was recognized to be independent of initial conditions. Another terminology has been adopted for this self-organization and is referred to as compositional patterning (Enrique and Bellon 2000). This phenomenon of self-organization is remarkable as it implies that alloy will maintain this steady state under prolonged irradiation and irrespective of its duration. As suggested by Nelson et al., this self-organization results from the competition between ballistic mixing (i.e. high energy displacement processes in cascade) and thermally derived diffusion (i.e. radiation-enhanced diffusion). The ballistic processes tend to homogenize the alloy, where assumption is that inside cascade the ballistic event energies are much higher compared to chemical bonds energy. On contrary, the thermal events (thermal diffusion) tend to restore the thermal equilibrium. A simple approach have been adapted by Nelson et al. to arrive at steady state radius of precipitates. Nelson's model assumes that the precipitate sputtering takes place at a rate which is proportional to r^3 (volume of precipitate), which implies that atoms liberated from precipitate recoil far from the precipitate and join the random distribution of solutes. Whereas, they join back a precipitate through irradiation enhanced diffusion and rate is proportional to r. The steady-state radius, which is manifested from these competing processes is mathematically formulated in (Nelson et al. 1972). Even though the model is predictive analytically, the steady-state length scale is always the same for compositional pattering. For example, the recoil distances of atoms from precipitates are typical of the order of one or two atomic distances, which would imply that these atoms should return to their parent precipitate either by ballistic mixing or irradiation to enhance diffusion. Many more models exist, giving more rigorous explanation by Heinig et al. (2003), Enrique and Bellon (2000) and Frost and Russell (1982).

A dynamic phase diagram shown in Figure 2 is constructed by Enrique and Bellon, which is based on the Cahn-Hilliard model. Here, two parameters are focused, γ_b, the ballistic to thermal jump ratio and the distance of the ballistic jump, R. Three distinct regions are clearly seen in Figure 2, macroscopic phase separation (MPS), compositional pattering (CP) and solid solution (SS). These different steady-state regions are strongly dependent on the value of γ_b. For $\gamma_b >> 1$, where ballistic mixing dominates over thermal diffusion, promoting random mixing and as a result of which solid solution of the alloy is attained. However, when $\gamma_b << 1$, thermal diffusion dominates and alloy restores the thermodynamic equilibrium and undergo macroscopic phase separation. Interestingly, compositional patterning is obtained at $\gamma_b \sim 1$. Here, both ballistic and thermal processes occur with similar strength but operate at different length scales. Thermal diffusion dominates at small distances, leading to phase separation and ballistic diffusion predominates at longer distances, governing the size of precipitating phase. These competing processes occurring at different length scales lead to compositional pattering.

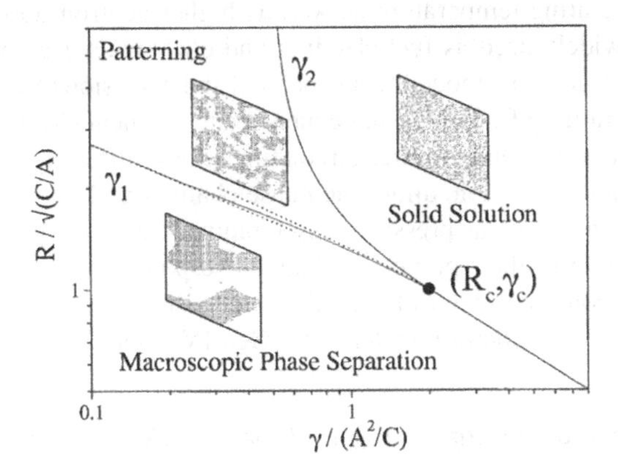

FIGURE 2 A phase space for steady-state as a function of R and γ(Enrique and Bellon 2000)

2.1.4 Irradiation Induced Creep (IIC)

Irradiation creep is of main concern associated with the design of materials for C-free energy production as it causes undesirable dimensional changes under prolonged irradiation. Generally, under stress and with thermal assistance, permanent deformation is activated by atomic displacement. Thermal creep is facilitated by the flow of lattice defect, contrarily the irradiation creep occurs by sustained production of point defects by energetic irradiation which feeds to dislocation evolution. Inevitably increasing the climb rate of dislocations. IIC is far from thermodynamic equilibrium and can occur at much lower temperatures required for thermal creep. This irradiation-induced permanent deformation and stress relaxation process has been well documented for metals and alloys (Ehrlich 1981). Unlike thermal creep, IIC has a weak dependence on temperature and is mostly governed by irradiation-induced microstructure changes, i.e. void swelling.

Typically, irradiation creeps strain rate is expressed as:

$$\dot{\varepsilon} = B_0 \sigma^n \varphi \tag{5}$$

where B_0 is irradiation creep compliance, σ is stress, n is stress exponent and ϕ the atomic displacement damage rate. From reported results, it can be inferred that $n = 1$ for metallic materials. Whereas, B_0 shows a strong dependence on fluence for transient creep regime. The irradiation creeps compliance for cold-worked 316 stainless steel is $\sim 3 \times 10^{-6}$ MPa^{-1} dpa^{-1} at 300-500°C for secondary creep (Grossbeck and Horak 1988). Whereas, for alloyed austenitic stainless, i.e. chromium ferritic and martensitic steels, it is in the range ~ 1×10^{-6} and 1×10^{-5} MPa^{-1} dpa^{-1} (Garner and Puigh 1991). With these references, it can be inferred that fewer alloyed metals will be subjected to greater creep deformation.

The main strategy to improve the long-term stability of these structural materials under irradiation is to introduce a high density of neutral defect sinks which can trap the irradiation-induced defects. For this purpose, some have shown the effectiveness of solutes (Braislford and Bullough 1978) and some advocate the supremacy of nanostructured materials. The main component of nanoscale materials which helps in mitigating the IIC is a large fraction of interfaces of incoherent and semi-coherent nature and grain boundaries.

2.2 Structural Material for the Gen-IV Reactor

Materials that are commonly used for commercial reactors, are not suitable for the Gen-IV design due to their higher operating temperature as well as higher neutron dose. For a typical example, zirconium alloys are widely used as fuel cladding and other reactor components due to their low neutron absorption cross-section, moderate mechanical and corrosion resistance at high temperature (at an operating temperature of ≈350°C) and aqueous environment (Murty and Charit 2008). With an increase in the operating temperature, zirconium alloys suffer from hydrogen embrittlement due to hydride formation, oxidation, allotropic phase changes and poor creep properties. Some of the outer core components (such as pressure vessel, piping, etc.) are typically made from low alloy steel, need to replace due to their poorer mechanical response and irradiation resistance at high temperatures. Several potential material candidates are listed in Table 2 and Figure 3 exhibiting promising performance as a structural material for Gen-IV reactors.

TABLE 2 List of candidate materials for Gen-IV reactors (Abram et al. 2002, Murty and Charit 2008).

Reactor System	F-M Steel	Austenitic SS	ODS Steel	Ni-Base Alloys	Graphite	Refractory Alloys	Ceramics
GFR	P	P	P	P	–	P	P
Pb-LFR	P	P	S	–	–	S	S
MSR	–	–	–	P	P	S	S
SFR	P	P	P	–	–	–	–
SCWR	P	P	S	S	–	–	–
VHTR	S	–	–	P	P	S	P

P = primary option, S = secondary option

For some of the listed materials, designated as primary, databases are already available. The only experiment related to its qualification is needed to be performed. Whereas, for secondary options, database generation along with qualification require the design of accelerated tests validated through materials modeling. Here, we will briefly discuss the performances of these materials, mostly irradiation resistance and mechanical properties which allow them to stand out as potential structural materials for Gen-IV reactors.

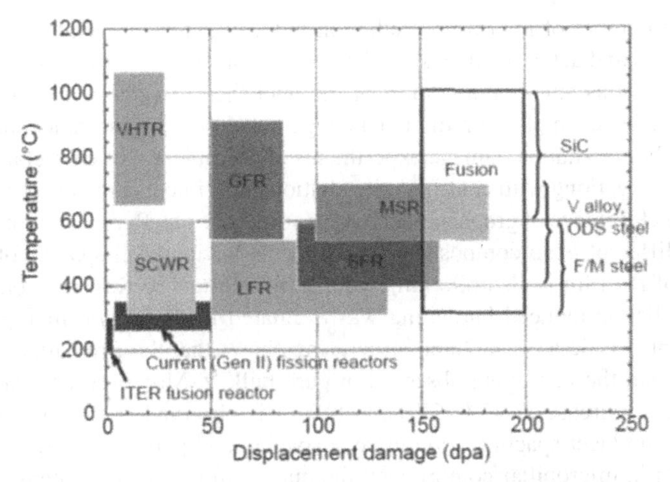

FIGURE 3 Displays the spectrum of temperature and displacement damage for Gen-II and IV reactors in combination with potential structural material.

2.3 Scope of Nanostructured Materials

Although the current material database gives an essential guideline, but designing and development of new materials for future generation reactor are almost unavoidable. Decades of expensive alloy designing methods yield nominal increments on alloy's radiation tolerance; however, advanced fusion and fission reactors need dramatic improvements in materials radiation tolerance. Therefore, the focus has been shifted toward tailoring existing materials with an innovative processing route to engineer high-performance materials. One of the key approaches in mitigating radiation damage is to control the density and structure of boundaries (interface) (Beyerlein et al. 2015). For example, in oxide dispersion strengthened (ODS) steel (Duan et al. 2017, Lu et al. 2017) and nanocrystalline metals [11] the high-density grain boundaries (GBs) and heterogeneous interfaces were found to be efficient sinks for irradiation-induced defects (Bai et al. 2010). Alternatively, multi-layer systems (such as Cu/Nb [11, 12], V/Ag [13], Cu/V [14, 15] and Fe/W [16]) and nanotwinned metals [17-19] with individual nano-meter-sized layer and particular interfacial structures possess excellent resistance to radiation damage and decent microstructural stability.

3. RADIATION DAMAGE OF GRAPHENE NANO-COMPOSITES

The core of a nuclear reactor, fuel cladding and tubes are subjected to heavy neutron irradiation of 10-150 displacement per atom (dpa) damage, resulting in the production of point defect, interstitial, dislocation loops and stacking faults, radiation enhance segregation at the grain boundaries as well as phase change through radiation enhanced diffusion (Zinkle and Was 2013). Such phenomena yield in hardening by means of providing restriction on dislocation movement causing embrittlement, severe loss of ductility, fast fracture along the grain boundary. Another important aspect for structural material is to manage a high amount of He (≈1,000 appm), derived as a transmutation product. He has very limited solubility in most of the metallic substances, resulting in precipitation of He bubbles at the grain boundary or interior (Odette et al. 2008). Conjugation of these bubbles results in void formation and its growth, which poses a serious threat to structural material. Unfortunately, a very limited amount of material database exists for the selection and qualification of materials for structural material in a nuclear reactor. Current understanding leads to design selection criteria of materials based on their neutron scattering cross-section to reduce swelling; left us with choices like high strength steels, Ni-base superalloys, W and V-based alloys. Another material selection criterion, which has recently gained a vast majority of attention, engineer material through deliberately designing a large number of interfaces. Radiation-induced defects typically get trap within these numerous interfaces and act as a defect sink (Beyerlein et al. 2015). Although this field is relatively new, it is getting serious attention and most of the published research to date is at the laboratory scale level. Here, a few examples of radiation damage in metal-graphene nanocomposites are listed, carried out using both radiation damage experiments and simulation of Gr-metal nano-composites.

Small-scale testing along with controlled irradiation experiments provide an initial indicator for the exceptional performance of graphene-derived nano-composite. Pure V film of 600 nm thickness along with two different nano-composite with graphene, having layer spacing of 300 and 110 nm (total thickness of the film is also 600 nm), were subjected to 120 KeV He^{+}-ion irradiation (Kim et al. 2016). Irradiation-induced hardening was evaluated using *in-situ* micropillar compression testing experiments inside a SEM. Nano-structuring shows that the irradiation hardening in pure V film was less than the hardening observed in pure bulk V. Moreover, an irradiation hardening induced flow stress increase of 25%, 61% and 88% was observed in the case of the graphene-V composite of 110 nm layer spacing, 300 nm layer spacing and pure V-film, respectively. Whereas, at 20% true strain in micropillar compression the pure V-film fails, nano-composite with 110 nm layer spacing deflects the crack propagation through V-graphene interface (Kim et al. 2016). Figure 4 displays the strain-strain curve and deformed microstructure of the composite.

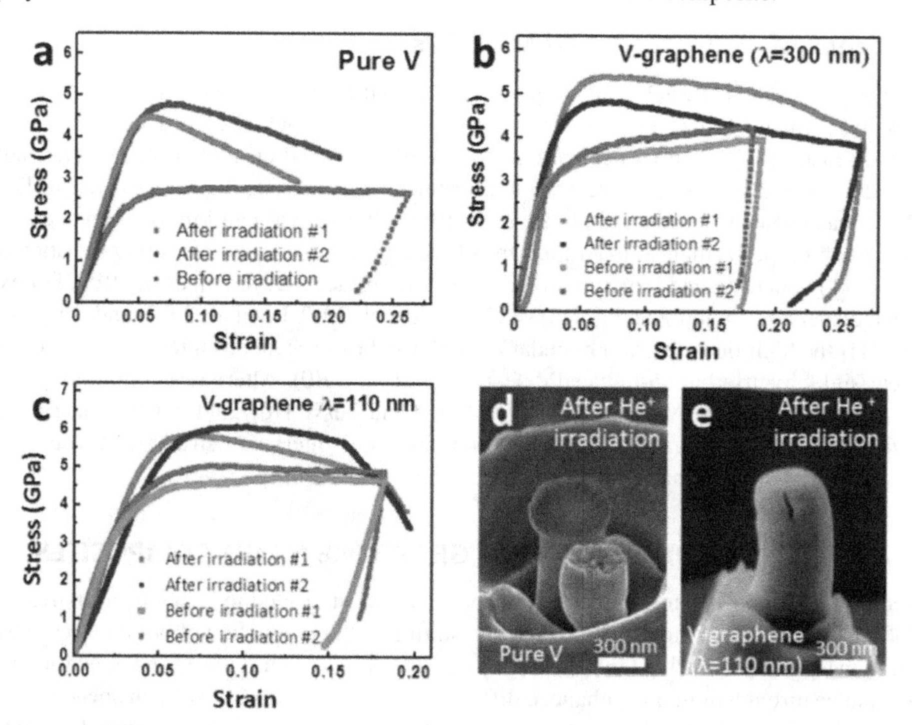

FIGURE 4 Strain versus strain curve obtained from the micro-pillar experiments along with deformed microstructure, reproduced from the reference of Kim et al. (2016).

As in the reactor structural material, He evolves as a transmutation product and due to its insolubility in most of the metals, it precipitates out at the grain boundaries. Reacting with the other radiation-induced defects it starts to producing He bubbles, which migrate and conjugate to produce a large void after reaching a critical size (Odette et al. 2008). In the case of the graphene-V nano-composite, due to the impermeable nature of atomic layer graphene toward He, manages to capture He bubbles at the graphene-V interface (Bunch et al. 2008). Moreover, graphene reduces the tendency of He migration to mitigate larger bubble formation and reduce the probability of brittle failure. The nano-structure of the film along with the V-Gr hetero-interface helps in retaining the bubble at the interface below the critical size. It was further substantiated that a very high number density and smaller sizes of the He bubbles is one of the primary reasons for the low hardening behavior of V-Gr nanocomposite (Kim et al. 2016). Molecular dynamics study of the irradiation process of the same V-Gr nanocomposite delineates that Gr-hetero-interface acts as an excellent sink for vacancy-interstitial defects by consuming PKA that annihilates radiation-induced defects. More the number of interfaces more the defect annihilation, resulting in better radiation resistance (Kim et al. 2016). Similarly, a radiation damage study on Ni/Gr nano-composite (prepared by electrodeposition) was performed using 300 keV He^{2+} ion irradiation with a fluence of 1×10^{17} ions/cm^2 at 823 K (Huang et al. 2018a). At peak damage, the He ions concentration within the composite was 5.3 at.%. Ni is typically used in molten salt and super-critical-water cooled reactors (Jin et al. 2012, Huang et al. 2016). Though after the fabrication of the composite, the Gr shows the disordered structure, aggravated by thermal annealing and irradiation but imparts high radiation damage resistance, compared to its pure Ni counterpart. The radiation-induced He bubble formation is lesser in size in the Ni/Gr composite with less damage in the matrix. Even after high-temperature irradiation, the Ni-Gr interface acts as defect sinks with an intrinsic self-healing mechanism of the Gr structure. A schematic illustration of the effectiveness of Ni/Gr interface on pining the He bubble at a smaller size is shown in Figure 5.

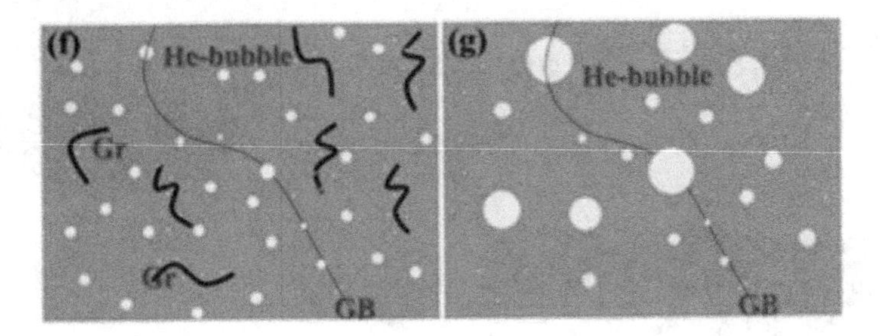

FIGURE 5 A schematic illustration of the role of Ni/Gr interface and Gb on the growth of He bubble upon irradiation (Huang et al. 2018b).

The role of Ni/Gr interface acting as a sink for radiation-induced defect has recently been unfolded by Huang et al. using molecular dynamics, molecular statics and climbing-image nudged elastic band simulation method (Huang et al. 2018b). Simulation study results in accumulation of point-defects at the Ni/Gr interface and tends to recombine or get trapped by the interface. Ni/Gr interface preferentially absorbs more interstitial than vacancy and forms a special defect structure through a major reduction in the formation energy (0.92 eV) (Huang et al. 2018a). This loaded Ni/Gr interface has an excellent sink strength which helps in self-healing the matrix defects and enhances radiation resistances. A more or less similar type of results was obtained when He-ion irradiation of 400 KeV was performed on a bulk-Al/reduced graphene oxide (RGO) composite (Liu et al. 2020). The Al/RGO composite shows a low hardening compare to the pure Al matrix and reduced irradiation-induced defect density. Small-scale micro-pillar tension and compression testing reveal the loss of ductility in the composite was less than the pure matrix (Liu et al. 2020). The excellent void swelling and irradiation resistance of the metal/Gr interface is derived from the excellent sink strength of Gr which promotes effective recombination of vacancy-interstitial, leading to trapping He in nm scale bubbles (Huang et al. 2015, 2017, Huang et al. 2018a, Liu et al. 2020). The mechanism of He trapping via the metal/Gr interface is shown in Figure 6.

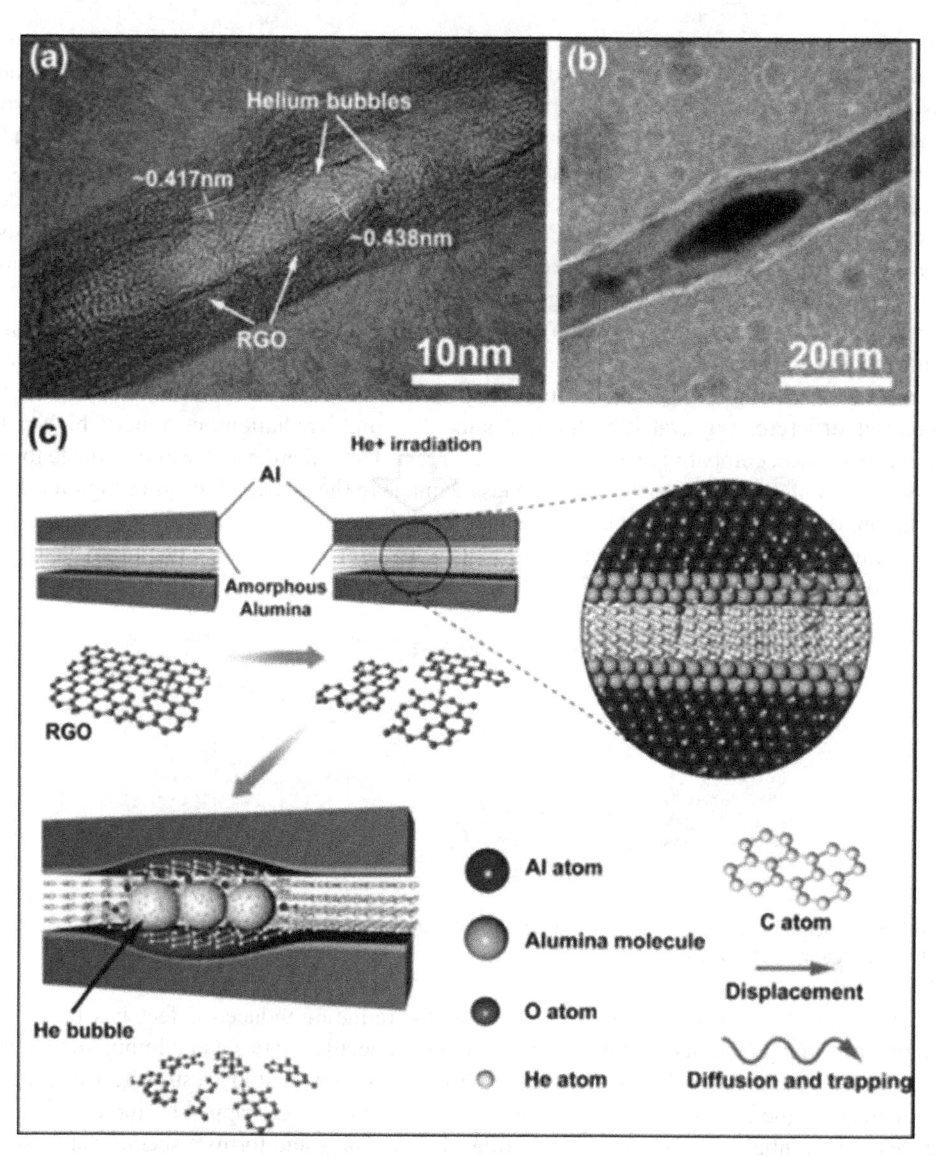

FIGURE 6 (a) and (b) representative TEM and STEM image of He bubble at the Al/Gr interface, (c) He bubble formation mechanism (Liu et al. 2020).

4. NOVEL PROCESSING OF GRAPHENE NANO-COMPOSITES

Graphene (Gr) nano-composites with various metal show a higher degree of radiation tolerance, swelling resistance and are capable of managing the high amount of He bubble at their interfaces but cost-effective processing of these materials are still a challenge. One of the first attempts to make this type of composite was carried out using sputtering. Stacking of W/Gr multilayer was prepared with a periodicity of ≈ 15, 30 and 40 nm. Monolayer graphene was transferred onto the W film sputtered on Si/SiO_2 substrate. Then, a W-nanofilm of the same thickness was deposited on top of the graphene. Repetition of the process creates a W/Gr composite of the total thickness of 3L (W/Gr), 5L (W/Gr), and 7L (W/Gr), where the numeric represents the number of cycles of G transferring and W sputtering (Si et al. 2017). Figure 7 shows TEM images of the successful formation of the multilayer structure along with the process flow.

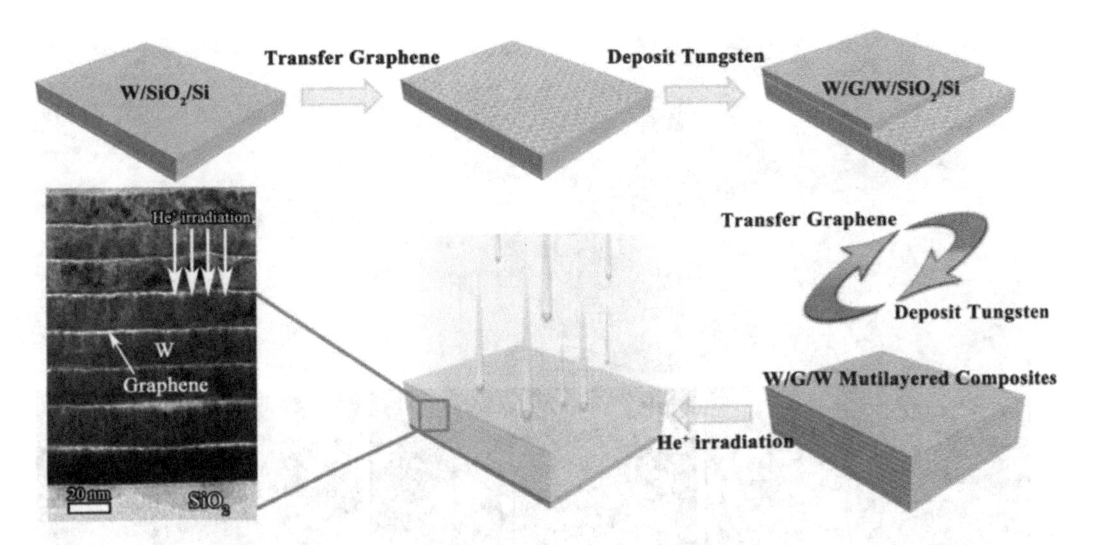

FIGURE 7 Schematic of W/Gr multilayer fabrication, the process flow of sputtering and Gr deposition, He^+ ion irradiation, TEM image of the multilayer structure (Si et al. 2017).

A similar type of V/Gr multilayer nano-composite was prepared using a combination of RF-sputtering and chemical vapor deposition. RF-sputtering of V nanolayer of 100 and 300 nm thickness were deposited on top of (625 μm/30 nm) Si/SiO_2 substrate under a vacuum of 5×10^{-6} Torr with Ar gas flow of 5 mTorr at 10 sccm, and RF power of 200 W. Then monolayer graphene was deposited using CVD technique (Kim et al. 2016). Although sputtering and CVD are able to produce hetero-interface structure, keeping a monolayer structure of the graphene. In comparison, electroplating can be a cost-effective and viable technology alternative. For electroplating, the incorporation of Gr was carried out through reduction of Gr-oxide by means of chemical or thermal reaction (Pei and Cheng 2012, Lorenzoni et al. 2015). However, several high-density topological defects and structural disorders, such as vacancies, dislocations, GBs, and holes, can be introduced into RGO (reduced graphene oxide) via GO reduction or composite preparation (Gómez-Navarro et al. 2010, Pavithra et al. 2014, Rabchinskii et al. 2016). The graphene is uniformly distributed in the electrolyte solution, having a size range ≈ 1-2 μm and uniform thickness of 0.9 nm of monolayer structure. By maintaining a pH of 4 of the solution and 50°C temperature, a 40 mm X 30 mm X 2 mm Graphene-Ni composite was prepared (Huang et al. 2018b). Figure 8 demonstrates a top surface view of the Gr-Ni nanocomposite showing the uniform distribution of Gr-oxide.

The aforementioned processing routes are quite successful in preparing thin films or thick coating, but the bulk scale production of this nano-composite is still under investigation. Very recently, a bulk processing route of Gr-metal (Ni, Al, and Cu) has been invented employing a conventional powder metallurgy route to produce bulk nanolaminates (Li et al. 2015).

5. FUTURE TRENDS AND SUMMARY

Material challenges for operating in a harsh environment in a nuclear reactor are quite severe, from maintaining high-temperature strength to exceptional radiation tolerance and void swelling resistance. Quest for higher conversion efficiency, the reactor design moves to generation-IV which demands withstanding higher radiation dose as well as higher temperature. The operating temperature often exceeds 600°C. Therefore, the design of new materials that can sustain the operating conditions and assure safety is one of the topmost priorities. Rather than going through a cumbersome process of new material/alloy design, microstructural engineering to produce

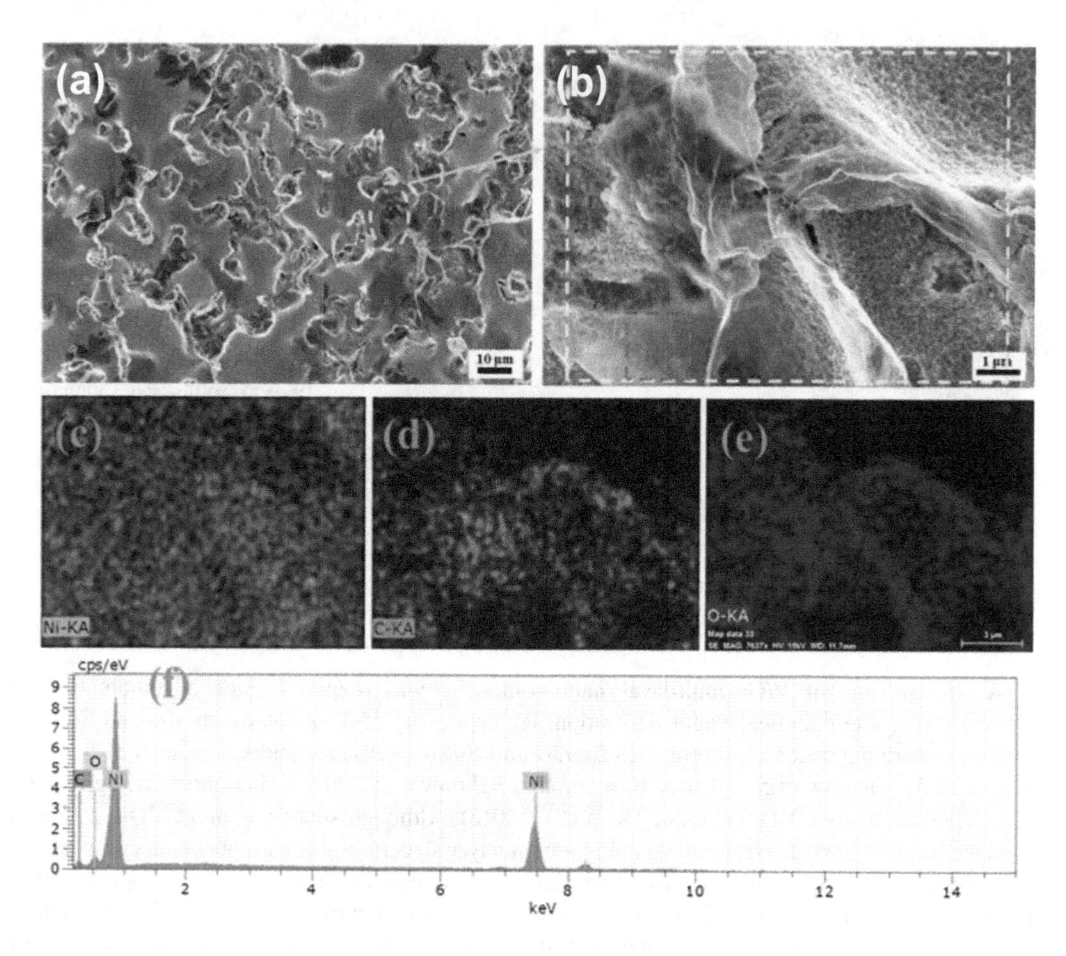

FIGURE 8 (a) SEM image of Ni/Gr representing the presence and uniform distribution of Gr in the matrix. In order to observe GO, the surface of Ni/Gr was pretreated by $FeCl_3$ solution. (b) Amplification of the highlighted area of the panel (a). (c–e) EDS elemental mapping of Ni/Gr in the highlighted area of the panel (b). Ni element (c), C element (d) and O element (e). (f) EDS spectrum of Ni/Gr in the small box area of the panel (b) (Huang et al. 2018b).

nano-structure of existing material candidates shows promising trends. Numerous heterogeneous interfaces can effectively serve as a sink to accommodate radiation-induced defects and trap He, produced as a transmutation product, in nm scale bubbles. Metal/Gr nano-composite in this regard demonstrates ample opportunities. The heterogeneous metal/gr interface not only increases the strength of material through the conventional Hall-Pitch strengthening effect but also provides high-temperature thermal stability. Metal/Gr interface provides excellent sink strength and demonstrates self-healing capability of the metal matrix from the radiation-induced defects. Nevertheless, the unique structure of the interface energetically favors the formation of He bubbles at the interface which are effectively pinned to a critical size and do not grow to form a void. Moreover, Gr is impermeable to He and does not allow He to migrate further to the grain boundaries to cause embrittlement. Although preliminary studies delineate exciting performances of these metal/gr nano-composite, a bulk scale production is still under investigation and a cost-effective processing route for large scale production is far away from the reality. However, some of the recent studies show shear mixing of the Gr within the metal nano-particle, consolidate and processed through conventional metallurgical routes has a great potential to produce bulk nanocomposite suitable with the industrial practice. Radiation damage studies on these composites are performed using mostly He or heavy ion irradiation that cannot be a true performance indicator. Neutron damage experiments and corresponding structure-property correlation is essential to qualify these new class of material as a substitute for existing structural material in a nuclear reactor. The mechanistic investigation, as well as a large database generation for different temperature-dpa damage, is the future need to predict the service life for this type of composite.

References

Abram, T., Crawford, D., Bart, G., Daeuwal, W., Ichimiya, M., Li, N., et al. 2002. Generation IV Roadmap Crosscutting Fuels and Materials R&D Scope Report.

Bai, X.-M., Voter, A.F., Hoagland, R.G., Nastasi, M. and Uberuaga, B.P. 2010. Efficient annealing of radiation damage near grain boundaries via interstitial emission. Science 327(5973): 1631-1634. doi: 10.1126/science.1183723.

Beyerlein, I.J., Demkowicz, M.J., Misra, A. and Uberuaga, B.P. 2015. Defect-interface interactions, Prog. Mater. Sci. 74: 125-210. doi: 10.1016/j.pmatsci.2015.02.001.

Braislford, A.D. and Bullough, R. 1978. Void growth and its relation to intrinsic point defect properties. J. Nucl. Mater. 69-70(C): 434-450. doi: 10.1016/0022-3115(78)90259-3.

Bunch, J.S., Verbridge, S.S., Alden, J.S. van der Zande, A.M., Parpia, J.M., Craighead, H.G., et al. 2008. Impermeable atomic membranes from graphene sheets. Nano Lett. 8(8): 2458-2462. doi: 10.1021/nl801457b.

Corwin, W.R. 2006. U.S. generation IV reactor integrated materials technology program. Nucl. Eng. Technol. 38: 591-618.

Demkowicz, M.J., Bellon, P. and Wirth, B.D. (no date). Atomic-scale design of radiation-tolerant nanocomposites Terms of Use Creative Commons Attribution-Noncommercial-Share Alike 3.0 Detailed Terms http://creativecommons.org/licenses/by-nc-sa/3.0/ 1 Atomic-scale design of radiation tolerant nanocomposites. doi: 10.1557/mrs2010.704.

Duan, B., Heintze, C., Bergner, F., Ulbricht, A., Akhmadaliev, S., Oñorbe, E., et al. 2017. The effect of the initial microstructure in terms of sink strength on the ion-irradiation-induced hardening of ODS alloys studied by nanoindentation. J. Nucl. Mater. 495: 118-127. doi: https://doi.org/10.1016/j.jnucmat.2017.08.014.

Ehrlich, K. 1981. Irradiation creep and interrelation with swelling in austenitic stainless steels. J. Nucl. Mater. 100(1-3): 149-166. doi: 10.1016/0022-3115(81)90531-6.

Enrique, R.A. and Bellon, P. 2000. Compositional patterning in systems driven by competing dynamics of different length scale. Phys. Rev. Lett. 84(13): 2885. doi: 10.1103/PhysRevLett.84.2885.

Frost, H.J. and Russell, K.C. 1982. Particle stability with recoil resolution. Acta Metall. 30(5): 953-960. doi: 10.1016/0001-6160(82)90202-4.

Garner, F.A. and Puigh, R.J. 1991. Irradiation creep and swelling of the fusion heats of PCA, HT9 and 9Cr-1Mo irradiated to high neutron fluence. J. Nucl. Mater. 179-181(PART 1): 577-580. doi: 10.1016/0022-3115(91)90153-X.

Grossbeck, M.L. and Horak, J.A. 1988. Irradiation creep in type 316 stainless steel and us PCA with fusion reactor He/dpa levels. J. Nucl. Mater. 155-157(PART 2): 1001-1005. doi: 10.1016/0022-3115(88)90457-6.

Gómez-Navarro, C., Meyer, J.C., Sundaram, R.S., Chuvilin, A., Kurasch, S., Burghard, M., et al. 2010. Atomic structure of reduced graphene oxide. Nano Lett. 10(4): 1144-1148. doi: 10.1021/nl9031617.

Heinig, K.H., Müller, T., Schmidt, B., Strobel, M. and Möller, W. 2003. Interfaces under ion irradiation: Growth and taming of nanostructures, Appl. Phys. A Mater. Sci. Process. 77(1): 17-25. doi: 10.1007/s00339-002-2061-9.

Hirvonen, J.M.J.P. and Nastasi, M. 1986. Microstructure of ion-bombarded Fe–Ti and Fe–Ti–C multilayered films. J. Appl. Phys. 60(3): 980-984.

Huang, H., Tang, X., Chen, F., Yang, Y., Liu, J., Li, H., et al. 2015. Radiation damage resistance and interface stability of copper–graphene nanolayered composite. J. Nucl. Mater. 460: 16-22. doi: https://doi.org/10.1016/j.jnucmat.2015.02.003.

Huang, H., Tang, X., Chen, F., Liu, J., Chen, D. 2017. Role of graphene layers on the radiation resistance of copper–graphene nanocomposite: Inhibiting the expansion of thermal spike. J. Nucl. Mater. 493: 322-329. doi: https://doi.org/10.1016/j.jnucmat.2017.06.023.

Huang, H., Tang, X., Chen, F., Gao, F., Peng, Q., Ji, L., et al. 2018a. Self-healing mechanism of irradiation defects in nickel–graphene nanocomposite: An energetic and kinetic perspective. J. Alloys Compd. 765: 253-263. doi: 10.1016/j.jallcom.2018.06.162.

Huang, H., Tang, X., Chen, F., Liu, J., Sun, X. and Ji, L. 2018b. Radiation tolerance of nickel–graphene nanocomposite with disordered graphene. J. Nucl. Mater. 510: 1-9. doi: 10.1016/j.jnucmat.2018.07.051.

Huang, H.F., Zhang, W., De Los Reyes, M., Zhou, X.L., Yang, C., Xie, R., et al. 2016. Mitigation of He embrittlement and swelling in nickel by dispersed SiC nanoparticles. Mater. Des. 90: 359-363. doi: https://doi.org/10.1016/j.matdes.2015.10.147.

Hung, J.M.L., Nastasi, M. and Gyulai, J. 1983. Ion-induced amorphous and crystalline phase formation in Al/Ni, Al/Pd, and Al/Pt thin films. Appl. Phys. Lett. 42(8): 672-674.

Jin, S., He, X., Li, T., Ma, S., Tang, R. and Guo, L. 2012. Microstructural evolution in nickel alloy C-276 after Ar-ion irradiation at elevated temperature. Mater. Charact. 72: 8-14. doi: https://doi.org/10.1016/j.matchar.2012.06.010.

Kim, Y., Baek, J., Kim, S., Ryu, S., Jeon, S. and Han, S.M. 2016. Radiation resistant vanadium-graphene nanolayered composite. Sci. Rep. 6: 1-9. doi: 10.1038/srep24785.

Li, Z., Guo, Q., Li, Z., Fan, G., Xiong, D.-B., Su, Y., et al. 2015. Enhanced mechanical properties of graphene (reduced graphene oxide)/aluminum composites with a bioinspired nanolaminated structure. Nano Lett. 15(12): 8077-8083. doi: 10.1021/acs.nanolett.5b03492.

Liu, Y., Zeng, Y., Guo, Q., Zhang, J., Li, Z., Xiong, D.B., et al. 2020. Bulk nanolaminated graphene (reduced graphene oxide)–aluminum composite tolerant of radiation damage. Acta Mater. 196: 17-29. doi: 10.1016/j.actamat.2020.06.018.

Lorenzoni, M., Giugni, A., Di Fabrizio, E., Pérez-Murano, F., Mescola, A. and Torre, B. 2015. Nanoscale reduction of graphene oxide thin films and its characterization. Nanotechnology 26(28): 285301. doi: 10.1088/0957-4484/26/28/285301.

Lu, C., Lu, Z., Wang, X., Xie, R., Li, Z., Higgins, M., et al. 2017. Enhanced radiation-tolerant oxide dispersion strengthened steel and its microstructure evolution under helium-implantation and heavy-ion irradiation. Sci. Rep. 7(1): 40343. doi: 10.1038/srep40343.

Martin, G. 1984. Phase stability under irradiation: Ballistic effects. Phys. Rev. B. 30(3): 1424.

Mayer, L.H.J., Tsaur, B. and Lau, S. 1981. Ion-beam-induced reactions in metal-semiconductor and metal-metal thin film structures. Nucl. Instruments Methods 182: 1-13.

Murty, K.L. and Charit, I. 2008. Structural materials for Gen-IV nuclear reactors: Challenges and opportunities. J. Nucl. Mater. 383(1-2): 189-195. doi: 10.1016/j.jnucmat.2008.08.044.

Nastasi, J.K.H.M. and Mayer, J. 1996. Ion-solid Interactions: Fundamentals and Applications. Cambridge University Press.

Naundorf, H.W.V. and Macht, M.P. 1992. Production rate of freely migrating defects for ion irradiation. J. Nucl. Mater. 186(3): 227-236.

Nelson, R.S., Hudson, J.A. and Mazey, D.J. 1972. The stability of precipitates in an irradiation environment. J. Nucl. Mater. 44(3): 318-330. doi: 10.1016/0022-3115(72)90043-8.

Norgett, I.T.M. and Robinson, M. 1975. A proposed method of calculating displacement dose rates. Nucl. Eng. Des. 33(1): 50-54.

Nuclear electricity generation worldwide from 1985 to 2020, © Statista (2021). https://www.statista.com/statistics/275048/gloobal-nuclcar-power-generation/.

Odette, G.R., Alinger, M.J. and Wirth, B.D. 2008. Recent developments in irradiation-resistant steels. Annu. Rev. Mater. Res. 38(1): 471-503. doi: 10.1146/annurev.matsci.38.060407.130315.

Ordered, Q. and Variation, D. 2014. Terms and Terms and. (April). pp. 353-362.

Pavithra, C.L.P., Sarada, B.V., Rajulapati, K.V., Rao, T.N. and Sundararajan, G. 2014. A new electrochemical approach for the synthesis of copper-graphene nanocomposite foils with high hardness. Sci. Rep. 4(1): 4049. doi: 10.1038/srep04049.

Pease, R.S. and K.G. 1955. The displacement of atoms in solids by radiation. Reports Prog. Phys. 18(1).

Pei, S. and Cheng, H.-M. 2012. The reduction of graphene oxide. Carbon 50(9): 3210-3228. doi: https://doi.org/10.1016/j.carbon.2011.11.010.

Rabchinskii, M.K., Shnitov, V.V., Dideikin, A.T., Aleksenskii, A.E., Vul', S.P., Baidakova, M.V., et al. 2016. Nanoscale perforation of graphene oxide during photoreduction process in the argon atmosphere. J. Phys. Chem. C. 120(49): 28261-28269. doi: 10.1021/acs.jpcc.6b08758.

Si, S., Li, W., Zhao, X., Han, M., Yue, Y., Wu, W., et al. 2017. Significant radiation tolerance and moderate reduction in thermal transport of a tungsten nanofilm by inserting monolayer graphene. Adv. Mat. Comm. 1-7. https://doi.org/10.1002/adma.201604623.

Siegel, R.W., Chang, S.M. and Balluffi, R.W. 1980. Vacancy loss at grain boundaries in quenched polycrystalline gold. Acta Metall. 28(3): 249-257. doi: 10.1016/0001-6160(80)90159-5.

Starikov, S.V., Insepov, Z., Rest, J., Kuksin, A.Y., Norman, G.E., Stegailov, V.V., et al. 2011. Radiation-induced damage and evolution of defects in Mo. Phys. Rev. B. 84(10): 101109.

Tsaur, J.M.B. 1981. Metastable Au-Si alloy formation induced by ion-beam interface mixing. Philos. Mag. A. 43(2): 345-361.

Various 2002. Generation IV Roadmap Crosscutting Fuels and Materials R&D Scope Report. Group, (December). pp. 1-76.

Was, G.S. 2007. Fundamentals of Radiation Materials Science: Metals and Alloys. Springer Science & Business Media.

Wei, R.A.L., Lang, E. and Flynn, C. 1999. Freely migrating defects in ion-irradiated Cu_3Au, Appl. Phys. Lett. 75(6): 805-810.

Williamson, P.J.W.D., Ozturk, O. and Wei, R. 1994. Metastable phase formation and enhanced diffusion in fcc alloys under high dose, high flux nitrogen implantation at high and low ion energies. Surf. Coatings Technol. 65(1-3): 15-23.

World Nuclear Power Reactors & Uranium Requirements 2020. https://world-nuclear.org/information-library/facts-and-figures/world-nuclear-power-reactors-and-uranium-requireme.aspx.

Ziegler, J.P.B.J.F. and Ziegler, M.D. 2010. SRIM–The stopping and range of ions in matter. Nucl. Instruments Methods Phys. Res. Sect. B: Beam Interact. with Mater. Atoms. 268(11-12): 1818-1823.

Zinkle, S.J. 2005. Fusion materials science: Overview of challenges and recent progress. Phys. Plasmas. 12(5): 58101. doi: 10.1063/1.1880013.

Zinkle, S.J. and Was, G.S. 2013. Materials challenges in nuclear energy. Acta Mater. 61(3): 735-758. doi: 10.1016/j.actamat.2012.11.004.

Emerging Pollutants in Aquatic Systems and Recent Advances in Its Removal Techniques

Deepika Saini,[1] Ruchi Aggarwal,[1] Prashant Dubey,[2]
Kumud Malika Tripathi[3,*] and Sumit Kumar Sonkar[1,*]

1. INTRODUCTION

The heavy emergence of pollutants in the aquatic environment is just a consequence of drastic and speedy anthropogenic activities (Benotti et al. 2009, Verlicchi and Zambello 2015). For few decades, diverse pollutants are introduced in the aquatic system of our Earth as a consequence of various outputs, such as waste sludges from wastewater treatment plants and domestic outlets, industrial units, sewer leakage, agriculture and surface runoffs from rural and urban zones and manure waste (Tijani et al. 2016, Zhang et al. 2015). Predominantly growing wastewater treatment plants are of major concern because they are directly dumping the emerging pollutants and micropollutants into water bodies as a result of overlooking proper treatment (Geissen et al. 2015, Rice and Westerhoff 2017). Additionally, the occurrence, transportation and metabolism of emerging pollutants in humans are still unknown and need to be taken care of thoroughly. A case study by Alves et al. (2017) reported the detection of four different emerging pollutants in the urine and fingernails of humans. This serious issue has alarmed the research fraternity engineers and civil authorities to look into the matter and to take some quick action as the level of these hazardous pollutants (Hao et al. 2019) has trepassed their permissible limit in the aquatic system. In the current time, the supply of clean and safe water to urban as well as rural populations is a very challenging task, and even it is more challenging to maintain careful treatment of wastewater along with good reusability without compromising environmental sustainability. Ironically, conventional wastewater treatment plants were much sensitive toward water nutrients, bacteria, heavy metals, pesticides and petroleum hydrocarbons (Pal et al. 2014). Whereas, present water treatment plant only focuses on primary pollutant whose impact on human health and animal is known (Rice and Westerhoff 2017). The issue of emerging pollutants is even more worse in developing countries that is due to the lack of systematic studies and overlooking the concerns (Rehman et al. 2015).

The occurrence of pollutant contaminants (generally known as organic and inorganic) in wastewater is increasing in non-regulated manner above the safe limits. These are not only harmful

[1] Department of Chemistry, Malaviya National Institute of Technology, Jaipur, Jaipur-302017, India.
[2] Centre of Material Sciences Institute of Interdisciplinary Studies, Nehru Science Complex, University of Allahabad, Prayagraj-211002, India.
[3] Department of Chemistry, Indian Institute of Petroleum and Energy, Visakhapatnam-530003, Andhra Pradesh, India.
[*] Corresponding authors: kumud20010@gmail.com, sksonkar.chy@mnit.ac.in.

to aquatic flora and fauna but also for terrestrial lives and recently have been observed by advanced analytical techniques. These trace compounds are termed emerging pollutants (Benotti et al. 2009). Generally, the organic and inorganic trace pollutants found in wastewater have the following broad category from which they are originating.

I. Pharmaceuticals (PhACs)
II. Antibiotics
III. Personal Care Products (PCPs)
IV. Ionic Liquids
V. Endocrine Disrupting Compounds (EDCs)
VI. Nanomaterials (NMs)
VII. Organic Pollutants and Heavy Metal Ions

Additionally, functional nanomaterials such as inorganic metal oxides or polymer nanoparticles cause malfunctioning in bacterial activity and make it difficult to remove NMs by using biological techniques in wastewater treatment plants (Wang et al. 2012). Consequently, emerging pollutants are easily circulated over surface water and even reach the groundwater from open resources of water. Moreover, disposal of municipal sludge is also a very challenging task as it is considered as the principal source of the liberation of emerging pollutants from in open aquatic ecosystem where industries and domestic outlets (as point sources) are directly connected without any significant treatment of wastewater to make it free of emerging pollutants. Emerging pollutants can enter water bodies in a number of routes as demonstrated in Figure 1.

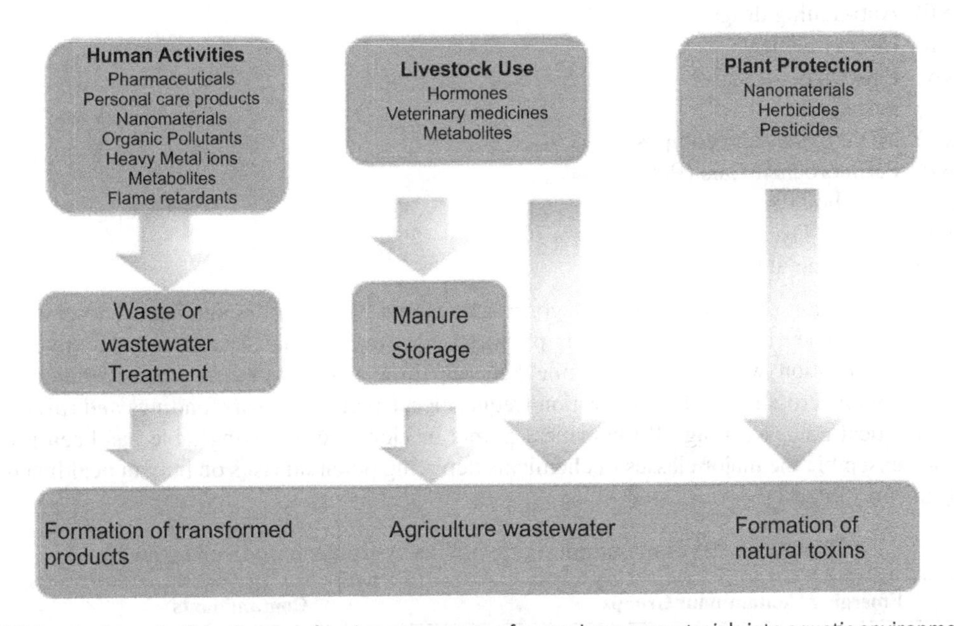

FIGURE 1 A schematic flow diagram showing entry route of emerging nanomaterials into aquatic environment.

2. EMERGING POLLUTANTS AND THEIR CLASSIFICATION

The emerging pollutants are categorized into six above said major groups of chemicals. Firstly, pharmaceutical compounds which are further subdivided into different categories viz. antibiotics, analgesics, steroids and non-steroid, etc. Pharmaceuticals are widely used to cure human and animal diseases (Ebele et al. 2017). Secondly, personal care products (PCPs) are identified as cosmetic and health care chemicals with large numbers and diverse chemical nature available in the market; for example, triclosan, galaxolide, paracetamol, musks, tonalide, iopamidol, carbamazepine, diclofenac (Ebele et al. 2017). Endocrine disrupting compounds (EDCs) are groups of compounds or chemicals that show an adverse impact on human endocrine systems, fertility problems, and a higher risk of cancer even at low concentrations of consumption (Falconer et al. 2006). Several chemicals including EDs and PCPs when exposed to pregnant women even at very low concentrations have been identified to adversely affect natural development and neural disorder in the human fetus (Falconer et al. 2006). Not only humans but pharmaceutical compounds can also reduce fertility in some aquatic species including fish (Falconer et al. 2006).

After executing a thorough review of reports on emerging pollutants found in water and wastewater, the following major classes falling in the above broad discussed categories are formed as per their abundant quantity in the aquatic system (Luo et al. 2014, Semblante et al. 2017, Subedi et al. 2015):

(i) Antibiotic
(ii) Antifungal/antimicrobial agents
(iii) Nonsteroidal anti-inflammatory drugs (NSAIDs)
(iv) Anticonvulsants/antidepressants
(v) Artificial sweeteners
(vi) Beta-adrenoceptor blocking agents
(vii) Lipid regulating drugs
(viii) Steroidal hormones
(ix) X-ray contrast media
(x) UV filters
(xi) Stimulants
(xii) Anti-itching drugs
(xiii) Insect repellents
(xiv) Plasticizers
(xv) Pesticides
(xvi) Metal oxides and composites
(xvii) Perfluorochemicals (PFCs)
(xviii) Ionic liquids
(xix) Organic Dyes
(xx) Heavy metal ions

The formation of the above subcategories Laurent is based on the number of observations by researchers. They have examined their pros and cons over human health and the environment. This classification was performed while keeping three main points into consideration, i.e. (i) consumption across the globe observation frequencies, (ii) effect on surroundings and (iii) quantity of analytical data. Keeping all the critical points in view, the following table has been prepared which ensemble the major classes of chemicals depicting potential risks on human health and other organisms (Table 1).

TABLE 1 List of major emerging pollutant groups found in wastewater.

Emerging Contaminant Groups	Contaminants
Pharmaceuticals	
Human antibiotics (Gurke et al. 2015, Tran et al. 2010)	Trimethoprim, erytromycine, amoxicillin, lincomycin, sulfamethaxozole, chloramphenicol and Triclosan
Analgesics, anti-inflammatory drugs (Tran et al. 2014a)	Ibuprofene, diclofenac, paracetamol, codein, acetaminophen, acetylsalicilyc acid, fenoprofen and Tramadol naproxen
Psychiatric drugs (Anticonvulsant) (Özbek and Gürdere 2020)	Diazepam, carbamazepine, primidone and salbutamol
β-blockers (Mohapatra et al. 2016)	Metoprolol, propanolol, timolol, atenolol and sotalol

Contd.

Lipid regulators (Sui et al. 2011, Sui et al. 2009, Sui et al. 2010, Sun et al. 2016)	Bezafibrate, clofibric acid, fenofibric acid, etofibrate and gemfibrozil
X-ray contrasts (Yang et al. 2017)	Iopromide, iopamidol and diatrizoate
Stimuants (Koelega 1993)	Caffeine and nicotine
Antihypertensive	Triamterene, Valsartan and Hydrochlorothiazide
Antidepressant (Beydoun and Backonja 2003)	Amitriptyline, Desmethyl-venlafaxine, desipramine, Temazepam, Hydroxybupropion and imipramine
Antiepileptic (Beydoun and Backonja 2003)	Phenytoin, Pregabalin, Lamotrigine, Gabapentin and oxcarbazepine
Antihyperlipidemic (Laurent 2017)	Gemfibrozil
Antiarrhythmic (Beydoun and Backonja 2003)	Lidocaine
Personal Care Products (PCPs)	
Fragrances (Arribas et al. 2012)	Benzyl alcohol, Cinnamyl alcohol, Benzyl salicylate Linalool, Limonene, Butylphenyl, methylpropional and Hydroxycitronella
Sun-screen agents (Rodil et al. 2009, Roscher et al. 1994)	Benzophenone and methylbenzylidene camphor
Insect repellents (Tran and Gin 2017)	N, N-diethyltoluamide
Endocrine Disrupting Chemicals (EDCs)	
Hormones and steroids (Scanes et al. 1979), (Lenzen et al. 1984)	Estradiol, estrone, estriol, diethylstilbestrol (Fatta-Kassinos et al.) and 17α-Ethinylestradiol
Perfluoronated and surfactants compounds (Nguyen et al. 2011)	Perfluorooctane sulfonate (PFOs), perfluorooctanoic acid (PFOA)
Flame retardants (Rahman et al. 2001)	Polybrominated diphenyl ethers (PBDEs): polybromonated biphenyls (PBBs) – polybromonated dibenzo-*p*-dioxins (PBDDs) – polybromonated dibenzofurans (PBDFs)
Industrial additives and agents (Nguyen et al. 2011)	Aromatic sulfonates
Gasoline additives (Yao et al. 2009)	Dialkyl ethers and Methyl-t-butyl ether (MTBE)
Antiseptics (McDonnell and Russell 1999)	Triclosan and chlorophene
Plastic precursor (Sun et al. 2016)	Bisphenol A and Phenol
Food additives (Subedi and Kannan 2014)	Sucralose, aspartame, saccharin
Herbicides (Helfrich et al. 2009)	2,4-D, Atrazine, Bromacil, Diuron, MCPP, Metolachlor, Triclopyr, Endothall, fluridone, glyphosate and Dichlobenil
Insecticide (Anju et al. 2010)	Imidacloprid, Dichlorodiphenyltrichloroethane, Pyrethroids Carbamates and Hexachlorocyclohexane
Artificial sweeteners (Tran et al. 2014b)	Acesulfame, Cyclamate, Saccharin and Sucralose

Contd.

Nanomaterials (Wiesner et al. 2009)	Inorganic metal oxides
UV-filters (Li et al. 2007)	Octocrylene and Oxybenzone
Organic Dyes (Aksu 2005, Gupta 2009, Zhou et al. 2015)	Cationic and anionic dyes such as Congo-red, Crystal-violet, Azure A, Azure B, Methanene yellow, Acid Red, P-nitro phenol, Methylene basic Blue, Rhodamine-B, etc.
Heavy Metal Ions (Fu and Wang 2011, Ngah and Hanafiah 2008)	Hexavalent Chromium, Lead, Mercury, Iron, Cadmium, etc.

2.1 Emerging Contaminants

2.1.1 Pharmaceuticals (PhACs)

PhACs are a set of compounds with diverse chemical nature, including nonsteroidal anti-inflammatory drugs (Fatta-Kassinos et al. 2011) (NSAIDs) to cure human and animal medication, cosmetic and sanitary applications. PhACs are emerging micropollutants ranging from ppb to ppm level concentration. PhACs include a diverse array of synthetic chemicals with complex anionic aromatic backbone and are highly soluble in water. The application of medicines is drastically increasing with the population across the globe. Our nation stands in the second position in the queue of highest populated countries after China. Amazingly, India has now become a highly growing pharmaceutical market. In fact, India is now the major exporter of generic drugs nearly (20% of gross global exports) as generics rules over more than 70% of the total market share in the Indian pharmaceutical sector, which is further scaled to be enhanced in near future. Moreover, multinational companies have been attracted to this giant market of pharmaceutics in the greed of cheap-cost production and flexible environment tribunal guidelines which is further contributed by efficient and fast techniques, quality research, the abundance of educated personnel and skilled laboratories making the Indian marketplace a fascinating choice for the overseas pharmaceutics majors for their drug manufacture and outsourcing. Today, the Indian pharmaceutical industry has become one of the leading industries of the nation, which has a very high caliber to make right from analgesics to medicine for any fatal disease like cancer and cardiac treatment. These all way too much high production and their assimilation have boosted the emergence of pharmaceutical pollutants in water and wastewater. Unfortunately, our conventional wastewater treatment plants are super-specialized to decontaminate the water from these pollutants.

Additionally, these pharmaceutical pollutants are found to be bio-accumulating in the different aquatic environments as they are very much active in the biological system. Ecological subsidy through life cycle facilitate the dispersion of aquatic biota to terrestrial food web to animals and finally in humans through the food chain (Richmond et al. 2018). The fate of PhACs contaminants and potential exposure to aquatic invertebrates to selected drugs, which are akin to human prescription is reported by Richmond et al. and demonstrated in Figure 2 (Richmond et al. 2018). They can remain in the body for a long time attributed to a specific mode of action and high chemical stability. Thus, these can be straightforwardly traced in wastewater, surface water and drinking water which ultimately cause bio-resistant properties in bacteria present anywhere. Furthermore, not only the pharmaceutically active compounds but also their derived or decomposed ingredients have been affecting the entire surroundings by causing lethal impact on the ecosystem gradually.

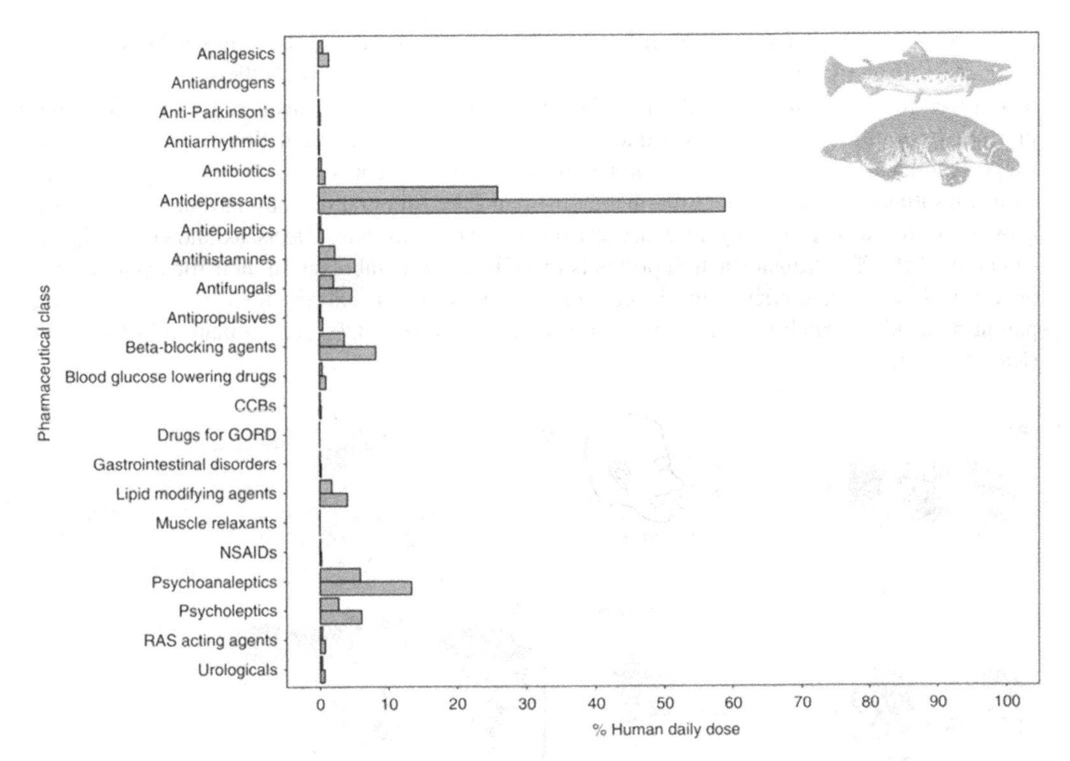

FIGURE 2 Estimated dietary intake of pharmaceuticals by two representative invertebrate predators compared to recommended human pharmaceutical doses. Dietary intake rates as a percentage of recommended human pharmaceutical daily doses by therapeutic class for platypus and brown trout in Brushy Creek. (CCBs calcium channel blockers, GORD gastroesophageal reflux disease, NSAID non-steroidal anti-inflammatory drugs, RAS renin angiotensin system). Adapted with permission from (Richmond et al. 2018), Copyright 2018, American Chemical Society.

Due to their very minute concentration, they have been unnoticed and never been prioritized by government guideline makers to treat wastewater for their removal. That is why they have been termed 'micropollutants' or 'pseudo-pollutant'. This situation has been continuing for few decades and no assessment has been executed seriously, except a few years back some studies noticed for the risk assessment of pharmaceutical ingredients and their derivatives. Recently, more than 160 different pharmaceutical compounds have been detected in water bodies, mainly wastewater treatment plant effluent in the concentration range of ng L^{-1} to low μg L^{-1}. These pharmaceutical drugs are not only affecting human beings but also surface water bodies in a tremendous way (Archana et al. 2017). So far, very inadequate information is collected about the toxicological effect of pharmaceutical drugs on terrestrial and aquatic lives.

Analgesics, anticonvulsants, antidepressants, antiepileptics, antihypertensives and beta-blockers are commonly observed in water systems owing to higher solubility in water and slower metabolic rate (Brown and Winterstein 2019). The antiepileptic gabapentin has been recorded with the highest median concentration in the aquatic system followed by metformin, lamotrigine, desmethylvenlafaxine, hydrochlorothiazide, sulfamethoxazole and hydroxycarbamazepine with (>100 ng L^{-1}) concentration.

2.1.2 Personal Care Products (PCPs)

PCPs generally consist of a diverse range of heterogenous organic compounds employed in cosmetic and healthcare products, which include deodorants, artificial hormones, cleaning agents, steroids, perfumes, hair care products, skincare agents, UV filters, sanitizers, cleaning products and other products with estrogenic activity (Noguera-Oviedo and Aga 2016). PCPs is most commonly recognized micropollutant not only in surface water but also in groundwater, even too detected in marine environment and sediments due to their uninterrupted disposal linked with accelerating usage (García et al. 2013). The major lethal effect causing concern is the lipophilic nature of their constituents, which hinders the metabolism and the lifecycle of flora and fauna in the water ecosystem and also proclivity to cause estrogenic and endocrine effects as shown in Figure 3 (Fu et al. 2019). The transformation products of PCPs after metabolism in their life cycle often are proven to have more tendency to bioaccumulation, more persistent and toxic in contrast to their parent molecules, which indicates the serious anxieties of the PCPs contamination in wastewater (Kar et al. 2020).

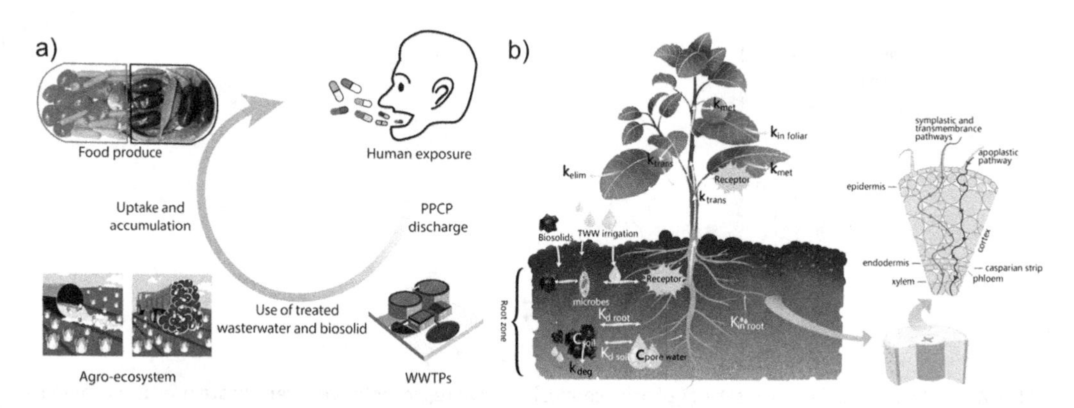

FIGURE 3 (a) A schematic representation of PCPs life cycle. (b) PPCPs transport and fate routes in the soil-plant system. Adapted with permission from Fu et al. (2019). Copyright 2019, American Chemical Society.

Fisher et al. investigated that exposure to Phthalates and Bisphenol A, which are commonly used PCPs in food processing and packaging and pregnant women could also impact the infant this way (Fisher et al. 2019). These are pervasive and have been detected in saliva, dust, blood, amniotic fluid, human milk, urine, meconium and follicular fluid (Fisher et al. 2019). Five transformation products of 2,6-di-tert-butyl-4-methylphenol (BHT), a commonly used synthetic chemical in cosmetics, foodstuffs, mineral oil or fuel additive, plastics and rubbers, were detected in greywater discharge after controlled dental exposure to humans (Liu and Mabury 2019).

2.1.3 Endocrine Disruptors (EDCS)

EDCs are xenobiotic chemicals that include both natural and engineered hormones and steroids, plastics and plasticizers, flame retardants, antiseptics, perfluoronated compounds, pesticides, such as dichloro diphenyl trichloroethane (DDT), surfactants and surfactant metabolites, food and gasoline additives, herbicides, insecticide and artificial sweeteners, which manifest carcinogenic effects on the human endocrine system due to their androgenic or estrogenic activities even at trace level (Lymperi and Giwercman 2018). Their emergence in aquatic bodies can cause disruptive physiological processes, sexual impairment, lesser fertility, cancer and antibiotic resistance in bacteria (Wang and Chen 2020). Overall, EDCs can increase the aquatic toxicity manifold and make the water system inhabitable for its dwellers. EDCs are way too fatal even to mock or alter the functioning of hormones and distort the body fluid. EDs contaminants can mess up with the human body's metabolism and adversely impact the cell regulation process, reproductive system, development of organs and increase metabolic disorder and pose the high risk of multiple cancer and epigenetic dysfunction (Salehi et al. 2017, Soto and Sonnenschein 2010). Moreover, EDs can interrupt the transport and metabolism of natural hormones and impact female sex organs and sexual differentiation (Vilela et al. 2018). EDs are hydrophobic and almost pervasive in the aquatic, common dietary environment.

There are three main classes of EDCs, estrogenic mocks the functioning of the body's natural estrogens, androgenic duplicates natural testosterone and last thyroidal disrupts the functioning of the thyroid (Lenzen and Bailey 1984). The emergence of EDCs is predominantly accelerated by above mentioned domestic and industrial waste products dumped directly into the water which finally reaches surface water or groundwater because wastewater is further poured into water bodies without EDCs specific treatment in water treatment plants (Rice and Westerhoff 2017, Salehi et al. 2017). This is quite unfortunate and in turn, is critical to sustaining wildlife habitat. Although they are found in immeasurably very low concentrations in wastewater, they can be very hazardous in long-term exposure to human beings and other organisms and can even pose ecological threats.

2.1.4 Nanomaterials

Increasing use of engineered nanomaterials (ENMs) includes diverse metal nanoparticles (NPs), metal oxide NPs, different allotropes of carbon and composite NPs creating new means of revolutionizations in a wide range of applications including wastewater treatment, (Das et al. 2019a, Khaksar et al. 2019) but at the same time, it also raises the concern of waste ENMs toward

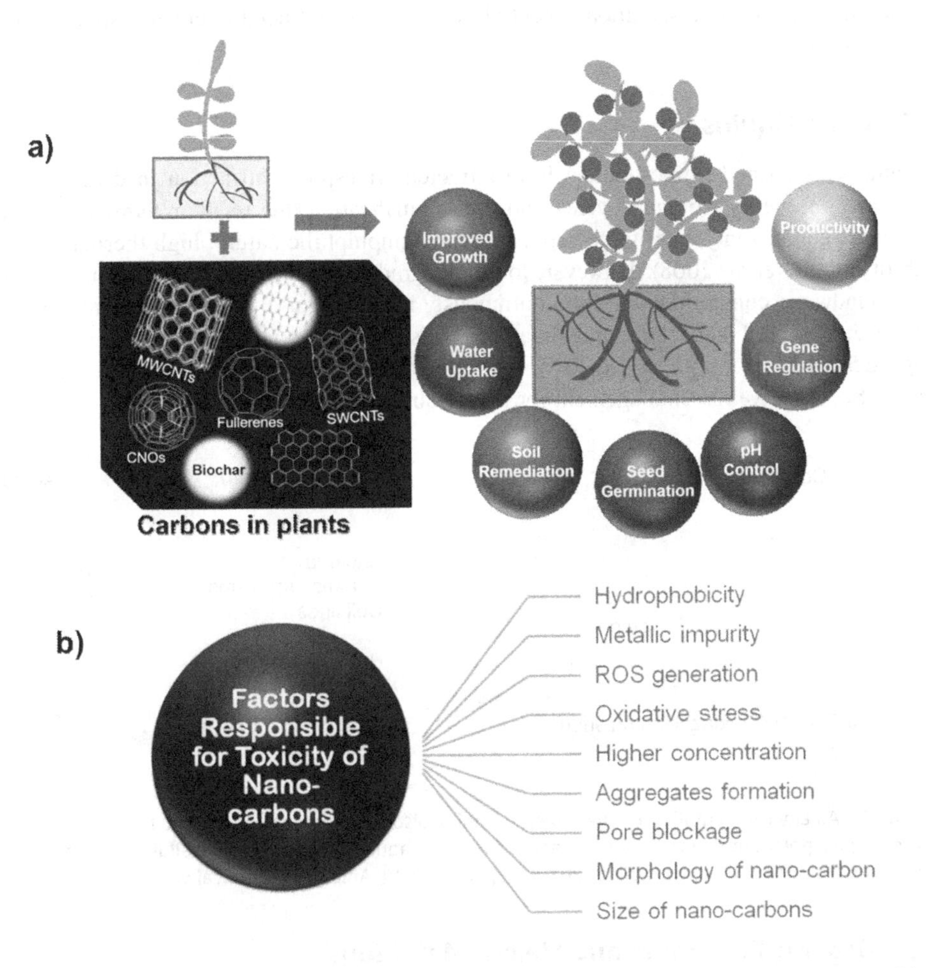

FIGURE 4 A schematic diagram of (a) effects of nano-carbons in plant system. (b) Factor responsible for measuring the toxicity of nano carbons. Adapted with permission from Bhati et al. (2018). Copyright 2018, Royal Society of Chemistry.

environment safety as being a new class of emerging pollutants. The small size of ENMs in many shapes and forms can be considered as truly emerging pollutants and are hard to monitor by traditional techniques and demands for new detection techniques. The environmental risk allied with ENMs is greatly dependent upon their synthetic route, source of origin, physiochemical properties and biological transformation after exposure (Khaksar et al. 2019). Although the fate and associated risk of ENMs are not well defined and the removal process is ill-understand. Very few studies regarding the fate and potentials of waste ENMs or wastewater discharged ENMs through nanotechnology-based treatment processes are reported. An accurate assessment of the presence, transport, bioaccumulation, toxicity and acute health effects of ENMs needs a long way to travel (Carl Englert 2007). Bhati et al. in a recent review investigated the effect of different carbon-based NPs on plant systems (Figure 4a) and concluded that insoluble forms of nano-carbons exhibit toxicity toward plant growth due to biological incompatibility, while water-soluble nano-carbons derived from biomass can enhance plant growth, retain soil fertility, increased crop productivity without and adverse effect (Figure 4b) (Bhati et al. 2018). The use of ENMs for the recognition or treatment of pollutants looks fancy, but it generates nano-waste that needs to be either dispose or recycle (Gao et al. 2008). The impact of these nano-waste could be even more adverse when exposed to other contaminants in wastewater than their own. The factors and process of adverse impacts of ENMs to environment and ecosystem are seem to be complex. Further, their agglomeration, surface properties and dissolution potential are complex and need detail investigation for proper risk assessment.

2.2 Ionic-Liquids

Organic salts termed 'ionic-liquids' have attracted widespread attraction in diverse applications especially in energy storage devices and water purification due to their flame-retardant nature, high ionic conductivity, negligible vapor pressure, amphiphilic nature, high thermal and chemical stability (Stolte et al. 2008). However, low biodegradability and low sorption to soil pose a risk for groundwater contamination with continuously increasing use. Although severe adverse effects have been observed on bacteria and aquatic organisms, their toxicologic assessment is very limited (Figure 5) (Oskarsson and Wright 2019). The toxicity is linked with their chemical nature and chain length. So, effective risk management for ionic liquids is highly needed.

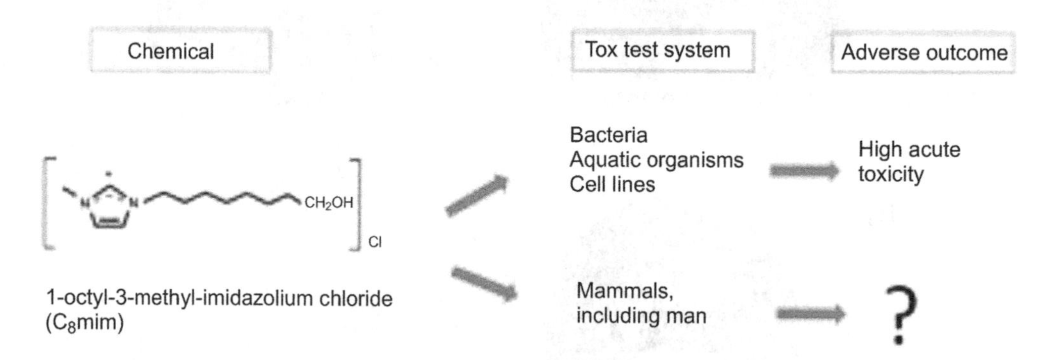

FIGURE 5 An environmentally friendly solvent, C_8 mim, also a well-known ionic liquid has been categorized as a new emerging pollutant due to their great acute toxicity in aquatic organisms and cellular systems. Adapted with permission from Oskarsson and Wright (2019). Copyright 2019, American Chemical Society.

2.3 Organic Pollutants and Heavy Metal Ions

With the rapid industrial growth, technologies and combustion activities, the waste effluent discharge from them into the aquatic system also increases with the straining of freshwater assets. These effluents mainly consist of toxic organic dyes and heavy metal ions, which are non-biodegradable, carcinogenic and mutagenic to living organisms. Dye is extremely used in leather, textile, clothes, furniture and paper industry, which have significant contribution toward water pollution and human life (Zhou et al. 2015, Aksu 2005, Gupta 2009). While doing the dying process, not all the dye molecules are consumed and hence leave unused dye molecules in the environment that needs further treatment. Along with this, heavy metal ions have strong tendency to bind with biological matter and lead to change in complex protein structures.

However, there are many heavy metal ions, such as copper(II), zinc (II), iron (III), aluminum (III) and chromium (III), which are essential for healthy enzymatic activity in organisms but their higher concentration can lead to toxicity. But generally heavy metal ions, like chromium [Cr(VI)], lead(II), arsenic(III), cadmium(II) and mercury(II), are very toxic even at trace amounts. The contamination of water by organic pollutants, such as phenol, chlorinated organic compounds and dyes, are used in textile industries and are also common effluents in water bodies, which cause harm to both environment and human health (Fu and Wang 2011, Ngah and Hanafiah 2008).

3. ADVANCES IN DETECTION AND REMEDIATION TECHNIQUES

In the past decades, advances and development in analytical techniques have demonstrated extraordinary performance from detection to degradation and removal of contaminants from wastewater. High-performance liquid-chromatography-coupled mass spectrometry (HPLC-MS), size exclusion chromatography (SEC), gel permeation chromatography (GPC), graphite furnace atomic absorption spectrometry (GFAAS), nuclear magnetic resonance (NMR), transmission electron microscopy (TEM), X-ray crystallography (XRD) and high-resolution scanning electron microscopy (HRSEM) can detect diverse micropollutants and emerging pollutant even at trace level. However, a single detection technique is not enough for the effective detection and quantification of a wide array of emerging pollutants and needs to be coupled together (Das et al. 2019b). The limited solubility of some emerging pollutants such as ENMs is also a critical issue for their detection. The state-of-the-art detection method is targeting only specific emerging pollutants and cannot be used for routine monitoring. (Abu-Danso et al. 2020, Aggarwal et al. 2019, Anand et al. 2019b, Bhati et al. 2019, Gunture et al. 2020a, Santhosh et al. 2020, Tripathi et al. 2017).

Most of the available techniques are extremely used for the quantification of emerging pollutants for research purposes only not in practice for complex environmental samples and wastewater remediation. Moreover, above listed techniques are used in a tandem fashion and have an inherent amount of error for the recognition of specific emerging pollutants due to interference with a complex matrix of environmental samples (Carl Englert 2007). The complex wastewater sample also requires a hard extraction process and demand for the investigation of new approaches for their separation. Advanced and ultra-sensitive techniques, which simplify sample preparation and simultaneously detect a number of diverse emerging pollutants are the need of the hour. Few investigations have detected emerging pollutants in wastewater by using the electrochemical technique (Torrinha et al. 2020). A recent review by Torinha et al. summarized the utilization of nano-carbons for the fabrication of electrochemical sensors towards the detection of pharmaceuticals in water systems (Torrinha et al. 2020). Also, nowadays, fluorescent-based sensors have been emerged as a new and efficient technique for the detection of various heavy metal ions, which are non-biodegradable pollutants, that are under consideration in order to maintain their permissible limit in water (Anand et al. 2019a, Chauhan et al. 2019). In light of the inadequacy of long-used traditional water or wastewater treatment technologies, the execution of alternative advanced technologies for the optimized recognition and uptake of emerging pollutants is growing continuously (de Andrade et al. 2018).

Among the available water treatment strategies adsorption is assumed to be environment friendly and the most efficient approach due to simple operation and no by-product generation (Gunture et al. 2020b). Sustainable adsorbents derived from biomass or biowaste are further increasing attention due to their low environmental footprint (Myung et al. 2019). Even adsorption seems promising for the removal of emerging pollutants due to its diverse nature (Figure 6) (Bajpai et al. 2019, de Andrade et al. 2018, Lin et al. 2018, Shukla et al. 2019). However, different kinds of emerging pollutants are regarded as the major concern due to low absorption efficiency attributed to small surface area and lack of surface binding sites. Batch scale absorption is a popular technique and is generally used for the removal of a range of pollutants, including emerging pollutants. With respect to the removal of dye-based pollutants from contaminated water bodies, Gunture et al. reported a unique methodology of converting diesel exhaust pollutant into a potent adsorbent black carbon for removal of cationic organic dyes, such as methylene blue, crystal violet, and rhodamine *b* and further utilized the treated water into plant growth (Gunture et al. 2020b). Wetland plant materials with the help of microorganisms are reported to remove the emerging pollutants, including ENMs, estrogen, para-chlorobenzoic acid with 60-70 percentage of removal efficiency (Sharif et al. 2013).

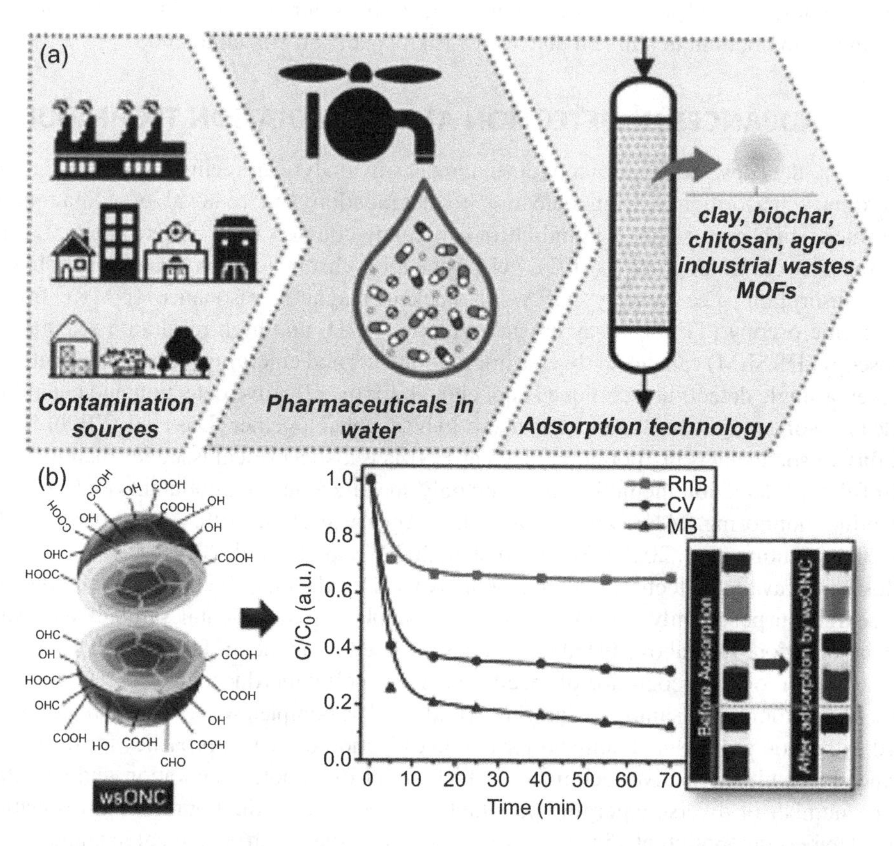

FIGURE 6 A scheme showing the removal of (a) pharmaceuticals from wastewater. Adapted with permission from de Andrade et al. (2018). Copyright 2018 American Chemical Society. (b) Organic pollutants using adsorption technology. Adapted with permission from Gunture et al. (2020b). Copyright 2020, American Chemical Society.

Photocatalysis is an advanced technique and provides a valuable alternative for the degradation of highly bio-resistance emerging pollutants as it generates highly reactive oxygen species (ROS) (Aggarwal et al. 2020). Mild operating conditions, no toxic byproducts generation and only requirement of light (without any chemicals) are the added advantage of photocatalysis (Das et al. 2020). The photocatalytic degradation of acetaminophen and antipyrine using TiO_2-ZnO as photocatalyst is reported under solar light irradiation (Tobajas et al. 2017). 2,4-Dinitrophenol (DNP) as a highly toxic micropollutant was degraded in a facile photocatalytic process with ZnO tetrapods-carbon nano onions composite induced by visible light. Further, the treated water was tested for ecological assessment as shown in Figure 7 (a and b) (Park et al. 2019). Being a cost-effective and very efficient method, photocatalysis was also explored in the degradation of complex organic dyes (Saini et al. 2019) and photoreduction of toxic hexavalent chromium ions to less toxic chromium (III) oxidation state in the presence of sunlight from wastewater (Bhati et al. 2019, Anand et al. 2019b, Saini et al. 2020). A schematic mechanism pathway for the reduction of Cr(VI) ions with NSCD under sunlight irradiation was shown in Figure 7 (c and d),

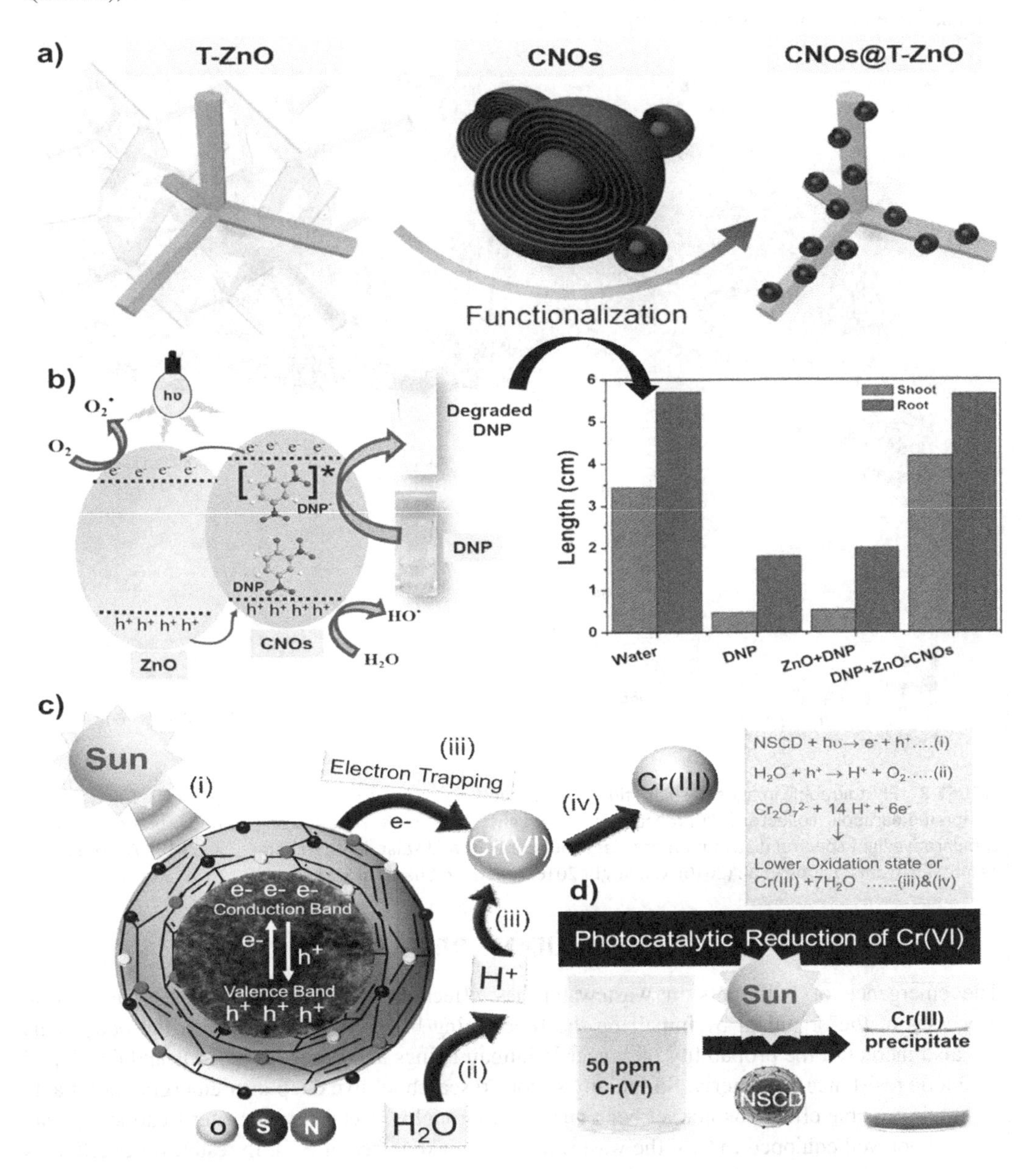

FIGURE 7 (a, b) Fabrication of ZnO tetrapods-carbon nano-onions for the photocatalytic degradation of DNP and ecological assessment of treated wastewater using gram plant. Adapted with permission from Park et al. (2019). Copyright 2019, Springer Nature. (c, d) Showing reduction mechanism of Cr(VI) to Cr(III) ions using NSCD in the sunlight. Adapted with permission from Saini et al. (2020). Copyright 2020, American Chemical Society.

Fenton-based processes are considering a promising approach for the emerging pollutants removal due to the utilization of common reagents and solar energy, but the generation of a high amount of waste and limited operational pH is the main barrier for practical applications (Mirzaei et al. 2017). In an innovative approach, Philippe et al. developed a solar simulator using P25 TiO_2 as an optical catalyst, the photograph of the device is shown in Figure 8 (Philippe et al. 2016). This solar simulator was used for the effective degradation of 13 different micropollutants and also able to inactivate microorganisms (Philippe et al. 2016).

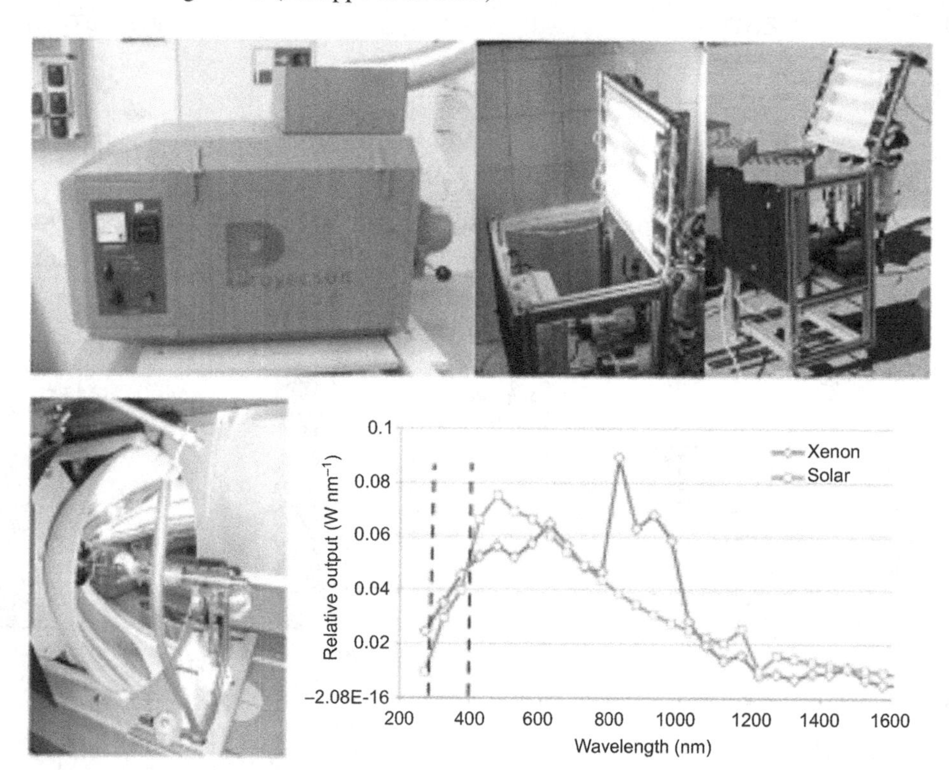

FIGURE 8 Photographic images of solar stimulator. Top left: the cinema project as a lamp holder. Top right: the compound parabolic collector (CPC) reactor and solar light. Bottom left: xenon lamp and reflector. Bottom right: comparative light spectral distribution of solar simulator light and solar light (Philippe et al. 2016). Adapted with permission from Philippe et al. (2016). Copyright 2016, American Chemical Society.

4. CONCLUDING REMARKS

The emergence of pollutants in wastewater has affected the physiological and reproduction processes of the organism by impairing the functioning biological system inside the body. This has also increased the probability of cancer in human beings and others organisms and developed antibiotic resistance in bacteria. So far, any serious research to dive deep over emerging pollutants and their possible effect has not yet been checked thoroughly. Hence, wastewater treatment plants are still not well equipped to treat the wastewater and make it free of ECs. In result, water effluents carrying an abundance of ECs have been carried out to natural water bodies causing a threat for wildlife as well as human beings.

Therefore, wastewater effluents after treatment from wastewater treatment plants must be motioned for the presence of ECs. In fact, wastewater treatment plants must be made to carry out the ECs removal task by developing effective detection and removal techniques to ensure the absence of ECs in wastewater effluents. After removal of ECs and mitigating much below the permissible limit wastewater effluents must be discharged to the aquatic system. Additionally, the long-term effect of ECs exposure to the ecosystem must be carefully analyzed to avoid any kind of casualties or incidence of aquatic pollution.

Acknowledgements

D.S. thanks the Department of Science and Technology (DST) for Inspire Doctoral Fellowship (IF180133), R.A. thanks MNIT Jaipur for doctoral fellowship and P. D. thanks UGC for funding. K.M.T thanks the Department of Biotechnology (DBT), India, for the Ramalingaswami Faculty Award (BT/RLF/Re-entry/45/2018) and financial assistance, and S.K.S. thanks DST, India (SB/EMEQ-383/2014) for funding.

References

Abu-Danso, E., Peräniemi, S., Leiviskä, T., Kim, T., Tripathi, K.M. and Bhatnagar, A. 2020. Synthesis of clay-cellulose biocomposite for the removal of toxic metal ions from aqueous medium. J. Hazard. Mater. 381: 120871.

Aggarwal, R., Anand, S.R., Saini, D., Gunture, Singh, R., Sonker, A.K. and Sonkar, S.K. 2019. Surface-passivated, soluble and non-toxic graphene nano-sheets for the selective sensing of toxic Cr(vi) and Hg(ii) metal ions and as a blue fluorescent ink. Nanoscale Adv. 1: 4481-4491.

Aggarwal, R., Saini, D., Singh, B., Kaushik, J., Garg, A.K. and Sonkar, S.K. 2020. Bitter apple peel derived photoactive carbon dots for the sunlight induced photocatalytic degradation of crystal violet dye. Sol. Energy. 197: 326-331.

Aksu, Z. 2005. Application of biosorption for the removal of organic pollutants: A review. Process Biochem. 40: 997-1026.

Alves, A., Giovanoulis, G., Nilsson, U., Erratico, C., Lucattini, L., Haug, L.S., et al. 2017. Case study on screening emerging pollutants in urine and nails. Environ. Sci. Technol. 51: 4046-4053.

Anand, S.R., Bhati, A., Saini, D., Gunture, Chauhan, N., Khare, P. and Sonkar, S.K. 2019a. Antibacterial nitrogen-doped carbon dots as a reversible "fluorescent nanoswitch" and fluorescent ink. ACS Omega 4: 1581-1591.

Anand, S.R., Aggarwal, R., Saini, D., Sonker, A.K., Chauhan, N. and Sonkar, S.K. 2019b. Removal of toxic chromium(VI) from the wastewater under the sunlight-illumination by functionalized carbon nano-rods. Sol. Energy. 193: 774-781.

Anju, A., Ravi, S.P. and Bechan, S. 2010. Water pollution with special reference to pesticide contamination in India. J. Water Resour. Prot. 2: 1793.

Archana, G., Dhodapkar, R. and Kumar, A. 2017. Ecotoxicological risk assessment and seasonal variation of some pharmaceuticals and personal care products in the sewage treatment plant and surface water bodies (lakes). Environ. Monit. Assess. 189: 446.

Arribas, M., Soro, P. and Silvestre, J. 2012. Allergic contact dermatitis to fragrances. Part 1. Actas Dermo-Sifiliográficas (English Edition) 103: 874-879.

Bajpai, V.K., Shukla, S., Khan, I., Kang, S.-M., Haldorai, Y., Tripathi, K.M., et al. 2019. A sustainable graphene aerogel capable of the adsorptive elimination of biogenic amines and bacteria from soy sauce and highly efficient cell proliferation. ACS Appl. Mater. Interfaces 11: 43949-43963.

Benotti, M.J., Trenholm, R.A., Vanderford, B.J., Holady, J.C., Stanford, B.D. and Snyder, S.A. 2009. Pharmaceuticals and endocrine disrupting compounds in U.S. drinking water. Environ. Sci. Technol. 43: 597-603.

Beydoun, A. and Backonja, M.-M. 2003. Mechanistic stratification of antineuralgic agents. J. Pain Symptom Manag. 25: S18-S30.

Bhati, A., Gunture, Tripathi, K.M., Singh, A., Sarkar, S. and Sonkar, S.K. 2018. Exploration of nano carbons in relevance to plant systems. New J. Chem. 42: 16411-16427.

Bhati, A., Anand, S.R., Saini, D., Gunture and Sonkar, S.K. 2019. Sunlight-induced photoreduction of Cr(VI) to Cr(III) in wastewater by nitrogen-phosphorus-doped carbon dots. npj Clean Water 2: 12.

Brown, J.D. and Winterstein, A.G. 2019. Potential adverse drug events and drug-drug interactions with medical and consumer cannabidiol (CBD) use. J. Clin. Med. 8: 989.

Carl Englert, B. 2007. Nanomaterials and the environment: Uses, methods and measurement. J. Environ. Monit. 9: 1154-1161.

Chauhan, N., Anand, S.R., Aggarwal, R., Kaushik, J., Shekhawat, S.S., Sonker, A.K., et al. 2019. Soluble non-toxic carbon nano-rods for the selective sensing of iron(iii) and chromium(vi). New J. Chem. 43: 10726-10734.

Das, G.S., Shim, J.P., Bhatnagar, A., Tripathi, K.M. and Kim, T. 2019a. Biomass-derived carbon quantum dots for visible-light-induced photocatalysis and label-free detection of Fe(III) and ascorbic acid. Sci. Rep. 9: 15084.

Das, G.S., Tripathi, K.M., Kumar, G., Paul, S., Mehara, S., Bhowmik, S., et al. 2019b. Nitrogen-doped fluorescent graphene nanosheets as visible-light-driven photocatalysts for dye degradation and selective sensing of ascorbic acid. New J. Chem. 43: 14575-14583.

Das, G.S., Bhatnagar, A., Yli-Pirilä, P., Tripathi, K.M. and Kim, T. 2020. Sustainable nitrogen-doped functionalized graphene nanosheets for visible-light-induced photocatalytic water splitting. Chem. Commun. 56: 6953-6956.

de Andrade, J.R., Oliveira, M.F., da Silva, M.G.C. and Vieira, M.G.A. 2018. Adsorption of pharmaceuticals from water and wastewater using nonconventional low-cost materials: A review. Ind. Eng. Chem. Res. 57: 3103-3127.

Ebele, A.J., Abou-Elwafa Abdallah, M. and Harrad, S. 2017. Pharmaceuticals and personal care products (PPCPs) in the freshwater aquatic environment. Emerg. Contam. 3: 1-16.

Falconer, I.R., Chapman, H.F., Moore, M.R. and Ranmuthugala, G. 2006. Endocrine-disrupting compounds: A review of their challenge to sustainable and safe water supply and water reuse. Environ. Toxicol. 21: 181-191.

Fatta-Kassinos, D., Meric, S. and Nikolaou, A. 2011. Pharmaceutical residues in environmental waters and wastewater: Current state of knowledge and future research. Anal. Bioanal. Chem. 399: 251-275.

Fisher, M., Arbuckle, T.E., MacPherson, S., Braun, J.M., Feeley, M. and Gaudreau, É. 2019. Phthalate and BPA exposure in women and newborns through personal care product use and food packaging. Environ. Sci. Technol. 53: 10813-10826.

Fu, F. and Wang, Q. 2011. Removal of heavy metal ions from wastewaters: A review. J. Environ. Manage. 92: 407-418.

Fu, Q., Malchi, T., Carter, L.J., Li, H., Gan, J. and Chefetz, B. 2019. Pharmaceutical and personal care products: From wastewater treatment into agro-food systems. Environ. Sci. Technol. 53: 14083-14090.

Gao, J., Bonzongo, J.-C.J., Bitton, G., Li, Y. and Wu, C.-Y. 2008. Nanowastes and the environment: Using mercury as an example pollutant to assess the environmental fate of chemicals adsorbed onto manufactured nanomaterials. Environ. Toxicol. Chem. 27: 808-810.

Geissen, V., Mol, H., Klumpp, E., Umlauf, G., Nadal, M., van der Ploeg, M., et al. 2015. Emerging pollutants in the environment: A challenge for water resource management. Int. Soil Water Conserv. Res. 3: 57-65.

Gunture, Dalal, C., Kaushik, J., Garg, A.K. and Sonkar, S.K. 2020a. Pollutant-soot-based nontoxic water-soluble onion-like nanocarbons for cell imaging and selective sensing of toxic Cr(VI). ACS Appl. Bio Mater. 3: 3906-3913.

Gunture, Kaushik, J., Garg, A.K., Saini, D., Khare, P. and Sonkar, S.K. 2020b. Pollutant diesel soot derived onion-like nanocarbons for the adsorption of organic dyes and environmental assessment of treated wastewater. Ind. Eng. Chem. Res. 59: 12065-12074.

Gupta, V. 2009. Application of low-cost adsorbents for dye removal: A review. J. Environ. Manage. 90: 2313-2342.

Gurke, R., Rößler, M., Marx, C., Diamond, S., Schubert, S., Oertel, R., et al. 2015. Occurrence and removal of frequently prescribed pharmaceuticals and corresponding metabolites in wastewater of a sewage treatment plant. Sci. Total Environ. 532: 762-770.

Hao, J., Zhang, Q., Chen, P., Zheng, X., Wu, Y., Ma, D., et al. 2019. Removal of pharmaceuticals and personal care products (PPCPs) from water and wastewater using novel sulfonic acid (–SO3H) functionalized covalent organic frameworks. Environ. Sci.: Nano 6: 3374-3387.

Helfrich, L.A., Weigmann, D.L., Hipkins, P.A. and Stinson, E.R. 2009. Pesticides and Aquatic Animals: A Guide to Reducing Impacts on Aquatic Systems. Virginia Cooperative Extension (VCE) 420-013.

Kar, S., Sanderson, H., Roy, K., Benfenati, E. and Leszczynski, J. 2020. Ecotoxicological assessment of pharmaceuticals and personal care products using predictive toxicology approaches. Green Chem. 22: 1458-1516.

Khaksar, M., Vasileiadis, S., Sekine, R., Brunetti, G., Scheckel, K.G., Vasilev, K., et al. 2019. Chemical characterisation, antibacterial activity, and (nano) silver transformation of commercial personal care products exposed to household greywater. Environ. Sci.: Nano 6: 3027-3038.

Koelega, H. 1993. Stimulant drugs and vigilance performance: A review. Psychopharmacology 111: 1-16.

Laurent, S. 2017. Antihypertensive drugs. Pharmacol. Res. 124: 116-125.

Lenzen, S. and Bailey, C.J. 1984. Thyroid hormones, gonadal and adrenocortical steroids and the function of the islets of Langerhans. Endocrine Rev. 5: 411-434.

Li, W., Ma, Y., Guo, C., Hu, W., Liu, K., Wang Y., et al. 2007. Occurrence and behavior of four of the most used sunscreen UV filters in a wastewater reclamation plant. Water Res. 41: 3506-3512.

Lin, S., Zhao, Y. and Yun, Y.-S. 2018. Highly effective removal of nonsteroidal anti-inflammatory pharmaceuticals from water by Zr(IV)-based metal–organic framework: Adsorption performance and mechanisms. ACS Appl. Mater. Interfaces 10: 28076-28085.

Liu, R. and Mabury, S.A. 2019. Synthetic phenolic antioxidants in personal care products in Toronto, Canada: Occurrence, human exposure, and discharge via greywater. Environ. Sci. Technol. 53: 13440-13448.

Luo, Y., Guo, W., Ngo, H.H., Nghiem, L.D., Hai, F.I., Zhang, J., et al. 2014. A review on the occurrence of micropollutants in the aquatic environment and their fate and removal during wastewater treatment. Sci. Total Environ. 473-474: 619-641.

Lymperi, S. and Giwercman, A. 2018. Endocrine disruptors and testicular function. Metabolism 86: 79-90.

McDonnell, G. and Russell, A.D. 1999. Antiseptics and disinfectants: Activity, action, and resistance. Clin. Microbiol. Rev. 12: 147-179.

Mirzaei, A., Chen, Z., Haghighat, F. and Yerushalmi, L. 2017. Removal of pharmaceuticals from water by homo/heterogonous Fenton-type processes: A review. Chemosphere 174: 665-688.

Mohapatra, S., Huang, C.-H., Mukherji, S. and Padhye, L.P. 2016. Occurrence and fate of pharmaceuticals in WWTPs in India and comparison with a similar study in the United States. Chemosphere 159: 526-535.

Myung, Y., Jung, S., Tung, T.T., Tripathi, K.M. and Kim, T. 2019. Graphene-based aerogels derived from biomass for energy storage and environmental remediation. ACS Sustainable Chem. Eng. 7: 3772-3782.

Ngah, W.W. and Hanafiah, M.M. 2008. Removal of heavy metal ions from wastewater by chemically modified plant wastes as adsorbents: A review. Bioresour. Technol. 99: 3935-3948.

Nguyen, V.T., Reinhard, M. and Karina, G.Y.-H. 2011. Occurrence and source characterization of perfluorochemicals in an urban watershed. Chemosphere 82: 1277-1285.

Noguera-Oviedo, K. and Aga, D.S. 2016. Lessons learned from more than two decades of research on emerging contaminants in the environment. J. Hazard. Mater. 316: 242-251.

Garcia, Q.D., Pinto, G.P., García-Encina, P.A. and Mata, R.I. 2013. Ranking of concern, based on environmental indexes, for pharmaceutical and personal care products: An application to the Spanish case. J. Environ. Manage. 129: 384-397.

Oskarsson, A. and Wright, M.C. 2019. Ionic liquids: New emerging pollutants, similarities with perfluorinated alkyl substances (PFASs). Environ. Sci. Technol. 53: 10539-10541.

Özbek, O. and Gürdere, M.B. 2020. A review on the synthesis and applications of molecules as anticonvulsant drug agent candidates. Med. Chem. Res. 29: 1553-1578.

Pal, A., He, Y., Jekel, M., Reinhard, M. and Gin, K.Y. 2014. Emerging contaminants of public health significance as water quality indicator compounds in the urban water cycle. Environ. Int. 71: 46-62.

Park, S.J., Das, G.S., Schütt, F., Adelung, R., Mishra, Y.K., Tripathi, K.M., et al. 2019. Visible-light photocatalysis by carbon-nano-onion-functionalized ZnO tetrapods: Degradation of 2,4-dinitrophenol and a plant-model-based ecological assessment. NPG Asia Mater. 11: 1-13.

Philippe, K.K., Timmers, R., van Grieken, R. and Marugan, J. 2016. Photocatalytic disinfection and removal of emerging pollutants from effluents of biological wastewater treatments, using a newly developed large-scale solar simulator. Ind. Eng. Chem. Res. 55: 2952-2958.

Rahman, F., Langford, K.H., Scrimshaw, M.D. and Lester, J.N. 2001. Polybrominated diphenyl ether (PBDE) flame retardants. Sci. Total Environ. 275: 1-17.

Rehman, M.S.U., Rashid, N., Ashfaq, M., Saif, A., Ahmad, N. and Han, J.-I. 2015. Global risk of pharmaceutical contamination from highly populated developing countries. Chemosphere 138: 1045-1055.

Rice, J. and Westerhoff, P. 2017. High levels of endocrine pollutants in US streams during low flow due to insufficient wastewater dilution. Nature Geosci. 10: 587-591.

Richmond, E.K., Rosi, E.J., Walters, D.M., Fick, J., Hamilton, S.K., Brodin, T., et al. 2018. A diverse suite of pharmaceuticals contaminates stream and riparian food webs. Nat. Commun. 9: 1-9.

Rodil, R., Moeder, M., Altenburger, R. and Schmitt-Jansen, M. 2009. Photostability and phytotoxicity of selected sunscreen agents and their degradation mixtures in water. Anal. Bioanal. Chem. 395: 1513-1524.

Roscher, N.M., Lindemann, M.K., Kong, S.B., Cho, C.G. and Jiang, P. 1994. Photodecomposition of several compounds commonly used as sunscreen agents. J. Photochem. Photobiol. A Chem. 80: 417-421.

Saini, D., Aggarwal, R., Anand, S.R. and Sonkar, S.K. 2019. Sunlight induced photodegradation of toxic azo dye by self-doped iron oxide nano-carbon from waste printer ink. Sol. Energy 193: 65-73.

Saini, D., Kaushik, J., Garg, A.K., Dalal, C. and Sonkar, S.K. 2020. N, S-codoped carbon dots for nontoxic cell imaging and as a sunlight-active photocatalytic material for the removal of chromium. ACS Appl. Bio Mater. 3: 3656-3663.

Salehi, A.S.M., Shakalli Tang, M.J., Smith, M.T., Hunt, J.M., Law, R.A., Wood, D.W., et al. 2017. Cell-free protein synthesis approach to biosensing hTRβ-specific endocrine disruptors. Anal. Chem. 89: 3395-3401.

Santhosh, C., Daneshvar, E., Tripathi, K.M., Baltrėnas, P., Kim, T., Baltrėnaitė, E., et al. 2020. Synthesis and characterization of magnetic biochar adsorbents for the removal of Cr(VI) and acid orange 7 dye from aqueous solution. Environ. Sci. Pollut. Res. 27: 32874-32887.

Scanes, C., Sharp, P., Harvey, S., Godden, P.M., Chadwick, A. and Newcomer, W. 1979. Variations in plasma prolactin, thyroid hormones, gonadal steroids and growth hormone in turkeys during the induction of egg laying and moult by different photoperiods. Br. Poult. Sci. 20: 143-148.

Semblante, G.U., Hai, F.I., McDonald, J., Khan, S.J., Nelson, M., Lee, D.-J., et al. 2017. Fate of trace organic contaminants in oxic-settling-anoxic (OSA) process applied for biosolids reduction during wastewater treatment. Bioresour. Technol. 240: 181-191.

Sharif, F., Westerhoff, P. and Herckes, P. 2013. Sorption of trace organics and engineered nanomaterials onto wetland plant material. Environ. Sci. Proces. 15: 267-274.

Shukla, S., Khan, I., Bajpai, V.K., Lee, H., Kim, T., Upadhyay, A., et al. 2019. Sustainable graphene aerogel as an ecofriendly cell growth promoter and highly efficient adsorbent for histamine from red wine. ACS Appl. Mater. Interfaces 11: 18165-18177.

Soto, A.M. and Sonnenschein, C. 2010. Environmental causes of cancer: Endocrine disruptors as carcinogens. Nat. Rev. Endocrinol. 6: 363-370.

Stolte, S., Abdulkarim, S., Arning, J., Blomeyer-Nienstedt, A.-K., Bottin-Weber, U., Matzke, M., et al. 2008. Primary biodegradation of ionic liquid cations, identification of degradation products of 1-methyl-3-octylimidazolium chloride and electrochemical wastewater treatment of poorly biodegradable compounds. Green Chem. 10: 214-224.

Subedi, B., Balakrishna, K., Sinha, R.K., Yamashita, N., Balasubramanian, V.G. and Kannan, K. 2015. Mass loading and removal of pharmaceuticals and personal care products, including psychoactive and illicit drugs and artificial sweeteners, in five sewage treatment plants in India. J. Environ. Chem. Eng. 3: 2882-2891.

Subedi, B. and Kannan, K. 2014. Fate of artificial sweeteners in wastewater treatment plants in New York state, U.S.A. Environ. Sci. Technol. 48: 13668-13674.

Sui, Q., Huang, J., Deng, S. and Yu, G. 2009. Rapid determination of pharmaceuticals from multiple therapeutic classes in wastewater by solid-phase extraction and ultra-performance liquid chromatography tandem mass spectrometry. Chin. Sci. Bull. 54: 4633-4643.

Sui, Q., Huang, J., Deng, S., Yu, G. and Fan, Q. 2010. Occurrence and removal of pharmaceuticals, caffeine and DEET in wastewater treatment plants of Beijing, China. Water Res. 44: 417-426.

Sui, Q., Huang, J., Deng, S., Chen, W. and Yu, G. 2011. Seasonal variation in the occurrence and removal of pharmaceuticals and personal care products in different biological wastewater treatment processes. Environ. Sci. Technol. 45: 3341-3348.

Sun, Q., Li, M., Ma, C., Chen, X., Xie, X. and Yu, C.-P. 2016. Seasonal and spatial variations of PPCP occurrence, removal and mass loading in three wastewater treatment plants located in different urbanization areas in Xiamen, China. Environ. Pollut. 208: 371-381.

Tijani, J.O., Fatoba, O.O., Babajide, O.O. and Petrik, L.F. 2016. Pharmaceuticals, endocrine disruptors, personal care products, nanomaterials and perfluorinated pollutants: A review. Environ. Chem. Lett. 14: 27-49.

Tobajas, M., Belver, C. and Rodriguez, J.J. 2017. Degradation of emerging pollutants in water under solar irradiation using novel TiO_2-ZnO/clay nanoarchitectures. Chem. Eng. J. 309: 596-606.

Torrinha, Á., Oliveira, T.M.B.F., Ribeiro, F.W.P., Correia, A.N., Lima-Neto, P. and Morais, S. 2020. Application of nanostructured carbon-based electrochemical (Bio) sensors for screening of emerging pharmaceutical pollutants in waters and aquatic species: A review. Nanomaterials 10: 1268.

Tran, N.H., Urase, T. and Kusakabe, O. 2010. Biodegradation characteristics of pharmaceutical substances by whole fungal culture *Trametes versicolor* and its Laccase. J. Water Environ. Technol. 8: 125-140.

Tran, N.H., Urase, T. and Ta, T.T. 2014a. A preliminary study on the occurrence of pharmaceutically active compounds in hospital wastewater and surface water in Hanoi, Vietnam. CLEAN – Soil, Air, Water 42: 267-275.

Tran, N.H., Nguyen, V.T., Urase, T. and Ngo, H.H. 2014b. Role of nitrification in the biodegradation of selected artificial sweetening agents in biological wastewater treatment process. Bioresour. Technol. 161: 40-46.

Tran, N.H. and Gin, K.Y.-H. 2017. Occurrence and removal of pharmaceuticals, hormones, personal care products, and endocrine disrupters in a full-scale water reclamation plant. Sci. Total Environ. 599: 1503-1516.

Tripathi, K.M., Tran, T.S., Kim, Y.J. and Kim, T. 2017. Green fluorescent onion-like carbon nanoparticles from flaxseed oil for visible light induced photocatalytic applications and label-free detection of Al(III) Ions. ACS Sustainable Chem. Eng. 5: 3982-3992.

Verlicchi, P. and Zambello, E. 2015. Pharmaceuticals and personal care products in untreated and treated sewage sludge: Occurrence and environmental risk in the case of application on soil: A critical review. Sci. Total Environ. 538: 750-767.

Vilela, C.L.S., Bassin, J.P. and Peixoto, R.S. 2018. Water contamination by endocrine disruptors: Impacts, microbiological aspects and trends for environmental protection. Environ. Pollut. 235: 546-559.

Wang, L. and Chen, Y. 2020. Luminescence-sensing Tb-MOF nanozyme for the detection and degradation of estrogen endocrine disruptors. ACS Appl. Mater. Interfaces 12: 8351-8358.

Wang, Y., Westerhoff, P. and Hristovski, K.D. 2012. Fate and biological effects of silver, titanium dioxide, and C60 (fullerene) nanomaterials during simulated wastewater treatment processes. J. Hazard. Mater. 201: 16-22.

Wiesner, M.R., Lowry, G.V., Jones, K.L., Hochella, J.M.F., Di Giulio, R.T., Casman, E., et al. 2009. Decreasing uncertainties in assessing environmental exposure, risk, and ecological implications of nanomaterials. Environ. Sci. Technol. 43: 6458-6462.

Yang, Y.-Y., Liu, W.-R., Liu, Y.-S., Zhao, J.-L., Zhang, Q.-Q., Zhang, M., et al. 2017. Suitability of pharmaceuticals and personal care products (PPCPs) and artificial sweeteners (ASs) as wastewater indicators in the Pearl River Delta, South China. Sci. Total Environ. 590: 611-619.

Yao, C., Yang, X., Roy Raine, R., Cheng, C., Tian, Z. and Li, Y. 2009. The effects of MTBE/ethanol additives on toxic species concentration in gasoline flame. Energy Fuels 23: 3543-3548.

Zhang, Q.-Q., Ying, G.-G., Pan, C.-G., Liu, Y.-S. and Zhao, J.-L. 2015. Comprehensive evaluation of antibiotics emission and fate in the river basins of China: Source analysis, multimedia modeling, and linkage to bacterial resistance. Environ. Sci. Technol. 49: 6772-6782.

Zhou, Y., Zhang, L. and Cheng, Z. 2015. Removal of organic pollutants from aqueous solution using agricultural wastes: A review. J. Mol. Liq. 212: 739-762.

Eradication of Personal Care Products by Liquid and Crystal Nanomaterials

Rachna[1], Uma Shanker[1,*] and Manviri Rani[2]

1. INTRODUCTION

Consumption of different types of organic compounds is part of our life today. The compounds consist of various personal care products (PCPs) and pharmaceuticals (Duca and Boldescu 2009, Ortiz de García et al. 2013). PCPs are a group of essential antibiotics, antimicrobial agents, drugs, food supplements and chemicals used in cosmetics, soaps, sunscreen and fragrances (Sören and Thiele-Bruhn 2003, Tan et al. 2015). Recently, wide range of publications have targeted PCPs and their occurrence in the environment owing to its presence in food (Dasenaki and Thomaidis 2015), marine life (Ramirez et al. 2007), water bodies (Yao et al. 2018, Borecka et al. 2015) and soil (Chen et al. 2011). Creation of these products and easy availability has raised their enrichment in the environment (Gorito et al. 2017). PCPs generally consist of ionic functional groups having different dissociation constants (Daughton and Ternes 1999). PCPs contain various subgroups, such as sulfonamides, tetracyclines and fluoroquinolones (Pan et al. 2009). Out of various compounds found in PCPs, bisphenol a is most commonly found endocrine compound (Novo et al. 2018). Many agriculture products, such as herbicides and fungicides, are part of personal care products (Tian et al. 2016, Zhang et al. 2013a). Disposal of PCPs is unregulated, specifically in developing countries with no proper regulations.

Removal of PCPs from the environment involves various conventional as well as advanced strategies. These include oxidation of PCPs through ozonation, Fenton's reagent, advanced oxidation using reactive species, adsorption on materials, degradation through catalysts, biological agents, and membrane filtration (Akbari et al. 2016, Matamoros et al. 2016, Liu et al. 2014, Mock et al. 2017, Schneider et al. 2014). However, each process has certain advantages and disadvantages associated with it. New wave of nanotechnology has fascinated the researchers through the extraordinary properties of nanomaterials, such as large surface area, semiconducting behavior and quantum effects. Rapid increase in the publications has been found on nanomaterial-based removal of PCPs through adsorption as well as photocatalysis.

Nanomaterials, particularly modified liquid crystals and other crystalline materials have fast response speed, thermal stability and better activity (Cui and Zhao 2004). Currently, modification of liquid crystal matrix through inorganic nanomaterials has gained attention (Shibaev 2009). Out

[1] Department of Chemistry, Dr B R Ambedkar National Institute of Technology, Jalandhar, Punjab, India-144011.
[2] Department of Chemistry, Malaviya National Institute of Technology, JLN Road, Malaviya Nagar, Rajasthan, India-302017.
[*] Corresponding author: shankeru@nitj.ac.in, umaorganic29@gmail.com.

of these materials, liquid crystals have been considered in the present study due to their exceptional involvement in display and photonics. Liquid crystals having photoconductivity, strong light absorption, advanced optical, thermal, and electrical properties are believed to be beneficial in the eradication of PCPs from the environment (Yadav et al. 2015). Liquid and crystalline materials may provide benefits to the electrical properties of the catalysts by entrapping and forming lines of tiny agglomerates in the defects. Coupling of crystalline materials with other nanostructured materials, such as nanorods, transition metal oxides, heterostructured materials, nanotubes, and polymers are subject of interest. However, selection of nanomaterials for the formation of coupled material should be appropriate to avoid the difficulties related to phase formation as well as agglomeration of phases.

Ordered graphene-based heterogeneous Fenton oxidation was employed for the eradication of PCPs from the environment at acidic as well as neutral pH (Divyapriya et al. 2018). Oxides of metals, like zinc, titanium, iron, and cadmium, have been widely employed for the photocatalytic removal of PCPs (Bagheri et al. 2016). PCPs such as antibiotics have been removed using metal doped CNTs as an adsorbent (Kang et al. 2018). Coupled materials of crystalline metals and bentonite have showed higher efficiency in the eradication of three antibiotics when targeted separately (Weng et al. 2018). Photocatalytic degradation of antibiotics with tin oxide nanomaterials showed good results with recyclability up to eight cycles. Recently, metal hexacyanoferrate-based nanocomposites have showed tremendous photocatalytic activity against phenols as well as dye removal under solar irradiation. Silver coupled ZnO nanoplates were utilized against the sunlight irradiated removal of antibiotics at neutral pH (Kaur et al. 2018). The ordered structure of crystalline materials is capable of enhancing their surface, conducting as well as light-responsive properties. Moreover, the presence of synergy between the constituents of composites can enhance their overall activity and stability against PCPs removal.

In the present chapter main focus has been given to the fate of PCPs in the environment and their removal through advanced processes involving the utilization of crystalline materials. The removal processes have been thoroughly discussed with an explanation of the mechanism of activity of materials. Reaction parameters play a vital role in the removal strategies, such as time, temperature and pH have been addressed.

2.1 Classification of Personal Care Products

Emerging micropollutants consist of a wide range of compounds, such as pesticides, antibiotics, PCPs, industrial compounds, and pharmaceuticals (Montes-Grajales et al. 2017). However, PCPs refer to compounds that are employed in maintaining health and hygiene and for cosmetic purposes. A class of anthropogenic chemicals having various daily use products (shampoo, cosmetic, toothpaste, and food) in it is called PCPs (Tolls et al. 2009).

2.1.1 Triclosan and Triclocarbon

Triclosan and triclocarban are used in soaps, toothpaste, deodorants, creams, and plastics to enable their antimicrobial and antifungal properties (McAvoy et al. 2002). Both these PCPs are listed among the top 10 commonly found wastewater compounds (Halden and Paull 2005). Their exposure to the general population is ubiquitous (Huang et al. 2016).

2.1.2 Fragrances

One of the most widely studied PCPs is fragrances. These habitually ubiquitous contaminants are generally synthetic musks. Deodorants, washing powder, and soaps include synthetic musks for fragrances. In the early eighties and mid-nineties, nitro musks and polycyclic musks were first introduced. Nitro musks and polycyclic musks are xylenes, ketons, tibetene, celestolide, and phantolide containing compounds. These are proven to be persistent and toxic to aquatic species (Peck 2006).

2.1.3 Insect Repellents

Chemicals used to limit the insects from impending on an applied surface are called insect repellents (Rodil and Moeder 2008). Insect repellants act as safe guards for some diseases in lack of any other protecting agent in tropical regions (Antwi et al. 2008). Significant amount of these have been detected in wastewaters, drinking, surface as well as ground water (Quednow and Puttmann 2009, Tay et al. 2009). N,N-diethyl-metatoluamide is broad-range spectrum used against mites, mosquitoes and tsetse flies formulated in 1946 (Costanzo et al. 2007, Antwi et al. 2008, Murphy et al. 2000).

2.1.4 Preservatives

In order to limit natural ripening and to prevent bacterial/fungal growth on plants, synthetic preservatives are used. Other than plants, these are widely utilized in soaps, cosmetics, food, and pharmaceuticals (Brausch and Rand 2011). Parabens are the most commonly used antimicrobial preservatives (Amin et al. 2019). Seven types of parabens having difference in word root are used as preservatives (Soni et al. 2005). Out of which methyl and propyl are one of the most commonly employed in makeup industry (Peck 2006). The 4-hydroxybenzoic acid is a low cost paraben having antifungal properties (Alvarez-Rivera et al. 2018).

2.1.5 Ultraviolet Filters

Recently, worldwide concern toward the harmful effect of ultraviolet (UV) radiation has enhanced the use of UV filters. These are used in cosmetics and sunscreen goods to safeguard against UV radiation. The UV filters make up to 10% of the total mass of cosmetic products (Schreurs et al. 2002). Out of various organic compounds, 16 are certified as sunscreen agents, while 27 are used as UV filters in plastics and cosmetics (Fent et al. 2008). Some of the estrogenic compounds found in UV filters are substitutes of benzophenone, cinnamate, and camphor (Zenker et al. 2008).

2.1.6 Biocide Compounds

Sterilization of health care and hospitals is done using various antiseptic and disinfectant compounds. These include chemicals such as iodine, chlorine, and alcohol. Triclosan, triclocarban, and benzotriazole are the most commonly used compounds. However, over-exposure of these is thought to cause microbial resistance (Brausch and Rand 2011).

2.1.7 Surfactants

Surfactants are one such PCPs that are utilized in the formation of herbicides, detergents, cosmetics, and the textile industry. Untreated disposal of surfactants into the environment has caused water pollution (da Silva et al. 2014). Surfactants being amphoteric are capable of accumulating in the soil, sludge, and sediments (Olmez-Hanci et al. 2011). These could be anionic, cationic, amphoteric, or nonionic (Reznik et al. 2010). Commonly, used surfactants in various industries are alkane sulfonates, nonylphenol, and nonylphenol ethoxylates (Baena-Nogueras et al. 2013, Bina et al. 2018).

2.1.8 Phthalates

Phthalates being flexible rigid polymers are used in various products such as polyvinylchloride. Production of products like toys, food wraps, medical devices, paints, and plastic goods involve phthalates in them. Other than these, cosmetics phthalates are used as solvents, suspension agents, skin emollients, elongators, plasticizers, antifoaming agents, or for fragrances (Api 2001, Hubinger and Havery 2006). However, overuse of these has caused their high load on the aquatic environment (Penalver et al. 2000).

2.1.9 Siloxanes

A new class of PCPs is siloxanes, having polymeric organic silicone with silicon-oxygen units as its backbone. These have low surface tension, smooth texture, good thermal stability, and physiologic inertness (Liu et al. 2014a). These properties make siloxanes extensively used in skin-care creams, color cosmetics, fuel additives, and automotive polishes. However, their toxic effects have raised concerns regarding their annual production (Horii and Kannan 2008). Their high Kow value has caused their adsorption onto the sludge and other water bodies (Sanchis et al. 2013). The group of siloxanes in PCPs includes various long-chain cyclic siloxanes (Richardson 2008).

2.2 Sources and Concentration of PCPs in Environment

The cosmetic industry alone in the early nineties had produced 550,000 metric tons of PCPs in European countries (Daughton and Ternes 1999). While, till 2010, the estimated sale was seen rising to 382.3 billion USD. Aging and sun protection has caused the growth of skincare products. Montes-Grajales et al. (2017) in their review, reported that around 72 PCPs have been listed as emerging micro-pollutants at a concentration ranging between 0.03 ngL^{-1} to 7.81 ngL^{-1}. Out of various PCPs, sunscreens, fragrances, and antiseptics were amongst the most detected groups in water bodies (Montes-Grajales et al. 2017). By the year 2020, the worldwide production of surfactants was estimated at 24,037.3 kilotons. Out of various types of surfactants, anionic have maximum volume (32%) in the year 2014, next was nonionic with 26% and 7.2% share was of cationic. Out of various markets of the world, Asia-Pacific is at number one for the highest production of surfactants (33.5%) (Reznik et al. 2010). Some of the calculated concentration of PCPs is listed in Table 1.

TABLE 1 PCPs with their type and range/lethal concentration (Brausch and Rand 2011).

Compound	Type	Range/Lethal Concentration
Triclosan	Disinfectant	<0.1-2,300 ngL^{-1}
Methyl triclosan	Disinfectant	0.5-74 ngL^{-1}
Triclocarbon	Disinfectant	19-1,425 ngL^{-1}
Musk ketone	Fragrance	4.8-390 ngL^{-1}
Musk xylene	Fragrance	1.1-180 ngL^{-1}
Celestolide	Fragrance	3.1-520 ngL^{-1}
Galaxolide	Fragrance	64-12,470 ngL^{-1}
Tonalide	Fragrance	52-6,780 ngL^{-1}
N,N-diethyl-metatoluamide	Insect repellant	13-660 ngL^{-1}
Paraben	Preservative	15-400 ngL^{-1}
4-methyl-benzilidine-camphor	UV filter	2.3-545 ngL^{-1}
benzophenone-3	UV filter	2.5-175 ngL^{-1}
2-ethyl-hexyl-4-trimethoxycinnamate	UV filter	2.7-224 ngL^{-1}
Octocrylene	UV filter	1.1-4,450 ngL^{-1}
Biphenylol	Antimicrobial	3.66 mgL^{-1}
Benzophenone	Fixative	56.8 mgL^{-1}
1,4-dichlorobenzene	Insect repellant	14 mgL^{-1}
Musk moskene	Nitro musk	>0.4 mgL^{-1}
Bezylparaben	Preservative	4.3, 5.7 mgL^{-1}
Butylparaben	Preservative	5.3, 7.3 mgL^{-1}
Ethylparaben	Preservative	25, 30 mgL^{-1}
Isobutylparaben	Preservative	7.6 mgL^{-1}
Isopropylparaben	Preservative	8.5 mgL^{-1}

Different behavior of triclosan at different pH values cause variation in its Dow value and hence the difference in bioaccumulation. Such that it accumulates more at higher pH values (Brausch and Rand 2011). Triclocarban with a concentration of around 6.8 μgL^{-1} has been observed in surface water. Till 2004, triclocarban concentration was in nanograms in the environment; however, its concentration has risen to significant levels over the last five years. Phenol and its substituted groups, such as 4-methylphenol and biphenylol, have been identified in surface water as high as 1.3 μgL^{-1} (Brausch and Rand 2011). Nitro musks have been found in more than 80% of river water samples and aquatic species of Japan with concentrations ranging from 26 to 36 ngL^{-1} (Yamagishi et al. 1983). Musk xylene and musk ketone were detected in 83 to 90% of effluents. In Germany, nitro (2 to 10 ngL^{-1}) and polycyclic musks (2 to 300 ngL^{-1}) were found in the river Elbe (Moldovan 2006). N, N-diethyl-metatoluamide has been used in more than 200 products in the US with annual usage ~1.8 million kg. It has been also found in waste water treatment plants and surface water (Glassmeyer et al. 2005, Sui et al. 2010). This PCP is persistent in water bodies but with lesser accumulation than fragrances and UV filters (Costanzo et al. 2007). The detected concentration of N, N-diethyl-metatoluamide is 0.2 μgL^{-1} in effluents and 55 ngL^{-1} in surface water. Beside this, 40% of surface water contained 1,4-dichlorobenzene (insect repellant) having 0.28 μgL^{-1} concentration in US (Glassmeyer et al. 2005).

In '80s, around 7,000 kg of parabens were utilized in cosmetics and sanitation products alone with a constant rise in number over time (Soni et al. 2005). Out of various parabens, methyl and propyl are often used in cosmetics to enhance preservative effects (Peck 2006). Their concentration in surface water is between 15 to 400 ng L^{-1} (Benijts et al. 2004). UV filters, such as 2-ethyl-hexyl-4-trimethoxycinnamate, benzophenone-3, 4-methyl-benzilidine-camphor, and octocrylene of mass 118g, 69g, 49g, and 28g, respectively, were detected per 10,000 people per day in Switzerland (Balmer et al. 2005). In one of a small lake in Switzerland, 966 kg of UV filters were thrown directly per year (Poiger et al. 2004). Alkylphenols, such as nonyl phenol and 4-tert-octylphenol, have been detected as intermediates in manufacturing industry. These phenols are also generated upon the degradation of nonionic surfactants. River and sludge water also contained these alkyl phenols (Bina et al. 2018, Inoue et al. 2003). Nonylphenol and its ethoxylates have been included in wide range of goods, like fabrics, coatings, paints and resins. Domestic products, cosmetics, and cleaning products are also formed of nonylphenol ethoxylates. Even though its use is restricted in European Union, regular use is still permitted worldwide (Jonkers et al. 2010). Phthalates have been detected in around 50 out of 70 cosmetic products ranging between 50 μgg^{-1} to 3% of whole product (Houlihan et al. 2002). Extensive use of phthalates have caused their distribution in aquatic bodies and sediments (Roslev et al. 2007, Chaler et al. 2004).

2.3 Environmental Fate and Challenges Caused by PCPs

Concern regarding environmental safety and human health has drawn the attention of researchers toward the PCPs fate. PCPs have been found far and wide in groundwater, surface and wastewater. Countries, such as Spain, US, Germany, England, China and Australia, have reported the occurrence of PCPs in water resources (Cabeza et al. 2012, Barnes et al. 2008, Reh et al. 2013, Peng et al. 2014, Liu et al. 2011). Residues of PCPs can enter into the food chain through waste water effluents or agricultural effluents (Rajapaksha et al. 2014). Other than direct exposure of PCPs, their metabolites could act as secondary pollutants (Yang et al. 2017). Their toxicity has been detected in invertebrates, aquatic animals, and plants. However, invertebrates are more sensitive to its toxicity. However, upon long term exposure aquatic animals can become resistant to the effects of PCPs, except algae and invertebrates. Disruption of lipid synthesis through PCPs by fatty acid synthesis or membrane destabilization causes their higher toxicity in algae and invertebrates (Brausch and Rand 2011). Musks are found to have acute toxicity limits (0.15 mgL^{-1}) to aquatic animals. Zebrafish upon longer exposure in its early life was detected to be most sensitive towards nitro musks (Carlsson and Norrgren 2004). However, the metabolites of nitro musks are highly toxic for aquatic species

(Daughton and Ternes 1999). Polycyclic musks also have adverse effects on invertebrates in comparison to other aquatic species (Dietrich and Chou 2001).

Parabens with a higher chain length have a considerable toxic effect on bacteria (Eklund 1980). However, the chronic effects of parabens are not well known till now on aquatic organisms (Dobbins et al. 2009). Out of various parabens, benzyl and butyl parabens are most hazardous to invertebrates and aquatic species, while lower members are least toxic. Chlorination of parabens can also increase their toxicity in bacteria (Terasaki et al. 2009). Higher stability and lipophilicity of UV filters has caused their bioaccumulation in fish (Balmer et al. 2005). In lipid tissue of fish, around 2 ppm of UV filters were detected (Nagtegaal 1997). Bioaccumulation factor of UV filters was more than 5,000 in fish (Hany and Nagel 1995). Zebrafish was most sensitive to short term exposure of UV filters (Fent et al. 2008). On exposure of UV filters, reproduction was effectively lowered in *P. antipodarum* and mortality was increased in Lumbriculus variegates upon exposure of 28 days (Schmitt et al. 2008). UV filters have also shown estrogenic activity in studies (Schlumpf and Lichtensteiger 2001, Kunz and Fent 2006). Fecundity and reproductivity was also affected adversely in some fish (Fent et al. 2008).

There are very few reports on the possible risks of PCPs on human health. However, it is well known that humans are under constant and close contact with PCPs (Cruz and Barcelo´ 2015). These have been detected in different human samples. For example, lipids in human breast and adipose tissue have ngg^{-1} concentration of fragrances (Yin et al. 2012, Kannan et al. 2005). Triclosan was detected at $ngmL^{-1}$ level in urine samples and at ngg^{-1} lipids in liver and brain tissue (Asimakopoulos et al. 2014, Geens et al. 2012). Triclosan can degrade into dioxins and exhibit toxicity to bacteria (Ricart et al. 2010). Parabens were found in urine, human milk, and breast tissue (Schlumpf et al. 2010, Darbre et al. 2004). UV filters have been detected in semen, urine, and human milk (Schlumpf et al. 2010). Triclosan causing bacterial resistance over antibiotics can initiate allergies in children (Bertelsen et al. 2013). Triclocarban can encourage the production of methemoglobin by transforming to primary amine in blood by heat (Johnson et al. 1963). Fragrances are associated with various health effects in humans, such as allergies, headaches, asthma and mucosal symptoms (Elberling et al. 2005). Long term exposure of fragrances can cause obesity, disrupt homeostasis, affect the immune, and enhance diabetes risk (Svensson et al. 2011). Other PCPs can result in seizures, coma, impaired cognitive functions, behavior change, and restlessness followed by death or neurotoxic effects (Abou-Donia 1996, Kim et al. 2009). Parabens have also been related to breast cancer, while phthalates to cancer in mice and rats (Darbre et al. 2004, Ito et al. 2007). All the above facts make it clear that there is a growing concern regarding the potential effect of PCPs on environment and life.

3. REMOVAL STRATEGIES OF PCPS

A recent concern of environmental hazards of PCPs has led to the development of various removal processes (Liang et al. 2014, Nakada et al. 2007). In general, conventional removal processes, such as filtration, screening, sedimentation, and aeration tanks are commonly used. However, these processes are ineffective in eliminating the PCPs completely from the effluents (Carballa et al. 2004). Eradication of PCPs in treatment plants is a complex process as it depends on the various properties (solubility, hydrophilicity, adsorption capacity degradability, and volatility) of pollutants (Evgenidou et al. 2015, Jones et al. 2005, Liu and Wong 2013). Moreover, these processes are capable of partial removal of PCPs. Hydrophilic nature of PCPs make their removal through the conventional processes a tricky task (Luo et al. 2014). Adsorption has been considered as a most relevant mechanism of PCPs removal (Suárez et al. 2008). PCPs such as fragrances can be effectively removed through primary processes due to its high partition capacity between different phases of matter (Stamatis and Konstantinou 2013). Advanced methods, such as nanofiltration, oxidation processes, and reverse osmosis are discovered for effective removal of PCPs (Liang et al. 2014, Nakada et al. 2007, Yoon et al. 2006).

Some of the advanced filtration techniques (ultra-filtration and microfiltration) were efficient in the removal of PCPs. However, the only limitation restricting the 100% success of this process was pore sizes (Watkinson et al. 2007). Some of the smaller pollutants were capable of passing through the membrane, thus causing low removal efficiency of the process (Schäfer et al. 2011). Nanofiltration, on the other hand, has a limitation of inefficiency under highly concentrated pollutants flow. Advanced processes like photocatalysis, Fenton's reaction, and ozonation can be highly effective in PCPs treatment (Klavarioti et al. 2009, Gerrity et al. 2010). These processes by changing the polarity and functional groups of PCPs can convert them into harmless products (Papageorgiou et al. 2014). Coupled or combined processes can further eliminate PCPs from the environment. The involvement of UV or UV/H_2O_2 is believed to enhance the efficiency of the overall removal process (Yang et al. 2016). A wide range of processes has been reported to combine the conventional process with advanced materials to improve the effectiveness of the eradication method (Li et al. 2015, Sun et al. 2014a, b).

3.1 Introduction to Nanomaterials

Nanotechnology has revolutionized recently and become a new wave of innovation. In the next two decades, it will become part of every individual on the planet through different industries. Nanomaterials are formed by the manipulation of atoms or clusters in a defined structure having new or different properties. Out of various ranges of nanomaterials, materials having dimensions between 1 to 100 nm are of more interest, owing to their extraordinary properties and advantages. The fascinating properties of nanomaterials occur because of their large surface activity and quantum effects. These properties have linked nanomaterials for the treatment of various environmental pollutants through techniques, like adsorption, filtration, and photocatalytic degradation. PCPs have been recently treated with various nanomaterials-based processes. Adsorption of PCPs through nanomaterials has shown high efficiency owing to their small size, higher surface area, and large number of active sites. Nanomaterials of graphene, metals, nanoclays, and metal oxides have been used as an alternative and advanced source for the removal of contaminants from wastewater. Nanomaterials through advanced oxidation processes can completely mineralize the pollutants. Oxides of various transition metals have been employed for photocatalytic removal of pollutants at a large scale (Bagheri et al. 2016). Nanoclays have been widely utilized in the adsorption of antibiotics (Rafati et al. 2018). Nanoparticles of hematite were highly effective in the removal of pharmaceutical wastes (Rajendran and Sen 2018). Photocatalysis has attracted attention for PCPs removal because of its enhanced effectiveness. Photocatalysts exhibit exclusive properties of surface area and surface states. Moreover, photocatalysts can be modified through coupling or functionalization of other materials. Upon coupling, the synergistic effect of coupled materials can break down the PCPs into smaller or safer products. Nanophotocatalysts, in presence of light, generate reactive species, such as hydroxyl radical, superoxide, and h+. These reactive species can further oxidize or break down the bulky molecules of pollutants into smaller products. In the present chapter, the removal of PCPs using nanomaterials having crystalline properties is described.

3.2 Liquid and Crystal Nanomaterials

The nanostructured materials have undoubtedly acquired attention in the academia and industrial sector owing to their excellent magnetic, electrical, chemical, surface, and semiconducting properties (Kumar 2001, Yang and Wu 2006, Coles and Morris 2010, Lagerwall and Scalia 2012, Beeckman et al. 2011, Shibaev et al. 2015, Gajanan and Tijare 2018). However, other than commonly known nanomaterials, liquid and crystal nanomaterials (LCN) can lead to an abundance of new and potential applications. The composite materials of LCN have improved macroscopic properties and exhibit high performance. The main objective of the formation of LCN is to amend the properties of materials and to take benefits out of them that are absent in basic materials (Yang and Xiang

2008, Xu and Lee 2014, Shiraishi et al. 2002). A comparison of different phases of matter has been shown in Figure 1.

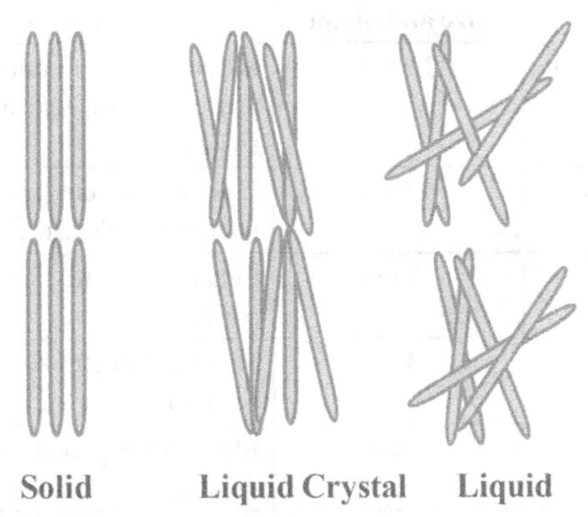

FIGURE 1 Schematic representation of different phases of liquid crystals.

The long-range orientation and connections in mesophases have shown an impact on LCN as a whole (Lee and Shih 2010). However, the properties, such as thermal, catalytic, electrical, mechanical, optical, and electrochemical, depend on the dopant material and its morphology and method of preparation of LCN. Generally, the formation of LCN involves metals, semiconducting materials, carbon, and ferroelectric materials as dopants. Liquid and crystal materials are referred to as the fourth state of matter. These are soft materials exhibiting order and mobility on different molecular levels. Their unique identity has found application in information, energy, environment, health care, and many other fields. The mesophases are due to the presence of both crystalline and an isotropic phase in these. Thermodynamic stability of LCN is due to organic or inorganic steady units used in its formation (Goodby et al. 2014). In general, materials exhibiting greater anisotropy can be referred to as LCN. These can have different shapes consisting of units of rods or discs orienting around an axis (Kumar 2001). Many researchers are focusing on the development of hybrid materials, like LCN, to increase their functionality beyond one particular sector.

Photothermal properties of nanoparticles are in limelight because of the distant and noncontact release of energy to different systems to show diverse functions. For instance, nanorods of gold on experiencing laser light absorb the light (matching surface plasmon resonance) and convert it into heat. In this regard, functionalized nanorods have been used to attain near-infrared beam-driven phase changes in LCN (Jayalakshmi et al. 2007). Such materials are gaining attention due to their potential and convenience. In order to increase the existing properties of materials, carbon nanotubes, gold nanoparticles, and graphene derivatives have been coupled with liquid crystalline polymers (Yang et al. 2015, Sun et al. 2012a, b, Wang et al. 2012, Wang et al. 2016, Kohlmeyer and Chen 2013, Li et al. 2012, Liu et al. 2015, Koerner et al. 2004, Ahir et al. 2005, Li et al. 2011). In Table 2, recent developments in the area of LCN have been listed.

Liquid crystalline elastomers of CNT fabricated that exhibited reversible actuation in infrared light (Yang et al. 2015). This phenomenon resulted from the absorption of IR radiation and the generation of heat by CNT and elastomer matrix, respectively. Orientation of material dispersed in the matrix is a critical requirement to exhibit anisotropic properties. Due to this particular requirement polymers and elastomers are mostly employed as the matrix (Wang et al. 2012). Other than CNTs, materials, such as graphene oxide, can also exhibit photomechanical responses (Yu et al. 2015, Yang et al. 2015a, Yu et al. 2014).

LCN can evolve as an independent venture depending upon its synthesis and design. LCN whose properties can be reversed through the light via dynamic gathering and dispersion may bring in a new direction in the field of nanomaterials. The utilization of alteration of nanoparticles and anisotropic materials seems a bright scenario. In this regard, chemists can play a critical role in the formation of engineered light-responsive LCN.

TABLE 2 Recent developments in the area of liquid and crystal nanomaterials.

LCN	Size and Shape	Properties	Reference
$CuInS_2$-ZnS and Nematic liquid crystal	2.2 nm	Superior electro-optical properties with controlled device switching	Lee et al. 2013
$CuInS_2$-ZnS and Ferroelectric liquid crystal	6.5 nm	Low spontaneous polarization, nonlinear dielectric properties	Singh et al. 2018
InP/ZnS and Nematic liquid crystal	~3 nm	Enhanced birefringence and conductivity	Roy et al. 2018
Co-ZnO/ZnO and Ferroelectric liquid crystal	~4 nm	Enhanced electro-optical properties	Doke et al. 2020
ZnO quantum dots and Ferroelectric liquid crystal	9 nm	Blue-shift in photoluminescence spectra	Joshi et al. 2014
ZnS/Mn-doped ZnS and Ferroelectric liquid crystal	~3 nm	Non-linear photoluminescence response	Singh et al. 2016
ZnS quantum dots and Ferroelectric liquid crystal	-	Enhanced photoluminescence	Vimal et al. 2016
Carbon dots and Phenylpyrimidine-based ferroelectric liquid crystal	10 nm	Spontaneous polarization, dc conductivity and dielectric constant	Shukla et al. 2015
Carbon dots and ferroelectric liquid crystal	4-5 nm	Modulated dielectric properties	Gangwar et al. 2019
$CsSnI_3$ QDs and Nematic liquid crystal	3-5 nm	Long term stability	Chen et al. 2018
$CsSn_{0.25}Pb_{0.75}Br_3$	Orthorhombic 3D	Band gap = 1.65 eV	Xian et al. 2020
$MASnI_3$	Cubic, 3D	Band gap = 1.15 eV	Dang et al. 2016a
$FASnI_3$	Cubic, 3D	Band gap = 1.4 eV	Dang et al. 2016a
Bi-doped Cs_2SnCl_6	Cubic	Band gap = 3 eV	Tan et al. 2018
$NH(CH_3)_3SnBr_3$	Orthorhombic	Band gap = 2.75 eV	Dang et al. 2016b
$NH(CH_3)_3SnCl_3$	Monoclinic	Band gap = 3.3 eV	Dang et al. 2016b
$Cs_3Bi_2I_9$	Hexagonal	Band gap = 2 eV	McCall et al. 2017
$Ru_3Bi_2I_9$	Monoclinic	Band gap = 1.9 eV	McCall et al. 2017
$MA_3Bi_2I_9$	Centrosymmetric	Band gap = 1.96 eV	Liu et al. 2020
$Rb_7Bi_3Cl_{16}$	Trigonal	Band gap = 3.27 eV	Xie et al. 2019
$MA_3Sb_2I_9$	Hexagonal	Band gap = 1.92 eV	Ju et al. 2018
$Cs_3Sb_2Br_9$	Trigonal	Band gap = 2.36 eV	Zhang et al. 2017
$Cs_3Sb_2I_9$	Defect structure	Band gap = 1.89 eV	McCall et al. 2017
$Ru_3Sb_2I_9$	Monoclinic	Band gap = 2.03 eV	McCall et al. 2017
$Cs_2AgBiBr_6$	Cubic	Band gap = 1.95 eV	Yuan et al. 2019
$(BA)_2CsAgBiBr_6$	Monoclinic	Band gap = 2.38 eV	Xu et al. 2019
$Cs_3Bi_2Br_9$	Trigonal	Emission peak = 468 nm	Yang et al. 2017
$Rb_7Bi_3Cl_{16}$	Trigonal	Emission peak = 437 nm	Xie et al. 2019
$Cs_3Sb_2Br_9$	Trigonal	Emission peak = 410 nm	Zhang et al. 2017
Cs_2PdBr_6	Cubic	Emission peak = 767 nm	Zhou et al. 2018
Cs_2ZnX_4 (X= Cl/ Br-Cl/ Cl-I)	Orthorhombic	Emission peak = 391 nm	Shankar et al. 2020

3.3 Liquid and Crystal Nanomaterials in Eradication of PCPs

Conventional removal techniques of PCPs are not fully successful due to complication strategies, cost, low efficiency and formation of hazardous by-products. Photocatalysis could be the one solution to neglect the above restrictions having reactive species formed *in situ* to degrade the pollutant (Pendergast and Hoek 2011). Earlier, adsorption use to work under dark conditions where a catalyst with active surface sites was allowed to absorb the pollutant molecules.

PCPs, such as triclosan and triclocarban, were recently removed effectively from water using core-shell iron oxide coupled covalent organic framework nanocomposites. The adsorption of targeted PCPs was monitored using liquid chromatography. Results showed high adsorption of PCPs following multilayer adsorption through Van der Waals force and π–π stacking. Triclosan was removed up to 82% and triclocarban to 92% with 10 times reusability of the process (Li et al. 2019). Removal of ibuprofen was tested using Fe/C granules. The composite of granules and sodium persulfate successfully removed 96% of ibuprofen involving the formation of eight intermediates. The intermediates are shown in Figure 2 (Yu et al. 2020).

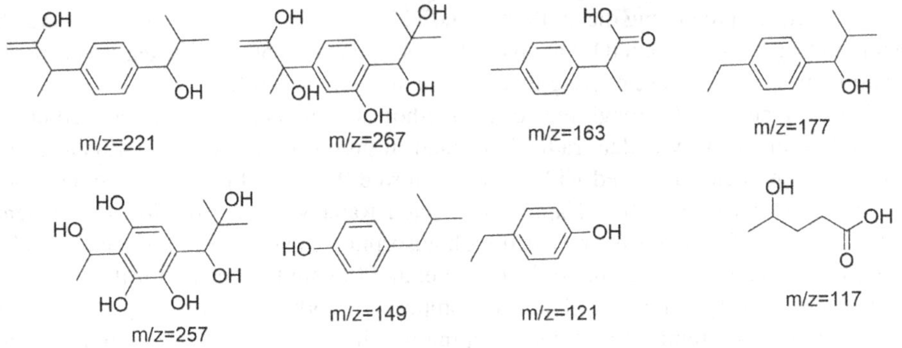

m/z=221 m/z=267 m/z=163 m/z=177 m/z=257 m/z=149 m/z=121 m/z=117

FIGURE 2 Intermediates detected after ibuprofen removal.

Meloxin and Naproxen were recently investigated for adsorption using Fe_3O_4@MIL-100(Fe) under different pH conditions. The adsorption followed pseudo first-order kinetics showing that the process was influenced by the chemical reaction. Adsorption was multilayer, spontaneous and endothermic (Liu et al. 2019). Photochemical degradation of four parabens was investigated using crystalline titanium dioxide under UV light. However, at higher concentrations of PCPs degradation efficiency of nanocatalyst was slow (Cuerda-Correa et al. 2016). Diclofenac was degraded using zinc-ferrite nanomaterial using the redox character of nanocatalyst. Within the two minutes of reaction, 90% of diclofenac was degraded through the reduction by electrons and radicals formed on catalysts surface. The degradation pathway of the contaminant has been shown in Figure 3 (Al-Anazi et al. 2020).

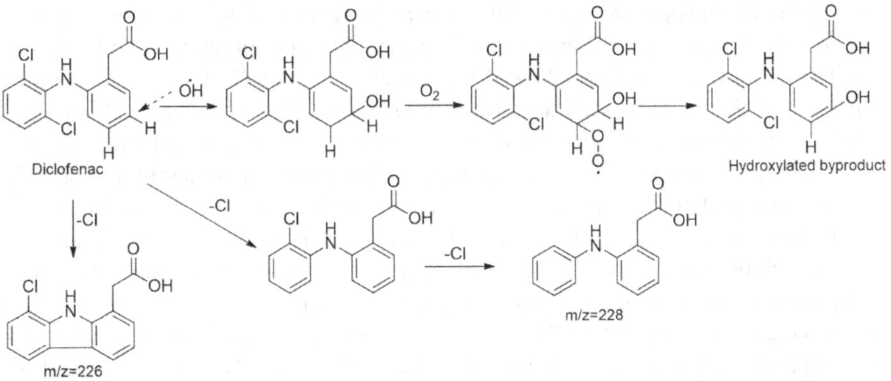

Diclofenac Hydroxylated byproduct m/z=228 m/z=226

FIGURE 3 Degradation pathway of diclofenac.

As shown in the figure, the phenylacetic acid ring having high electron density on the attack of hydroxyl radical resulted in the hydroxylated product. The other paths showed the different by-products formed upon the removal of chlorine. Ofloxacin was photo degraded under sunlight with silver-modified zinc oxide nanoparticles in 150 min (Kaur et al. 2018). Fe doped ZnO nanocatalyst was used for PCP removal under sunlight at three different concentrations of nanocatalyst. Maximum degradation ability of nanocatalyst was observed at moderate concentration due to the formation of reactive oxygen species under these conditions (Das et al. 2018). Similarly, ZnO-based nanocomposites, such as $ZnO\text{-}Ag_2O/pg\text{-}C_3N_4$, were investigated for degradation of PCPs. The lower bandgap of nanocomposite (2.6 eV) was helpful in the quick degradation of PCPs (Rong et al. 2016). A graphene-based heterogeneous catalyst was utilized for the oxidation of ciprofloxacin at neutral pH. The process showed many times higher activity than the bare graphene. The 99% of PCP was removed in 120 minutes at neutral pH (Divyapriya et al. 2018). Antibiotic, sulfachloropyridazine, was adsorbed on N/CNT/Ni nanocatalyst, such that doping of nitrogen and nickel enhanced the overall stability of the catalyst (Kang et al. 2018). Antibiotics, namely amoxicillin, ampicillin, and penicillin, were removed using bentonite-supported iron and nickel-based nanocatalyst. The catalyst was active in removing the antibiotics individually; however, lower efficiency was found in a mixture (Weng et al. 2018). SnO_2 nanoparticles showed excellent activity against carbamazepine degradation for up to eight cycles (Begum and Ahmaruzzaman 2018).

$Co/BiFeO_3$ synthesized through the sol-gel method with higher surface area decomposed 81% of tetracycline in water by sulfate radicals formed in presence of persulfate (Zhang et al. 2018). Iron oxide nanoparticles activated with carbon removed 97% of 10 ppm ceftriaxone at acidic pH in 90 minutes (Badi et al. 2018). Naproxen affected water was treated with Si/Graphene oxide nanoparticles at pH 5. The results revealed high adsorption capacity (83%) of nanoparticles in the process (Mohammadi Nodeh et al. 2018). Rafati et al. 2018 studied the removal of ibuprofen with functionalized nanoclay composite. The nanocomposite adsorbed 95% of 10 ppm of PCP within 2 hours. Hematite-based removal of carbamazepine was investigated by Rajendran and Sen (2018). Removal of PCPs was also studied by transition metals doped TiO_2 nanocatalyst in visible light (Inturi et al. 2014). In one more report non-metal doped TiO_2 was utilized in the process (Lin and Yang 2014).

Fiber membranes-based nanofiltration cross-linked with polymers having high energy conservation and separation efficiency was also employed for the PCPs removal (Yee et al. 2014). The nanofilters were capable of adsorbing small PCP molecules through their comparable pore size (Yu and Zhang 2015, Zhu and Tian 2015). These techniques were effective in removing carbamazepine, known as the most often detected PCP in the water (Roberts et al. 2016, Rohricht et al. 2010). Nanoparticle-based membranes can remove other PCP, such as bisphenol A and phthalates from water (Hu and Zhang 2011). Activated carbon and titanium dioxide-based composites have been widely studied for PCPs degradation (Basha et al. 2011, Gar Alalm et al. 2016, Khraisheh et al. 2013, Quinones et al. 2014, Rosa et al. 2017). The uniform pore structure and stability of these catalysts make their utilization apt for the purpose (Ma et al. 2017). Khraisheh et al. (2013) fabricated activated charcoal using coconut shell material, a cheap source of carbon with superior quality and formed composite material. The as-formed composite showed highest efficiency (90%) for carbamazepine removal. The properties of the catalyst, such as surface area, active sites, pore size and crystal structure, influenced the overall removal process. It was suggested that pharmaceuticals, like amoxicillin and diclofenac, can be completely removed within 180 minutes under sunlight (Gar Alalm et al. 2016). The overall mechanism involves the transfer of absorbates on activated carbon to photoactive metal oxide through the interface (Leary and Westwood 2011). In one more study, amoxicillin and a b-lactam antibiotic were photos degraded using activated carbon/titanium dioxide (Basha et al. 2011). The chemisorption took place between amoxicillin and active groups on the nanocatalyst surface involving electron exchange (Basha et al. 2010). Nitrogen-doped carbon nanotubes and metal oxide-based photocatalyst degraded ibuprofen under visible light. The involvement of nitrogen in the catalyst was to lower the bandgap energy. The overall synergy of the catalyst resulted into strong binding (Yuan et al. 2016, Murgolo et al. 2015). Similarly, CNT based metal oxide catalyst was employed for the light-assisted degradation of 4-chlorophenol. The CNT lowered the recombination of active species by acting as an electron sink (Zouzelka et al. 2016). The LCN based processes are beneficial because of (a) easy adaptation for continuous process, (b) limit catalysts agglomeration, and (c) no special separation required (Dong et al. 2014).

$BiFeO_3$ and CdSe coupled catalyst through photo Fenton's reaction degraded phenol effectively under visible light. The synergistic effect of the constituents resulted in quick degradation of phenol from 42% to 98% with one hour (Huang et al. 2020). Sunlight irradiated phenol degradation was investigated using Ag/TiO_2 nanofibers. Findings revealed that degradation depends on four factors, such as contaminants concentration, pH, catalyst dosage, and interaction between both catalyst and phenol. Phenol concentration of 5.6 ppm was removed up to 92% at neutral pH (Norouzi et al. 2020).

Biochar-based composite using $CoFe_2O_4$ was activated with peroxydisulfate and utilized for bisphenol a (BPA) degradation. BPA having 100 ppm concentration was degraded up to 95% using this strategy within eight minutes. The overall reaction of the process was eight times faster than without a catalyst. Synergism in the constituents of the composite activated its overall efficiency via a free radical-based pathway (Li et al. 2020).

Spinel of cobalt ferrite was applied for the removal of BPA from the water. It was observed that 45 μM of BPA was eliminated up to 97% within 40 minutes from water. The overall activity of the catalyst was attributed to its higher surface area and higher dose of cobalt content. Mineralization of BPA was also observed up to 81%. The primary species, like hydroxyl radical, encouraged the degradation of BPA as shown in Figure 4 (Long et al. 2021).

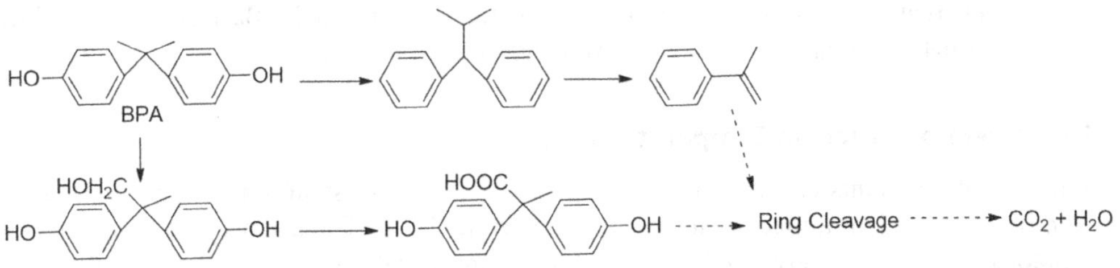

FIGURE 4 Degradation pathway of Bisphenol A.

Titanium dioxide is one of the crystal nanomaterials highly applied for PCPs degradation. TiO_2@ nanodiamond under UV light showed complete removal of 10 ppm BPA in 100 minutes of reaction. The degradation pathway of BPA has been shown in Figure 5 (Hunge et al. 2021).

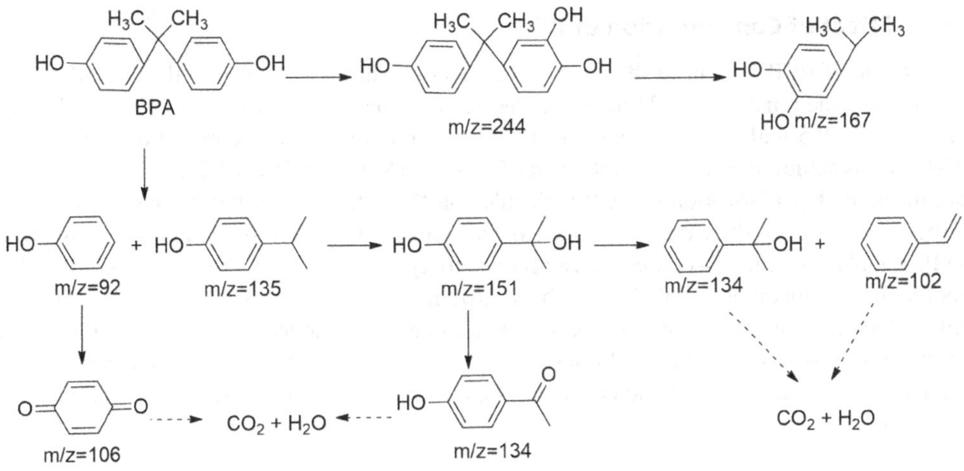

FIGURE 5 Proposed mechanism of BPA degradation.

3.4 Factors Affecting Eradication of PCPs

As discussed above the degradation PCPs depends on various catalysts depending as well as system depending factors. Various factors affecting the eradication of PCPs are discussed below.

3.4.1 Effect of pH

The nature of the surface of the catalyst has a lot to do with the overall process. The surface charge depends on the pH of the catalyst making it the most significant operating parameter (Chong et al. 2010). The overall charge of catalysts could be positive or negative owing to the amphoteric nature of most of the metal oxides (Mirzaei et al. 2016). The effect of pH is generally optimized by the point of zero charges of the catalyst. The point of zero charges refers to the pH at which the charge on the material surface is zero. It mostly depends on the surface chemistry of the catalyst. Various reports have found the point of zero charge of catalyst ranging between 4-5, 4-6, 6-8, and 7-9 for CNT-TiO_2, AC-TiO_2, TiO_2, and Graphene-TiO_2, respectively (Ahmadi et al. 2016, Hua et al. 2016, Andronic et al. 2014, Malekshoar et al. 2014). The reaction of the catalyst leading to its protonation and deprotonation depends on the pH. At acidic pH photogenerated holes dominate as oxidizing agents; however, at neutral or basic pH, hydroxyl radicals become dominant in oxidation (Lee et al. 2016). The photocatalytic reaction could be inhibited under basic pH due to hydroxyl anions scavenging the holes. The highest removal activity of CNT-TiO_2-Urea was obtained at pH 5 for PCP (Yuan et al. 2016). The point of zero charges of the catalyst was six, such that under acidic conditions PCP could deprotonate to be negatively charged conjugate base (Hashim and Khan 2011). Change in the pH can create electrostatic attraction amongst the PCP and LCN (Basha et al. 2011). Hence, adjustment of pH is critical in order to achieve valuable PCP removal.

3.4.2 Effect of Reaction Temperature

Eradication of pollutants is affected by the temperature of the system. Photocatalysis is one such system that can work at ambient temperature (Lee et al. 2016). The average reaction temperature for photocatalysis ranges between 20 to 80°C (Herrmann 1999). In photocatalysis, the temperature influences the electron and hole diffusion/recombination rates as well as the extent of adsorption (Sarkar et al. 2014). In case the temperature is greater than 800°C, system adsorption can then decrease with the recombination of charge carriers (Chong et al. 2010, Gaya and Abdullah 2008). On the other hand, under lower temperature conditions activity of the catalyst may decrease owing to the large activation energy required to excite the catalyst (Herrmann 1999).

3.4.3 Effect of Concentration of PCP

The amount of PCP present in the reaction has a significant role to play in the overall activity of the method. One report on PCP removal suggested that the highest efficiency (95%) of the catalyst was at low as 0.5 mgL^{-1} concentration of PCP (Ahmadi et al. 2016). Removal of metoprolol with AC-TiO_2 was achieved at a concentration was 50 mgL^{-1} (Rey et al. 2012). Higher PCP concentration lead to the higher adsorption of PCP molecules on the catalyst and that ultimately inhibited the further adsorption on the catalyst surface (Mirzaei et al. 2016). Moreover, light scattering through the PCP molecules can also lower the catalytic activity of the catalyst (Ahmad et al. 2016). Earlier reports on the remediation of PCPs have been experimenting with a concentration as high as 1,000 mgL^{-1}. Although, high concentration-based studies may not be appropriate to predict the activity of the method applied under real conditions (Gulyas et al. 2016). It is important to learn the catalytic activity of the catalyst at realistic concentrations (mg L^{-1}) to limit the restraining effects of high PCP load.

3.4.4 Effect of Dissolved Oxygen

Photocatalysis depends upon the content of dissolved oxygen in the water sample. Dissolved oxygen acts as an acceptor of electrons and manages the formation of hydroxyl ions (Hoffmann et al. 1995). The overall rate of photocatalysis tends to boost with the rise in dissolved oxygen amount. It helps in the formation of superoxide ions and limits the recombination of charge carriers (Martínez et al. 2011). However, the formation of hydroxyl ions still mainly depends on the water content in the reaction (Kondrakov et al. 2016). Other than water, amount of hydroxyl ion depends on the available holes rather than electrons to react with oxygen to form superoxide ions.

3.4.5 Effect of Natural Organic Matter

Organic matter in the reaction is a mixture of substances, such as proteins, lipids, acids, humic substances, amino acids, and carbohydrates (Ateia et al. 2017, Shimizu et al. 2018). The performance of catalysts depends upon the organic matter as studied in various reports (Van Doorslaer et al. 2015, Drosos et al. 2015). It was observed that at high organic matter content overall activity of the LCN could be lower, while its minimal amount has a negligible effect on the overall process (Drosos et al. 2015). The species formed from the organic matter at their low concentration, such as hydroxyl ion, superoxide ion, and solvated electrons help in enhancing the degradation of PCPs through energy transfer (Doll and Frimmel 2003). However, their ability to absorb and scatter light can lower the degradation rate for PCPs (Frimmel 1994). Lower organic matter content can also create p-p interactions between LCN and PCP, while higher content can prevent PCP from reaching the surface of LCN (Drosos et al. 2015). The inhibitory effect of organic content could be counteracted by the phosphate and higher dissolved oxygen content (Long et al. 2021, Ren et al. 2018).

3.4.6 Effect of Inorganic Species

Other than organic matter, the amount or presence of inorganic species can also affect the photocatalytic activity of LCN in PCP removal. Inorganic species, such as hydrogen peroxide, peroxydisufate ion, bromate ion, and sulfite ion, have a beneficial role in enhancing the removal of PCPs by acting as a scavenger for electrons, thus boosting the production of hydroxyl radicals/ other species (Lee et al. 2016). Although unnecessary concentration rise of inorganic species can limit the degradation rate by scavenging the hydroxyl radicals. In one report activity of titanium dioxide was highly affected by the presence of multiple inorganic species, such as NaCl, $CaCl_2$, $FeCl_3$, $Fe_2(SO_4)_3$, $NaHCO_3$, $Al_2(SO_4)_3$, Na_2SO_4, $FeCl_2$, and Na_2CO_3) because of the struggle for free radicals, catalyst blockage, and change in the pH of reaction (Rioja et al. 2016). The presence of a solution of iron in the reaction causes the attenuation of UV light and forms the overall solution of yellowish color. Scavenging effect of sulfates/carbonates was found to be stronger than chlorides. However, the photocatalytic activity can be controlled by ionic strength, such that at higher ionic strength repulsion between the contaminants could be lowered and adsorption raised (Aguedach et al. 2008). Thus, it can be concluded that the overall effect of inorganic species on the activity of the catalyst cannot be ignored and pretreatment is required to remove these for effective treatment.

3.5 Degradation Mechanism of PCPs

Yu et al. (2020) showed that the degradation rate of catalyst could be enhanced by activating it with secondary materials. The coupled catalyst can degrade the PCPs through an oxidation reaction involving various active species. The active species can destroy the bulky molecules of PCP and finally convert them into CO and H_2O. The degradation mechanism of PCP using Fe/C granules and peoxydisulfate has been shown in Figure 6.

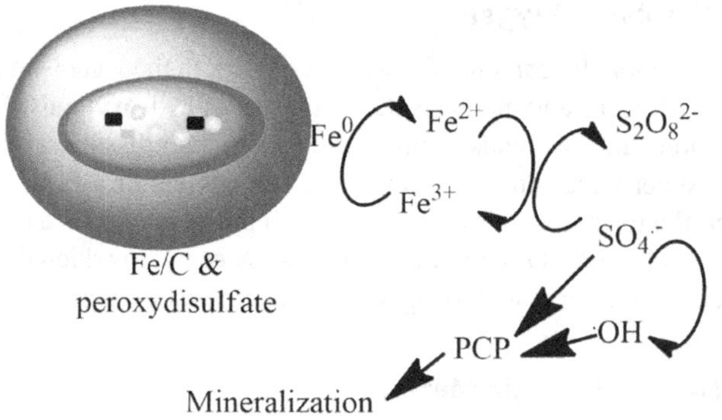

FIGURE 6 Degradation mechanism of personal care product with nanocatalyst.

The oxidizing species, such as hydroxyl radicals, superoxide ions, and other radical species, are consumed during the reaction with not only PCPs but other organic matter present in the reaction mixture. Other than this, the scavenging ions of inorganic species can also limit the role of reactive species. The findings were reported by various scientific communities. In one study, TiO_2 nanoparticles were used to degrade clofibric acid from the aqueous matrix. The removal efficiency of catalysts was highly affected by the unwanted complexes (NaCl, $NaHCO_3$, humic acids, and surfactants t) in water samples (Cuerda-Correa et al. 2016).

Degradation of ofloxacin with silver doped ZnO included the silver as electron sink helping in better charge separation and thus increasing the number of radical species in the reaction mixture. The more number hydroxyl radicals present in the reaction can help in the quick degradation of PCPs. Similar degradation mechanisms were obtained for other PCPs (Kaur et al. 2018, Das et al. 2018). The surface chemistry of coupling materials has also an important role in the formation of oxidants. In some cases, activated carbon acted as a photocatalyst and promoted the electrons for the generation of hydroxyl ion in the presence of light source. Various concluding factors for such studies reveal that (a) higher surface oxygen can limit the generation of hydroxyl ion, (b) under light source formation of superoxide ion could be enhanced through physisorbed water, (c) higher the number of carboxyl groups, higher is the concentration of hydroxyl ion, (d) aromatic ring and presence of oxygen group can stabilize the photogenerated electrons, and (e) lower the bandgap, higher the production of hydroxyl ion. Based on the above factors, the degradation rate of the LCN could be enhanced for PCPs removal. The degradation mechanism of one such factor depending system is shown in Figure 7 (Huang et al. 2020).

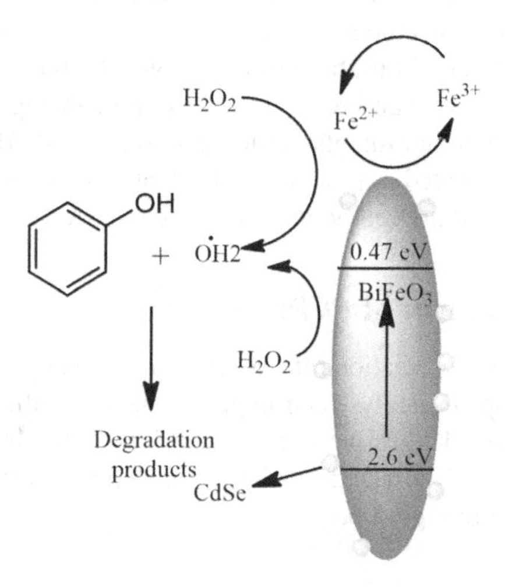

FIGURE 7 Radicals based degradation mechanism of personal care product.

4. CONCLUSIONS AND FUTURE SCOPES

This chapter provides the thorough applicability of LCN materials in the removal of personal care products. There are very few reports on the exhaustive literature survey of degradation of PCP with the LCN materials. Hence, the chapter provides the required one-hand information of LCN in the remediation of toxic PCPs. Various conclusions and future research suggestions have been drawn from this review:

- Liquid and crystalline nanomaterials are capable of increasing the overall available features of nanotechnology and can play a vital role in the remediation of environmental pollutants.
- A lot of liquid and crystalline nanomaterials (coupled with metal oxides, graphene oxide, carbon and other non-metals) were fabricated and applied in various applications of environmental concern.
- Doping and synergism play important roles in better properties of LCN, such as beneficial bandgap energy, limit charge recombination, increase the overall surface active sites and charge carriers.
- Mostly oxides of transition metal oxides, like zinc, iron, and titanium, were used in environmental applications, such as degradation of pollutants from water, soil and sediments. More attention is needed to be paid to the other metal oxides and nonconventional materials to form LCN so that bio-saturation of one kind of material could be restricted.
- Liquid and crystalline materials and the various factors affecting the efficiency of the materials were investigated in the present chapter.
- The fabricated materials were able to remove pollutants, such as parabens, pesticides, and phenols from the environment completely.
- Limited reports were able to thoroughly investigate the degradation mechanism of personal care products and provide the possible pathway of reactive species formation.
- A very few reports were available on the utilization of non-conventional metal oxides in the formation of liquid and crystalline nanomaterials.
- Limited reports were available on the degradation of personal care products with the LCN materials.
- Although efforts have been made to explore the various application of LCN in the environmental industry, yet a lot of work is required to have a full profit from the excellent properties provided by the LCN materials.
- Complete knowledge of the LCN materials fabrication is lacking and simple/cost-effective methods development is required.
- To achieve the desired results from the LCN materials, a better understanding of its features is required. Materials stability under various environmental factors, such as temperature, pH, and the solvent is needed to be known.
- Limited data is available on the application of LCN materials in PCPs removals such as fragrances, parabens, and screening materials. Moreover, most of the results were based on the lab reports, field-based investigation of LCN activity should be investigated to receive the real-life profits of the materials.
- Expansion of more simplistic, one-step, and cost-effective processes for industrial application is highly recommended.

Acknowledgments

Authors are grateful to the Ministry of Human Resources Development for financial assistance.

References

Abou-Donia, M.B. 1996. Neurotoxicity resulting from coexposure to pyridostigmine bromide, DEET, and permethrin: Implications of Gulf War chemical exposures. J. Toxicol. Environ. Health Part. A 48: 35-56.

Aguedach, A., Brosillon, S., Morvan, J. and Lhadi, E.K. 2008. Influence of ionic strength in the adsorption and during photocatalysis of reactive black 5 azo dye on TiO_2 coated on non woven paper with SiO_2 as a binder. J. Hazard Mater. 150: 250-256. https://doi.org/10.1016/j.jhazmat.2007.04.086.

Ahir, S.V. and Terentjev, E.M. 2005. Photomechanical actuation in polymer-nanotube composites. Nat. Mater. 4: 491-495.

Ahmad, R., Ahmad, Z., Khan, A.U., Mastoi, N.R., Aslam, M. and Kim, J. 2016. Photocatalytic systems as an advanced environmental remediation: Recent developments, limitations and new avenues for applications. J. Environ. Chem. Eng. 4: 4143-4164. https://doi.org/10.1016/j.jece.2016.09.009.

Ahmadi, M., Ramezani Motlagh, H., Jaafarzadeh, N., Mostoufi, A., Saeedi, R., Barzegar, G., et al. 2016. Enhanced photocatalytic degradation of tetracycline and real pharmaceutical wastewater using MWCNT/TiO_2 nano-composite. J. Environ. Manage. 186: 55-63. https://doi.org/10.1016/j.jenvman.2016.09.088.

Akbari, A., Sheath, P., Martin, S.T., Shinde, D.B., Shaibani, M., Banerjee, P.C., et al. 2016. Large-area graphene-based nanofiltration membranes by shear alignment of discotic nematic liquid crystals of graphene oxide. Nat. Commun 7: 10891.

Al-Anazi, A., Abdelraheem, W.H., Scheckel, K., Nadagouda, M.N., O'Shea, K. and Dionysiou, D.D. 2020. Novel franklinite-like synthetic zinc-ferrite redox nanomaterial, for degradation of diclofenac in water. Appl. Catal. B. Environ. 270: 119098.

Alvarez-Rivera, G., Llompart, M., Lores, C. and Garcia–Jares, C. 2018. Preservatives in cosmetics: Regulatory aspects and analytical methods. pp. 175-224. *In*: Salvador, A. and Chisvert, A. [eds.]. Analysis of Cosmetic Products (Second Edition). Elsevier. https://doi.org/10.1016/B978-0-444-63508-2.00009-6.

Amin, M.M., Tabatabaeian, M., Chavoshani, A., Amjadi, E., Hashemi, M., Ebrahimpour, K., et al. 2019. Paraben content in adjacent normal-malignant breast tissues from women with breast cancer. Biomed. Environ. Sci. 32: 893-904.

Andronic, L., Enesca, A., Cazan, C. and Visa, M. 2014. TiO_2-active carbon composites for wastewater photocatalysis. J. Sol-Gel Sci. Technol. 71: 396-405. https://doi.org/10.1007/s10971-014-3393-6.

Antwi, F.B., Shama, L.M. and Peterson, R.K. 2008. Risk assessments for the insect repellents DEET and picaridin. Regul. Toxicol. Pharmacol. 51: 31-36.

Api, A. 2001. Toxicological profile of diethyl phthalate: A vehicle for fragrance and cosmetic ingredients. Food Chem. Toxicol. 39: 97-108.

Asimakopoulos, A.G., Thomaidis, N.S. and Kannan, K. 2014. Widespread occurrence of bisphenol A diglycidyl ethers, p-hydroxybenzoic acid esters (parabens), benzophenone type-UV filters, triclosan, and triclocarban in human urine from Athens, Greece. Sci. Total. Environ. 470: 1243-1249.

Ateia, M., Ran, J., Fujii, M. and Yoshimura, C. 2017. The relationship between molecular composition and fluorescence properties of humic substances. Int. J. Environ. Sci. Technol. 14: 867-880. https://doi.org/10.1007/s13762-016-1214-x.

Badi, M.Y., Azari, A., Pasalari, H., Esrafili, A. and Farzadkia, M. 2018. Modification of activated carbon with magnetic Fe_3O_4 nanoparticle composite for removal of ceftriaxone from aquatic solutions. J. Mol. Liquids. 261: 146-154.

Baena-Nogueras, R.M., Gonzälez-Mazo, E. and Lara-Martín, P.A. 2013. Determination and occurrence of secondary alkane sulfonates (SAS) in aquatic environments. Environ. Pollut. 176: 151-157.

Bagheri, H., Afkhami, A. and Noroozi, A. 2016. Removal of pharmaceutical compounds from hospital wastewaters using nanomaterials: A review. Anal. Bioanal. Chem. Res. 3: 1-18.

Balmer, M.E., Buser, H.-R., Muller, M.D. and Poiger, T. 2005. Occurrence of some organic UV filters in wastewater, in surface waters, and in fish from Swiss lakes. Environ. Sci. Technol. 39: 953-962.

Barnes, K.K., Kolpin, D.W., Furlong, E.T., Zaugg, S.D., Meyer, M.T. and Barber, L.B. 2008. A national reconnaissance of pharmaceuticals and other organic wastewater contaminants in the United States—I) Groundwater. Sci. Total. Environ. 402: 192-200.

Basha, S., Keane, D., Morrissey, A., Nolan, K., Oelgem€oller, M. and Tobin, J. 2010. Studies on the adsorption and kinetics of photodegradation of pharmaceutical compound, indomethacin using novel photocatalytic adsorbents (IPCAs). Ind. Eng. Chem. Res. 49: 11302-11309. https://doi.org/10.1021/ie101304a.

Basha, S., Barr, C., Keane, D., Nolan, K., Morrissey, A., Oelgem€oller, M., et al. 2011. On the adsorption/photodegradation of amoxicillin in aqueous solutions by an integrated photocatalytic adsorbent (IPCA): Experimental studies and kinetics analysis. Photochem. Photobiol. Sci. 10: 1014-1022. https://doi.org/10.1039/c0pp00368a.

Beeckman, J., Neyts, K. and Vanbrabant, P.J.M. 2011. Liquid crystal photonic applications. Opt. Eng. 50: 081202.

Begum, S. and Ahmaruzzaman, M. 2018. CTAB and SDS assisted facile fabrication of SnO_2 nanoparticles for effective degradation of carbamazepine from aqueous phase: A systematic and comparative study of their degradation performance. Water Res. 129: 470-485.

Benijts, T., Lambert, W. and De Leenheer, A. 2004. Analysis of multiple endocrine disruptors in environmental waters via wide-spectrum solid-phase extraction and dual-polarity ionization LC-ion trap-MS/MS. Anal. Chem. 76: 704-711.

Bertelsen, R.J., Longnecker, M.P., LøVik, M., Calafat, A.M., Carlsen, K.H., London, S.J., et al. 2013. Triclosan exposure and allergic sensitization in Norwegian children. Allergy 68: 84-91.

Bina, B., Mohammadi, F., Amin, M.M., Pourzamani, H.R. and Yavari, Z. 2018. Determination of 4-nonylphenol and 4-tert-octylphenol compounds in various types of wastewater and their removal rates in different treatment processes in nine wastewater treatment plants of Iran. Chin. J. Chem. Eng. 26: 183-190.

Borecka, M., Siedlewicz, G., Haliński, L.P., Sikora, K., Pazdro, K., Stepnowski, P., et al. 2015. Contamination of the southern Baltic Sea waters by the residues of selected pharmaceuticals: Method development and field studies. Mar. Pollut. Bull. 94: 62-71.

Brausch, J.M. and Rand, G.M. 2011. A review of personal care products in the aquatic environment: Environmental concentrations and toxicity. Chemosphere 82: 1518-1532.

Cabeza, Y., Candela, L., Ronen, D. and Teijon, G. 2012. Monitoring the occurrence of emerging contaminants in treated wastewater and groundwater between 2008 and 2010. The Baix Llobregat (Barcelona, Spain). J. Hazard. Mater. 239: 32-39.

Carballa, M., Omil, F., Lema, J.M., Llompart, M.A., García-Jares, C., Rodríguez, I., et al. 2004. Behavior of pharmaceuticals, cosmetics and hormones in a sewage treatment plant. Water Res. 38: 2918-2926.

Carlsson, G. and Norrgren, L. 2004. Synthetic musk toxicity to early life stages of zebrafish (Danio rerio). Arch. Environ. Contam. Toxicol. 46: 102-105.

Chaler, R., Cantón, L., Vaquero, M. and Grimalt, J.O. 2004. Identification and quantification of *n*-octyl esters of alkanoic and hexanedioic acids and phthalates as urban wastewater markers in biota and sediments from estuarine areas. J. Chromatogr. A 1046: 203-210.

Chen, F., Ying, G.-G., Kong, L.-X., Wang, L., Zhao, J.-L., Zhou, L.-J., et al. 2011. Distribution and accumulation of endocrine-disrupting chemicals and pharmaceuticals in wastewater irrigated soils in Hebei, China. Environ. Pollut. 159: 1490-1498.

Chen, L.J., Dai, J.H., Lin, J.D., Mo, T.S., Lin, H.P., Yeh, H.C., et al. 2018. Wavelength-tunable and highly stable perovskite-quantum dots-doped lasers with liquid crystal lasing cavities. ACS Appl. Mater. Interfaces 10: 33307-33315.

Chong, M.N., Jin, B., Chow, C.W.K. and Saint, C. 2010. Recent developments in photocatalytic water treatment technology: A review. Water Res. 44: 2997-3027. https://doi.org/10.1016/j.watres.2010.02.039.

Coles, H. and Morris, S. 2010. Liquid-crystal lasers. Nat. Photon 4: 676-685.

Costanzo, S., Watkinson, A., Murby, E., Kolpin, D.W. and Sandstrom, M.W. 2007. Is there a risk associated with the insect repellent DEET (N, N-diethyl-m-toluamide) commonly found in aquatic environments? Sci. Total. Environ. 384: 214-220.

Cruz, S. and Barceló, D. 2015. Personal Care Products in the Aquatic Environment. Springer, Switzerland.

Cuerda-Correa, E.M., Domínguez-Vargas, J.R., Muñoz-Peña, M.J. and González, T. 2016. Ultraviolet-photoassisted advanced oxidation of parabens catalyzed by hydrogen peroxide and titanium dioxide. Improving the system. Ind. Eng. Chem. Res. 55: 5152-5160

Cui, L. and Zhao, Y. 2004. Azopyridine side chain polymers: An efficient way to prepare photoactive liquid crystalline materials through self-assembly. Chem. Mater. 16: 2076-2082. https://doi.org/10.1021/cm0348850.

Dang, Y., Zhong, C., Zhang, G., Ju, D., Wang, L., Xia, S., et al. 2016a. Crystallographic investigations into properties of acentric hybrid perovskite single crystals $NH(CH_3)_3SnX_3$ (X = Cl, Br). Chem. Mater. 28: 6968-6974. https://doi.org/10.1021/acs.chemmater.6b02653.

Dang, Y., Zhou, Y., Liu, X., Ju, D., Xia, S., Xia, H., et al. 2016b. Formation of hybrid perovskite tin iodide single crystals by top-seeded solution growth. Angew. Chem. 55: 3447-3450. https://doi.org/10.1002/anie.201511792.

Darbre, P., Aljarrah, A., Miller, W., Coldham, N., Sauer, M. and Pope, G. 2004. Concentrations of parabens in human breast tumours. J. Appl. Toxicol.: An. Int. J. 24: 5-13.

Das, S., Ghosh, S., Misra, A., Tamhankar, A., Mishra, A., Lundborg, C., et al. 2018. Sunlight assisted photocatalytic degradation of ciprofloxacin in water using Fe doped ZnO nanoparticles for potential public health applications. Int. J. Environ. Res. Public Health 15: 2440. https://doi.org/10.3390/ijerph15112440.

Dasenaki, M.E. and Thomaidis, N.S. 2015. Multi-residue determination of 115 veterinary drugs and pharmaceutical residues in milk powder, butter, fish tissue and eggs using liquid chromatography-tandem mass spectrometry. Anal. Chim. Acta 880: 103-121.

da Silva, S.S., Chiavone-Filho, O., De Barros Neto, E.L., Mota, A.L., Foletto, E.L. and Nascimento, C.A. 2014. Photodegradation of non-ionic surfactant with different ethoxy groups in aqueous effluents by the photo-Fenton process. Environ. Technol. 35: 1556-1564.

Daughton, C.G. and Ternes, T.A. 1999. Pharmaceuticals and personal care products in the environment: Agents of subtle change? Environ. Health Perspect. 107: 907-938.

Dietrich, D.R. and Chou, Y.-J. 2001. Ecotoxicology of musks. pp. 156-167. *In*: Daughton, Christian G. and Jones-Lepp, Tammy L. [eds.]. ACS Symposium Series. ACS Publications, US.

Divyapriya, G., Nambi, I. and Senthilnathan, J. 2018. Ferrocene functionalized graphene based electrode for the electro—fenton oxidation of ciprofloxacin. Chemosphere 209: 113-123.

Dobbins, L.L., Usenko, S., Brain, R.A. and Brooks, B.W. 2009. Probabilistic ecological hazard assessment of parabens using Daphnia magna and Pimephales promelas. Environ. Toxicol. Chem. 28: 2744-2753.

Doke, S., Ganguly, P. and Mahamuni, S. 2020. Improvement in molecular alignment of ferroelectric liquid crystal by Co-ZnO/ZnO core/shell quantum dots. Liq. Cryst. 47: 309-316.

Doll, T.E. and Frimmel, F.H. 2003. Fate of pharmaceuticals—photodegradation by simulated solar UV-light. Chemosphere 52: 1757-1769. https://doi.org/10.1016/S0045-6535(03)00446-6.

Dong, Y., Tang, D. and Li, C. 2014. Photocatalytic oxidation of methyl orange in water phase by immobilized TiO_2-carbon nanotube nanocomposite photocatalyst. Appl. Surf. Sci. 296: 1-7. https://doi.org/10.1016/j.apsusc.2013.12.128.

Drosos, M., Ren, M. and Frimmel, F.H. 2015. The effect of NOM to TiO_2: Interactions and photocatalytic behavior. Appl. Catal. B Environ. 165: 328-334. https://doi.org/10.1016/j.apcatb.2014.10.017.

Duca, G. and Boldescu, V. 2009. Pharmaceuticals and Personal Care Products in the Environment. Springer, Netherlands, Dordrecht, pp. 27-35.

Eklund, T. 1980. Inhibition of growth and uptake processes in bacteria by some chemical food preservatives. J. Appl. Bacteriol. 48: 423-432.

Elberling, J., Linneberg, A., Dirksen, A., Johansen, J., FRøLund, L., Madsen, F., et al. 2005. Mucosal symptoms elicited by fragrance products in a population-based sample in relation to atopy and bronchial hyper-reactivity. Clin. Exp. Allergy 35: 75-81.

Evgenidou, E.N., Konstantinou, I.K. and Lambropoulou, D.A. 2015. Occurrence and removal of transformation products of PPCPs and illicit drugs in wastewaters: A review. Sci. Total Environ. 505: 905-926.

Fent, K., Kunz, P.Y. and Gomez, E. 2008. UV filters in the aquatic environment induce hormonal effects and affect fertility and reproduction in fish. CHIMIA Int. J. Chem. 62: 368-375.

Frimmel, F.H. 1994. Photochemical aspects related to humic substances. Environ. Int. 20: 373-385. https://doi.org/10.1016/0160-4120(94)90123-6.

Gajanan, K. and Tijare, S.N. 2018. Applications of nanomaterials. Mater. Today: Proc. 5: 1093-1096.

Gangwar, L.K., Kumar, A., Singh, G., Choudhary, A., Singh Rajesh, S.P. and Biradar, A.M. 2019. Probing the impact of carbon quantum dots on partially unwound helical mode in ferroelectric liquid crystals. J. Appl. Phys. 125: 125108.

Gar Alalm, M., Tawfik, A. and Ookawara, S. 2016. Enhancement of photocatalytic activity of TiO_2 by immobilization on activated carbon for degradation of pharmaceuticals. J. Environ. Chem. Eng. 4: 1929-1937. https://doi.org/10.1016/j.jece.2016.03.023.

Gaya, U.I. and Abdullah, A.H. 2008. Heterogeneous photocatalytic degradation of organic contaminants over titanium dioxide: A review of fundamentals, progress and problems. J. Photochem. Photobiol. C Photochem. Rev. 9: 1-12. https://doi.org/10.1016/j.jphotochemrev.2007.12.003.

Geens, T., Neels, H. and Covaci, A. 2012. Distribution of bisphenol-A, triclosan and nnonylphenol in human adipose tissue, liver and brain. Chemosphere 87: 796-802.

Gerrity, D., Stanford, B.D., Trenholm, R.A. and Snyder, S.A. 2010. An evaluation of a pilot-scale nonthermal plasma advanced oxidation process for trace organic compound degradation. Water Res. 44: 493-504.

Glassmeyer, S.T., Furlong, E.T., Kolpin, D.W., Cahill, J.D., Zaugg, S.D., Werner, S.L., et al. 2005. Transport of chemical and microbial compounds from known wastewater discharges: Potential for use as indicators of human fecal contamination. Environ. Sci. Technol. 39: 5157-5169.

Goodby, J.W., Collings, P.J., Kato, T., Tschierske, C., Glesson, H. and Raynes, P. 2014. Handbook of Liquid Crystals, 2nd ed.; Eds.; Wiley-VCH: Weinheim.

Gorito, A.M., Ribeiro, A.R., Almeida, C.M.R. and Silva, A.M.T. 2017. A review on the application of constructed wetlands for the removal of priority substances and contaminants of emerging concern listed in recently launched EU legislation. Environ. Pollut. 227: 428-443.

Gulyas, H., Ogun, M.K., Meyer, W., Reich, M. and Otterpohl, R. 2016. Inadequacy of carbamazepine-spiked model wastewaters for testing photocatalysis efficiency. Sci. Total Environ. 542: 612-619. https://doi.org/10.1016/j.scitotenv.2015.10.116.

Halden, R.U. and Paull, D.H. 2005. Co-occurrence of triclocarban and triclosan in US water resources. Environ. Sci. Technol. 39: 1420-1426.

Hany, J. and Nagel, R. 1995. Detection of sunscreen agents in human breast-milk. Dtsch. Lebensmittel-Rundschau. 91: 341-345.

Hashim, N.H. and Khan, S.J. 2011. Enantioselective analysis of ibuprofen, ketoprofen and naproxen in wastewater and environmental water samples. J. Chromatogr. A 1218: 4746-4754. https://doi.org/10.1016/j.chroma.2011.05.046.

Herrmann, J. 1999. Heterogeneous photocatalysis: Fundamentals and applications to the removal of various types of aqueous pollutants. Catal. Today 53: 115-129. https://doi.org/10.1016/S0920-5861(99)00107-8.

Hoffmann, M.R., Martin, S., Choi, W. and Bahnemannt, D.W. 1995. Environmental applications of semiconductor photocatalysis. Chem. Rev. 95: 69-96. https://doi.org/10.1021/cr00033a004.

Horii, Y. and Kannan, K. 2008. Survey of organosilicone compounds, including cyclic and linear siloxanes, in personal-care and household products. Arch. Environ. Contam.Toxicol. 55: 701.

Houlihan, J., Brody, C., Schwan, B., Meunick, R., Doyle, M.B., Patterson, J., et al. 2002. Not too pretty: Phthalates, beauty products and the FDA. https://static.ewg.org/reports/2002/NotTooPretty.pdf.

Hua, Z., Zhang, J., Bai, X., Ye, Z., Tang, Z., Liang, L., et al. 2016. Aggregation of TiO_2- graphene nanocomposites in aqueous environment: Influence of environmental factors and UV irradiation. Sci. Total Environ. 539: 196-205. https:// doi.org/10.1016/j.scitotenv.2015.08.143.

Huang, C.L., Abass, O.K. and Yu, C.P. 2016. Triclosan: A review on systematic risk assessment and control from the perspective of substance flow analysis. Sci. Total Environ. 566-567: 771-785.

Huang, T., Zhu, J., Ge, S., Guo, T., Jiang, C. and Xie, L. 2020. Synthesis of novel CdSe QDs/$BiFeO_3$ composite catalysts and its application for the photo-Fenton catalytic degradation of phenol. J. Environ. Chem. Eng. 8: 104384.

Hubinger, J.C. and Havery, D.C. 2006. Analysis of consumer cosmetic products for phthalate esters. J. Cosmet. Sci. 57: 127-137.

Hunge, Y.M., Yadav, A.A., Khan, S., Takagi, K., Suzuki, N., Teshima, K., et al. 2021. Photocatalytic degradation of bisphenol A using titanium dioxide@nanodiamond composites under UV light illumination. J. Colloid. Interf. Sci. 582: 1058-1066.

Inoue, K., Kawaguchi, M., Okada, F., Takai, N., Yoshimura, Y., Horie, M., et al. 2003. Measurement of 4-nonylphenol and 4-tert-octylphenol in human urine by column switching liquid chromatography–mass spectrometry. Anal. Chim. Acta. 486: 41-50.

Inturi, S.N.R., Boningari, T., Suidan, M. and Smirniotis, P.G. 2014. Visible-light-induced photodegradation of gas phase acetonitrile using aerosol-made transition metal (V, Cr, Fe, Co, Mn, Mo, Ni, Cu, Y, Ce, and Zr) doped TiO_2. Appl. Catal. B 144: 333-342.

Ito, Y., Yamanoshita, O., Asaeda, N., Tagawa, Y., Lee, C.-H., Aoyama, T., et al. 2007. Di (2-ethylhexyl) phthalate induces hepatic tumorigenesis through a peroxisome proliferator-activated receptor α-independent pathway. J. Occup. Health 49: 172-182.

Jayalakshmi, V., Nair, G.G. and Prasad, S.K. 2007. Effects of aerosol dispersions on the photoinduced nematic-isotropic transition. J. Phys.: Condens. Matter. 19: 226213.

Johnson, R.R., Navone, R. and Larson, E.L. 1963. An unusual epidemic of methemoglobinemia. Pediatrics 31: 222-225.

Jones, O.A.H., Voulvoulis, N. and Lester, J.N. 2005. Human pharmaceuticals in wastewater treatment processes. Crit. Rev. Environ. Sci. Technol. 35: 401-427.

Jonkers, N., Sousa, A., Galante-Oliveira, S., Barroso, C.M., Kohler, H.-P.E. and Giger, W. 2010. Occurrence and sources of selected phenolic endocrine disruptors in Ria de Aveiro, Portugal. Environ. Sci. Pollut. Res. 17: 834-843.

Joshi, T., Ganguly, P., Haranath, D., Singh, S. and Biradar, A.M. 2014. Tuning the photoluminescence of ferroelectric liquid crystal by controlling the size of dopant ZnO quantum dots. Mat. Lett. 114: 156-158.

Ju, D., Jiang, X., Xiao, H., Chen, X., Hu, X. and Tao, X. 2018. Narrow band gap and high mobility of lead-free perovskite single crystal Sn-doped MA3Sb2I9. J. Mater. Chem. A 6: 20753-20759. https://doi.org/10.1039/c8ta08315k.

Kang, J., Duan, X., Wang, C., Sun, H., Tan, X., Tade, M.O., et al. 2018. Nitrogen-doped bamboo-like carbon nanotubes with Ni encapsulation for persulfate activation to remove emerging contaminants with excellent catalytic stability. Chem. Eng. J. 332: 398-408.

Kannan, K., Reiner, J.L., Yun, S.H., Perrotta, E.E., Tao, L., Johnson-Restrepo, B., et al. 2005. Polycyclic musk compounds in higher trophic level aquatic organisms and humans from the United States. Chemosphere 61: 693-700.

Kaur, A., Gupta, G., Ibhadon, A.O., Salunke, D.B., Sinha, A.S.K. and Kansal, S.K. 2018. A facile synthesis of silver modified ZnO nanoplates for efficient removal of ofloxacin drug in aqueous phase under solar irradiation. J. Environ. Chem. Eng. 6: 3621-3630. https://doi.org/10.1016/j.jece.2017.05.032.

Khraisheh, M., Kim, J., Campos, L., Al-Muhtaseb, A.H., Walker, G.M. and Alghouti, M. 2013. Removal of carbamazepine from water by a novel TiO_2-coconut shell powder/UV process: Composite preparation and photocatalytic activity. Environ. Eng. Sci. 30: 515-526. https://doi.org/10.1089/ees.2012.0056.

Kim, B.-N., Cho, S.-C., Kim, Y., Shin, M.-S., Yoo, H.-J., Kim, J.-W., et al. 2009. Phthalates exposure and attention-deficit/hyperactivity disorder in school-age children. Biol. Psychiatry 66: 958-963.

Klavarioti, M., Mantzavinos, D. and Kassinos, D. 2009. Removal of residual pharmaceuticals from aqueous systems by advanced oxidation processes. Environ. Int. 35(2): 402-417.

Koerner, H., Price, G., Pearce, N.A., Alexander, M. and Vaia, R.A. 2004. Remotely actuated polymer nanocomposites—stress-recovery of carbon-nanotube-filled thermoplastic elastomers. Nat. Mater. 3: 115-120.

Kohlmeyer, R.R. and Chen, J. 2013. Wavelength-selective, IR light-driven hinges based on liquid crystalline elastomer composites. Angew. Chem., Int. Ed. 52: 9234-9237.

Kondrakov, A.O., Ignatev, A.N., Lunin, V.V., Frimmel, F.H., Br€ase, S. and Horn, H. 2016. Roles of water and dissolved oxygen in photocatalytic generation of free OH radicals in aqueous TiO_2 suspensions: An isotope labeling study. Appl. Catal. B Environ. 182: 424-430. https://doi.org/10.1016/j.apcatb.2015.09.038.

Kumar, S. 2001. Liquid Crystals: Experimental Study of Physical Properties and Phase Transitions. Cambridge University Press.

Kunz, P.Y. and Fent, K. 2006. Multiple hormonal activities of UV filters and comparison of *in vivo* and *in vitro* estrogenic activity of ethyl-4-aminobenzoate in fish. Aquat. Toxicol. 79: 305-324.

Lagerwall, J.P.F. and Scalia, G. 2012. A new era for liquid crystal research: Applications of liquid crystal in soft matter nano-, bio- and microtechnology. Current Appl. Phys. 12: 1387-1412.

Leary, R. and Westwood, A. 2011. Carbonaceous nanomaterials for the enhancement of TiO_2 photocatalysis. Carbon N.Y. 49: 741-772. https://doi.org/10.1016/ j.carbon.2010.10.010.

Lee, C.W. and Shih, W.P. 2010. Quantification of ion trapping effect of carbon nano-materials in liquid crystal. Mater. Lett. 64: 466-468.

Lee, K.M., Lai, C.W., Ngai, K.S. and Juan, J.C. 2016. Recent developments of zinc oxide based photocatalyst in water treatment technology: A review. Water Res. 88: 428-448. https://doi.org/10.1016/j.watres.2015.09.045.

Lee, W.K., Hwang, S.J., Cho, M.J., Park, H.G., Han, J.W., Song, S., et al. 2013. CIS–ZnS quantum dots for self-aligned liquid crystal molecules with superior electro-optic properties. Nanoscale 5: 193-199.

Li, C., Liu, Y., Lo, C. and Jiang, H. 2011. Reversible white-light actuation of carbon nanotube incorporated liquid crystalline elastomer nanocomposites. Soft Matter. 7: 7511-7516.

Li, C., Liu, Y., Huang, X. and Jiang, H. 2012. Direct sun-driven artificial heliotropism for solar energy harvesting based on a photothermomechanical liquid-crystal elastomer nanocomposite. Adv. Funct. Mater. 22: 5166-5174.

Li, S., Wu, Y., Zheng, Y., Jing, T., Tian, J., Zheng, H., et al. 2020. Free-radical and surface electron transfer dominated bisphenol A degradation in system of ozone and peroxydisulfate co-activated by $CoFe_2O_4$-biochar. Appl. Surf. Sci. https://doi.org/10.1016/j.apsusc.2020.147887.

Li, W., Shi, Y., Gao, L., Liu, J. and Cai, Y. 2015. Occurrence, fate and risk assessment of parabens and their chlorinated derivatives in an advanced wastewater treatment plant. J. Hazard. Mater. 300: 29-38.

Li, Y., Zhang, H., Chen, Y., Huang, L., Lin, Z. and Cai, Z. 2019. Core–shell structured magnetic covalent organic framework nanocomposites for triclosan and triclocarban adsorption. ACS Appl. Mater. Interfaces 11: 22492-22500.

Liang, R., Hu, A., Hatat-Fraile, M. and Zhou, N. 2014. Fundamentals on adsorption, membrane filtration, and advanced oxidation processes for water treatment. Nanotechnology for Water Treatment and Purification. Springer, Anming Hu, Allen Apblett, Switzerland.

Lin, C.J. and Yang, W.T. 2014. Ordered mesostructured Cu-doped TiO_2 spheres as active visible-light-driven photocatalysts for degradation of paracetamol. Chem. Eng. J. 237: 131-137.

Liu, F.-f., Zhao, J., Wang, S., Du, P. and Xing, B. 2014. Effects of solution chemistry on adsorption of selected pharmaceuticals and personal care products (PPCPs) by graphenes and carbon nanotubes. Environ. Sci. Technol. 48: 13197-13206.

Liu, J.L. and Wong, M.H. 2013. Pharmaceuticals and personal care products (PPCPs): A review on environmental contamination in China. Environ. Int. 59: 208-224.

Liu, N., Shi, Y., Li, W., Xu, L. and Cai, Y. 2014a. Concentrations and distribution of synthetic musks and siloxanes in sewage sludge of wastewater treatment plants in China. Sci. Total. Environ. 476: 65-72.

Liu, S., Zhao, Y., Wang, T., Liang, N. and Hou, X. 2019. Core–shell Fe_3O_4@MIL-100(Fe) magnetic nanoparticle for effective removal of meloxicam and naproxen in aqueous solution. J. Chem. Eng. Data 64: 2997-3007.

Liu, X., Wei, R., Hoang, P.T., Wang, X., Liu, T. and Keller, P. 2015. Reversible and rapid laser actuation of liquid crystalline elastomer micropillars with inclusion of gold nanoparticles. Adv. Funct. Mater. 25: 3022-3032.

Liu, Y., Zhang, Y., Yang, Z., Cui, J., Wu, H., Ren, X., et al. 2020. Large lead-free perovskite single crystal for high-performance coplanar X-ray imaging applications. Adv. Opt. Mater. 8: 2000814. https://doi.org/10.1002/adom.202000814.

Liu, Y.-S., Ying, G.-G., Shareef, A. and Kookana, R.S. 2011. Simultaneous determination of benzotriazoles and ultraviolet filters in ground water, effluent and biosolid samples using gas chromatography–tandem mass spectrometry. J. Chromatogr. A 1218: 5328-5335.

Long, X., Yang, S., Qiu, X., Ding, D., Feng, C., Chen, R., et al. 2021. Heterogeneous activation of peroxymonosulfate for bisphenol A degradation using $CoFe_2O_4$ derived by hybrid cobalt-ion hexacyanoferrate nanoparticles. Chem. Eng. J. 404: 127052.

Luo, Y.L., Guo, W.S., Ngo, H.H., Nghiem, L.D., Hai, F.I., Zhang, J., et al. 2014. A review on the occurrence of micropollutants in the aquatic environment and their fate and removal during wastewater treatment. Sci. Total Environ. 473: 619-641.

Ma, R., Wang, X., Huang, J., Song, J., Zhang, J. and Wang, X. 2017. Photocatalytic degradation of salicylic acid with magnetic activated carbon-supported F-N codoped TiO_2 under visible light. Vacuum 141: 157-165. https://doi.org/10.1016/j.vacuum.2017.04.003.

Malekshoar, G., Pal, K., He, Q., Yu, A.P. and Ray, A.K. 2014. Enhanced solar photocatalytic degradation of phenol with coupled graphene-based titanium dioxide and zinc oxide. Ind. Eng. Chem. Res. 53: 18824-18832. https://doi.org/10.1021/ie501673v.

Martínez, C., Canle, L.M., Fernández, M.I., Santaballa, J.A. and Faria, J. 2011. Aqueous degradation of diclofenac by heterogeneous photocatalysis using nanostructured materials. Appl. Catal. B Environ. 107: 110-118. https://doi.org/10.1016/j.apcatb.2011.07.003.

Matamoros, V., Uggetti, E., García, J. and Bayona, J.M. 2016. Assessment of the mechanisms involved in the removal of emerging contaminants by microalgae from wastewater: A laboratory scale study. J. Hazard. Mater. 301: 197-205.

McAvoy, D.C., Schatowitz, B., Jacob, M., Hauk, A. and Eckhoff, W.S. 2002. Measurement of triclosan in wastewater treatment systems. Environ. Toxicol. Chem.: An. Int. J. 21: 1323-1329.

McCall, K.M., Stoumpos, C.C., Kostina, S.S., Kanatzidis, M.G. and Wessels, B.W. 2017. Strong electron–phonon coupling and self-trapped excitons in the defect halide perovskites $A_3M_2I_9$ (A = Cs, Rb; M = Bi, Sb). Chem. Mater. 29: 4129-4145. https://doi.org/10.1021/acs.chemmater.7b01184.

Mirzaei, A., Chen, Z., Haghighat, F. and Yerushalmi, L. 2016. Removal of pharmaceuticals and endocrine disrupting compounds from water by zinc oxide-based photocatalytic degradation: A review. Sustain. Cities Soc. 27: 407-418. https://doi.org/10.1016/j.scs.2016.08.004.

Mock, T., Otillar, R.P., Strauss, J., McMullan, M., Paajanen, P., Schmutz, J., et al. 2017. Evolutionary genomics of the cold adapted diatom Fragilariopsis cylindrus. Nature 541: 536.

Mohammadi Nodeh, M.K., Radfard, M., Zardari, L.A. and Rashidi Nodeh, A. 2018. Enhanced removal of naproxen from wastewater using silica magnetic nanoparticles decorated onto graphene oxide: Parametric and equilibrium study. Sep. Sci. Technol. 53(15): 1-10.

Moldovan, Z. 2006. Occurrences of pharmaceutical and personal care products as micropollutants in rivers from Romania. Chemosphere 64: 1808-1817.

Montes-Grajales, D., Fennix-Agudelo, M. and Miranda-Castro, W. 2017. Occurrence of personal care products as emerging chemicals of concern in water resources: A review. Sci. Total Environ. 595: 601-614.

Murgolo, S., Petronella, F., Ciannarella, R., Comparelli, R., Agostiano, A., Curri, M.L., et al. 2015. UV and solar-based photocatalytic degradation of organic pollutants by nano-sized TiO_2 grown on carbon nanotubes. Catal. Today 240: 114-124. https://doi.org/10.1016/j.cattod.2014.04.021.

Murphy, M.E., Montemarano, A.D., Debboun, M. and Gupta, R. 2000. The effect of sunscreen on the efficacy of insect repellent: A clinical trial. J. Am. Acad. Dermatol. 43: 219-222.

Nagtegaal, M. 1997. Detection of sunscreen agents in water and fish of the meerfelder maar, the eifel, Germany. Env. Sci. Eur. 9: 79-86.

Nakada, N., Shinohara, H., Murata, A., Kiri, K., Managaki, S., Sato, N., et al. 2007. Removal of selected pharmaceuticals and personal care products (PPCPs) and endocrine-disrupting chemicals (EDCs) during sand filtration and ozonation at amunicipal sewage treatment plant. Water Res. 41: 4373-4382.

Norouzi, N., Fazelia, A. and Tavakoli, O. 2020. Phenol contaminated water treatment by photocatalytic degradation on electrospun Ag/TiO_2 nanofibers: Optimization by the response surface method. J. Water Proc. Eng. 37: 101489.

Novo, M., Verdu, I., Trigo, D. and Martinez-Guitarte, J.L. 2018. Endocrine disruptors in soil: Effects of bisphenol A on gene expression of the earthworm Eisenia fetida. Ecotoxicol. Environ. Saf. 150: 159-167.

Olmez-Hanci, T., Arslan-Alaton, I. and Basar, G. 2011. Multivariate analysis of anionic, cationic and nonionic textile surfactant degradation with the H_2O_2/UV-C process by using the capabilities of response surface methodology. J. Hazard. Mater. 185: 193-203.

Ortiz de García, S., Pinto Pinto, G., García Encina, P. and Irusta Mata, R. 2013. Consumption and occurrence of pharmaceutical and personal care products in the aquatic environment in Spain. Sci. Total Environ. 444: 451-465.

Pan, B., Ning, P. and Xing, B. 2009. Part V-sorption of pharmaceuticals and personal care products. Environ. Sci. Pollut. Res. 16: 106-116.

Papageorgiou, A., Voutsa, D. and Papadakis, N. 2014. Occurrence and fate of ozonation byproducts at a full-scale drinking water treatment plant. Sci. Total Environ. 481: 392-400.

Peck, A.M. 2006. Analytical methods for the determination of persistent ingredients of personal care products in environmental matrices. Anal. Bioanal. Chem. 386: 907-939.

Penalver, A., Pocurull, E., Borrull, F. and Marce, R. 2000. Determination of phthalate esters in water samples by solid-phase microextraction and gas chromatography with mass spectrometric detection. J. Chromatogr. A 872: 191-201.

Pendergast, M.M. and Hoek, E.M. 2011. A review of water treatment membrane nanotechnologies. Energy Environ. Sci. 4: 1946-1971.

Peng, X., Ou, W., Wang, C., Wang, Z., Huang, Q., Jin, J., et al. 2014. Occurrence and ecological potential of pharmaceuticals and personal care products in groundwater and reservoirs in the vicinity of municipal landfills in China. Sci. Total. Environ. 490: 889-898.

Poiger, T., Buser, H.-R., Balmer, M.E., Bergqvist, P.-A. and Muller, M.D. 2004. Occurrence of UV filter compounds from sunscreens in surface waters: Regional mass balance in two Swiss lakes. Chemosphere 55: 951-963.

Quednow, K. and Puttmann, W. 2009. Temporal concentration changes of DEET, TCEP, terbutryn, and nonylphenols in freshwater streams of Hesse, Germany: Possible influence of mandatory regulations and voluntary environmental agreements. Environ. Sci. Pollut. Res. 16: 630-640.

Quinones, D.H., Rey, A., Alvarez, P.M., Beltrán, F.J. and Plucinski, P.K. 2014. Enhanced activity and reusability of TiO_2 loaded magnetic activated carbon for solar photocatalytic ozonation. Appl. Catal. B Environ. 144: 96-106. https://doi.org/10.1016/j.apcatb.2013.07.005.

Rafati, L., Ehrampoush, M.H., Rafati, A.A., Mokhtari, M. and Mahvi, A.H. 2018. Removal of ibuprofen from aqueous solution by functionalized strong nano-clay composite adsorbent: Kinetic and equilibrium isotherm studies. Int. J. Environ. Sci. Technol. 15: 513-524.

Rajapaksha, A.U., Vithanage, M., Lim, J.E., Ahmed, M.B.M., Zhang, M., Lee, S.S., et al. 2014. Invasive plant-derived biochar inhibits sulfamethazine uptake by lettuce in soil. Chemosphere 111: 500-504.

Rajendran, K. and Sen, S. 2018. Adsorptive removal of carbamazepine using biosynthesized hematite nanoparticles. Environ. Nanotechnol. Monit. Manage. 9: 122-127.

Ramirez, A.J., Mottaleb, M.A., Brooks, B.W. and Chambliss, C.K. 2007. Analysis of pharmaceuticals in fish using liquid chromatography-tandem mass spectrometry. Anal. Chem. 79: 3155-3163.

Reh, R., Licha, T., Geyer, T., Nödler, K. and Sauter, M. 2013. Occurrence and spatial distribution of organic micro-pollutants in a complex hydrogeological karst system during low flow and high flow periods, results of a two-year study. Sci. Total. Environ. 443: 438-445.

Ren, A.S., Chai, F., Xue, H., Anderson, D.M. and Chavez, F.P. 2018. A sixteen-year decline in dissolved oxygen in the central California current. Sci. Rep. 8: 7290.

Rey, A., Quiñones, D.H., Alvarez, P.M., Beltrán, F.J. and Plucinski, P.K. 2012. Simulated solar-light assisted photocatalytic ozonation of metoprolol over titania-coated magnetic activated carbon. Appl. Catal. B Environ. 111-112: 246-253. https:// doi.org/10.1016/j.apcatb.2011.10.005.

Reznik, G.O., Vishwanath, P., Pynn, M.A., Sitnik, J.M., Todd, J.J., Wu, J., et al. 2010. Use of sustainable chemistry to produce an acyl amino acid surfactant. Appl. Microb. Biotech. 86: 1387-1397.

Ricart, M., Guasch, H., Alberch, M., Barceló, D., Bonnineau, C., Geiszinger, A., et al. 2010. Triclosan persistence through wastewater treatment plants and its potential toxic effects on river biofilms. Aquat. Toxicol. 100: 346-353.

Richardson, S.D. 2008. Environmental mass spectrometry: Emerging contaminants and current issues. Anal. Chem. 80: 4373-4402.

Rioja, N., Zorita, S. and Peñas, F.J. 2016. Effect of water matrix on photocatalytic degradation and general kinetic modeling. Appl. Catal. B Environ. 180: 330-335. https://doi.org/10.1016/j.apcatb.2015.06.038.

Roberts, J., Kumar, A., Du, J., Hepplewhite, C., Ellis, D.J., Christy, A.G., et al. 2016. Pharmaceuticals and personal care products (PPCPs) in Australia's largest inland sewage treatment plant, and its contribution to a major Australian river during high and low flow. Sci. Total Environ. 541: 25-1637.

Rodil, R. and Moeder, M. 2008. Stir bar sorptive extraction coupled to thermodesorption–gas chromatography–mass spectrometry for the determination of insect repelling substances in water samples. J. Chromatogr. A 1178: 9-16.

Rohricht, M., Krisam, J. and Weise, U. 2010. Elimination of pharmaceuticals from wastewater by submerged nanofiltration plate modules. Desalination 250: 1025-1026.

Rong, X., Qiu, F., Jiang, Z., Rong, J., Pan, J., Zhang, T., et al. 2016. Preparation of ternary combined $ZnO-Ag_2O$/porous $g-C_3N_4$ composite photocatalyst and enhanced visible-light photocatalytic activity for degradation of ciprofloxacin. Chem. Eng. Res. Des. 111: 253-261. https://doi.org/10.1016/j.cherd.2016.05.010.

Rosa, S.M.C., Nossol, A.B.S., Nossol, E., Zarbin, A.J.G. and Peralta-Zamora, P.G. 2017. Nonsynergistic UV-A photocatalytic degradation of estrogens by nano-TiO_2 supported on activated carbon. J. Braz. Chem. Soc. 28: 582-588. https://doi.org/10.5935/0103-5053.20160201.

Roslev, P., Vorkamp, K., Aarup, J., Frederiksen, K. and Nielsen, P.H. 2007. Degradation of phthalate esters in an activated sludge wastewater treatment plant. Water Res. 41: 969-976.

Roy, A., Pathak, G., Herman, J., Inamdar, S.R., Srivastava, A. and Manohar, R. 2018. InP/ZnS quantum-dot-dispersed nematic liquid crystal illustrating characteristic birefringence and enhanced electro-optical parameters. Appl. Phys. A. 124: 273.

Sanchis, J., Martínez, E., Ginebreda, A., Farre, M. and Barceló, D. 2013. Occurrence of linear and cyclic volatile methylsiloxanes in wastewater, surface water and sediments from Catalonia. Sci. Total. Environ. 443: 530-538.

Sarkar, S., Das, R., Choi, H. and Bhattacharjee, C. 2014. Involvement of process parameters and various modes of application of TiO_2 nanoparticles in heterogeneous photocatalysis of pharmaceutical wastes – a short review. RSC Adv. 4: 57250-57266. https://doi.org/10.1039/C4RA09582K.

Schlumpf, M., Kypke, K., Wittassek, M., Angerer, J., Mascher, H., Mascher, D., et al. 2010. Exposure patterns of UV filters, fragrances, parabens, phthalates, organochlor pesticides, PBDEs, and PCBs in human milk: Correlation of UV filters with use of cosmetics. Chemosphere 81: 1171-1183.

Schlumpf, M. and Lichtensteiger, W. 2001. *In vitro* and *in vivo* estrogenicity of UV screens: Response. Environ. Health Perspect. A359-A361.

Schmitt, C., Oetken, M., Dittberner, O., Wagner, M. and Oehlmann, J. 2008. Endocrine modulation and toxic effects of two commonly used UV screens on the aquatic invertebrates Potamopyrgus antipodarum and Lumbriculus variegatus. Environ. Pollut. 152: 322-329.

Schreurs, R., Lanser, P., Seinen, W. and Burg, B. 2002. Estrogenic activity of UV filters determined by an *in vitro* reporter gene assay and an *in vivo* transgenic zebrafish assay. Arch. Toxic. 76: 257-261.

Schneider, J., Matsuoka, M., Takeuchi, M., Zhang, J., Horiuchi, Y., Anpo, M., et al. 2014. Understanding TiO_2 photocatalysis: Mechanisms and materials. Chem. Rev. 114: 9919-9986.

Schäfer, A.I., Akanyeti, I. and Semião, A.J. 2011. Micropollutant sorption to membrane polymers: A review of mechanisms for estrogens. Adv. Colloid Interf. Sci. 164: 100-117.

Shankar, H., Ghosh, S. and Kar, P. 2020. Highly stable blue fluorescent lead free allinorganic $Cs_2ZnX_4$2D perovskite nanocrystals. J. Alloys Compd. 844: 156148. https://doi.org/10.1016/j.jallcom.2020.156148.

Shibaev, P.V., Wenzlick, M., Murray, J., Tantillo, A. and Howard-Jennings, J. 2015. Rebirth of liquid crystals for sensoric applications: Environmental and gas sensors. Adv. Cond. Matter Phys. 2015: 1-8.

Shibaev, V.P. 2009. Liquid-crystalline polymers: Past, present and future. Polym. Sci. Ser. A. 1: 1131-1193. https://doi.org/10.1134/S0965545X09110029.

Shimizu, Y., Ateia, M. and Yoshimura, C. 2018. Natural organic matter undergoes different molecular sieving by adsorption on activated carbon and carbon nanotubes. Chemosphere 203: 345-352. https://doi.org/10.1016/j.chemosphere.2018.03.197.

Shiraishi, Y., Toshima, N., Maeda, K., Yoshikawa, H., Xu, J. and Kobayashi, S. 2002. Frequency modulation response of a liquid-crystal electro-optic device doped with nanoparticles. Appl. Phys. Lett. 81: 2845.

Shukla, R.K., Mirzaei, J., Sharma, A., Hofmann, D., Hegmann, T. and Haase, W. 2015. Electro-optic and dielectric properties of a ferroelectric liquid crystal doped with chemically and thermally stable emissive carbon dots. RSC Adv. 5: 34491.

Singh, D.P., Daoudi, A., Gupta, S.K., Pandey, S., Vimal, T., Manohar, R., et al. 2016. Mn^{2+} doped ZnS quantum dots in ferroelectric liquid crystal matrix: Analysis of new relaxation phenomenon, faster optical response, and concentration dependent quenching in photoluminescence. J. Appl. Phys. 119: 094101.

Singh, D.P., Vimal, T., Mange, Y.J., Varia, M.C., Nann, T., Pandey, K.K., et al. 2018. $CuInS_2$/ZnS QD-ferroelectric liquid crystal mixtures for faster electro-optical devices and their energy storage aspects. J. Appl. Phys. 123: 034101.

Soni, M., Carabin, I. and Burdock, G. 2005. Safety assessment of esters of p-hydroxybenzoic acid (parabens). Food Chem. Toxicol. 43: 985-1015.

Stamatis, N.K. and Konstantinou, I.K. 2013. Occurrence and removal of emerging pharmaceutical, personal care compounds and caffeine tracer in municipal sewage treatment plant in Western Greece. J. Environ. Sci. Health B 48: 800-813.

Sui, Q., Huang, J., Deng, S., Yu, G. and Fan, Q. 2010. Occurrence and removal of pharmaceuticals, caffeine and DEET in wastewater treatment plants of Beijing, China. Water Res. 44: 417-426.

Sun, P., Casteel, K., Dai, H., Wehmeyer, K.R., Kiel, B. and Federle, T. 2014a. Distributions of polycyclic musk fragrance in wastewater treatment plant (WWTP) effluents and sludges in the United States. Sci. Total Environ. 493: 1073-1078.

Sun, Q., Lv, M., Hu, A., Yang, X. and Yu, C.-P. 2014b. Seasonal variation in the occurrence and removal of pharmaceuticals and personal care products in a wastewater treatment plant in Xiamen, China. J. Hazard Mater. 277: 69-75.

Sun, X., Wang, W., Qiu, L., Guo, W., Yu, Y. and Peng, H. 2012a. Unusual reversible photomechanical actuation in polymer/nanotube composites. Angew. Chem., Int. Ed. 51: 8520-8524.

Sun, Y., Evans, J.S., Lee, T., Senyuk, B., Keller, P., He, S., et al. 2012b. Optical manipulation of shape-morphing elastomeric liquid crystal microparticles doped with gold nanocrystals. Appl. Phys. Lett. 100: 241901.

Suárez, S., Carballa, M., Omil, F. and Lema, J.M. 2008. How are pharmaceutical and personal care products (PPCPs) removed from urban wastewaters? Rev. Environ. Sci. Biotechnol. 7: 125-138.

Svensson, K., Hernández-Ramírez, R.U., Burguete-García, A., Cebrián, M.E., Calafat, A.M., Needham, L.L., et al. 2011. Phthalate exposure associated with self-reported diabetes among Mexican women. Environ. Res. 111: 792-796.

Sören, T. 2003. Pharmaceutical antibiotic compounds in soils: A review. J. Plant Nutr. Soil Sci. 166: 546.

Tan, Y., Guo, Y., Gu, X. and Gu, C. 2015. Effects of metal cations and fulvic acid on the adsorption of ciprofloxacin onto goethite. Environ. Sci. Pollut. Res. Int. 22: 609-617.

Tan, Z., Li, J., Zhang, C., Li, Z., Hu, Q., Xiao, Z., et al. 2018. Highly efficient blue-emitting Bi-doped Cs_2SnCl_6 perovskite variant: Photoluminescence induced by impurity doping. Adv. Funct. Mater. 28: 1801131. https://doi.org/10.1002/adfm.201801131.

Tay, K.S., Rahman, N.A. and Abas, M.R.B. 2009. Degradation of DEET by ozonation in aqueous solution. Chemosphere 76: 1296-1302.

Terasaki, M., Makino, M. and Tatarazako, N. 2009. Acute toxicity of parabens and their chlorinated by-products with Daphnia magna and Vibrio fischeri bioassays. J. Appl. Toxicol. 29: 242-247.

Tian, G., Wang, W., Zong, L., Kang, Y. and Wang, A. 2016. A functionalized hybrid silicate adsorbent derived from naturally abundant low-grade palygorskite clay for highly efficient removal of hazardous antibiotics. Chem. Eng. J. 293: 376-385.

Tolls, J., Berger, H., Klenk, A., Meyberg, M., Beiersdorf, A., Müller, R., et al. 2009. Environmental safety aspects of personal care products—a European perspective. Environ. Toxicol. Chem. 28: 2485-2489. http://dx.doi.org/10.1897/09-104.1.

Van Doorslaer, X., Dewulf, J., De Maerschalk, J., Van Langenhove, H. and Demeestere, K. 2015. Heterogeneous photocatalysis of moxifloxacin in hospital effluent: Effect of selected matrix constituents. Chem. Eng. J. 261: 9-16. https://doi.org/10.1016/j.cej.2014.06.079.

Vimal, T., Singh, D.P., Agrahari, K., Srivastava, A. and Manohar, R. 2016. Analysis of optical properties and mechanism of photoluminescence enhancement of a quantum dot – ferroelectric liquid crystal composite. Photonics. Lett. 8: 23-25.

Wang, M., Sayed, S.M., Guo, L.-X., Lin, B.-P., Zhang, X.-Q., Sun, Y., et al. 2016. Multi-stimuli responsive carbon nanotube incorporated polysiloxane azobenzene liquid crystalline elastomer composites. Macromolecules 49: 663-671.

Wang, W., Sun, X., Wu, W., Peng, H. and Yu, Y. 2012. Photoinduced deformation of crosslinked liquid-crystalline polymer film oriented by a highly aligned carbon nanotube sheet. Angew. Chem. 124: 4722-4725.

Watkinson, A.J., Murby, E.J. and Costanzo, S.D. 2007. Removal of antibiotics in conventional and advanced wastewater treatment: Implications for environmental discharge and wastewater recycling. Water Res. 41: 4164-4176.

Weng, X., Cai, W., Lan, R., Sun, Q. and Chen, Z. 2018. Simultaneous removal of amoxicillin, ampicillin and penicillin by clay supported Fe/Ni bimetallic nanoparticles. Environ. Pollut. 236: 562-569.

Xian, Y., Zhang, Y., Rahman, N.U., Yin, H., Long, Y., Liu, P., et al. 2020. An emerging all-inorganic $CsSn_xPb_{1-x}Br_3(0 \leq x \leq 1)$ perovskite single crystal: Insight on structural phase transition and electronic properties. J. Phys. Chem. C 124: 13434-13446. https://doi.org/10.1021/acs.jpcc.0c02423.

Xie, J.-L., Huang, Z.-Q., Wang, B., Chen, W.-J., Lu, W.-X., Liu, X., et al. 2019. New lead-free perovskite Rb7Bi3Cl16 nanocrystals with blue luminescence and excellent moisture-stability. Nanoscale 11: 6719-6726. https:// doi.org/10.1039/c9nr00600a.

Xu, T.T. and Lee, J. 2014. Nanomaterials: Electrical, magnetic, and photonic applications. JOM 66: 654.

Xu, Z., Liu, X., Li, Y., Liu, X., Yang, T., Ji, C., et al. 2019. Exploring lead-free hybrid double perovskite crystals of $(BA)_2CsAgBiBr_7$ with large mobility-lifetime product toward X-ray detection. Angew. Chem. 58: 15757-15761. https://doi.org/10.1002/anie.201909815.

Yadav, S.P., Pande, M., Manohar, R. and Singh, S. 2015. Applicability of TiO_2 nanoparticle towards suppression of screening effect in nematic liquid crystal. J. Mol. Liq. 208: 34-37. https://doi.org/10.1016/j.molliq.2015.04.031.

Yamagishi, T., Miyazaki, T., Horii, S. and Akiyama, K. 1983. Synthetic musk residues in biota and water from Tama River and Tokyo Bay (Japan). Arch. Environ. Contam. Toxicol. 12: 83-89.

Yang, B., Chen, J., Hong, F., Mao, X., Zheng, K., Yang, S., et al. 2017. Lead-free, air-stable all-inorganic cesium bismuth halide perovskite nanocrystals. Angew. Chem. 56: 12471-12475. https://doi.org/10.1002/anie.201704739.

Yang, D.K. and Wu, S.T. 2006. Fundamentals of Liquid Crystal Devices. John Wiley and Sons, New York, NY, USA.

Yang, H., Liu, J.-J., Wang, Z.-F., Guo, L.-X., Keller, P., Lin, B.-P., et al. 2015. Near-infrared-responsive gold nanorod/liquid crystalline elastomer composites prepared by sequential thiol-click chemistry. Chem. Commun. 51: 12126-12129.

Yang, J. and Xiang, H. 2008. Low dimensional nanomaterials for spintronics. One dimensional nanostructures. pp. 247-271. *In*: Wang, Z.M. [ed.]. Lecture Notes in Nanoscale Science and Technology, vol 3. Springer, New York, NY.

Yang, X., Sun, J., Fu, W., Shang, C., Li, Y., Chen, Y., et al. 2016. PPCP degradation by UV/chlorine treatment and its impact on DBP formation potential in real waters. Water Res. 98: 309-318.

Yang, Y., Zhan, W., Peng, R., He, C., Pang, X., Shi, D., et al. 2015a. Graphene-enabled superior and tunable photomechanical actuation in liquid crystalline elastomer nanocomposites. Adv. Mater. 27: 6376-6381.

Yang, Y., Ok, Y.S., Kim, K.-H., Kwon, E.E. and Tsang, Y.F. 2017. Occurrences and removal of pharmaceuticals and personal care products (PPCPs) in drinking water and water/sewage treatment plants: A review. Sci. Total. Environ. 596: 303-320.

Yao, L., Zhao, J.L., Liu, Y.S., Zhang, Q.Q., Jiang, Y.X., Liu, S., et al. 2018. Personal care products in wild fish in two main Chinese rivers: Bioaccumulation potential and human health risks. Sci. Total Environ. 621: 1093-1102.

Yee, K.O., Fu, Y.L. and Shi, P.S. 2014. Nanofiltration hollow fiber membranes for textile wastewater treatment: Lab-scale and pilot-scale studies. Chem. Eng. Sci. 114: 51-57.

Yin, J., Wang, H., Zhang, J., Zhou, N., Gao, F., Wu, Y., et al. 2012. The occurrence of synthetic musks in human breast milk in Sichuan, China. Chemosphere 87: 1018-1023.

Yoon, Y., Westerhoff, P., Snyder, S.A. and Wert, E.C. 2006. Nanofiltration and ultrafiltration of endocrine disrupting compounds, pharmaceuticals and personal care products. J. Membr. Sci. 270: 88-100.

Yu, D., Cui, J., Wang, Y. and Pei, Y. 2020. Removal of ibuprofen by using a novel Fe/C granule-induced heterogeneous persulfate system at near neutral pH. Ind. Eng. Chem. Res. 59: 1073-1082.

Yu, L., Cheng, Z., Dong, Z., Zhang, Y. and Yu, H. 2014. Photomechanical response of polymer-dispersed liquid crystals/graphene oxide nanocomposites. J. Mater. Chem. C. 2: 8501-8506.

Yu, L. and Yu, H. 2015. Light-powered tumbler movement of graphene oxide/polymer nanocomposites. ACS Appl. Mater. Interfaces 7: 3834-3839.

Yu, L. and Zhang, Y.T. 2015. High flux, positively charged loose nanofiltration membrane by blending with poly (ionic liquid) brushes grafted silica spheres. J. Hazard. Mater. 287: 373-383.

Yuan, C., Hung, C.H., Li, H.W. and Chang, W.H. 2016. Photodegradation of ibuprofen by TiO_2 co-doping with urea and functionalized CNT irradiated with visible light-effect of doping content and pH. Chemosphere 155: 471-478. https://doi.org/ 10.1016/j.chemosphere.2016.04.055.

Yuan, W., Niu, G., Xian, Y., Wu, H., Wang, H., Yin, H., et al. 2019. *In situ* regulating the orderedisorder phase transition in $Cs_2AgBiBr_6$ single crystal toward the application in an X-ray detector. Adv. Funct. Mater. 29: 1900234. https://doi.org/10.1002/adfm.201900234.

Zenker, A., Hansruedi, S. and Fent, K. 2008. Simultaneous trace determination of nine organic UV-absorbing compounds (UV filters) in environmental samples. J. Chromat. A. 1202: 64-74

Zhang, G., Liu, X., Sun, K., He, Q., Qian, T. and Yan, Y. 2013a. Interactions of simazine, metsulfuron-methyl, and tetracycline with biochars and soil as a function of molecular structure. J. Soils Sediments 13: 1600-1610.

Zhang, H., Cheng, S., Li, B., Cheng, X. and Cheng, Q. 2018. Fabrication of magnetic Co/$BiFeO_3$ composite and its advanced treatment of pharmaceutical waste water by activation of peroxysulphate. Sep. Purif. Technol. 202: 242-247.

Zhang, J., Yang, Y., Deng, H., Farooq, U., Yang, X., Khan, J., et al. 2017. High quantum yield blue emission from lead-free inorganic antimony halide perovskite colloidal quantum dots. ACS Nano 11: 9294-9302. https:// doi.org/10.1021/acsnano.7b04683.

Zhang, S., Pelligra, C.I., Feng, X. and Osuji, C.O. 2018. Directed assembly of hybrid nanomaterials and nanocomposites. Adv. Mater. 30: 1705794.

Zhou, L., Liao, J.-F., Huang, Z.-G., Wang, X.-D., Xu, Y.-F., Chen, H.-Y., et al. 2018. All-inorganic lead-free Cs_2PdX_6 (X = Br, I) perovskite nanocrystals with single unit cell thickness and high stability. ACS Energy Lett. 3: 2613-2619. https://doi.org/10.1021/acsenergylett.8b01770.

Zhu, J.Y. and Tian, M.M. 2015. Fabrication of a novel "loose" nanofiltration membrane by facile blending with chitosan–montmorillonite nanosheets for dyes purification. Chem. Eng. J. 265: 184-193.

Zouzelka, R., Kusumawati, Y., Remzova, M., Rathousky, J. and Pauporté, T. 2016. Photocatalytic activity of porous multiwalled carbon nanotube-TiO_2 composite layers for pollutant degradation. J. Hazard. Mater. 317: 52-59. https://doi.org/ 10.1016/j.jhazmat.2016.05.056.

Supramolecular Gels as Smart Sorbent Materials for Removal of Pollutants From Water

Bhagwati Sharma

1. INTRODUCTION

Clean water is one of the most basic requirements for all living beings. However, the scarcity of clean water has been recognized as one of the most serious issues for mankind in the twenty-first century (Mishra and Clark 2013, Okesola and Smith 2016). The main reasons for such scarcity of water are increased population, climate changes and rapid industrialization (Harrison 2006, Albelda et al. 2012, Wang et al. 2019, Kummu et al. 2016). It is almost impossible to imagine life without industrialization in the present world, but its pace has severely impacted the quality of water across the globe. The discharge of hazardous waste materials, such as metal cations and anions, toxic dyes, pesticides, fertilizers, etc., is the most prominent reason for water pollution (Okesola and Smith 2016, Wang et al. 2019, Lim et al. 2019, Sharma et al. 2013, Marimuthu et al. 2020, Zhang et al. 2016). The presence of these hazardous materials in water can have several deleterious effects on both human and aquatic lives. The consumption of water containing such toxic materials can lead to cancer, liver damage, neurological disorder as well as cognitive dysfunction (Okesola and Smith 2016). Therefore, the development of biological as well as physicochemical methods to fight water pollution has been a focal point of research. Several physicochemical methods, such as sedimentation, flocculation as well as photocatalysis, have been used for the treatment of polluted water, but these methods have proven to be less effective in the treatment of water containing micropollutants, such as toxic household as well as industrial chemicals (Lim et al. 2019). Adsorption technique offers advantages, such as simplicity, easy operation, economic feasibility and regeneration, which has led to its use in the uptake of such organic/inorganic micropollutants (Okesola and Smith 2016, Lim et al. 2019). Adsorbent materials, such as activated carbon, zeolites, minerals and organoclays, noble metal nanoparticles and inorganic oxides, have been reported to be used for the removal of micropollutants (Hager 1967, Wong et al. 2018, Wang and Peng 2010, Srinivasan 2011, Beall 2003, Pradeep and Anshup 2009, Lin et al. 2020, Maleki 2016). Nevertheless, the majority of such adsorbent materials suffer from disadvantages, such as poor bio-degradability, lack of selectivity, costly regeneration processes, etc. (Zhao et al. 2014, Smith et al. 2016).

Gels are a class of soft materials that consist of a three-dimensional cross-linked solid-like network capable of trapping a large volume of water (hydrogels) or organic solvents (organogels)

Materials Research Centre, Malaviya National Institute of Technology, Jaipur-302017, India.
E-mail: bhagwati.mrc@mnit.ac.in.

(Sangeetha and Maitra 2005, Mondal et al. 2020, Nebot and Smith 2013). Gels have found widespread applications in cosmetics, pharmaceuticals, lubricants as well as the food industry. Most of the gels used in the everyday application are primarily polymeric in nature, wherein the polymers are entangled into a three-dimensional solid, like the network (Appel et al. 2012). However, in recent times, a new class of gels derived from low molecular weight organic gelators (molecular mass <3,000) have gained immense interest. Such gels, known as supramolecular gels, form by the self-assembly of the gelator molecules under the influence of various non-covalent interactions, such as H-bonding, π-π stacking, dipole-dipole interactions, solvophobic effects and coordination interactions (Sangeetha and Maitra 2005, Du et al. 2015, Babu et al. 2014, Tam and Yam 2013, Sharma et al. 2016, Sharma et al. 2018a, b, Banerjee et al. 2009). The self-assembly of the gelator molecules within the solvent leads to the formation of solid-like entangled networks, mostly composed of nanofibers/microfibers (Figure 1). Supramolecular gels, although highly solvated, have rheological properties similar to that of solids but possess highly porous structures, which allow the percolation of solvent molecules and significant internal diffusion kinetics (Okesola and Smith 2016). These gels have a very high exposed surface area, which allows the solid network in the gel materials to be in intimate contact with the liquid phase (Lim et al. 2019). The material properties of these gels, such as stiffness, sol to gel transformation, etc., can easily be tuned by varying the molecular constituents and functional groups on the gelator molecules. Furthermore, the supramolecular gels are also stimuli-responsive and self-healing in nature (Häring and Dìaz Dìaz 2016, Sharma et al. 2018c, Rahim et al. 2016, Thakur et al. 2018, Qi and Schalley 2014). Owing to the presence of weak, non-covalent interactions that can easily be broken and reformed (Häring and Dìaz Dìaz 2016, Qi and Schalley 2014), supramolecular gels can undergo reversible gel to sol transition under the influence of various stimuli, such as heat, pH, light, chemicals and sonication (Sharma et al. 2018c, Rahim et al. 2016, Thakur et al. 2018).

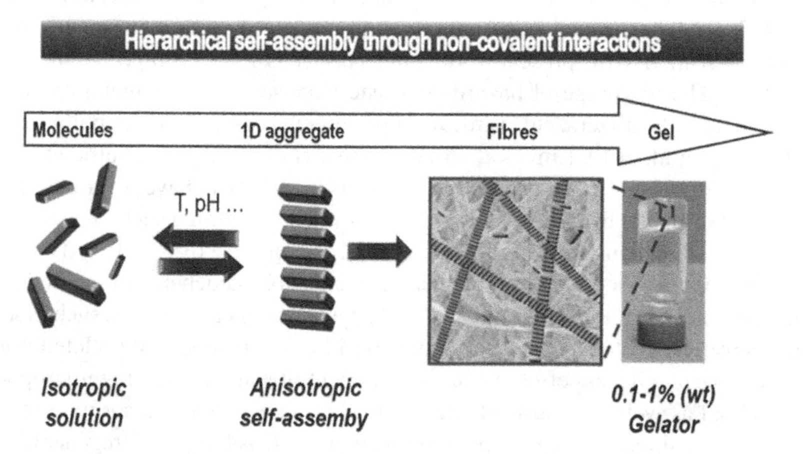

FIGURE 1 Schematic representation for the bottom up self-assembly of low molecular weight compounds leading to gel formation. Reproduced with permission from Nebot and Smith 2013. Copyright 2013 The Royal Society of Chemistry.

Despite finding applications in numerous areas including but not limited to drug delivery, tissue engineering, energy storage, molecular sensing and catalysis (Hirst et al. 2008, Skilling et al. 2014, Saunders and Ma 2019, Thakur et al. 2019), the potential of supramolecular gels for environmental remediation and wastewater treatment had not been realized. Only in the past few years, there has been a surge in the exploitation of supramolecular gels for the treatment of wastewater. The highly porous nature and high surface area in gel materials allow the immobilization of a range of diverse chemical entities and the well-established chemistry allows stimuli responsiveness, self-healing, programmability and biocompatibility. When combined, the gel materials exhibit unique properties, which are not usually observed in the traditional materials, are used for water treatment. Moreover, most of the supramolecular gels are environmentally friendly, and they do not act as a source of pollution. Due to their robust nature, they can easily be recycled and they do not require the use of sophisticated equipment for operation (Okesola and Smith 2016, Lim et al. 2019). In this chapter, the application of supramolecular gels for the removal or degradation of pollutants from water is discussed. Specifically, the chapter focuses on the use of supramolecular gels for oil spill recovery from water, adsorption and degradation of pollutant dyes and uptake and removal of toxic metal cations and anions.

2. SUPRAMOLECULAR GELS FOR TREATMENT OF WASTEWATER

2.1 Supramolecular Gels for Remediation of Oil Spills

The intentional or accidental discharge of crude oil and other petroleum products into water is the main reason for marine pollution and poses serious environmental concerns (Doshi et al. 2018). The discharge of large quantities of crude oil and other such products into seawater not only causes economic loss by the wastage of oil but also has a lasting effect on sea animals and humans who consume the seafoods obtained from the polluted water region (Okesola and Smith 2016). Moreover, such release or discharge of crude oil into the water also leads to the accumulation of harmful volatile hydrocarbons in the atmosphere, which seriously impacts the climate (Guterman 2009). The methods usually employed for the remediation of oil spills in water, include absorption, dispersion and solidification by the formation of a gel (Dave and Ghaly 2011, Prenderghast and Gschwend 2014). While the absorption method involves the use of a sorbent material to absorb the oil, the dispersion method leads to emulsification of the oil and the solidification method involves the use of polymeric materials, which can selectively gel the oil from an oil-water mixture (Okesola and Smith 2016). Nevertheless, these methods suffer from disadvantages, such as high cost or even they can themselves leave behind certain impurities, which can bioaccumulate through food chains. Supramolecular gels that are economically viable, environmentally benign, thermo-reversible and reusable are excellent materials for remediation of oil spills. Recently, phase selective organogelators (PSOGs) which can selectively gel oil in presence of water have come to the forefront for oil spill remediation. These PSOGs can be applied in powdered form to selectively gel the oil phase without the use of any toxic co-solvent. Several groups are working on the use of supramolecular gels for oil spill remediation. However, some of the reported gelators, which show the ability to gel oil selectively from oil-water biphasic mixture require heating-cooling cycles for gelation (Bhattacharya and Krishnan-Ghosh 2001, Trivedi et al. 2003, Trivedi et al. 2004), which is not feasible for practical applications and are not discussed here. Some of the recent works which have employed PSOGs for oil spill remediation and do not require heating-cooling to form the gels are discussed below.

A glucose-based phase selective organogelator (1) was synthesized in 2016 by substituting the O-alkyl group at the anomeric position of 4,6-O-benzylidene-glucopyranosides by a thioaryl or thioalkyl group (Vibhute et al. 2016). The motive behind modifying the compound was to make it more hydrophobic without compromising the gelation ability of the gelator. The authors synthesized five gelators by substituting the O-alkyl group with five different thioalkyl or thioaryl groups (Figure 2). All the gelators showed the ability to gel non-polar liquids, such as silicon oil, pump oil and diesel. Two of the five gelators which formed stable and strong gels in aromatic solvents and crude oil were then used for gelation of oil in an oil-water biphasic system. Although the biphasic critical gelation concentration (0.7 and 0.9 wt% respectively) were slightly higher than their critical gelation concentration (0.5 and 0.6 wt% respectively), the gelators showed excellent phase selectivity by selectively gelling the oil phase from the oil-water biphasic system. To show the practical utility of their gelator, the authors dispersed the powdered gelator 1-(i) in the crude oil-sea water mixture (20:200 v/v) in a glass bath. The mixture was gently stirred and after some time they observed that the oil layer was gelled and stopped swirling, while the water layer continued to swirl, suggesting that the gelator could diffuse through the oil phase to selectively gelate the oil phase.

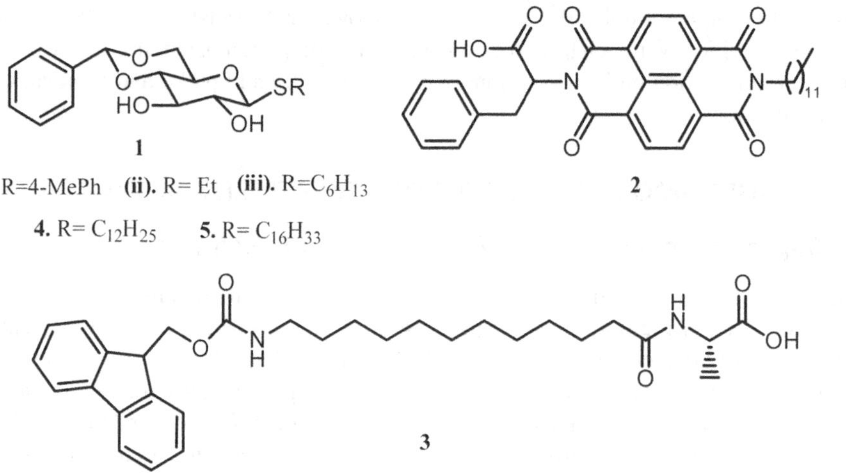

(i.) R=4-MePh **(ii).** R= Et **(iii).** R=C_6H_{13}

4. R= $C_{12}H_{25}$ **5.** R= $C_{16}H_{33}$

FIGURE 2 Chemical structures of the gelators used for oil spill recovery by different research groups. (1) Vibhute et al. 2016, (2) Datta et al. 2018 and (3) Basu et al. 2017 respectively.

The same group also developed a novel sorbent material by impregnation of cellulose pulp with a mannitol-derived oleogelator and 1,2:5,6-di-O-cyclohexylidene mannitol (Prathap and Sureshan 2017). The mannitol derivative is a cheap phase selective gelator and can be synthesized easily. Moreover, the gelator when impregnated with cellulose pulp could mask the exposed hydroxyl groups of the cellulose fibers by forming H-bonding and their hydrophobic parts make the otherwise hydrophilic cellulose fibrils hydrophobic. The gelator adsorbed cellulose pulp (GACP) was prepared by adding cellulose pulp globules into a hot hexane solution of the gelator followed by drying. SEM images confirmed that the adsorption of the gelator onto the cellulose pulp did not affect the morphology of the cellulose pulp. The hydrophobicity of GACP was studied by dropping a drop of water onto a glass slide coated with GACP that showed a contact angle of 110°C, confirming the hydrophobic nature of GACP. The authors further illustrated the hydrophobicity of the GACP by adding GACP and cellulose pulp separately in two beakers containing water. While the pure cellulose pulp absorbed water and started sinking in the beaker, the GACP was unaffected and floated at the surface, which further confirmed their hydrophobic nature. Oil absorption studies were performed by separately treating crude oil-water mixture (80 mL crude oil + 15 mL seawater) with 100 mg of the gelator, 500 mg of cellulose pulp and 500 mg of GACP. The gelator formed a semi-solid by absorbing the oil, but it could not be separated from the water phase. On the other hand, the cellulose pulp alone absorbed some oil along with water and submerged in water. Upon removal from water, the oil oozed out of the cellulose pulp. The inefficient absorption by the cellulose pulp suggested that the absorption of oil was only due to the spongy nature of the pulp, and there was no adhesive interaction between the cellulose pulp and the crude oil. Interestingly, the GACP absorbed all the oil and floated on the water surface and unlike cellulose pulp, it retained the oil even after separation from water (Figure 3). The GACP after the absorption of the oil could easily be scooped off, leaving clean water behind. Moreover, the GACP could congeal oil and could form a stronger gel at a concentration lower than the critical gelation concentration.

FIGURE 3 Demonstration of the oil spill recovery using the mannitol derivative-cellulose pulp composite. (a) The sorbent to be used to congeal oil selectively from marine water (3% NaCl). (b) Biphasic system formed by introducing crude oil into marine water. (c) The composite sorbent is spread over the oil surface in the biphasic mixture. (d) Absorption of the oil phase by the composite instantaneously. (e) The congealed oil is scooped using a sieve spatula, and (f) the recovered oil from the biphasic system. Reproduced with permission from Prathap and Sureshan 2017. Copyright 2017 John Wiley and Sons.

A naphthalene diimide (NDI) based supramolecular gelator (2) was used to congeal heavy crude oil from water in seconds (Datta et al. 2018). The gelator was composed of a phenylalanine unit that served as an H-bonding motif at one end of the NDI core, which could be used for pi-stacking. Apart from these two groups that allowed two different non-covalent interactions, the other terminal of NDI was functionalized with an *n*-dodecyl group to allow for the dispersibility of the gelator in an organic medium, such as oil. The functionalization with the non-polar group assured almost complete insolubility in water, which decreased any detrimental effect on marine life. The FESEM image of the gel in methylcyclohexanone revealed that the gel was composed of fibrous aggregates that were hierarchically assembled into porous globular microstructures. The ability of the gel to congeal oil was studied by pouring heavy crude oil (25 mL) in water (150 mL) containing 3.5% salt, followed by a sprinkling of the dry powder of the gelator (2.5 wt%) over the oil layer. Interestingly, it was observed that the crude oil layer congealed within 20 seconds, which could be scooped off from the water.

Banerjee and co-workers synthesized a peptide-based gelator (Basu et al. 2017), which comprised of various zones of non-covalent interactions, like π-π stacking, H-bonding, Van der Waal's interactions and a polar head group (3). The gelator was ambidextrous and could gelate various solvents, such as water (pH 7.5-8.0), as well as aromatic and aliphatic organic solvents, such as benzene, toluene, *o*-xylene, hexane, octane and petroleum ether. The ability of the gelator to remove oil from the oil-water mixture was studied by injecting 100 μL of the gelator in ethyl acetate (10% w/v) to 4 mL of 1:1 mixture of saltwater and oil. Within 20 seconds, the oil part was gelled by the gelator solution, leaving behind water. Furthermore, the oil could be removed from the gel using a rotavapor and the dried gelator could be reused for subsequent cycles.

Recent advancements in the field of PSOGs have addressed several issues, which are critical in solving the oil spill problem. PSOGs have been developed, which do not require a heat-cool cycle to form the gel and thus can easily be applied in the form of powder to form gels at room temperature. Some of the parameters, which are critical to the design of PSOGs are: (i) the gelator should possess H-bonding groups for gel formation; (ii) it should also possess groups with π-π stacking interactions that can assist in the self-assembly process; (iii) it should consist of non-polar alkyl groups to allow solubility in the oil medium, which is essential for rapid gelation and use of the gelator in powdered form for removal of oil from water.

2.2 Supramolecular Gels for Dye Removal From Water

Commercial dyes are widely used in textile industries. Their use, however, is not limited to textile industries and they have also found widespread use in food, paint, drugs, cosmetics, etc. (Okesola and Smith 2016, Anwer et al. 2019, Routoula and Patwardhan 2020). Being non-biodegradable in nature, the discharge of even significantly lower concentrations of these dyes in lakes and rivers can significantly affect the aquatic environment. These dyes affect the immune and reproductive systems of animals and also exhibit potential genotoxicity and cardiotoxicity (Ratna and Padhi 2012). Therefore, the removal of toxic dyes from the biosphere has been a focal point of research. Supramolecular gels due to their highly porous fibrous structure provide a large surface area in the gel matrix to trap toxic pollutant dyes. While the enhanced surface area permits direct contact of the gel with the dyes in water, the reversibility and stimuli-responsive nature of the gels allows for its recyclability (Okesola and Smith 2016). Moreover, the gels can be used for trapping the dye molecules either in the solvated (gel) state or in the dried state (xerogel). Further, the amphiphilic nature of the gels assists them in forming interactions with the dye molecules, and the structural design flexibility in supramolecular gels may allow for the selective adsorption of dyes with specific charges.

In one of the initial works on the use of gels for removal of dyes, a pH-responsive peptide-based bola-amphiphile (4), which formed a supramolecular gel upon sonication in the presence of divalent metal ions (pH 6.5-7.2) was used for the removal of crystal violet (cationic), naphthol blue-black (anionic) and pyrene (non-ionic) dyes from water Ray et al. 2007). The dried xerogel was submerged into an aqueous solution containing the various dyes for the removal of the dyes and within a few hours, all the dye content was adsorbed by the xerogel, leaving clear water (Figure 4b). Ionic dyes were adsorbed more efficiently and the maximum adsorption was calculated to be 84 mg g^{-1} for naphthol blue-black. TEM and EDX studies of the xerogels after dye adsorption indicated that the dye particles were adsorbed on the nanofibrous network of the gel (Figures 4c and d).

In another work, the same group synthesized a library of tripeptide-based gelators (5) that could gel water (Adhikari et al. 2009). The hydrogels in the gel state were used for the removal of rhodamine B, reactive blue 4 and direct red 80 from water. Within 48 hours, all three dyes could almost completely be adsorbed by the gel. The hydrogelators, however, self-assembled to form stable hydrogels only at a pH greater than 11.5, below which they were insoluble and precipitated out in the water. This property of the hydrogels was utilized for the regeneration of the gelators, whereby upon decreasing the pH, the gelators precipitated out in the water, leaving the toxic dyes in water. The gelator was then filtered and washed, dried and could be reused.

A two-component supramolecular gel comprising of adenine and tricarboxylic acids was reported in 2013 (Sukul and Malik 2013). The hydrogels were thermoreversible and demonstrated modest adsorption capability for methylene blue, Rhodamine 6G and crystal violet from their aqueous solutions. However, the maximum uptake of the dyes was only 6-8 mg g^{-1}.and the adsorption process was quite slow. The process required nearly 48 hours for adsorption of the dyes. Hydrophobic interactions between the gelators and the dye molecules were suggested to be responsible for the adsorption of the dyes.

A two-component metal-organic gel comprising of pyridine-3,5-bis(benzimidazole-2-yl) ligand (6) as the organic part and divalent cadmium/copper ions as the metal component was fabricated (Samai and Biradha 2012). The coordination interaction between the two components resulted in the formation of self-assembled intertwined three-dimensional fibrous networks, which could gelate several aliphatic alcohols. Since the ligand is comprised of both hydrophilic and hydrophobic domains (Figure 5), therefore its potential to adsorb toxic dyes was studied. The dye adsorption studies were performed by the addition of small quantities of the dried xerogels to an aqueous solution of methyl orange. After nearly 24 hours, almost all of the dye was adsorbed by the xerogel and nearly clear water could be observed. Microscopic studies confirmed that the dye was entrapped within the fibrillar network of the gel. However, the dye absorption efficiency of these gels was

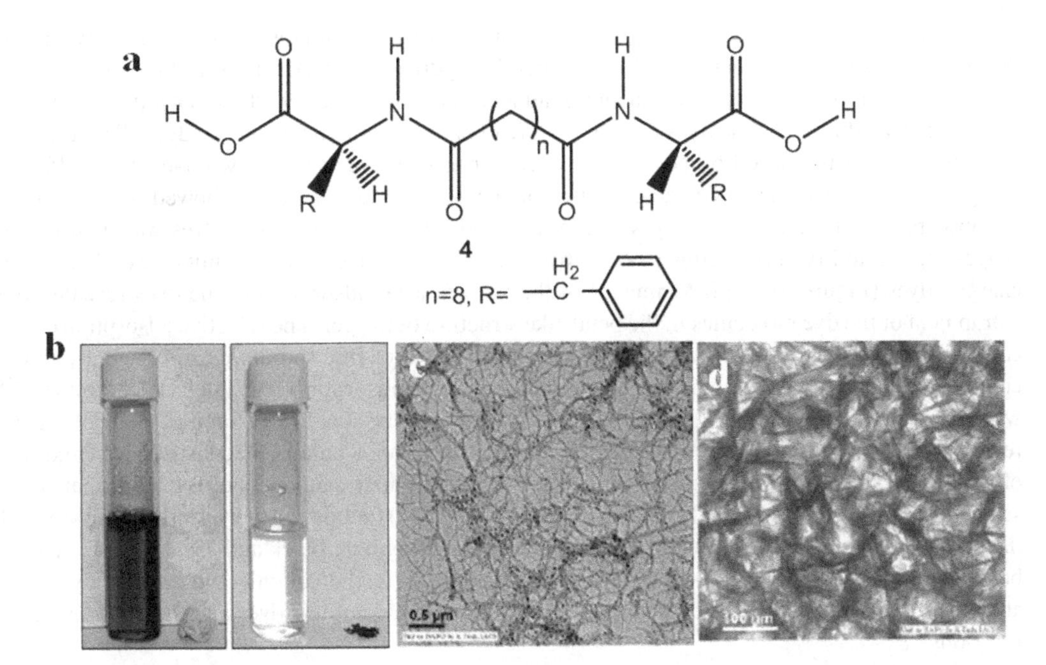

FIGURE 4 (a) Chemical structure of the peptide-based gelator employed for dye adsorption by Ray et al. 2007. (b) Digital image showing the adsorption of crystal violet dye by the xerogel formed by drying the gel of (4) with Cu^{2+} ions. (c), (d) TEM images of the xerogels after adsorption of Naphthol blue-black and crystal violet respectively. Reproduced with permission from Ray et al. 2007. Copyright 2007 American Chemical Society.

significantly lower and a maximum dye uptake of 10.7 mg per gram of the gel was observed. The group further extended the scope of the organic ligand and substituted the N-H group of benzimidazole to afford the ligand pyridine-3,5-bis(1-methylbenzimidazole-2-yl) (**7**) (Dey et al. 2013). The interaction of the ligand with halides of Cu^{2+} and Cd^{2+} in alcohol led to the self-assembly of the gelator into fibrous structures, which could trap the solvent molecules in the matrix of fibers, leading to the formation of gels. The presence of both hydrophilic (imine N-atom) and hydrophobic groups (N-Me and aromatic rings) in the ligand assisted in the use of these xerogels for the removal of the dyes MO and Rh B from water.

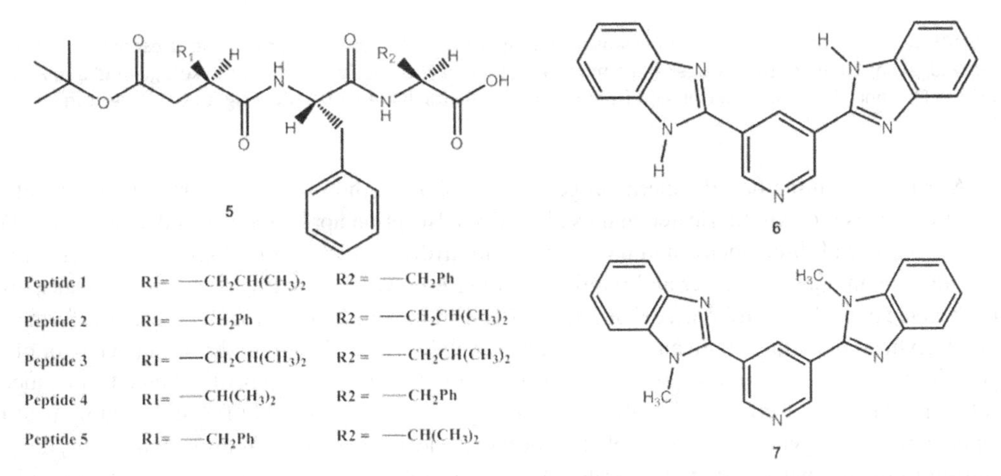

FIGURE 5 Chemical structures of the phenylalanine-based tripeptide gelators (5) synthesized by Samai and Biradha 2012 and chemical structures of the pyridine-based organic ligands (6) and (7) used to synthesize metallogels with cadmium/copper ions by Dey et al. 2013.

Another series of metal-organic gels were fabricated by coordination interaction between the disodium salt of carboxymethyl-(3,5-di-*tert*-butyl-2-hydroxy-benzyl)-amino acetic acid (Na_2HL) with Cd(II) and Zn(II) halides (Karan and Bhattacharjee 2016). The gels showed a unique ability to selectively adsorbing cationic dyes from a mixture of cationic, anionic or neutral dyes. The selective dye adsorption was studied by adding the dried xerogel to a mixture of two dyes with different charges. Time-dependent UV-visible studies of the aqueous dye mixture showed a decrease in the absorption band for cationic dyes, while the band for anionic or neutral dyes was unaffected, suggesting the ability of the xerogel to selectively adsorb cationic dyes from a mixture of differently charged dyes (Figure 6). FESEM images of the xerogel after adsorption of the dyes revealed the entrapment of the dye molecules in the petal-like structure of the gel. The selective adsorption of the cationic dyes by the xerogels was attributed to the presence of free sodium ions in the gels, which could be replaced by the cationic dyes. Although moderate adsorption of crystal violet (56 mg g^{-1}) was observed, the uptake of Rhodamine B and methylene blue was low (9.6 mg g^{-1} and 7.9 mg g^{-1} respectively). Davis and co-workers recently reported supramolecular gels consisting of mixtures of guanosine and 8-aminoguanosine in presence of stoichiometric amounts of Ba^{2+} ions that could selectively absorb anionic dyes from water (Plank et al. 2017). While this gel absorbed only 15% of the cationic dye safranin O, it could remove 89% of naphthol blue-black and 75% of Rose Bengal, both of which are anionic dyes. The difference in affinity to the cationic and anionic dyes was attributed to the stronger electrostatic interactions between the anionic dyes and the gel containing Ba^{2+} ions.

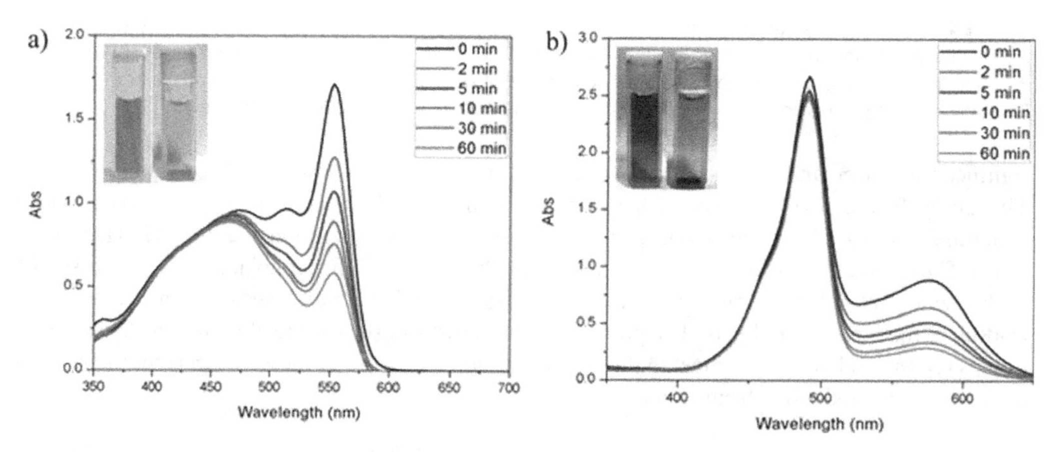

FIGURE 6 Time-dependent UV-visible absorption spectrum of a dye mixture in the presence of Cd-Na_2HL metallogel. (a) 10 mg of the xerogel and a mixture of RB and MO and (b) 10 mg of the xerogel and a mixture of CV and FL. Reproduced with permission from Karan and Bhattacharjee 2016. Copyright 2016 American Chemical Society.

A triphenylalanine-based superhydrogel (8) that showed the unique property of self-shrinking (syneresis) was recently fabricated and used for dye adsorption application (Basak et al. 2017). Due to the presence of three phenylalanine moieties with hydrophobic, π-core (Figure 7) the gelator was insoluble in an aqueous buffer and required heating followed by cooling to afford the transparent hydrogel in phosphate buffer at a pH of 7.4. The FESEM images of the xerogel revealed the formation of intertwined nanofibers with an average diameter of 25-45 nm. The self-shrinking behavior of the gel with time was exploited for the removal of toxic methylene blue and brilliant blue from aqueous solutions. The dyes were added to the solution of gelator in phosphate buffer followed by heating. Upon cooling, the gelator self-assembled to form dye encapsulated gels. Within 60 minutes, the gels started to shrink by expelling clear water. The UV-visible spectrum of the expelled water did not show any peaks corresponding to the dyes, suggesting all the dye molecules were trapped by the gel in the shrunken state. Using these gels, high uptake of methylene blue (420 mg g^{-1} of the gelator) and brilliant blue (120 mg g^{-1} of the gelator) could be achieved.

Takashita et al. synthesized a tris-urea based gelator (9) that self-assembled into fibers upon the addition of NaOH (Takashita et al. 2017). The addition of HCl or $CaCl_2$ to the above solution afforded a self-standing supramolecular gel (1-HCl or 1-$CaCl_2$). The dye adsorption ability of these gels was studied by submerging a cylindrical supramolecular gel into an aqueous solution of cationic dyes such as Rhodamine 6G and methylene blue. It was observed that the maximum uptake of Rh6G and MB was 604 and 1,243 mg respectively per gram of the gel using 5 mM 1-HCl gel. On the other hand, under similar conditions and concentration, the 1-$CaCl_2$ gel could adsorb 892 and 1,261 mg Rh 6G and MB respectively.

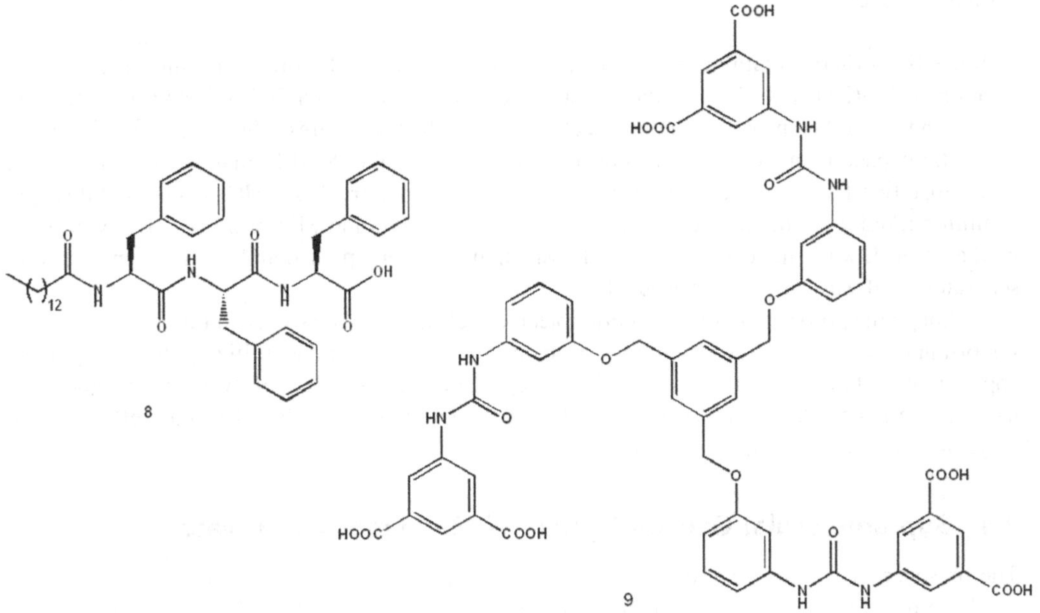

FIGURE 7 Chemical structures of the triphenylalanine-based superhydrogel (8) that showed syneresis property (Basak et al. 2017) and tris urea based gelator (9) synthesized by Takashita et al. 2017.

A novel dibenzylidene sorbitol (DBS) functionalized with hydrazide (DBS-$CONHNH_2$) gelator (10) was synthesized by Smith's group in 2013 (Okesola and Smith 2013). The gelator self-assembled in water-DMSO mixture under simple heating-cooling into fibrous nanostructures with an average diameter of 10 nm. The solution upon cooling led to the trapping of water molecules to afford a self-standing hydrogel. This gel was stable in a broad pH range from 2 to 11.5. Due to its ability to withstand changes in pH, it was utilized for the removal of high concentrations (700-1,050 mg g^{-1}) of various dyes at different pH values. The dye adsorption studies were performed by pipetting out the dye solutions on top of the hydrogel and allowing for the dye to adsorb in the pores of the hydrogel. Under basic conditions, this hydrogel was capable of removing cationic dye (MB) from an aqueous solution. The anionic dyes acid blue 25 and naphthol blue-black were better adsorbed in acidic pH (Figure 8b). Moreover, the gels could also adsorb a mixture of anionic dyes from water under acidic conditions, which is a real-world scenario where a mixture of different dyes is present in wastewater.

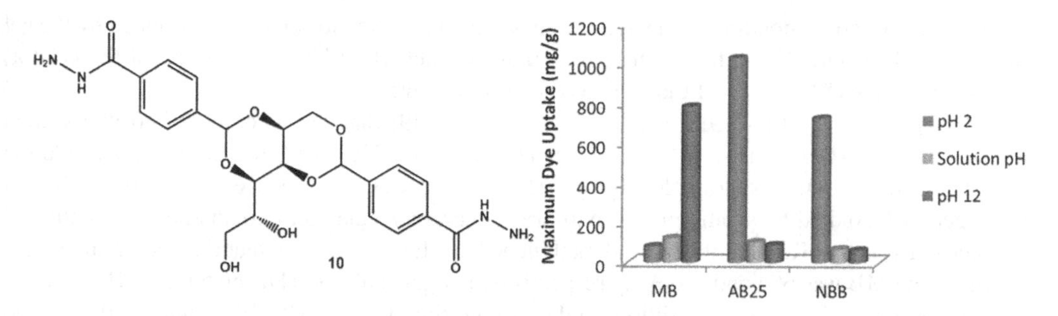

FIGURE 8 Chemical structure of the DBS functionalized hydrazide gelator (10) and maximum dye adsorption by the hydrogel (10) at different pHs. Reproduced with permission from Okesola and Smith 2013. Copyright 2013 The Royal Society of Chemistry.

Recently, a diethanolamine-based metallogel was fabricated by direct mixing of an aqueous solution of $Mg(NO_3)_2{\cdot}6H_2O$ with diethanolamine (Alam and Sarma 2020). Electron microscopy studies indicated that the gel was composed of 2D sheet-like structures. The dried xerogel was used as a column packing material and was utilized for the removal of RhB from a mixture of dyes with high separation factors. When a mixture of RhB + congo red or RhB + MO was passed through a column packed with the xerogel, the anionic dyes (congo red and MO) were selectively adsorbed by the xerogel, while the cationic dye RhB was unaffected and passed and the solution containing separated RhB was collected as effluent.

Although the porous structure of supramolecular gels has shown promise for dye adsorption, it is important to understand that further improvements are necessary before utilizing them for practical applications. Although some gels have shown high and fast adsorption of dyes, while some others have shown good selectivity toward dyes of a particular charge, yet the adsorption efficiency and recyclability require further improvements.

2.3 Supramolecular Gels for Uptake of Metal Ions From Water

The presence of various toxic metal ions in concentrations above the permissible limits in water can have deleterious effects on human health (Chonamada et al. 2019). Metal ions such as Hg^{2+}, Pb^{2+} and Cd^{2+} have a very strong binding affinity toward various biomolecules, which can lead to either modification in the activity of the biomolecules or their complete activity loss. Therefore, these toxic metal ions when consumed directly or indirectly above the permissible limits can cause severe diseases, including kidney failure, damage to the central nervous system and immune system (Aragay et al. 2011). Unfortunately, these hazardous metal ions are a constituent of industrial waste, which is discharged in water regularly. Supramolecular gels can be used to remove these metal ions via two different strategies. One of the strategies utilizes the coordination ability of the metal ions to bind to specific ligands leading to the formation of a metal-organic gel, hence effectively separating the toxic metal ions from water into the nanoscale gel network. The second strategy exploits the porous structure of supramolecular gels for the adsorption of these toxic metal ions into the matrix of the gel. Some of the examples of supramolecular gels for uptake and removal of toxic metal ions are discussed below:

A novel pyridine-pyrazole-based amide molecule was synthesized recently and used to sequester three toxic metal ions, viz Pb^{2+}, Hg^{2+} and Cd^{2+} (De and Mondal 2018). The molecule, N^2, N^6-bis (5-(3)-(pyridine-2-yl)-1H-pyrazole-3(5)-carbonyl)-pyridine-2,6-dicarbohydrazide (11) (BP3D) is unique in the sense that it has multiple metals chelating and solvent immobilizer functional groups. Initially, BP3D alone was tested to see if it could self-assemble to form a gel with fibrous morphology in a DMSO-water mixture. Unsurprisingly, it led to the formation of a stable and transparent self-assembling gel. The ability of the BP3D hydrogel to capture the toxic metal ions was then studied by adding an aqueous solution of the metal salts on top of the BP3D gel (Figure 9b). Within a few minutes, the color of the top layer of the gel which was in contact with the metal ion started showing a color change. The color change was not just confined to the interface, but gradually with time, the color started spreading throughout the gel, suggesting the possible coordination of the ligand BP3D with the metal ions. However, the gel did not undergo a gel-sol transition, as evidenced by the tube inversion method. The TEM image of the metallogel obtained by the addition of the metal ions to the hydrogel showed the formation of nanoscale metal-organic particles originating from the nanofibers in all three cases. The BP3D gel was capable of capturing 85.63% of Pb^{2+} ions and 99.95% of Hg^{2+} ions from water in 3 months. Although the process of metal sequestration is significantly slow, nevertheless it opens up avenues for developing gel-based material for sequestration of toxic metal ions from wastewater using hydrogel columns as filters.

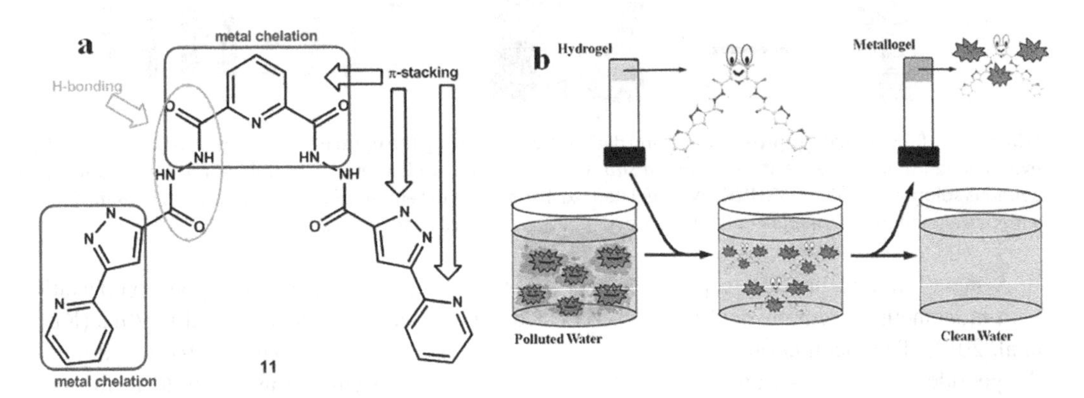

FIGURE 9 (a) Chemical structure of BP3D gelator with multiple metals chelating, H-bonding and π-π stacking interaction sites. (b) Schematic representation for the capture and uptake of toxic metal ions from contaminated water by BP3D leading to metallogel formation. Reproduced with permission from De and Mondal 2018. Copyright 2018 American Chemical Society. Source: https://pubs.acs.org/doi/10.1021/acsomega.8b00758. Further permissions related to the material excerpted should be directed to the ACS.

Smith and co-workers used the DBS-$CONHNH_2$ (10) supramolecular gel to selectively extract precious metal ions via the reduction of metal ions to form metal nanoparticles (Okesola et al. 2016). Water contaminated with gold and silver cations was added on top of the preformed hydrogel and allowed to stand for 48 hours to allow for the diffusion of the cations in the gel. ICP-MS analysis revealed that the maximum uptake of the gold and silver ions was 2,000 mg g^{-1} and 900 mg g^{-1} (metal: gelator), respectively, which was higher than many reported adsorbents. As the metal ions started diffusing into the gel, the gel started changing its color from colorless to ruby red and yellow, suggesting the formation of Au and Ag nanoparticles, respectively. While Au nanoparticles started to form within 2 seconds, Ag nanoparticles were formed in 30 minutes. XPS analysis confirmed the formation of Au^0 and Ag^0 in the hydrogel. TEM images showed Au and Ag NPs with an approximate average size of 5 and 10 nm decorated along the fibers of the gel, rather than the solvent pockets in the gel. The hydrazide groups in the gelator were responsible for the reduction of the metal ions to metal NPs. The selectivity of the hydrogel toward metal ions was tested by adding an aqueous mixture containing $FeCl_2$, $NiCl_2$, $CuCl_2$, $ZnCl_2$, $PtCl_2$, $PdCl_2$, $AuCl_3$ and $AgNO_3$ (all 100 mg L^{-1}) to the hydrogel. ICP-MS analysis of the supernatant indicated that the gel had a higher affinity toward precious metal ions and the metal ions with the highest reduction potential were extracted better.

While 100% of the precious metal ions (Au^{3+}, Ag^{+}, Pt^{2+}, Pd^{2+}) were extracted by the gel, only 25-50% uptake of the other metal ions was observed (Fig. 10b). The hydrogel-NP-composite was stable in a wide pH range (2-12) and at temperatures from 65-85°C depending upon the concentration of NPs. Further, the nanocomposite was highly conductive and showed conductance similar to gel doped with single-wall carbon nanotubes.

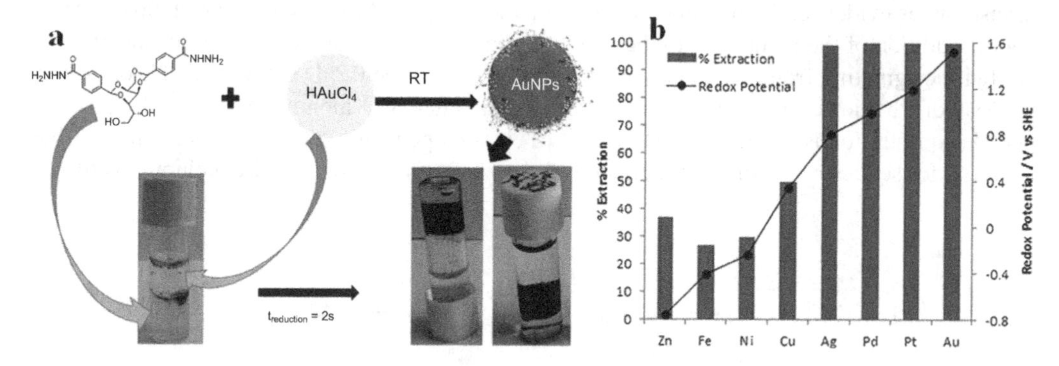

FIGURE 10 (a) Schematic representation and digital images showing the uptake of Au^{3+} ions by the DBS-$CONHNH_2$ hydrogel and their subsequent reduction *in situ* to form Au NPs. (b) Percentage metal ion recovery from mixed aqueous solutions by DBS-$CONHNH_2$ hydrogel (left axis) and redox potentials of the metals (right axis). Reproduced with permission from Okesola et al. 2016. Copyright 2016 John Wiley and Sons.

A metal-binding peptide, capable of inducing the formation of hydrogel upon complexation with monomethyl arsenous acid (MMA), Zn^{2+}, Cd^{2+}, Hg^{2+} and Pb^{2+} was designed in 2012 (Knerr et al. 2012). The metal coordination upon addition of stoichiometric amounts of the metal ions to the peptide resulted in an amphiphilic β-hairpin, which underwent self-assembly to form a self-standing gel. The binding of the metal ion to the two cysteinyl thiols flanking the β-turned off the peptide and initiated the peptide folding. Circular dichroism and mass spectroscopy indicated a 1:1 metal:peptide coordination and electron microscopic analysis revealed the formation of twisted, elongated fibers. Although the metal-induced gelation for uptake of the metal ions was an encouraging result, the tedious synthetic methods and high costs of the peptide limited the use of the method for real-world applications.

Hg^{2+} ion-based complexes (12a-c) were synthesized in excellent yields by King and McNeil (King and McNeil 2010) using commercially available quinoxalines (11a-c). It was interesting to note that, one of the three synthesized Hg^{2+}-quinoxaline complexes (12a), could form a gel in methanol:H_2O mixtures. Furthermore, it was also observed that gelation could also be induced *in situ* by the addition of aqueous $Hg(OAC)_2$ to a hot methanolic solution of the quinoxaline 11a. Control experiments in absence of Hg^{2+} ions under similar conditions only resulted in precipitation, indicating the importance of Hg^{2+} coordination to the ligand. The use of other metal ions, such as Co^{2+}, Ni^{2+}, Zn^{2+}, Cd^{2+}, Ba^{2+}, Ag^{+}, under similar conditions led to precipitation, suggesting that the gel formation was selective for Hg^{2+} ions. The specificity for gel formation with Hg^{2+} ions was attributed to the linear geometry of the resulting complex. The authors then demonstrated the removal of Hg^{2+} ions from water using the ligand through an *in situ* gel formation process. It was observed that out of 3,800 ppm Hg^{2+} ions contaminated water, only 289 ppm of the toxic metal ion could be detected in the solution after gel formation, suggesting that the gel formation process could be used to remove significantly higher quantities of Hg^{2+} ions from water. Thus, the gel formation process could be used both for the detection of Hg^{2+} ions as well as their uptake. However, the gel was not tolerant toward Cl^- ions and disrupted into sol upon addition of Cl^- ions, which prevented their use for practical applications.

The same group later synthesized five new Hg^{2+} complexes (13-17) (Figure 11) which could form gels in at least one of the studied solvent systems (Carter et al. 2014). The authors performed controlled experiments in methanol, ethanol and ethanol:acetic acid mixture, which led to the conclusion that the gel formation was largely dependent on the nature and position of the substituents, rather than the quinoxaline frameworks. Of the various gelators synthesized, one of these should excellent tolerance against Cl^- ions and could also be used for the removal of 99% of Hg^{2+} ions from contaminated water. The robustness of the gel for detection of Hg^{2+} ions was studied by using bottled water, tap water and Huron river water, by deliberately adding Hg^{2+} ions externally. While the addition of the quinoxaline (15a) to all the contaminated water samples resulted in gel formation, its addition to the uncontaminated water (not containing externally added Hg^{2+}) did not result in gelation consistent with the formation of gel triggered by Hg^{2+} ions.

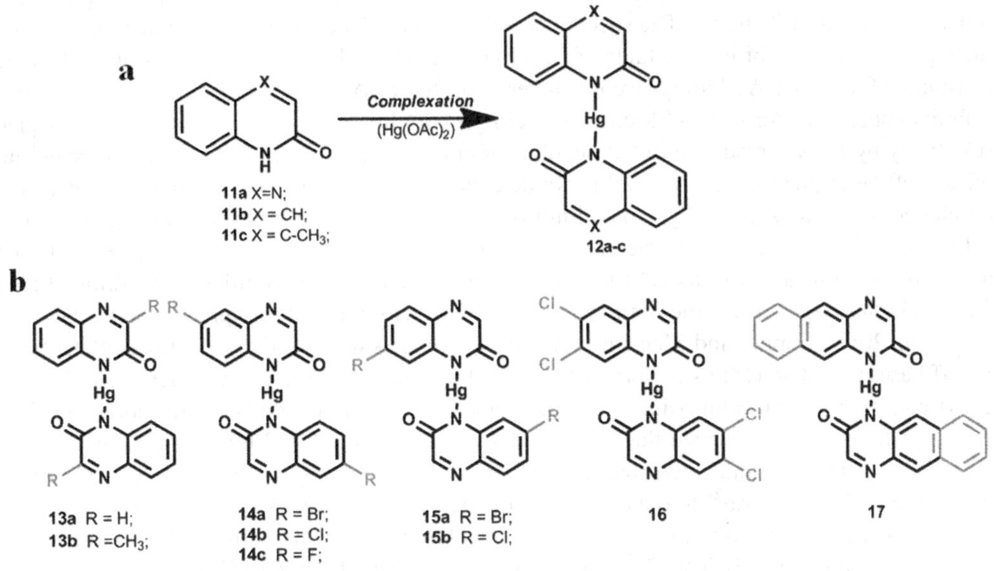

FIGURE 11 (a) Chemical structures of the Hg-complexes synthesized by King and McNeil 2010 and Carter et al. 2014.

Although nanomaterials have shown huge potential in a variety of applications, the release of nanoparticulate metals into the environment might have associated health hazards. Therefore, it is important to develop methods for the removal of nanoparticles from water in addition to the metal ions. Keeping this in mind, Barthèlèmy and co-workers used a glycosylated-nucleoside fluorinated (GNF) amphiphile as a trapping scaffold for the entrapment of different nanoparticles (Patwa et al. 2015). The GNF hydrogels were separately incubated with water samples containing quantum dots, Au NPs and TiO_2 NPs. After incubation for 48 hours, a gel formed at the bottom of the tube. Fluorescence and UV-visible studies were employed to monitor the uptake of the nanoparticles. The supernatant solutions did not show any signatures of quantum dots and Au or TiO_2 NPs in the fluorescence or UV-visible spectrum, respectively, indicating the trapping of the nanoparticles in the gel matrix. The entrapment of the nanoparticles in the gel matrix was attributed to the more favorable interactions of the nanoparticles with the 3D network of the gel, which might lead to a more stable state.

2.4 Supramolecular Gels for Sensing and Removal of Anions From Water

Anions although are present in several biological systems, but their presence in an appreciable amount in water can severely pollute water and negatively impact human health and the marine system (Okesola and Smith 2016, Wang et al. 2019). For instance, while anions, such as phosphates and nitrates, can cause eutrophication of lakes and rivers, increased concentrations of nitrite and fluoride can result in carcinogenesis and fluorosis, respectively (Busschaert et al. 2015). On the other hand anions, such as perchlorate are explosive and pertechnate anion, exhibit radioactivity (Busschaert et al. 2015). Industry effluents, refineries and chemical storage sites are the main sources through which anions are exposed to the environment. The stimuli-responsive nature of the supramolecular gels makes them attractive materials for the sensing and remediation of anions.

One of the earliest works on the use of supramolecular gels for selective recognition of anions was reported in 2009 (Shen et al. 2009). They used Ag (I)-glutathione (Ag-GSH) coordination polymer gel for the selective detection of I^- ions. The Ag-GSH gel exhibited a gel-to-sol transition within 60 minutes of the introduction of I^- ions, making it a naked eye sensor for I^- ions (Figure 12a). The use of other halides, such as F^-, Cl^- or Br^- and $H_2PO_4^-$, could not induce such a gel to sol transition even after 24 hours. The specificity of I^- ions to induce the gel to sol transition was due to the higher value of stability constants of Ag-I than the Ag-thiolates, whereas the values of stability constants of all other Ag-halides were smaller than that of Ag-thiolates. As a result of the higher stability constant of Ag-I, the –SR component in the coordination polymer backbone was displaced selectively by I^- ions, resulting in the depolymerization of Ag-GSH polymeric gel and subsequent release of the trapped water. A quantitative detection method for I^- ions in model wastewater was developed using the Ag-GSH gel. The limit of detection was as low as 0.423 mM. Although this gel was an effective sensor for selective detection of I^- ions in water, it could not be used for the uptake of the same as it lost its gel-like nature in presence of I^- ions. Similar competition of anions for metal ions was used for the optical detection of the environmentally problematic cyanide ions (Sun et al. 2016). Copper and Zinc based metal-organic gels were fabricated using L-glutamic acid Schiff base derivative (18) as the organic ligand. Two gels containing single metal ions (ZnG and CuG) and another gel comprising both the metal components (ZnCuG) were fabricated. While ZnG gel showed excellent blue fluorescence with a maximum of 457 nm, the CuG gel was non-fluorescent. The result indicated that the fluorescence could easily be modulated by the rational introduction of Cu^{2+} or Zn^{2+} to the corresponding gel. The addition of Cu^{2+} ions over the ZnG gel, followed by heating led to a transition to a green-colored sol. Upon cooling, a Cu containing ZnG gel was formed (Zn-CuG). The Cu^{2+} ions could competitively bind with G in ZnG and release the Zn^{2+} ions. This resulted in complete fluorescence quenching of the ZnG gel and the color of the gel also changed from colorless to green. The anion response of the mixed metal gel was studied by the addition of different anions to the gel. Of the various anions added, only CN^- ions could generate fluorescence in the Zn-CuG, which was similar to that of ZnG gel. The results confirmed that CN^- ions coordinated to Cu^{2+} competitively resulting in stable $[Cu(CN)_x]^{n-}$ and the Zn^{2+} ions again coordinated to G. Further, the fluorescence could again be quenched by the addition of Cu^{2+} to the CN^- treated gel.

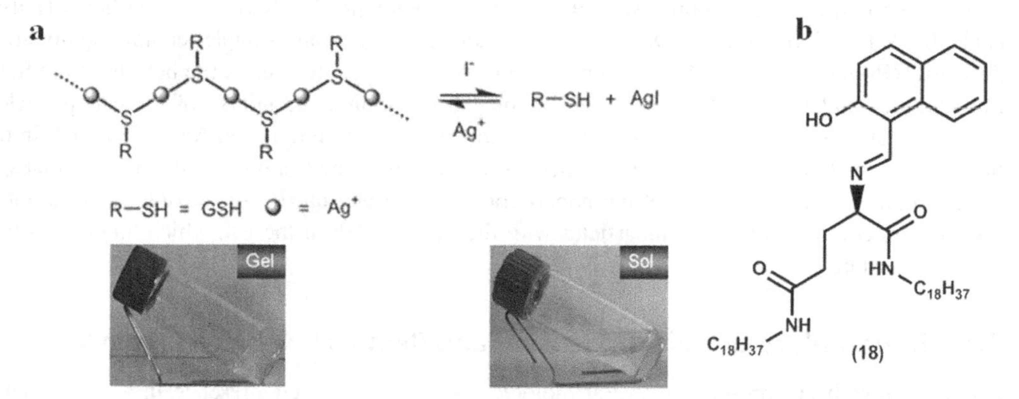

FIGURE 12 (a) Schematic illustration for the reversible gel to sol transition of Ag-GSH coordination polymer gel upon addition of I^- ions and Ag^+ ions respectively. Reproduced with permission from Shen et al. 2009. Copyright 2009 The Royal Society of Chemistry. (b) The chemical structure of the ligand is used to form a mixed metal gel (Sun et al. 2016).

A two-component organogel that showed the ability to colorimetrically detect NO_2^- ions were fabricated by mixing the components naphthalimide undecanoic acid (19) and diaminoanthraquinone (DAQ) (20) in various organic solvents (Xia et al. 2014). The formed gel was red due to absorption at 517 nm by DAQ. Upon the addition of NO_2^- ions, a suspension was formed, which indicated that the gel underwent a transition to sol state. Moreover, the red color of the gel faded and the emission of the gel was also quenched significantly (Figure 13). Mass spectroscopic studies revealed that the observed changes were due to the reaction between nitrite and DAQ resulting in the formation of benzotriazole. The gel was highly selective only toward NO_2^- ions as under similar conditions, the use of F^-, HCO_3^-, SO_3^{2-} or PO_4^{3-} did not induce any appreciable change.

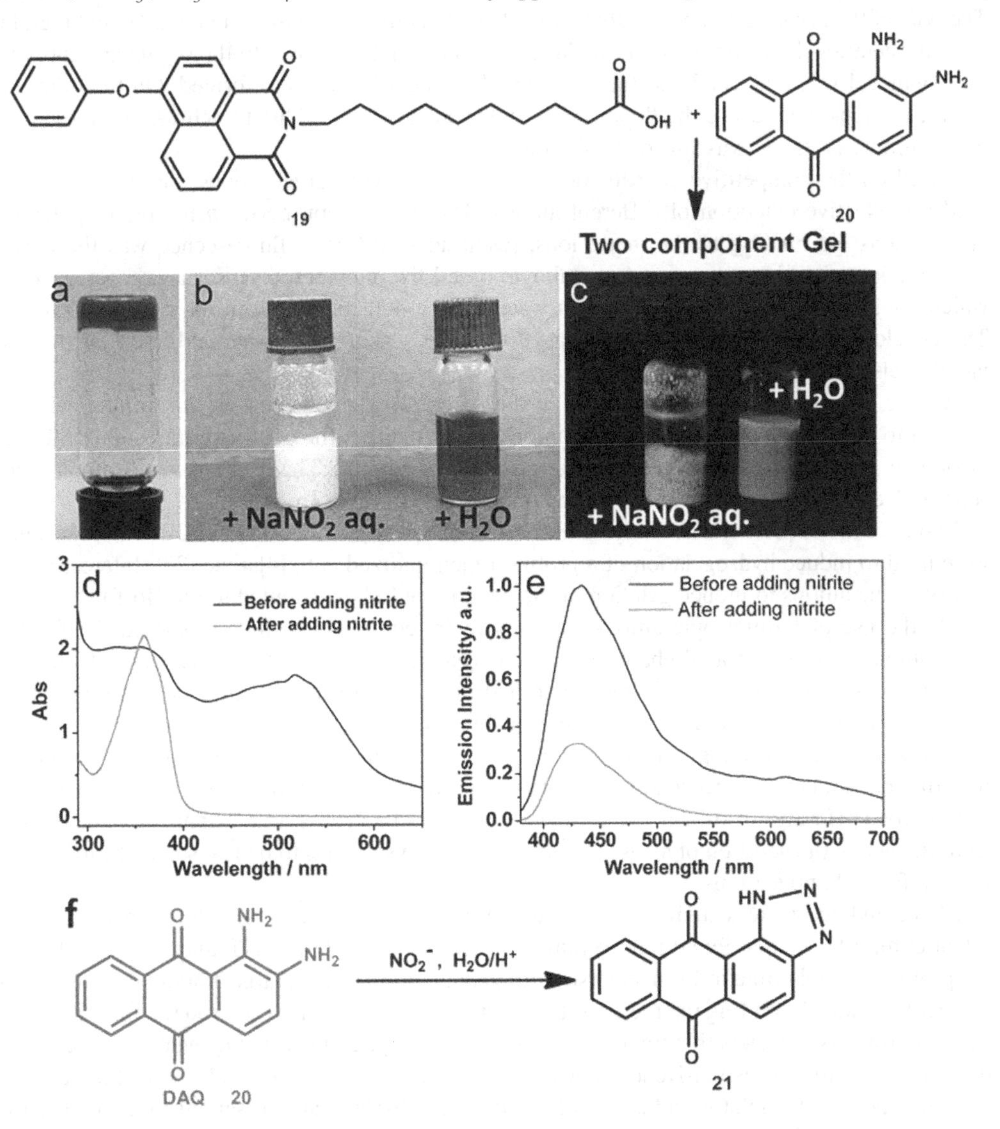

FIGURE 13 Chemical structure of the two-component gel formed by naphthalimide undecanoic acid 19 and diaminoanthraquinone 20. Photograph of gel (19 + 20) (25 mg mL^{-1} in acetonitrile) (a), and the gel with added $NaNO_2$ solution (left) and pure water (right), (b) and with 365 nm light irradiation (c), (d) Absorption and (e) emission spectra of 19 + 20 before and after treatment with NO_2^-. (f) Reaction scheme between 20 and NO_2^-. Reproduced with permission from Xia et al. 2014. Copyright 2014 The Royal Society of Chemistry.

Detection of multiple analytes using a single sensor is an interesting area of research, as the single material can be used to detect several analytes, which can significantly reduce the cost issues associated with the development of separate sensors for each analyte. Zhang and co-workers developed gel-based sensor arrays, which can detect multiple anions (Lin et al. 2015). They first carefully designed a gelator (22) that had multiple self-assembling driving forces, coordination sites and fluorescent signal groups (Figure 14). The gelator was capable of gelating DMF and formed a stable organogel in DMF. Interestingly, a hot solution of gelator in DMF did not show any fluorescence, but upon cooling the solution to a temperature below the gelation temperature of the gel, a drastic increase in fluorescence was observed due to aggregation-induced emission (AIE). The AIE of the organogel could easily be tuned by the addition and diffusion of different metal ions to form metallogels. Upon addition of Cu^{2+}, Hg^{2+}, Fe^{3+} or Cr^{3+} ions into the organogel, the AIE of the organogel was quenched and the corresponding metallogels were formed. On the other hand, when Zn^{2+} ions were used, the fluorescence color changed from blue to yellow, and the emission maximum was red and was shifted by 40 nm.

Based on the competitive coordination of metal ions with target anions, the metallogels were used for selective detection of different anions. The competition between the organogelator and other anions for binding to the metal ions, resulting in different fluorescence was the basis for the detection of different anions. A metallogel based five-membered sensor array was developed, which could detect CN^-, SCN^-, S^{2-}, I^- in water with a detection limit in the range of 0.1-10 μM. The development of the practical and easy-to-handle sensor array was facilitated by the use of metallogels.

All the examples discussed above show the potential of gel as sensors for anions, which can be used to detect environmentally toxic anions through either the gel to sol transition or a change in optical properties of the gel upon the addition of anions. There are fewer examples where supramolecular gels have been used for the remediation of toxic anions.

In one of the earliest works for remediation of anions by gels (Becker et al. 2008), NO_3^- and Br^- were used to induce hydrogelation of a proline functionalized calix[4]arene (23). Interestingly, the ability of the anions to induce gelation of the gelator could be attributed to the Hoffmeister series. While the use of kosmotropic anions (strongly hydrated), such as SO_4^{2-} promoted the formation of solution, the less hydrated chaotropic anions (NO_3^-, Br^-) induced self-assembly of the gelator into fibrous structure resulting in the stable hydrogel. Nonetheless, if too chaotropic anions, like ClO_4^- and I^- are used, rapid gelation followed by crystallization, which leads to the collapse of the gel is observed. Although the authors did not explore the use of this gelator for environmental remediation, it meets several conditions that are required for this application. The gel can be formed over a broad pH range of 0-7 in a single step. Moreover, the gelator and the anion can be recovered from the gel by the addition of a base. Nonetheless, the system has to be investigated further before using it for real applications.

Zhang and co-workers studied the interaction of protonated melamine (24) with several anions (Shen et al. 2010). They found that oxyanions, such as NO_3^-, PO_4^{3-}, ATP and SO_4^{2-}, could trigger the protonated melamine to form superstructures that could gelate a large amount of water through electrostatic and H-bonding interactions. Electron microscopic images showed the formation of 1-D nano/microfibers or sheets that trapped the water molecules, giving a stable hydrogel. The hydrogel demonstrated stimuli responsiveness toward pH and temperature. The addition of the base led to deprotonation of the gelator and a loss of the positive charge that is essential to interact with the anion in a competitive aqueous solution. Other anions, such as OAc^-, F^- or Cl^-, did not result in the formation of hydrogels. Such aqueous gelation induced by anions, like NO_3^- and PO_4^{3-}, may be useful in the removal of these anions from aqueous waste.

In another excellent work, a gelator whose gelation could be switched on in presence of NO_2^- ions was used for the detection of nitrite ions (Zurcher et al. 2014). The method employed the Griess reaction, in which the aromatic amine (26) was reacted with NO_2^- ions to form a diazonium ion, which in turn was reacted with sodium 6-hydroxynaphthalene-2-sulfonate (28) to form the azobenzene derivative (29), that could function as a gelator and trap water molecules. To demonstrate the ability of the system to capture nitrite ions from aqueous waste, nitrite anion spiked water samples were added to vials containing 26 suspended in 4M H_2SO_4 and allowed to react for 10 minutes, followed by the addition of compound 28 in the borax buffer. The resulting mixture was then subjected to heating-cooling. It was observed that the samples in which nitrite ions were spiked all formed a red-colored gel due to the formation of the red azobenzene-based gelator (Figure 15b). On the other hand, the control samples without nitrite ions did not result in any color change or gelation. This gelator system can, therefore, be used not only for the detection of the toxic nitrite ions through a color change observable to the naked eye but could also be used for the uptake of the nitrite ions through the formation of the gel.

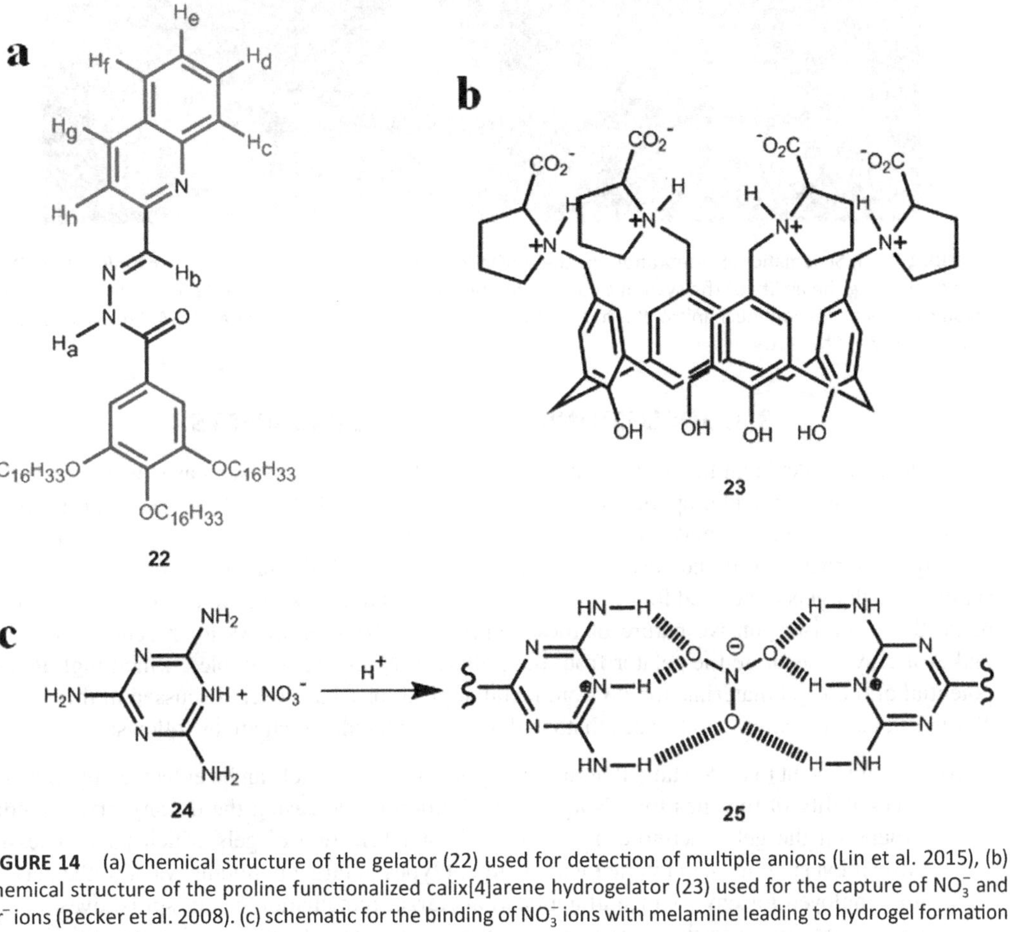

FIGURE 14 (a) Chemical structure of the gelator (22) used for detection of multiple anions (Lin et al. 2015), (b) chemical structure of the proline functionalized calix[4]arene hydrogelator (23) used for the capture of NO_3^- and Br^- ions (Becker et al. 2008). (c) schematic for the binding of NO_3^- ions with melamine leading to hydrogel formation (Shen et al. 2010).

b

FIGURE 15 (a) Schematic representation for the nitrite ion induced formation of gelator scaffold 29. (b) Digital images showing the ability of the system to detect the presence of nitrite anions, and potentially remediate the anion by a gel formation mechanism. Reproduced with permission from Zurcher et al. 2014. Copyright 2014 The Royal Society of Chemistry.

3. CONCLUSIONS AND FUTURE PROSPECTS

In conclusion, it is evident that self-assembled supramolecular systems, such as gels, are promising materials for environmental applications, like wastewater remediation. The intrinsic properties of this class of soft materials make them unique for the removal and uptake of pollutants from water. The highly porous nature and solvent compatibility, along with the fast response time enable the supramolecular gels to be used for oil spill remediation and dye as well as ion capture. On the other hand, the stimuli-responsive nature of these supramolecular gels allows for regeneration, which makes it easy to separate the gelator from the pollutant. Numerous examples, which highlight the potential of these gel materials for environmental applications, have been discussed in this chapter. Nevertheless, there are several issues that need to be addressed, which are as follows:

(i) It is important to understand that the adsorption ability of a gel can be enhanced by increased accessibility of the sites for adsorption in addition to increasing the density of the binding groups on the gel. Therefore, there is need for fabrication of gels which have accessible adsorption sites and can interact with different types of target pollutant. Most of the gels that are employed for environmental applications target remediation of oil, spills, dyes, cations and anions. However, there are also other classes of pollutants, such as pharmaceutical drugs, pesticides, insecticides, cosmetic compounds as well as chemical compounds (chlorinated compounds), which are all known contaminants for water. There is a need for the development of supramolecular gel systems that can sequester these important pollutants from the water as well.

(ii) Enhanced design and control over the material morphology and property together with a decreased cost is to be focussed.

(iii) The selectivity of the gel for uptake of pollutants needs to be improved. For instance, most of the examples discussed herein, are studied under laboratory conditions, which are different from real practical conditions, where the gels become less effective in uptake of target pollutants in presence of various other ions.
(iv) Supramolecular gel systems should be integrated with effective water treatment infrastructures to exploit their potential in the industry. Technological advances are necessary to incorporate gels within filtration systems.

The use of supramolecular gels for environmental remediation is an interesting field and is increasing every day. The development of novel self-assembled systems through smarter supramolecular chemistry will definitely play a defining role in providing advanced solutions to the grave issue of environmental pollution. It is expected that in the coming few years, supramolecular gel-based wastewater systems will be incorporated within filtration systems for practical use.

References

Adhikari, B., Palui, G. and Banerjee, A. 2009. Self-assembling tripeptide based hydrogels and their use in removal of dyes from waste-water. Soft Matter 5: 3452-3460.

Alam, N. and Sarma, D. 2020. A thixotropic supramolecular metallogel with 2D sheet morphology: Iodine sequestration and column based dye separation. Soft Matter 16: 10620-10627.

Albelda, M.T., Frias, J.C., Garcia-Espana, E. and Schneider, H.J. 2012. Supramolecular complexation for environmental control. Chem. Soc. Rev. 41: 3859-3877.

Anwer, H., Mahmood, A., Lee, J., Kim, K.-H., Park, J.-W. and Yip, A.C.K. 2019. Photocatalysts for degradation of dyes in industrial effluents: Opportunities and challenges. Nano. Res. 12: 955-972.

Appel, E.A., del Barrio, J., Loh, X.J. and Scherman, O.A. 2012. Supramolecular polymeric hydrogels. Chem. Soc. Rev. 41: 6195-6214.

Aragay, G., Pons, J. and Merkoçi, A. 2011. Recent trends in macro-, micro-, and nanomaterial-based tools and strategies for heavy-metal detection. Chem. Rev. 111: 3433-3458.

Babu, S.S., Praveen, V.K. and Ajayaghosh, A. 2014. Functional π-gelators and their applications. Chem. Rev. 114: 1973-2129.

Banerjee, S., Das, R.K. and Maitra, U. 2009. Supramolecular gels 'in action'. J. Mater. Chem. 19: 6649-6687.

Basak, S., Nandi, N., Paul, S., Hamley, I.W. and Banerjee, A. 2017. A tripeptide-based self-shrinking hydrogel for waste-water treatment: removal of toxic organic dyes and lead (Pb^{2+}) ions. Chem. Commun. 53: 5910-5913.

Basu, K., Nandi, N., Mondal, B., Dehsorki, A., Hamley, I.W. and Banerjee, A. 2017. Peptide-based ambidextrous bifunctional gelator: Applications in oil spill recovery and removal of toxic organic dyes for waste water management. Interface Focus 7: 20160128.

Beall, G.W. 2003. The use of organo-clays in water treatment. Appl. Clay Sci. 24: 11-20.

Becker, T., Goh, C.Y., Jones, F., Mclldowie, M.J., Mocerino, M. and Ogden, M.I. 2008. Proline-functionalised calix[4]arene: An anion-triggered hydrogelator. Chem. Commun. 3900-3902.

Bhattacharya, S. and Krishnan-Ghosh, Y. 2001. First report of phase selective gelation of oil from oil/water mixtures. Possible implications toward containing oil spills. Chem. Commun. 185-186.

Busschaert, N., Caltagirone, C., Van Rossom, W. and Gale, P.A. 2015. Applications of supramolecular anion recognition. Chem. Rev. 115: 8038-8155.

Carter, K.K., Rycenga, H.B. and McNeil, A.J. 2014. Improving Hg-triggered gelation via structural modifications. Langmuir 30: 3522-3527.

Chonamada, T.D., Sharma, B., Nagesh, J., Shibu, A., Das, S., Bramhaiah, K., et al. 2019. Origin of luminescence-based detection of metal ions by Mn–doped zns quantum dots. ChemistrySelect 4: 13551-13557.

Datta, S., Samanta, S. and Chaudhuri, D. 2018. Near instantaneous gelation of crude oil using naphthalene diimide based powder gelator. J. Mater. Chem. A 6: 2922-2926.

Dave, D. and Ghaly, A.E. 2011. Remediation technologies for marine oil spills: A critical review and comparative analysis. Am. J. Environ. Sci. 7: 423-440.

De, A. and Mondal, R. 2018. Toxic metal sequestration exploiting an unprecedented low-molecular-weight hydrogel-to-metallogel transformation. ACS Omega 3: 6022-6030.

Dey, A., Mandal, S.K. and Biradha, K. 2013. Metal–organic gels and coordination networks of pyridine-3,5-bis(1-methyl-benzimidazole-2-yl) and metal halides: Self-sustainability, mechano, chemical responsiveness and gas and dye sorptions. CrystEngComm 15: 9769-9778.

Doshi, B., Silanpää, M. and Kalliola, S. 2018. A review of bio-based materials for oil spill treatment. Water. Res. 135: 262-277.

Du, X., Zhou, J., Shi, J. and Xu, B. 2015. Supramolecular hydrogelators and hydrogels: From soft matter to molecular biomaterials. Chem. Rev. 115: 13165-13307.

Guterman, L. 2009. Exxon Valdez Turns 20. Science 323: 1558-1559.

Hager, D.G. 1967. Activated carbon used for large scale water treatment. Environ. Sci. Technol. 1: 287-291.

Harrison, R.M. 2006. The world's waters: A chemical contaminant perspective. pp. 77-121. *In*: Harrison, R.M. [ed.]. An Introduction to Pollution Science. RSC, Cambridge.

Hirst, A.R., Escuder, B., Miravet, J.F. and Smith, D.K. 2008. High-tech applications of self-assembling supramolecular nanostructured gel-phase materials: From regenerative medicine to electronic devices. Angew. Chem. Int. Ed. 47: 8002-8018.

Häring, M. and Díaz Díaz, D. 2016. Supramolecular metallogels with bulk self-healing properties prepared by *in situ* metal complexation. Chem. Commun. 52: 13068-13071.

Karan, C.K. and Bhattacharjee, M. 2016. Self-healing and moldable metallogels as the recyclable materials for selective dye adsorption and separation. ACS Appl. Mater. Interfaces 8: 5526-5535.

King, K.N. and McNeil, A.J. 2010. Streamlined approach to a new gelator: Inspiration from solid-state interactions for a mercury-induced gelation. Chem. Commun. 46: 3511-3513.

Knerr, P.J., Branco, M.C., Nagarkar, R., Pochan, D.J. and Schneider, J.P. 2012. Heavy metal ion hydrogelation of a self-assembling peptide *via* cysteinyl chelation. J. Mater. Chem. 22: 1352-1357.

Kummu, M., Guillaume, J.H.A., de Moel, H., Eisner, S., Flörke, M., Porkka, M., et al. 2016. The world's road to water scarcity: Shortage and stress in the 20th century and pathways towards sustainability. Sci. Rep. 6: 38495.

Lim, J.Y.C., Goh, S.S., Liow, S.S., Xue, K. and Loh, X.J. 2019. Molecular gel sorbent materials for environmental remediation and wastewater treatment. J. Mater. Chem. A 7: 18759-18791.

Lin, Q., Lu, T.-T., Zhu, X., Sun, B., Yang, Q.-P., Wei, T.-B., et al. 2015. A novel supramolecular metallogel-based high-resolution anion sensor array. Chem. Commun. 51: 1635-1638.

Lin, Y., Cao, Y., Yao, Q., Chai, O.J.H. and Xie, J. 2020. Engineering noble metal nanomaterials for pollutant decomposition. Ind. Eng. Chem. Res. 59: 20561-20581.

Maleki, H. 2016. Recent advances in aerogels for environmental remediation applications: A review. Chem. Eng. J. 300: 98-118.

Marimuthu, S., Antonisamy, A.J., Malayandi, S., Rajendran, K., Tsai, P.-C., Pugazhendi, A., et al. 2020. A review on synthesis, treatment methods, mechanisms, photocatalytic degradation, toxic effects and mitigation of toxicity. J. Photochem. Photobiol. B 205: 111823.

Mishra, A. and Clark, J.H. 2013. Greening the blue: How the world is addressing the challenge of green remediation of water. pp. 1-10. *In*: Mishra, A. and Clark, J.H. [eds.]. Green Materials for Sustainable Water Remediation and Treatment. RSC, Cambridge,

Mondal, S., Das, S. and Nandi, A.K. 2020. A review on recent advances in polymer and peptide hydrogels. Soft Matter 16: 1404-1454.

Nebot, V.J. and Smith D.K. 2013. Techniques for the characterisation of molecular gels. pp. 30-66. *In*: Escuder, B. and Miravet, J.F. [eds.]. Functional Molecular Gels. RSC, Cambridge.

Okesola, B. and Smith D.K. 2016. Applying low-molecular weight supramolecular gelators in an environmental setting – self assembled gels as smart materials for pollutant removal. Chem. Soc. Rev. 45: 4226-4251.

Okesola, B.O. and Smith, D.K. 2013. Versatile supramolecular pH-tolerant hydrogels which demonstrate pH-dependent selective adsorption of dyes from aqueous solution. Chem. Commun. 49: 11164-11166.

Okesola, B.O., Suravaram, S.K., Parkin, A. and Smith, D.K. 2016. Selective extraction and *in situ* reduction of precious metal salts from model waste to generate hybrid gels with embedded electrocatalytic nanoparticles. Angew. Chem. Int. Ed. 55: 183-187.

Patwa, A., Labille, J., Bottero, J.-Y., Thiery, A. and Barthèlèmy, P. 2015. Decontamination of nanoparticles from aqueous samples using supramolecular gels. Chem. Commun. 51: 2547-2550.

Plank, T.N., Skala, L.P. and Davis, J.T. 2017. Supramolecular hydrogels for environmental remediation: G4-quartet gels that selectively absorb anionic dyes from water. Chem. Commun. 53: 6235-6238.

Pradeep, T. and Anshup. 2009. Noble metal nanoparticles for water purification: A critical review. Thin Solid Films 517: 6441-6478.

Prathap, A. and Sureshan, K.M. 2017. Organogelator – cellulose composite for practical and ecofriendly marine oil spill recovery. Angew. Chem. Int. Ed. 56: 9405-9409.

Prenderghast, D.D. and Gschwend, P.M. 2014. Assessing the performance and cost of oil spill remediation technologies. J. Cleaner Prod. 78: 233-242.

Qi, Z. and Schalley, C.A. 2014. Exploring macrocycles in functional supramolecular gels: From stimuli responsiveness to systems chemistry. Acc. Chem. Res. 47: 2222-2233.

Rahim, Md. A., Björnmalm, M., Suma, T., Faria, M., Ju, Y., Kempe, K., et al. 2016. Metal–phenolic supramolecular gelation. Angew. Chem. Int. Ed. 55: 13803-13807.

Ratna and Padhi, B.S. 2012. Pollution due to synthetic dyes toxicity & carcinogenicity studies and remediation. Int. J. Environ. Sci. 3: 940-955.

Ray, S., Das, A.K. and Banerjee, A. 2007. pH-responsive, bolaamphiphile-based smart metallo-hydrogels as potential dye-adsorbing agents, water purifier, and vitamin B_{12} carrier. Chem. Mater. 19: 1633-1639.

Routoula, E. and Patwardhan, S.V. 2020. Degradation of anthraquinone dyes from effluents: A review focusing on enzymatic dye degradation with industrial potential. Environ. Sci. Tech. 54: 647-664.

Samai, S. and Biradha, K. 2012. Chemical and mechano responsive metal–organic gels of bis(benzimidazole)-based ligands with Cd(II) and Cu(II) halide salts: Self sustainability and gas and dye sorptions. Chem. Mater. 24: 1165-1173.

Sangeetha, N.M. and Maitra, U. 2005. Supramolecular gels: Functions and uses. Chem. Soc. Rev. 34: 821-836.

Saunders, L. and Ma, P.X. 2019. Self-healing supramolecular hydrogels for tissue engineering applications. Macromol. Biosci. 19: 1800313.

Sharma, B., Mahata, A., Mandani, S., Sarma, T.K. and Pathak, B. 2016. Coordination polymer hydrogels through Ag(I)-mediated spontaneous self-assembly of unsubstituted nucleobases and their antimicrobial activity. RSC Adv. 6: 62968-62973.

Sharma, B., Mahata, A., Mandani, S., Thakur, N., Pathak, B. and Sarma, T.K. 2018a. Zn(II)–nucleobase metal–organic nanofibers and nanoflowers: Synthesis and photocatalytic application. New. J. Chem. 42: 17983-17990.

Sharma, B., Mandani, S., Thakur, N. and Sarma, T.K. 2018b. Cd(ii)–nucleobase supramolecular metallo-hydrogels for *in situ* growth of color tunable CdS quantum dots. Soft Matter 14: 5715-5720.

Sharma, B., Singh, A., Sarma, T.K., Sardana, N. and Pal, A. 2018c. Chirality control of multi-stimuli responsive and self-healing supramolecular metallo-hydrogels. New. J. Chem. 42: 6427-6432.

Sharma, R.K., Adholeya, A., Das, M. and Puri, A. 2013. Green materials for sustainable remediation of metals in water. pp. 11-29. *In*: Mishra, A. and Clark, J.H. [eds.]. Green Materials for Sustainable Water Remediation and Treatment. RSC, Cambridge.

Shen, J.-S., Li, D.-H., Cai, Q.-G. and Jiang, Y.-B. 2009. Highly selective iodide-responsive gel–sol state transition in supramolecular hydrogels. J. Mater. Chem. 19: 6219-6224.

Shen, J.-S., Cai, Q.-G., Jiang, Y.-B. and Zhang, H.-W. 2010. Anion-triggered melamine based self-assembly and hydrogel. Chem. Commun. 46: 6786-6788.

Skilling, K.J., Citossi, F., Bradshaw, T.D., Ashford, M., Kellam, B. and Marlow, M. 2014. Insights into low molecular mass organic gelators: A focus on drug delivery and tissue engineering applications. Soft Matter 10: 237-256.

Smith, C.R., Hatcher, P.G., Kumar, S. and Lee, J.W. 2016. Investigation into the sources of biochar water-soluble organic compounds and their potential toxicity on aquatic microorganisms. ACS Sustain. Chem. Eng. 4: 2550-2558.

Srinivasan, R. 2011. Advances in application of natural clay and its composites in removal of biological, organic, and inorganic contaminants from drinking water. Adv. Mater. Sci. Eng. 872531.

Sukul, P.K. and Malik, S. 2013. Removal of toxic dyes from aqueous medium using adenine based bicomponent hydrogel. RSC Adv. 3: 1902-1915.

Sun, J., Lin, Y., Jin, L., Chen T. and Yin, B. 2016. Coordination-induced gelation of an L-glutamic acid Schiff base derivative: The anion effect and cyanide-specific selectivity. Chem. Commun. 52: 768-771.

Takashita, J., Hasegawa, Y., Yanai, K., Yamamoto, A., Ishii, A., Hasegawa, M., et al. 2017. Organic dye adsorption by amphiphilic tris-urea supramolecular hydrogel. Chem. Asian. J. 12: 2029-2032.

Tam, A.Y.-Y. and Yam, V.W.-W. 2013. Recent advances in metallogels. Chem. Soc. Rev. 42: 1540-1567.

Thakur, N., Sharma, B., Bishnoi, S., Jain, S., Mishra, S.K., Nayak, D., et al. 2018. Multifunctional inosine monophosphate coordinated metal–organic hydrogel: Multistimuli responsiveness, self-healing properties, and separation of water from organic solvents. ACS Sustain. Chem. Eng. 6: 8659-8671.

Thakur, N., Sharma, B., Bishnoi, S., Jain, S., Nayak, D. and Sarma, T.K. 2019. Biocompatible Fe^{3+} and Ca^{2+} dual cross-linked G-quadruplex hydrogels as effective drug delivery system for pH-responsive sustained zero-order release of doxorubicin. ACS Appl. Bio Mater. 2: 3300-3311.

Trivedi, D.R., Ballabh, A. and Dastidar, P. 2003. An easy to prepare organic salt as a low molecular mass organic gelator capable of selective gelation of oil from oil/water mixtures. Chem. Mater. 15: 3971-3973.

Trivedi, D.R., Ballabh, A., Dastidar, P. and Ganguly, B. 2004. Structure–property correlation of a new family of organogelators based on organic salts and their selective gelation of oil from oil/water mixtures. Chem. – Eur. J. 10: 5311-5322.

Vibhute, A.M., Muvvala, V. and Sureshan, K.M. 2016. A sugar-based gelator for marine oil-spill recovery. Angew. Chem. Int. Ed. 55: 7782-7785.

Wang, H., Ji, X., Ahmed, M., Huang, F. and Sessler, J.L. 2019. Hydrogels for anion removal from water. J. Mater. Chem. A 7: 1394-1403.

Wang, S. and Peng, Y. 2010. Natural zeolites as effective adsorbents in water and wastewater treatment. Chem. Eng. J. 156: 11-24.

Wong, S., Ngadi, N., Inuwa, I.M. and Hassan, O. 2018. Recent advances in applications of activated carbon from biowaste for wastewater treatment: A short review. J. Clean. Prod. 175: 361-375.

Xia, Q., Mao, Y., Wu, J., Shu, T. and Yi, T. 2014. Two-component organogel for visually detecting nitrite anion. J. Mater. Chem. C 2: 1854-1861.

Zhang, Y., Su, Z., Li, B., Zhang, L., Fan, D. and Ma, H. 2016. Recyclable magnetic mesoporous nanocomposite with improved sensing performance toward nitrite. ACS Appl. Mater. Interfaces 8: 12344-12351.

Zhao, J., Wang, Z., White, J.C. and Xing, B. 2014. Graphene in the aquatic environment: Adsorption, dispersion, toxicity and transformation. Environ. Sci. Technol. 48: 9995-10009.

Zurcher, D.M., Adhia, Y.J., Romero, J.D. and McNeil, A.D. 2014. Modifying a known gelator scaffold for nitrite detection. Chem. Commun. 50: 7813-7816.

12

CHAPTER

Liquid and Crystal Nanomaterials for Water Remediation: Synthesis, Application and Environmental Fate

Jigneshkumar V. Rohit[1,*] and Vaibhavkumar N. Mehta[2]

1. INTRODUCTION

Water is the most abundant molecule on the earth for living organisms to carry out essential functions, like drinking, food preparation and processing and sanitation for sustainable life. Recently, due to the increase in the global population, the industrial revolution for economic development and climate change resulted in the water pollution of drinking, ground and surface water, which led to the scarcity of clean water for human consumption and escalated to be a great challenge among global scientists (Grey et al. 2013, Levin et al. 2002). In this connection, various physical (filtration, adsorption and sedimentation), chemical (oxidation, reduction, ion-exchange and flocculation) and biological water remediation techniques have been already utilized to remove the chemical and biological contaminants from water (Yuan 2008, Malaviya and Singh 2011, Wei et al. 2016). The quality of water is highly decreased due to presence of the heavy metals, agricultural chemicals, pesticides, industrial fuel and solvents and microbes in water (Nie et al. 2015, Kim and Benjamin 2004). The traditional available methods for the purification of water are economically costly processes, which require high energy and suffer from the incomplete removal of pollutants and generation of secondary pollutants (Musico et al. 2014). Therefore, there is a need to develop a cost-effective and efficient alternative water remediation technique for the production of clean water to meet the requirement of the global population.

Recently, the development in the area of nanoscience and nanotechnology could provide a promising solution for water remediation with high efficiencies and more cost-effectiveness over the available traditional technique due to their unique physico-chemical properties, such as high surface area, high reactivity, ease of functionalization, high porosity, hydrophobicity and high adsorption capacities. In recent years, the incorporation of nanomaterials (NMs) has been extensively studied to improve the water quality using nanosorbents, bioactive nanoparticles, catalytic membrane based on the nanoparticles (NPs) and nanocatalysis, which ultimately reduced the concentration of toxic elements via filtration and desalination of seawater (Qu et al. 2013). Khan et al. reviewed the use of engineered NMs, such as titanium dioxide (TiO_2), zinc Oxide (ZnO), silver NPs (Ag NPs), iron NPs (Fe NPs), graphene and nanocomposites, for the removal of waterborne pathogen, organic and

[1] Department of Chemistry, National Institute of Technology, Srinagar-190006, J&K, INDIA.
[2] ASPEE Shakilam Biotechnology Institute, Navsari Agricultural University, Surat-395007, Gujarat, INDIA.
[*] Corresponding author: Jignesh@nitsri.ac.in.

inorganic pollutants (Khan and Malik 2019). Adeleye and co-workers also reviewed and compared the conventional and nanotechnology-based technologies for water purification and environmental remediation based on the classification of pollutants (Adeleye et al. 2016). Furthermore, Lu and Astruc provided a comprehensive literature review for the removal of heavy metals from tap water, groundwater and wastewater by using transition metal (Au, Ag and Fe), metal oxides (Iron oxide and TiO_2), sulfide nanoparticles (MoS_2, ZnS and Ag_2S) and carbon/silicon-based nanomaterials (Lu and Astruc 2018). Moreover, the applications of inorganic NMs, metal-organic framework (MOFs), nanomembranes and organic polymer-based nanomaterials for the removal of organic pollutants are briefly discussed by the same group based on their cost-effectiveness and environmental concerns (Lu and Astruc 2018). The application of chemically synthesized NMs for water remediation and purification is of great point of discussion in the view of the environmental impact of NPs due to their toxicity as well as economical aspects. Nasrollahzadeh and co-workers provided a detailed review regarding recent advancements in the use of green synthesized nanomaterials and nanocatalysis for water remediation from groundwater, drinking water and wastewater (Nasrollahzadeh et al. 2021). The applications of various biological nanofactories and their mechanism for the removal of organic and inorganic contaminants have been briefly discussed by Das and co-workers (Das et al. 2018).

Here, we compiled the available data in the state-of-the-art to provide a comprehensive review to researchers working in the field NMs based water purification techniques. In this chapter, we discussed the fabrication of various NMs with controlled size, shape and suitable functionality for the remediation of organic, inorganic and nuclear waste from wastewater. Also, we highlighted the mechanism and removal/degradation efficiency of NMs toward hazardous water pollutants. In the concluding remark, we emphasize on challenges and prospects of NMs-based water purification techniques.

2. SYNTHESIS OF LIQUID AND CRYSTAL NANOMATERIALS

The selection of fabrication method is of great importance to obtain the desired size, shape and functionality of NMs for a wide variety of applications. For instance, NMs can be mainly fabricated via top-down and bottom-up approaches. The bottom-up approach is considered a constructive and commonly accepted approach for the formation of NMs from atoms to clusters to NPs, whereas the top-down approach is a destructive method to obtain NMs from the bulk product (Ealia and Saravanakumar 2017). The bottom-up approach includes the co-precipitation, chemical reduction, sol-gel method, chemical vapor deposition and spray pyrolysis, and the top-down approach includes the sputtering, etching, ball-milling and lithographic techniques (Cao 2004).

2.1 Synthesis of Liquid Nanomaterials

NMs in liquid/dispersion form (Ag NPs, Au NPs, Pt NPs, Fe NPs, Cu NPs, etc.) have been widely used for water remediation. Chemical reduction is one of the potential techniques for the synthesis of liquid NMs with the desired shape, size and functionality. In the first step, the development of zero valence metal was done by process of chemical reduction and nucleation with subsequent growth leads to the preparation of liquid NMs (Yu et al. 2008). Synthesis of liquid NMs can be done by using many precursors in presence of reducing and stabilizing agents. In this connection, Chauhan et al. prepared Au NPs via reduction of chloroauric acid with the help of strong reducing agent sodium tetrahydroborate (Chauhan et al. 2011) and prepared well-dispersed Au NPs with an average size of 8 nm. Similarly, Zhou et al. reported the synthesis of silver Ag NPs by the reduction of silver nitrate using Sodium borohydride (Zhou et al. 2011).

Also, the synthesis of liquid NMs carried out using micelles/microemulsion methods. In this type of synthesis, micelles or immiscible microemulsion of organic solvent with water was used in the preparation of liquid NMs. Also, liquid NMs prepared using this method was showing a very fine distribution of particles with controlled size. In this lineup, Sine and Comninellis prepared Pt NPs using the water-in-oil microemulsion method and as-synthesized Pt NPs showed the narrow size particle distribution in the range of 2-5 nm (Sine and Comninellis 2005).

2.2 Synthesis of Crystal Nanomaterials

Crystal NMs using for the removal/degradation of water pollutants, which includes semiconductor NPs, quantum dots, metal oxide NPs, core-shell NPs, etc. The most common approach to synthesize crystal NMs is pyrolysis of precursor (organometallic) with a stabilizing agent at the desired temperature and in an inert atmosphere. The formed crystal NMs using this approach showed uniform size distribution and dispersion. In this lineup, Tamrakar et al. synthesized crystalline ZnS NPs using chemical precipitation techniques in which an aqueous solution of $ZnCl_2$ and the aqueous solution of Na_2S were mixed in the presence of mercaptoethanol as a stabilizer (Tamrakar et al. 2008). Similarly, Wang et al. prepared cadmium sulfide quantum dots (CdS QDs) using $CdCl_2$ and Na_2S in presence of sodium citrate as a stabilizer (Wang et al. 2011).

Synthesis of oxide NPs was generally carried out using the sol-gel method because steps involved in the preparation of oxide NPs were comparatively more complex than the steps involved in the synthesis of liquid and semiconductor NMs. As the sol-gel method is basically a polymerization technique, it involved steps like initiation, propagation and termination. In this connection, Oskamand and Poot synthesized crystalline TiO_2 NPs by using Ti(O-*i*Pr)$_4$ and deionized water in presence of concentrated HNO_3 (Oskamand and Poot 2006). Another research group prepared crystalline ZnO nanoparticles via the sol-gel process with narrow size distribution (Sakohara et al. 1998).

Core/shell NMs having core coated with shell material in which core and shell are of completely different material with different properties. Also, it was observed that prepared core/shell NMs showing diverse properties than the material used for the synthesis of these NMs. Many researchers developed various core/shell NMs by varying material of core and shell. In this connection, Selvakannan et al. used Au as core Ag as shell material for the preparation of Au/Ag core-shell NPs with an average particle size of 22 nm (Selvakannan et al 2004).

3. APPLICATION OF LIQUID AND CRYSTAL NANOMATERIALS FOR WATER REMEDIATION

3.1 Removal/Degradation of Heavy Metal Ions

Over the last many years, heavy metals are the major cause of water pollution as they are non-degradable, highly poisonous and have considerably hazardous effects on human health and the environment (Gola et al. 2018). Trace level contamination of heavy metal ions (lead, cadmium, arsenic, copper, chromium, mercury, cobalt, silver, barium, manganese and aluminum) in water bodies can harm aqua-culture by changing their natural habitats (Sekar et al. 2004). So, detection and removal of heavy metal ions from water is a very important task to get heavy metals free water for humans and animals. NMs have found a wide range of applications according to their shapes, sizes, and compositions. Also, the surface chemistry of NMs plays an important role to make them promising tools to remove pollutants from water.

In this regard, many researchers developed methods to remove heavy metal ions from water using NMs (Table 1). In this connection, Kaur et al. synthesized Ag NPs using leaf extract of mulberry and incorporated it into SnZrMoP ion exchanger (Kaur et al. 2020). This method is highly selective to detect and remove Ba^{2+} and Sr^{2+} ions from industrial effluents. Also, this method showed a good recovery by removing more than 88% of metal ions and detoxifies industrial water. One of the most hazardous metal ions, mercury was detected and removed from water using Au NPs modified filter paper (Figure 1) (Lei et al. 2020). This method is capable to detect Hg^{+2} with a limit of detection (LOD) 0.09 mg L^{-1} using Au NPs-based removal method. Similarly, Hg^{+2} was detected and separated from water samples using poly(*N*-isopropylacrylamide) (PNIPAm) stabilized Ag NPs (Zhu et al. 2019). This reported method was very simple and detection of Hg^{+2} was observed by a color change of Ag NPs from yellow to brown with the formation of Ag-Hg core-shell NPs. The above method showed high sensitivity with LOD 75 nM and 95.40% efficiency to remove Hg^{+2} from tap water samples. Furthermore, Hg^{2+} and Ag^{+} ions were removed effectively from the water sample using magnetic silica sphere (MSS)@Au NPs functionalized with oligonucleotide (Liu et al. 2014). One of the major advantages of the above method was about 80% of the MSS@Au NPs were recycled without any difficulty which made it one of the low-cost techniques.

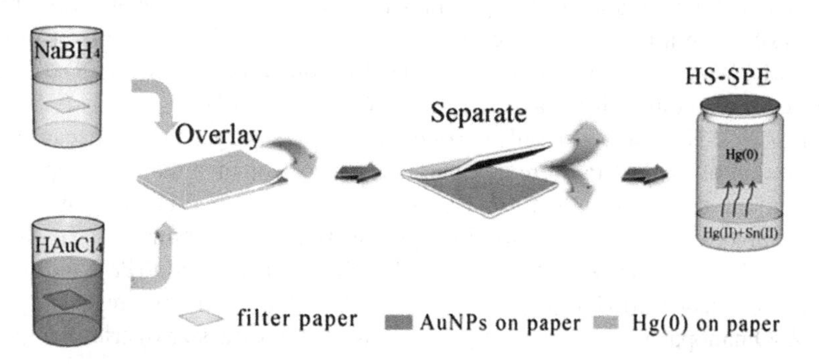

FIGURE 1 The preparation route of the AuNP-modified filter paper sensor (Reprinted with permission from Lei et al., *New J. Chem.*, 44(2020) 14299-14305).

Also, many researchers investigated the use of crystal NMs are in the analysis and removal of heavy metal ions from water (Table 1). In this lineup, Sharma et al. prepared copper oxide/zinc oxide-tetrapods (CuO/ZnO-T) and successfully removed Cr^{6+} and Pb^{2+} from wastewater (Sharma et al. 2020). This CuO/ZnO-T showed superior efficiency to remove 99% of Cr^{6+} and 97% of Pb^{2+} at a concentration of 10 ppm. Similarly, another analytical method for the removal of heavy metal ions Pb^{2+} and Cd^{2+} was developed using thio-(3-glycidyloxopropyl) trimethoxysilane (GLYMO(S)) and ethylenediamine (en) modified Fe_3O_4@SiO_2 magnetic NPs (Masjedi et al. 2020). The core-shell structure of Fe_3O_4@SiO_2 NPs with the presence of amino and thiol groups surfaces made it possible to detect and absorb Pb^{2+} and Cd^{2+} with 90% efficiency to remove both the heavy metal ions from a water sample. Moreover, an analytical method for the removal of Cr^{6+} was developed by using zerovalent iron nanoparticles (ZVNI) (Madhavi et al. 2013). The leaf extracts of *Eucalyptus globules* were used to synthesized ZVNI. The reduction and stabilization of the ZVNI were done by polyphenolic compounds present in the leaf extract of *Eucalyptus globules*. This method was proficient to remove Cr^{6+} with an adsorption efficiency of 98.1%, which was highly comparable and even better than many other techniques which were used to remove Cr^{6+} from water.

3.2 Removal/Degradation of Pesticides

Pesticides are the agrochemical used to remove or manage pests. So, pesticides are widely used for crops protection from diseases and damages caused by various pests in the agriculture field (Rani et al. 2017). However, the largest use of these chemicals shows harmful effects on humans and the environment. In humans and animals, upon exposure pesticides caused eye and skin irritation, problems appear related to the nervous system and sometimes even cause death (Rawtani et al. 2018). So, it is in great demand to develop a method to remove pesticides from water as they decrease the quality of water due to their hazardous nature and non-degradability. Over the last few years,

NMs proved as promising tools to detect, degrade and remove toxic pesticides from water, soil and food samples. So, by taking the advantage of NMs, many researchers developed a method for the removal of pesticides from water (Table 1). In this connection, Sun et al. developed a method for the removal of carbaryl insecticide using *p*-amino benzenesulfonic acid (PABSA) functionalized Au NPs (Sun et al. 2013). This reported method was highly selective, capable to detect carbaryl among other carbamate pesticides with similar structures and separation efficiency of up to 44.8%. Similarly, Zhang et al. prepared imidazole ionic liquid (I-IL) functionalized Au NPs and used them in the detection and removal of imidacloprid pesticides (Zhang et al. 2014). This method was able to detect imidacloprid up to a very low concentration (5×10^{-7} M) and showed high efficiency (89.7%) to remove imidacloprid from water.

To remove pesticides from various sample matrices, various crystal NMs were used since the last decade. Pesticide permethrin was removed with an efficiency of 99% using chitosan-loaded zinc oxide nanoparticles (CS-ZnONPs) (Dehaghi et al. 2014). In this method, CS-ZnONPs beads were prepared using a polymer-based method, and these big beads having the advantage of showing a regeneration capacity of 56% after three cycles. Another pesticide fenarimol was removed with the help of iron oxide-palygorskite nanoparticles (Fe_2O_3-Pal NPs) and this method was demonstrated by Ouali et al. (Ouali et al. 2015). Algerian Fe_2O_3-Pal NPs were synthesized using the chemical co-precipitation method and having very good efficiency to remove fenarimol pesticide. Furthermore, Liu et al. developed the method to remove six organochlorine pesticides (Hydrochlorobenzene HCB, *trans*-chlordane, *cis*-chlordane, *o,p*-DDT, *p,p*-DDT and mirex) using TiO_2 nanoparticles (TiO_2 NPs) (Liu et al. 2015). In this method, as-prepared TiO_2 NPs provided a good alternative to the commercial fibers, which were used for the determination and removal of organochlorine pesticides from water samples. Then, Kaur et al. showed pesticides removal efficiency comparison between bare and functionalized Zinc Oxide nanoparticles (ZnO NPs) (Kaur et al. 2017). Naphthalene a widely used pesticide was successfully removed by 1-butyl-3-methylimidazolium tetrafluoroborate (BMTF) functionalized ZnO NPs with adsorption capacity 148.3 mg g^{-1}, which was higher than the 66.80 mg g^{-1} of without functionalized ZnO NPs. Furthermore, the extraction and removal method for pyrathroid pesticides (Cyhalothrin, Deltamethrin, Ethofenprox and Bifenthrin) was developed by Fan et al. using ionic liquid (IL) functionalized Fe_3O_4 NPs (Fan et al. 2017). As shown in Figure 2, IL [P4448][Br] was used to enrich pyrethroids and further reaction with $Na[N(CN)_2]$ forming hydrophobic IL as a removal/extraction agent. Then, magnetic Fe_3O_4 NPs were added to the above solutions and elution was done using acetonitrile. Also, the developed method has many benefits, like short pretreatment time, good recoveries and very low solvent consumption.

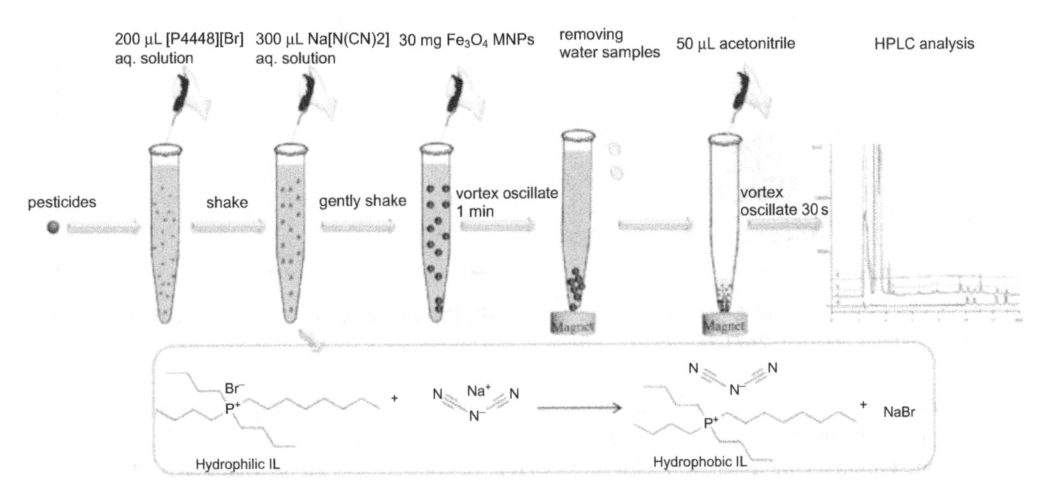

FIGURE 2 Scheme of pretreatment procedure under optimum conditions. (Reprinted with permission from Fan et al., *Anal. Chim. Acta* 975(2017) 20-29).

Also, NMs show their potential as alternative reductive tools for degradation of pesticides and process is having various benefits, like easy to use, inexpensive and selective. In this lineup, many research groups developed methods for the degradation of pesticides using liquid and crystal NMs. Degradation of chlorpyrifos (CP) one of the widely used organophosphorus pesticides was showed by Bootharaju et al. using Ag NPs (Bootharaju and Pradeep 2012). As shown in Figure 3, after adsorption of CP on Ag NPs surfaces it decomposed to 3,5,6-trichloro-2-pyridinol (TCP) and diethyl thiophosphate (DETP), which led to the aggregation of Ag NPs. Furthermore, the degradation of sulfentrazone pesticides to their less toxic compound was done using bimetallic Fe/Ni NPs (Nascimento et al. 2016). The degradation products were examined by mass spectrometry, and toxicity analysis was done on water fleas *Daphnia Similis*. Similarly, bimetallic silver/copper nanoparticles (Ag/Cu NPs) were used to degrade chlorpyrifos pesticide (Rosbero and Camacho 2017). Au/Cu NPs were prepared from leaf extracts of *Carica papaya* via green synthetic route and catalytic degradation of chlorpyrifos was completed after its adsorption on Au/Cu NPs surfaces. Moreover, the degradation method of the same pesticide chlorpyrifos was developed using magnetic ion nanoparticles (MNPs) with 99% of degradation efficiency (Das et al. 2017). This method showed great reusability potential as MNPs gave 95% activity even after five washing. Furthermore, bimetallic Fe/Ni NPs were used as a catalyst to degrade organophosphorus pesticide profenofos (Mansouriieh et al. 2019). After optimization of various parameters, this method showed 94.51% degradation efficiency for profenofos pesticide.

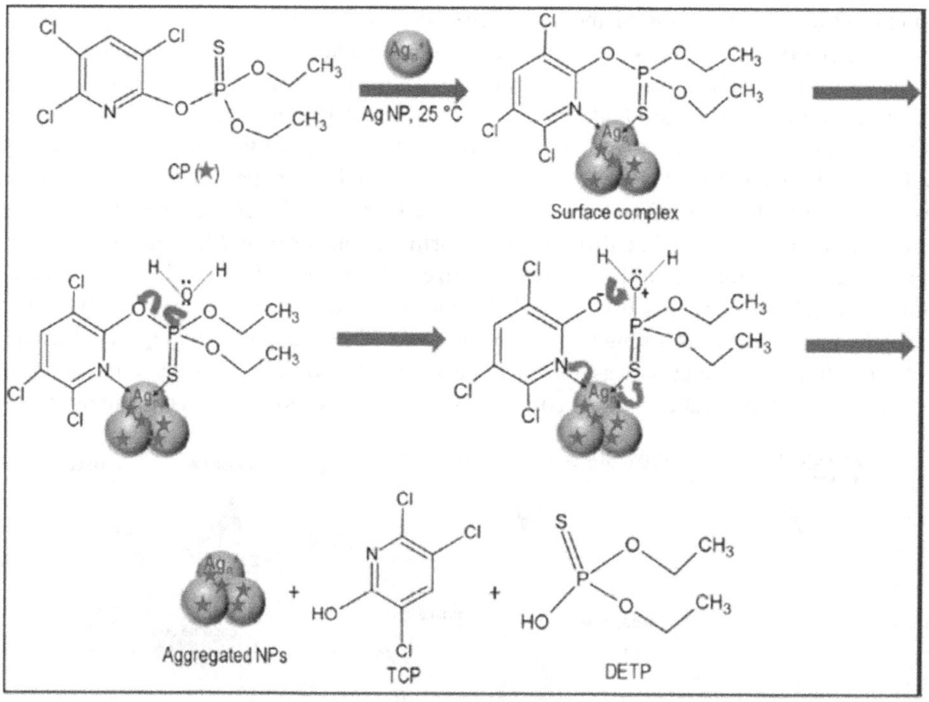

FIGURE 3 Representation of Degradation of CP on Ag NPs (Reprinted with permission from Bootharaju and Pradeep, Langmuir 28(2012) 2671-2679).

As with liquid NMs, crystal NMs show good efficiency to degrade pesticides. So far, many methods have been developed for the degradation of pesticides using crystal NMs. In this connection, Gupta et al. prepared $CoFe_2O_4@TiO_2$ decorated graphene oxide for photocatalytic degradation of chlorpyrifos pesticide (Gupta et al. 2015). This method was successfully applied to remove chlorpyrifos from wastewater and as prepared $CoFe_2O_4$@TiO_2/graphene oxide showed good reusability during experiments. Similarly, another method was developed for photocatalytic degradation of two pesticides monocrotophos and chlorpyrifos using TiO_2 NPs as photocatalysts (Amalraj and Pius 2015). After optimization of various experimental conditions, authors concluded that prepared TiO_2 photocatalyst showing higher efficiency for degradation of monocrotophos and chlorpyrifos pesticides under irradiation of UV light. Furthermore, two types of nanoscale zero-valent iron (nZVI) (T-type prepared using reduction of iron oxides in the gas phase under H_2 and B-type prepared using borohydride precipitation) for degradation of organochlorine pesticide DDT (El-Temsah et al. 2016). In this method, authors attended degradation efficiency > 90% and 78% using nZVI-B and nZVI-T, respectively, for DDT pesticide.

3.3 Removal/Degradation of Dyes

Dyes are largely used in paper, textile, plastic and other industries. Among commercially used dyes more than 15% are released in the environment without any pre-treatment, and it is a leading example of environmental pollution (Ramasamy et al. 2018). These dyes having a highly stable structure, so potential and economical methods are requiring for the removal/degradation of such a stable compound. Liquid and crystal NMs-based techniques have the efficiency to remove/degrade dyes from environmental resources. In this connection, many research groups made efforts to develop a method based on liquid/crystal NMs for the removal/degradation of dyes (Table 1).

To remove/degrade methylene blue dye from an aqueous solution, Ag NPs were decorated on surfaces of carbon nanotubes (CNTs) (Wu et al. 2020). In this method, catalyst Ag NPs in the networks of CNTs were responsible for the chemical degradation of methylene blue provided an efficient and simple approach for removal of methylene blue dye. Also, Ag NPs/CNTs composites were proven as efficient recyclable NMs, which was a major advantage of this method. Recently, another research group synthesized biogenic Ag NPs using *Terminalia bellerica* kernel extract (K-Ag NPs) and applied them for catalytic reduction of dyes (methylene blue, eosin yellow and methyl orange) and other pollutants (4-nitrophenol) (Sherin and et al. 2020). The authors' concluded that only 1.5% reduction of pollutants occurred in presence of strong reducing agent sodium borohydride but it reached 87% after addition of catalytic amount of K-Ag NPs to the sodium borohydride (Figure 4). Furthermore, β-csyclodextrin-based nanoparticles (β-CDN) were synthesized and studied their adsorption capacity to pollutants (dye methyl orange and heavy metal Pb^{+2}) in wastewater (Liu et al. 2020). As prepared NPs showed pollutants removal efficiency more than 95%, which was higher than many other adsorbent compounds/materials.

The use of crystal NMs was explored by many scientists to check their capability to act as a tool for the removal/degradation of dyes. To contribute toward method development based on crystal NMs for dyes degradation, Ramasamy et al. prepared Ca doped CeO_2 quantum dots for photodegradation of methylene blue dye (Ramasamy et al. 2018). This method showed 84% of efficiency for the degradation of methylene blue under direct sunlight conditions. Similarly, Sheshmani et al. developed a method for the removal of dye acid orange 7 using nanosized ZnS/chitosan/graphene oxide (ZnO/Ch/GO) (Sheshmani and Kazemi 2020). After optimization of various experimental conditions, the method showed the efficiency of adsorption (95.81%) and photodegradation (71.3%) for removal/degrade dye acid orange 7. Furthermore, Sharma et al. prepared copper oxide/zinc oxide-tetrapods (CuO/ZnO-T) nanocomposite for the treatment of wastewater (Sharma et al. 2020). As prepared CuO/ZnO-T nanocomposite was examined for the removal of dyes (Reactive yellow-145 [RY-145], Basic violet-3 [BV-3]) and heavy metal ions (Chromium [VI] and Lead [II]) and found a good removal efficiency against these pollutants.

TABLE 1 Removal/Degradation of water pollutants using NMs.

Pollutant	NMs	Surface Modification of NMs	Removal/Degradation Mechanism	Removal/Degradation Efficiency (%)	Reference
Heavy metal ions					
Ba^{2+} and Sr^{2+}	Ag NPs	–	Ion-exchange	88	Kaur et al. 2020
Hg^{+2}	Au NPs	–	Absorption	–	Lei et al. 2020
Hg^{+2}	Ag NPs	PNIPAm	Core-Shell formation	95.40	Zhu et al. 2019
Hg^{2+} and Ag^{+}	Au NPs	Oligonucleotide	Magnetic affinity	–	Liu et al. 2014
Cr^{6+} and Pb^{2+}	CuO/ZnO-T	–	Adsorption	99 (Cr^{6+}) and 97 (Pb^{2+})	Sharma et al. 2020
Pb^{2+} and Cd^{2+}	Fe_3O_4@SiO_2	GLYMO(S)-en	Adsorption	90	Masjedi et al. 2020
Cr^{6+}	ZVNI	–	Adsorption	98.1	Madhavi et al. 2013
Pesticides					
Carbaryl	Au NPs	PABSA	Aggregation	44.8	Sun et al. 2013
Imidacloprid	Au NPs	I-IL	Aggregation	89.7	Zhang et al. 2014
Permethrin	ZnO NPs	CS	Adsorption	99	Dehaghi et al. 2014
Fenarimol	Fe_2O_3 NPs	Pal	Adsorption	70	Ouali et al. 2015
HCB, *trans*-chlordane, *cis*-chlordane, *o,p*-DDT, *p,p*-DDT and mirex	TiO_2 NPs	–	Adsorption	–	Liu et al. 2015
Naphthalene	ZnO NPs	BMTF	Adsorption	–	Kaur et al. 2017
Cyhalothrin, Deltamethrin, Ethofenprox, Bifenthrin	Fe_3O_4	IL	Adsorption	–	Fan et al. 2017
Chlorpyrifos	Ag NPs	Citrate	Adsorption/Aggregation	–	Bootharaju et al. 2012
Sulfentrazone	Fe/Ni NPs	–	Adsorption	–	Nascimento et al. 2016

Contd.

Chlorpyrifos	Ag/Cu NPs	–	Adsorption	–	Rosbero et al. 2017
Chlorpyrifos	MNPs	–	Adsorption	99	Das et al. 2017
Profenofos	Fe/Ni NPs	–	Adsorption	94.51	Mansouriieh et al. 2019
Chlorpyrifos	Graphene oxide	$CoFe_2O_4@TiO_2$	Adsorption	–	Gupta et al. 2015
Monocrotophos and Chlorpyrifos	TiO_2 NPs	–	Adsorption	–	Amalraj et al. 2015
DDT	nZVI-B and nZVI-T	–	Adsorption	>90 and 78	El-Temsah et al. 2016
Dyes					
Methylene blue	CNTs	Ag NPs	Adsorption	–	Wu et al. 2012
Methylene blue, Eosin yellow, Methyl orange, and 4-nitrophenol	Ag NPs	*Terminalia bellerica* kernel extract	Adsorption	87	Sherin et al. 2020
Methyl orange and Pb^{+2}	β-CDN	Polyethyleneimine (PEI) and phosphonitrilic chloride trimer	Adsorption/ Photodegradation	95 for both (Methyl orange and Pb^{+2})	Liu et al. 2020
Methylene blue	CeO_2 quantum dots	Ca	Adsorption/ Photodegradation	84	Ramasamy et al. 2018
Acid orange 7	ZnO/Ch/GO	Chitosan	Adsorption/ Photodegradation	95.81 *via* adsorption and 71.3 *via* Photodegradation	Sheshmani et al. 2019
Reactive yellow-145 (RY-145), Basic violet-3 (BV-3), Chromium (VI) and Lead (II)	CuO/ZnO-T	–	Adsorption/ Photodegradation	80 for RY-145, 86 for BV-3, 99 for Chromium (VI) and 97 for Lead (II)	Sharma et al. 2020
Nuclear Waste					
$^{125}I^-$	Au NPs	Dextran	Adsorption	99	Choi et al. 2016

Contd.

$^{125}I^-$	Au NPs	*Deinococcus radiodurans R1*	Adsorption	99	Choi et al. 2017
UO_2^{2+}	Au NPs	Carbon nanotubes	Adsorption	99	Omidi et al. 2017
$^{125}I^-$	Ag NPs	Cellulose acetate	Adsorption	99	Shim et al. 2018
$^{60}Co(II)$	TiO_2 NPs	–	Complexation or adsorption	–	Wang et al. 2013
Ba(II) and Cs(I)	TiO_2 NPs	PVA and alginate	Adsorption	99 and 9	Majidnia et al. 2015
U(VI)	TiO_2 NPs	Polydopamine and Fe_3O_4	Adsorption	96.45	Zhang et al. 2018
I^- and UO_2^{2+}	Ag_2O	$Mg(OH)_2$	Surface precipitation	96.82 and 99.39	Chen et al. 2018
I^-	Fe_3O_4	Polypyrrole	Adsorption	–	Harijan et al. 2018
Microbial Contaminations					
E. coli and *B. subtili*	Ag NPs	–	Adsorption/Cellular Interaction	99.99	Prabhakar et al. 2020
E. coli and *Bacillus*	Ag NPs	SnZrMoP	Adsorption/Cellular Interaction	–	Kaur et al. 2020
E. coli and *S. aureus*	β-CDN	Polyethyleneimine and phosphonitrilic chloride	Adsorption/Cellular Interaction	99.99	Liu et al. 2020
S. aureus and *P. aeruginosa*	ZnO NPs	Ethylene glycol, gelatin, polyvinyl alcohol and polyvinylpyrrolidone	Photocatalytic degradation/ Cellular Interaction	25	Akhil et al 2016
Escherichia coli	Ag@TiO_2 NPs	–	Photocatalytic degradation/ Cellular Interaction	42	S. Sreeja and K.V. Shetty, 2016
Enterococcus faecalis and *Micrococcus luteus*	ZnO QDs and TiO_2 QDs	–	Electrostatic Interaction	–	Fakri et al. 2018

Contd.

Staphylococcus aureus and *Pseudomonas aeruginosa*	CdO NPs	Starch, Methyl cellulose, tween 20, Polyvinylpyrrolidone, Ethylene glycol, Gelatin and Polyvinyl alcohol	Photocatalytic degradation/ Cellular Interaction	~40	Janani et al. 2020
Other Toxic Molecules					
4-Nitrophenol	Ag NPs	Poly (acrylic acid)	Photocatalytic degradation	–	Kastner and Thunemann 2016
4-Nitrophenol	Co NPs	Tetrabutyl ammonium bromide	Photocatalytic degradation	–	Mondal et al. 2017
Picric acid, 4-nitroaniline, 4-nitrophenol and 2- nitrophenol.	Ag NPs	*Colocasia esculenta*	Photocatalytic degradation	90 (Picric acid), 96.3 (4-nitroaniline), 97.83 (4-nitrophenol) and 92.97 (2-nitrophenol)	
4-Nitrophenol	Ag NPs	*Piper chaba*	Photocatalytic degradation	–	Mahiuddin et al. 2020
Acenaphthene, Phenanthrene and Fluorene	TiO_2 NPs	–	Photocatalytic degradation	96 (Acenaphthene), 95 (Phenanthrene) and 93 (Fluorine)	Rachna et al. 2020
4-Nitrophenol	Co@NC NPs	–	Photocatalytic degradation	~100	Wang et al. 2020
p-arsanilic acid, 2-aminophenylarsonic acid, 2-nitrophenylarsonic acid, Roxarsone, 4-hydroxyphenylarsonic acid and Phenylarsonic acid	$CoFe_2O_4$ NPs	–	Adsorption	–	Liu et al. 2020

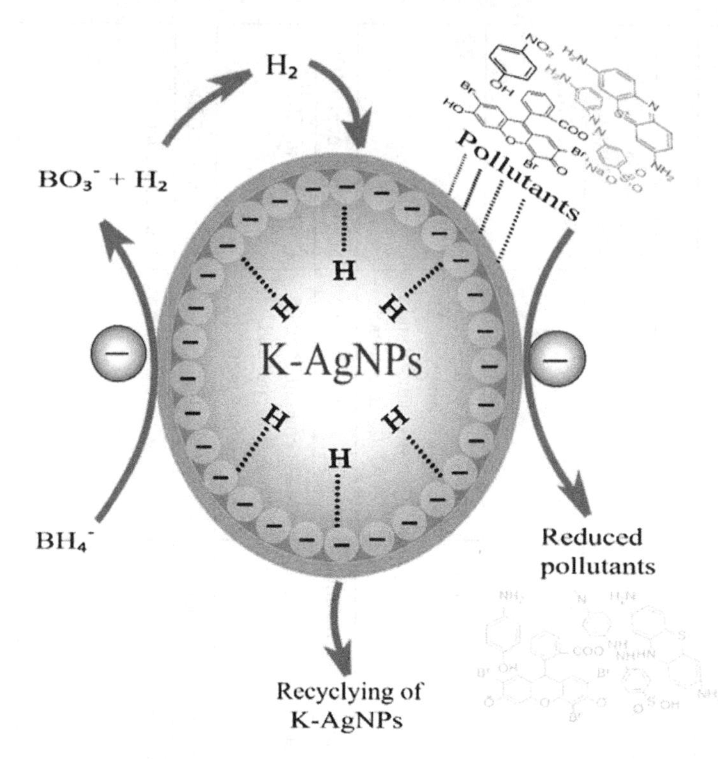

FIGURE 4 Schematic mechanism of K-AgNPs catalysis. (Reprinted with permision from Sherin et al., *Colloids Interface Sci. Commun.* 37(2020) 100276).

3.4 Removal/Degradation of Nuclear Waste

As an alternative renewable energy tool, nuclear energy received great importance globally. Also, it is less expensive and clean energy used in nuclear power plants, nuclear weapon testing, mining industry and medical applications (Zhang and Liu 2020). But at the same time, radioactive pollution caused during the generation/use of nuclear energy is a major concern for the environment and health point of view. Water, air and soil contamination with various radioisotopes (e.g. ^{125}I, ^{129}I, ^{131}I, ^{137}Cs, ^{152}Eu, ^{106}Ru, ^{241}Am, ^{90}Sr, ^{144}Ce, ^{99m}Tc, ^{99}Mo, ^{60}Co and ^{75}Se) cause health hazards and long time genetic disorders in living organisms. So, regulation and remediation of nuclear waste is a big task for nuclear energy nations around the world. To handle radioactive waste, presently available waste treatment techniques are not efficient as radioactive contamination is comparatively low than the other contaminants. Thus, it needs to develop sensitive and selective degradation/removal methods for nuclear waste. To solve this issue, recently many methods were developed based on NMs to remove contaminations of nuclear waste as nanometer-scale material shows stronger adsorption ability and higher chemical affinity with radioactive ions compared with traditional other materials (Table 1).

Liquid NMs, such as Ag NPs and Au NPs, acted as excellent adsorbents for the removal of radioactive ions. In this connection, Choi et al. developed the method for removal of radioactive $^{125}I^-$ based on fact that gold atoms having a good affinity toward iodine atoms (Choi et al. 2016). Dextran gel complex (DC) was immobilized on Au NPs and this system has the capacity to remove 99% radioactive $^{125}I^-$ (Fig. 5). The same group developed another method using Au NPs containing bacteria *Deinococcus radiodurans R1* for removal of radioactive $^{125}I^-$ (Choi et al. 2017). This method also showed good affinity toward $^{125}I^-$ and efficiently remove more than 99% of $^{125}I^-$ within 15 minutes. In a similar way, radioactive UO_2^{2+} ions were removed from a water sample using carbon nanotubes (CNTs) functionalized Au NPs (Omidi et al. 2017). At optimized experimental conditions, this method is able to remove 99% of radioactive UO_2^{2+} within 5 minutes followed by simple sonication. Au NPs and Ag NPs also remove radioactive I^- ions due to their high affinity between Ag and I^- atoms. Shim et al. prepared Ag NPs immobilized with cellulose acetate membrane (CAM) and studied its efficiency to remove radioactive I^- (Shim et al. 2018). Their study revealed that as-prepared CAM-Ag NPs were capable to remove radioactive I^- about 99% with a simple continuous filtration method.

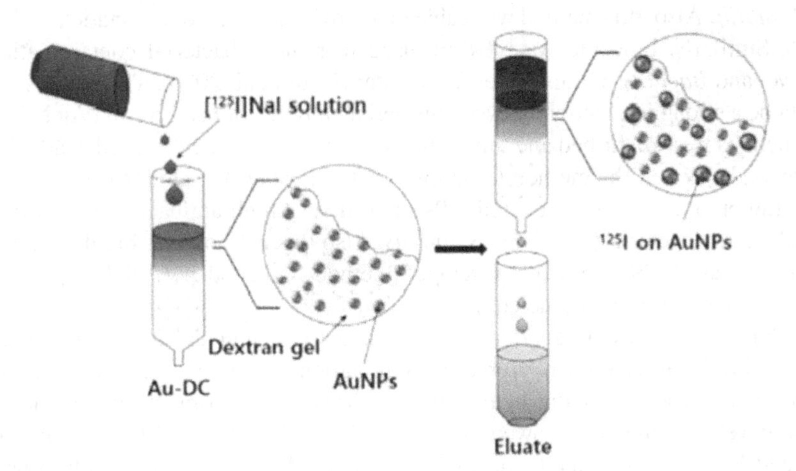

FIGURE 5 Schematic illustration of the desalination produce using Au-DC (Reprinted with permission from Choi et al., *ACS Appl. Mater. Interfaces* 8(2016) 29227-29231).

Crystal NMs also have a good affinity toward radioactive anions and cations. For example, TiO_2 NPs were used to remove radioactive $^{60}Co(II)$ using complexation or chemical adsorption (Wang et al. 2013). Similarly, TiO_2 NPs surface-functionalized with PVA and alginate were found effective to remove radioactive ions (Majidnia et al. 2015). This developed method was very simple and was able to remove Ba(II) and Cs(I) under sunlight with the efficiency of 99% and 9%, respectively. Another researcher, prepared core-shell NPs using polydopamine (PDA) functionalized TiO_2 NPs deposited on the surface of Fe_3O_4 for removal of radioactive U(VI) (Zhang et al. 2018). Under the optimized experimental conditions, these core-shell NPs showed the highest removal efficiency of 96.45% for U(VI) using the adsorption mechanism. TiO_2 NPs and Ag_2O can also effectively remove radioactive I^- and UO_2^{2+} ions from wastewater samples (Chen et al. 2018). This developed method was highly efficient and capable to remove I^- and UO_2^{2+} ions 96.82% and 99.39%, respectively, within 5 minutes. In a similar way, polypyrrole(PPy) encapsulated magnetite(Fe_3O_4) NPs are used to capturing radioactive I^- from water (Harijan et al. 2018). This method is very simple and there is no need for any complex separation techniques as it is based on a simple magnetic separation principle.

3.5 Removal/Degradation of Microbial Contaminants

The availability and supply of safe drinking water to the community are major aspects of any developing country. But the presence of bacteria, viruses and pathogens in drinking water increases concern about the quality of water and associated health risk (Ashbolt 2015). The presence of these microorganisms in drinking water can cause vomiting, diarrhea, fever, cramps, headaches, nausea, fatigue and sometimes serious disease, like cholera. Also, low immunity personals, especially children and the elderly get easily infected with these waterborne diseases. The use of NMs for the removal of microbial contaminations from water has been explored for many years. As NMs having a high surface-to-volume ratio, higher chemical reactivity and recyclability, they are widely used to remove/degrade waterborne microbes (bacteria, protozoa and viruses) (Kokkinos et al. 2020). Also, NMs-based water disinfectants technology is in high demand as ease of fabrication, low cost and low energy requirement and reliability.

Liquid NMs showed excellent disinfectant properties, especially Ag and Au NPs (Table 1). Prabhakar et al. developed a device to remove microbial contamination from water using Ag NPs and UV-light (Prabhakar et al. 2020). This device was highly effective against common bacteria *E. coli* and *B. subtili.* Also, this method was cable to remove bacterial contamination with an efficiency of 99.99%. Similarly, biogenic Ag NPs are used to remove bacterial contamination of *E. coli* (Gram (–) ve) and *Bacillus* (Gram (+) ve) from water (Kaur et al. 2020). The authors suggested that there might be a strong interaction between the cell membrane of bacteria and Ag NPs due to which cellular functions were disturbed and this could be the reason for bacterial cell death. Another group of researchers developed the method to remove bacteria using β-cyclodextrin-based nanoparticles (β-CDN) (Liu et al. 2020). As prepared NPs showed very high antibacterial activity against both G(–)ve and G(+)ve bacteria *E. coli* and *S. aureus*. Also this method was highly efficient to remove bacteria more than 99.99%. So, the developed method has good potential to purify microbially contaminated water with great efficiency.

Crystal NMs also showed good antimicrobial activity and capable of removing/degrading pathogens (Table 1). To remove microbial contaminations from water, Akhil et al. prepared ZnO NPs capped with various capping ligands and recorded their antimicrobial activity (Akhil et al. 2016). The developed method showed good antimicrobial activity ~100% against both bacteria *S. aureus* and *P. aeruginosa* and degraded 25% bacteria. Another group of authors used Ag@TiO_2 NPs to study photocatalytic disinfection of *E. coli* contaminated water (Sreeja and Shetty 2016). This study revealed that the method was very efficient and capable of 100% disinfect water and degraded about 42% of bacteria. In a similar way, Fakhri et al. prepared ZnO QDs decorated CuO nanosheets and TiO_2 QDs decorated WO_3 nanosheets (Fakhri et al. 2018). Both the composites were shown antimicrobial activity against *Enterococcus faecalis* and *Micrococcus luteus*. The possible mechanism for antibacterial activity was electrostatic interaction between +ve charge metal oxide and the -ve charge of the bacteria. Furthermore, CdO NPs were prepared and functionalized with various capping ligands to check their antimicrobial activity against *Staphylococcus aureus* and *Pseudomonas aeruginosa* bacteria (Janani et al. 2020). This developed method showed good antibacterial activity with degradation of the pathogen with an efficiency of ~40%.

3.6 Removal/Degradation of Other Toxic Molecules

Organic pollutants are products/byproducts of chemical industries and major sources of water pollution. As they are highly toxic molecules, their degradation/removal from water resources is a necessary task. 4-Nitrophenol and its derivatives are commonly found to toxic molecules in water bodies as they are used by many industries as an intermediate in the preparation of many drugs, dyes and pesticides. But this nitrogen-based organic compounds are highly health hazardous and cause skin irritation, dermatitis, diarrhea, conjunctivitis, colic, corneal and liver damage (Bamba et al. 2017). Also, nitro-containing aromatic molecules show mutagenic and carcinogenic activity due to their easy conversion into their N-hydroxy and nitroso derivatives. The use of NMs for the conversion of toxic substances into environmentally friendly molecules is the simplest way to remove/degrade 4-nitrophenol and their derivatives from water samples (Chen et al. 2016).

Liquid NPs have high catalytic activity and is able to reduce/degrade nitro-based aromatic compounds using photocatalytic degradation (Table 1). In this connection, Kastner and Thunemann synthesized poly (acrylic acid) (PAA) functionalized Ag NPs and used them for the reduction of 4-nitrophenol (Kastner and Thunemann 2016). As prepared PAA stabilized, Ag NPs showed particles good catalytic activity and were able to reduce 4-nitrophenol to 4-amionophenol. Another group of researchers used surfactant stabilized cobalt (Co) NPs as a metal catalysts to reduce 4-nitrophenol (Mondal et al. 2017). Similarly, Ismail et al. synthesized biogenic Ag NPs using plant extract of *Colocasia esculenta* and used them to reduce picric acid, nitroaniline and nitrophenols (Ismail et al. 2018). This method showed high efficiency and capacity to reduce 90% of picric acid, 96.3% of 4-nitroaniline, 97.83% of 4-nitrophenol and 92.97% of 2-nitrophenol. Recently, Mahiuddin et al. prepare Ag NPs used steam extract of *Piper chaba* and used them for the reduction of anthropogenic water pollutants 4-nitrophenol (Mahiuddin et al. 2020). This method was very simple, cost-effective and required 9 min to reduce 4-nitrophenol pollutants.

For last many years, crystal NMs have also been used as an effective photo-catalyst to degrade/remove toxic organic molecules (Table 1). In this connection, Rachna et al. used TiO_2 NPs based zinc hexacyanoferrate (ZnHCF) framework for degradation of organic pollutants (acenaphthene, phenanthrene and fluorene) (Rachna et al. 2020). This developed method showed good efficiency and successfully removed 96% of acenaphthene, 95% of phenanthrene and 93% of fluorine from water samples. Similarly, Wang et al. prepared highly efficient nitrogen-doped carbon-coated cobalt (Co@NC) NPs for the reduction of 4-nitrophenol (Wang et al. 2020). The method was able to reduce nearly 100% of pollutant 4-nitrophenol, which showed high capability developed NMs. Furthermore, Liu et al. synthesized flower-like $CoFe_2O_4$ NPs and used them as a sorbent to remove aromatic organoarsenicals (*p*-arsanilic acid, 2-aminophenylarsonic acid, 2-nitrophenylarsonic acid, roxarsone, 4-hydroxyphenylarsonic acid and phenylarsonic acid) (Liu et al. 2020). Under the optimized experimental conditions, as-prepared $CoFe_2O_4$ NPs act as high efficient sorbents to adsorb and remove hazardous aromatic organoarsenicals from water samples.

4. ENVIRONMENTAL FATE OF NANOMATERIALS-BASED TECHNOLOGY

As NMs have outstanding physico-chemical properties, like tunable shape, nanometer range size and high surface to volume ratio, which allows them to substitute traditional materials/techniques used for water purification. By taking the advantage of these excellent properties, NMs based techniques are proven as one of the advanced tools in the field of water cleaning and purification methods. But there are many disputes and challenges for the use of NMs in the treatment of polluted water and their commercialization on large scale. The experimental data revealed that many NMs showed a toxic effect on living organisms and may be harmful to the environment (Liu et al. 2014). The level of toxicity is not exactly clear, but NMs release into water resources may lead to a hazardous effect on the human and marine ecosystem. Also, it is suggested that NMs were non-toxic in short exposure but maybe having adverse effects in long-term exposure (Usmani et al. 2017). For example, Ag, ZnO, TiO_2, Cu and Fe_2O_3 NPs can damage DNA, proteins and cellular membrane (Brandelli et al. 2017). Also, many researchers noticed that metal NPs not only affect the microorganism but showed an adverse effect on plants and animals also (Brinkmann et al. 2020). Furthermore, the presence of NMs in water affected humans and the environment via various possible routes of uptakes and causes hazardous effects (Kamali et al. 2019).

To avoid possible environmental and health hazards, NMs used in water purification techniques are subjected to toxicological tests (Bodzek et al. 2020). To take actual advantage of NMs based technologies, there is a need to investigate the adverse effects of NMs instead of only demonstrating their benefits. So, during the method/technique development process, researchers also have to consider the health and environmental hazard effects of NMs that are used in their developed techniques.

5. CONCLUSIONS AND FUTURE PROSPECTS

The use of liquid and crystal NMs for the purification of water has been one of the most explored research fields in the last decade. Here, we collected and compiled research data and provided a common platform to researchers working in the field of NMs based water purification techniques. The synthesis of liquid and crystal NMs with controlled size, shape and surface chemistry, using appropriate reducing and stabilizing agents, made them the best suitable material for water purification. Also, functionalizing ligands on surfaces of NMs act as a linker to bind with targeted molecules/ions, which helps in the easy removal of pollutants from water. The effective removal/ degradation of pollutants was processed through binding between NMs surfaces and pollutants via various interaction mechanisms and understanding of these mechanisms is the base of NMs based water purification method development. The NMs based water purification techniques are highly efficient, cost-effective, simple and can purify water from any kind of pollutants (heavy metal ions, pesticides, dyes, nuclear waste and microbial contamination, etc.).

In the end, still, there are lots of possibilities in the development of new water purification methods/techniques using NMs by preparing more efficient new NMs or tuning the properties of existing NMs. But at the same time, future research and development on the use NMs for water purification techniques should consider the environmental and health hazardous effects of NMs. The preparation of environment-friendly NMs and their use in water purification techniques are the future change and give a scope of work to researchers from all around the world.

References

Adeleye, A.S., Conway, J.R., Garner, K., Huang, Y., Su, Y. and Keller, A.A. 2016. Engineered nanomaterials for water treatment and remediation: Costs, benefits, and applicability. Chem. Eng. J. 286: 640-662.

Akhil, K., Jayakumar, J., Gayathri, G. and Khan, S.S. 2016. Effect of various capping agents on photocatalytic, antibacterial and antibiofilm activities of ZnO nanoparticles. J. Photochem. Photobiol. B Biol. 160: 32-42.

Amalraj, A. and Pius, A. 2015. Photocatalytic degradation of monocrotophos and chlorpyrifos in aqueous solution using TiO_2 under UV radiation. J. Water Process Eng. 7: 94-101.

Ashbolt, N.J. 2015. Microbial contamination of drinking water and human health from community water systems. Curr. Envir. Health Rpt. 2: 95-106.

Bamba, D., Coulibaly, M. and Robert, D. 2017. Nitrogen-containing organic compounds: Origins, toxicity and conditions of their photocatalytic mineralization over TiO_2. Sci. Total Environ. 580: 1489-1504.

Bodzek, M., Konieczny, K. and Kwiecińska-Mydlak, A. 2020. Nanotechnology in water and wastewater treatment. Graphene – the nanomaterial for next generation of semipermeable membranes. Crit. Rev. Env. Sci. Tec. 50: 1515-1579.

Bootharaju, M.S. and Pradeep, T. 2012. Understanding the degradation pathway of the pesticide, chlorpyrifos by noble metal nanoparticles. Langmuir 28: 2671-2679.

Brandelli, A., Ritter, A.C. and Veras, F.F. 2017. Antimicrobial activities of metal nanoparticles. pp. 337-363. *In*: Cravo-Laureau, C., Cagnon, C., Lauga, B. and Duran, R. [ed.]. In Microbial Ecology. Springer, New York, USA.

Brinkmann, B.W., Koch, B.E.V., Spaink, H.P., Peijnenburg, W.J.G.M. and Vijver, M.G. 2020. Colonizing microbiota protect zebrafish larvae against silver nanoparticle toxicity. Nanotoxicology 14: 725-739.

Cao, G. 2004. Nanostructures & Nanomaterials: Synthesis, Properties and Applications. Imperial College Press, London.

Chauhan, N., Gupta, S., Singh, N., Singh, S., Islam, S.S., Sood, K.N., et al. 2011. Aligned nanogold assisted one step sensing and removal of heavy metal ions. J. Colloid. Interface Sci. 363: 42-50.

Chen, P., Xing, X., Xie, H., Sheng, Q. and Qu, H. 2016. High catalytic activity of magnetic $CuFe_2O_4$/graphene oxide composite for the degradation of organic dyes under visible light irradiation. Chem. Phys. Lett. 660: 176-181.

Chen, Y.-Y., Yu, S.-H., Yao, Q.-Z., Fu, S.-Q. and Zhou, G.-T. 2018. One-step synthesis of $Ag_2O@Mg(OH)_2$ nanocomposite as an efficient scavenger for iodine and uranium. J. Colloid. Interf. Sci. 510: 280-291.

Choi, M.H., Jeong, S.-W., Shim, H.E., Yun, S.-J., Mushtaq, S., Choi, D.S., et al. 2017. Efficient bioremediation of radioactive iodine using biogenic gold nanomaterial-containing radiation-resistant bacterium, Deinococcus radiodurans R1. Chem. Commun. 53: 3937-3940.

Choi, M.H., Shim, H.-E., Yun, S.-J., Park, S.-H., Choi, D.S., Jang, B.-S., et al. 2016. Gold-nanoparticle-immobilized desalting columns for highly efficient and specific removal of radioactive iodine in aqueous media. ACS Appl. Mater. Interfaces 8: 29227-29231.

Das, A., Singh, J. and Yogalakshmi, K.N. 2017. Laccase immobilized magnetic iron nanoparticles: Fabrication and its performance evaluation in chlorpyrifos degradation. Int. Biodeterior. Biodegrad. 117: 183-189.

Das, S., Chakraborty, J., Chatterjee, S. and Kumar, H. 2018. Prospects of biosynthesized nanomaterials for the remediation of organic and inorganic environmental contaminants. Environ. Sci. Nano 5: 2784-2808.

Dehaghi, S.M., Rahmanifar, B., Moradi, A.M. and Azar, P.A. 2014. Removal of permethrin pesticide from water by chitosan–zinc oxide nanoparticles composite as an adsorbent. J. Saudi Chem. Soc. 18: 348-355.

Ealia, S.A.M. and Saravanakumar, M.P. 2017. A review on the classification, characterisation, synthesis of nanoparticles and their application. IOP Conf. Ser. Mater. Sci. Eng. 263: 032019.

El-Temsah, Y.S., Sevcu, A., Bobcikova, K., Cernik, M. and Joner, E.J. 2016. DDT degradation efficiency and ecotoxicological effects of two types of nano-sized zero-valent iron (nZVI) in water and soil. Chemosphere 144: 2221-2228.

Fakhri, A., Azad, M., Fatolahi, L. and Tahami, S. 2018. Microwave-assisted photocatalysis of neurotoxin compounds using metal oxides quantum dots/nanosheets composites: Photocorrosion inhibition, reusability and antibacterial activity studies. J. Photochem. Photobiol. B Biol. 178: 108-114.

Fan, C., Liang, Y., Dong, H., Ding, G., Zhang, W., Tang, G., et al. 2017. *In situ* ionic liquid dispersive liquid-liquid microextraction using a new anion-exchange reagent combined Fe_3O_4 magnetic nanoparticles for determination of pyrethroid pesticides in water samples. Anal. Chim. Acta 975: 20-29.

Gola, D., Malik, A., Namburath, M. and Ahammad, S.Z. 2018. Removal of industrial dyes and heavy metals by *Beauveriabassiana*: FTIR, SEM, TEM and AFM investigations with Pb(II). Environ. Sci. Pollut. Res. 25: 20486-20496.

Grey, D., Garrick, D., Blackmore, D., Kelman, J., Muller, M. and Sadoff, C. 2013. Water security in one blue planet: Twenty-first century policy challenges for science. Phil. Trans. R. Soc. A. 371: 20120406.

Gupta, V.K., Eren, T., Atar, N., Yola, M.L., Parlak, C. and Karimi-Maleh, H. 2015. $CoFe_2O_4$@TiO_2 decorated reduced graphene oxide nanocomposite for photocatalytic degradation of chlorpyrifos. J. Mol. Liq. 208: 122-129.

Harijan, D.K.L., Chandra, V., Yoon, T. and Kim, K.S. 2018. Radioactive iodine capture and storage from water using magnetite nanoparticles encapsulated in polypyrrole. J. Hazard. Mater. 344: 576-584.

Ismail, M., Khan, M.I., Khan, S.B., Akhtar, K., Khan, M.A. and Asiri, A.M. 2018. Catalytic reduction of picric acid, nitrophenols and organic azo dyes *via* green synthesized plant supported Ag nanoparticles. J. Mol. Liq. 268: 87-101.

Janani, B., Gayathri, G., Syed, A., Raju, L.L., Marraiki, N., Elgorban, A.M., et al. 2020. The effect of various capping agents on surface modifications of CdO NPs and the investigation of photocatalytic performance, antibacterial and antibiofilm activities. J. Inorg. Organomet. Polym. Mater. 30: 1865-1876.

Kamali, M., Persson, K.M., Costa, M.E. and Capela, I. 2019. Sustainability criteria for assessing nanotechnology applicability in industrial wastewater treatment: Current status and future outlook. Environ. Int. 125: 261-276.

Kastner, C. and Thunemann, A.F. 2016. Catalytic reduction of 4-nitrophenol using silver nanoparticles with adjustable activity. Langmuir 32: 7383-7391.

Kaur, R., Kaushal, S. and Singh, P.P. 2020. Biogenic synthesis of a silver nanoparticle–SnZrMoP nanocomposite and its application for the disinfection and detoxification of water. Mater. Adv. 1: 728-737.

Kaur, Y., Bhatia, Y., Chaudhary, S. and Chaudhary, G.R. 2017. Comparative performance of bare and functionalize ZnO nanoadsorbents for pesticide removal from aqueous solution. J. Mol. Liq. 234: 94-103.

Khan, S.T. and Malik, A. 2019. Engineered nanomaterials for water decontamination and purification: From lab to products. J. Hazard. Mater. 363: 295-308.

Kim, J. and Benjamin, M.M. 2004. Modeling a novel ion exchange process for arsenic and nitrate removal. Water Res. 38: 2053-2062.

Kokkinos, P., Mantzavinos, D. and Venieri, D. 2020. Current trends in the application of nanomaterials for the removal of emerging micropollutants and pathogens from water. Molecules 25: 2016-2047.

Kästner, C. and Thünemann, A.F. 2016. Catalytic reduction of 4-nitrophenol using silver nanoparticles with adjustable activity. Langmuir 32: 7383-7391.

Lei, Y., Zhang, F., Guan, P., Guo, P. and Wang, G. 2020. Rapid and selective detection of Hg(II) in water using AuNP *in situ*-modified filter paper by a head-space solid phase extraction Zeeman atomic absorption spectroscopy method. New J. Chem. 44: 14299-14305.

Levin, R., Epstein, P., Ford, T., Harrington, W., Olson, E. and Reichard, E. 2002. US drinking water challenges in the twenty-first century. Environ. Health Perspect. 110: 43-52.

Liu, J., Li, B., Wang, G., Qin, L., Ma, X., Hu, Y., et al. 2020. Facile synthesis of flower-like $CoFe_2O_4$ particles for efficient sorption of aromatic organoarsenicals from aqueous solution. J. Colloid Interface Sci. 568: 63-75.

Liu, M., Wang, Z., Zong, S., Chen, H., Zhu, D., Wu, L., et al. 2014. SERS detection and removal of mercury(ii)/silver(i) using oligonucleotide-functionalized core/shell magnetic silica sphere@Au nanoparticles. ACS Appl. Mater. Interfaces 6: 7371-7379.

Liu, S., Xie, L., Zheng, J., Jiang, R., Zhu, F., Luan, T., et al. 2015. Mesoporous TiO_2 nanoparticles for highly sensitive solid-phase microextraction of organochlorine pesticides. Anal. Chim. Acta 878: 109-117.

Liu, X.T., Xi, Y.M., Xiao, L.W., Li, X.M., Wen, B.G., Yong, Q.M., et al. 2014. Toxicity of multi-walled carbon nanotubes, graphene oxide, and reduced grapheme oxide to zebrafish embryos. Biomed. Environ. Sci. 27: 676-683.

Liu, Y., Jia, J., Gao, T., Wang, X., Yu, J., Wu, D., et al. 2020. One-pot fabrication of antibacterial β-cyclodextrin-based nanoparticles and their superfast, broad-spectrum adsorption towards pollutants. J. Colloid Interf. Sci. 576: 302-312.

Lu, F. and Astruc, D. 2018. Nanomaterials for removal of toxic elements from water. Coord. Chem. Rev. 356: 147-164.

Lu, F. and Astruc, D. 2020. Nanocatalysts and other nanomaterials for water remediation from organic pollutants. Coord. Chem. Rev. 408: 213180.

Madhavi, V., Prasad, T.N.V.K.V., Reddy, A.V.B., Reddy, B.R. and Madhavi, G. 2013. Application of phytogenic zerovalent iron nanoparticles in the adsorption of hexavalent chromium. Spectrochim. Acta Part A 116: 17-25.

Mahiuddin, M., Saha, P. and Ochiai, B. 2020. Green synthesis and catalytic activity of silver nanoparticles based on piper chaba stem extracts. Nanomaterials 10: 1777-1792.

Majidnia, Z., Idris, A., Majid, M., Zin, R. and Ponraj, M. 2015. Efficiency of barium removal from radioactive waste water using the combination of maghemite and titania nanoparticles in PVA and alginate beads. Appl. Radia. Isotopes 105: 105-113.

Malaviya, P. and Singh, A. 2011. Phsicochemical technologies for remediation of chromium-containing waters and waste waters. Cris. Rev. Environ. Sci. Technol. 41: 1111-1172.

Mansouriieh, N., Sohrabi, M.R. and Khosravi, M. 2019. Optimization of profenofos organophosphorus pesticide degradation by zero-valent bimetallic nanoparticles using response surface methodology. Arabian J. Chem. 12: 2524-2532.

Masjedi, A., Askarizadeh, E. and Baniyaghoob, S. 2020. Magnetic nanoparticles surface-modified with tridentate ligands for removal of heavy metal ions from water. Mater. Chem. Phys. 249: 122917.

Mondal, A., Adhikary, B. and Mukherjee, D.K. 2017. Cobalt nanoparticles as reusable catalysts for reduction of 4-nitrophenol under mild conditions. Bull. Mater. Sci. 40: 321-328.

Musico, Y.L.F., Santos, C.M., Dalida, M.L.P. and Rodrigues, D.F. 2014. Surface modification of membrane filters using graphene and graphene oxide-based nanomaterials for bacterial inactivation and removal. ACS Sustain. Chem. Eng. 2: 1559-1565.

Nascimento, M.A., Lopes, R.P., Cruz, J.C., Silva, A.A. and Lima, C.F. 2016. Sulfentrazone dechlorination by iron-nickel bimetallic nanoparticles. Environ. Pollut. 211: 406-413.

Nasrollahzadeh, M., Sajjadi, M., Iravani, S. and Varma, R.S. 2021. Green-synthesized nanocatalysts and nanomaterials for water treatment: Current challenges and future perspectives. J. Hazard. Mater. 401: 123401.

Nie, M.H., Yan, C.X., Li, M., Wang, X.N., Bi, W.L. and Dong, W.B. 2015. Degradation of chloramphenicol by persulfate activated by Fe^{2+} and zerovalent iron. Chem. Eng. J. 279: 507-515.

Omidi, M.H., Azad, F.N., Ghaedi, M., Asfaram, A., Azqhandi, M.H.A. and Tayebi, L. 2017. Synthesis and characterization of Au-NPs supported on carbon nanotubes: Application for the ultrasound assisted removal of radioactive UO_2^{2+} ions following complexation with Arsenazo III: Spectrophotometric detection, optimization, isotherm and kinetic study. J. Colloid Inter. Sci. 504: 68-77.

Oskamand, G. and Poot, F.J.P. 2006. Synthesis of ZnO and TiO_2 nanoparticles. J. Solgel Sci. Technol. 37: 157-160.

Ouali, A., Belaroui, L.S., Bengueddach, A., Galindo, A.L. and Peña, A. 2015. Fe_2O_3–palygorskite nanoparticles, efficient adsorbates for pesticide removal. Appl. Clay Sci. 115: 67-75.

Prabhakar, A., Agrawal, M., Mishra, N., Roy, N., Jaiswar, A., Dhwaja, A., et al. 2020. Cost-effective smart microfluidic device with immobilized silver nanoparticles and embedded UV-light sources for synergistic water disinfection effects. RSC Adv. 10: 17479-17485.

Qu, X., Alvarez, P.J.J. and Li, Q. 2013. Applications of nanotechnology in water and wastewater treatment. Water Res. 47: 3931-3946.

Rachna, Rani, M. and Shanker, U. 2020. Degradation of tricyclic polyaromatic hydrocarbons in water, soil and river sediment with a novel TiO_2 based heterogeneous nanocomposite. J. Environ. Manage. 248: 109340.

Ramasamy, V., Mohana, V. and Rajendran, V. 2018. Characterization of Ca doped CeO_2 quantum dots and their applications in photocatalytic degradation. Open Nano 3: 38-47.

Rani, M., Shanker, U. and Jassal, V. 2017. Recent strategies for removal and degradation of persistent & toxic organochlorine pesticides using nanoparticles: A review. J. Environ. Manage. 190: 208-222.

Rawtani, D., Khatri, N., Tyagi, S. and Pandey, G. 2018. Nanotechnology-based recent approaches for sensing and remediation of pesticides. J. Environ. Manage. 206: 749-762.

Rosbero, T.M.S. and Camacho, D.H. 2017. Green preparation and characterization of tentacle-like silver/copper nanoparticles for catalytic degradation of toxic chlorpyrifos in water. J. Environ. Chem. Eng. 5: 2524-2532.

Sakohara, S., Ishida, M. and Anderson, M.A. 1998. Visible luminescence and surface properties of nanosized zno colloids prepared by hydrolyzing zinc acetate. J. Phys. Chem. B 102: 10169-10175.

Sekar, M., Sakthi, V. and Rengaraj, S. 2004. Kinetics and equilibrium adsorption study of lead(II) onto activated carbon prepared from coconut shell. J. Colloid Interface Sci. 279: 307-313.

Selvakannan, P., Swami, A., Srisathiyanarayanan, D., Shirude, P.S., Pasricha, R., Mandale, A.B., et al. 2004. Synthesis of aqueous au core-ag shell nanoparticles using tyrosine as a ph-dependent reducing agent and assembling phase-transferred silver nanoparticles at the air-water interface. Langmuir 20: 7825-7836.

Sharma, M., Poddar, M., Gupta, Y., Nigam, S., Avasthi, D.K., Adelung, R., et al. 2020. Solar light assisted degradation of dyes and adsorption of heavy metal ions from water by CuO-ZnO tetrapodal hybrid nanocomposite. Mater. Today Chem. 17: 100336.

Sherin, L., Sohail, A., Amjad, U., Mustafa, M., Jabeen, R. and Ul-Hamid, A. 2020. Facile green synthesis of silver nanoparticles using *Terminalia bellerica* kernel extract for catalytic reduction of anthropogenic water pollutants. Colloids Interface Sci. Commun. 37: 100276.

Sheshmani, S. and Kazemi, A. 2020. Graphene oxide and chitosan co-modified ZnS as photocatalyst and adsorbent: Preparation, characterisation, removal of acid orange 7, kinetic studies, and adsorption isotherms. Int. J. Environ. Anal. Chem. 100: 1362-1375.

Shim, H., Yang, J., Jeong, S.-W., Lee, C., Song, L., Mushtaq, S., et al. 2018. Silver nanomaterial-immobilized desalination systems for efficient removal of radioactive iodine species in water. Nanomaterials 8: 660.

Sine, G. and Comninellis, C. 2005. Nafion®-assisted deposition of microemulsion-synthesized platinum nanoparticles on BDD: Activation by electrogenerated OH radicals. Electrochim. Acta. 50: 2249-2254.

Sreeja, S. and Shetty, K.V. 2016. Microbial disinfection of water with endotoxin degradation by photocatalysis using Ag@TiO_2 core shell nanoparticles. Environ. Sci. Pollut. Res. 23: 18154-18164.

Sun, Z., Cui, Z. and Li, H. 2013. *p*-Amino benzenesulfonic acid functionalized gold nanoparticles: Synthesis, colorimetric detection of carbaryl and mechanism study by zeta potential assays. Sens. Actuators B Chem. 183: 297-302.

Tamrakar, R., Ramrakhiani, M. and Chandra, B.P. 2008. Effect of capping agent concentration on photophysical properties of zinc sulfide nanocrystals. The Open Nanoscience Journal 2: 12-16.

Usmani, M.A., Khan, I., Bhat, A.H., Pillai, R.S., Ahmad, N., Haafiz, M.K.M., et al. 2017. Current trend in the application of nanoparticles for waste water treatment and purification: A review. Curr. Org. Synth. 14: 206-226.

Wang, D., Zeng, F., Hu, X., Li, C. and Su, Z. 2020. Synthesis of a magnetic 2D Co@NC-600 material by designing a MOF precursor for efficient catalytic reduction of water pollutants. Inorg. Chem. 59: 2672-12680.

Wang, G., Dong, Y., Yang, H. and Li, Z. 2011. Ultrasensitive cysteine sensing using citrate-capped CdS quantum dots. Talanta 83: 943-947.

Wang, S., Tan, L., Jiang, J., Chen, J. and Feng, L. 2013. Preparation and characterization of nanosized TiO_2 powder as an inorganic adsorbent for aqueous radionuclide Co(II) ions. J. Radioanal. Nucl. Chem. 295: 1305-1312.

Wei, X.Y., Gao, N.Y., Li, C.J., Deng, Y., Zhou, S.Q. and Li, L. 2016. Zero-valentiron (ZVI) activation of persulfate (PS) for oxidation of bentazon in water. Chem. Eng. J. 285: 660-670.

Wu, J., Zhang, J., Zhou, S., Yang, Z. and Zhang, X. 2020. Ag nanoparticle-decorated carbon nanotube sponges for removal of methylene blue from aqueous solution. New J. Chem. 44: 7096-7104.

Yu, C.H., Tam, K. and Tsang, E.S. 2008. Chemical methods for preparation of nanoparticles in solution. pp. 113-141. *In*: Blackman, J.A. [ed.]. Handbook of Metal Physics. Vol. 5. Elsevier publication, Amsterdam.

Yuan, G. 2008. Nanomaterials to the rescue. Nano Today 3: 61.

Zhang, H., Dai, Z., Sui, Y., Xue, J. and Ding, D. 2018. Adsorption of U(VI) from aqueous solution by magnetic core–dual shell Fe_3O_4@PDA@TiO_2. J. Radioanal. Nucl. Chem. 317: 613-624.

Zhang, X., Sun, Z., Cui, Z. and Li, H. 2014. Ionic liquid functionalized gold nanoparticles: Synthesis, rapidcolorimetric detection of imidacloprid. Sens. Actuators B Chem. 191: 313-319.

Zhang, X. and Liu, Y. 2020. Nanomaterials for radioactive wastewater decontamination. Environ. Sci. Nano 7: 1008-1040.

Zhou, Y., Zhao, H., He, Y., Ding, N. and Cao, Q. 2011. Colorimetric detection of Cu^{2+} using 4-mercaptobenzoic acid modified silver nanoparticles. Colloids. Surf. A. 391: 179-183.

Zhu, C., Bian, J., Li, Y., Liu, J., Liu, X., Gao, X., et al. 2019. A novel and ultrasensitive yellow to taupe brown colorimetric sensing and removal method for Hg(II) based on thermosensitive poly (N-isopropyl acrylamide) stabilized sliver nanoparticles. New J. Chem. 43: 15879-15885.

13 CHAPTER

Modern Applications and Current Status of Liquid and Crystal Nanomaterials in Environmental Industry

Rachna[1], Uma Shanker[1,*] and Manviri Rani[2]

1. INTRODUCTION

A liquid crystalline state is a state of highly ordered molecular orientation. The synthesis of materials of such properties is an important issue. There are generally three basic phases of liquid crystalline anisotropy. These are named as nematic phase, smectic and cholesteric phase (Bukowczan et al. 2020). Out of these three phases, cholesteric liquid crystals have a special place due to their consideration as nanocomponents having properties of size, structure and other parameters similar to nanomaterials. Liquid crystals are known to form nanostructures with unique construction. These are used to form substances of ordered textures, such as nano electrodes and mesoporous systems. The introduction of other material frameworks, such as carbon nanotubes can contribute to better electrical, thermal conductivity and thermal insulation (Kragelsky et al. 2017, Shen and Dierking 2019). Liquid and crystal materials are considered the most important part of nanotechnology due to their supramolecular type assembly of molecules (Myshkin and Goryacheva 2017). Anisotropy in magnetic properties is a commending property of liquid and crystalline nanomaterials. Thus, upon applying the magnetic field, the molecules are oriented under the field effects. The anisotropy of its dielectric and diamagnetism is governed by the orientation of major axes (Dierking 2003). Mostly known two classes of liquid crystals are thermotropic and lyotropic materials. Thermotropic liquid crystals are exclusively dependent on the variation of temperature. Lyotropic are formed by the variation of concentration of dopant materials in the host liquid (Neto and Salinas 2005).

Vanadium pentaoxide is the classic type of liquid and crystal material having tremendous anisotropy in it. Sonin has reviewed many other forms of liquid crystals having minerals, clays and inorganic materials in them (Sonin 2018). Single and multi-walled carbon-based liquid and crystal materials have shown lyotropic crystalline behavior in them. The formed materials have been utilized in electrical and magnetical molecular switches (Dierking 2004, 2005, Lagerwall et al. 2007, Kumar and Bisoyi 2007, Song et al. 2003, Badaire et al. 2005). The Kerr effect-based electro-optic applications have been observed in various graphene oxide-based liquid and crystalline nanomaterials (Kim et al. 2011, Xu and Gao 2011, Shen et al. 2014, Al-Zangana et al. 2016, Narayan

[1] Department of Chemistry, Dr. B R Ambedkar National Institute of Technology, Jalandhar, Punjab, India-144011.
[2] Department of Chemistry, Malaviya National Institute of Technology, JLN Road, Malaviya Nagar, Rajasthan, India-302017.
* Corresponding author: shankeru@nitj.ac.in, umaorganic29@gmail.com.

et al. 2016). There are various articles on the nanorods of inorganic materials, magnetic materials, ferroelectric particles and nematics of ferromagnetic platelets (Zhang et al. 2011, Ren et al. 2012, Li et al. 2002, Thorkelsson et al. 2015, Li et al. 2006, Basu 2014, Al-Zangana et al. 2017, Podoliak et al. 2012, Mertelj et al. 2013). Quick speed, superior actuator control and thermal stability are the beneficial properties of liquid and crystalline materials (Cui and Zhao 2004, Hayasaka et al. 2008, Shibaev et al. 2003). Modification of nanomaterials to achieve better crystalline properties and different materials matrixes are incorporated in it. These materials are formed by coupling with inorganic as well as crystalline nanostructures (Shibaev 2009).

The environmental applications of nanomaterials are one of the sustainable approaches to focus on. The natural environmental system can provide various opportunities for liquid and crystalline nanomaterials. These opportunities could be associated with the current environmental issues or proactive in providing and solving future problems. These applications include treatment of polluted water, remediation and environmental sensing. Practical applications mainly consist of green fabrication and cost-effective substitutes of sources. The advantages of new materials must also have implications of atmosphere, society and ethics issues. These implications can be achieved by selecting the safest design out of all the available sources. Presently, the baseline data for the providence of nanomaterials is lacking. Hence, the only possibility is to assess the impact related to the utilization of these materials in the environment and to track the adverse effects on living species. Liquid and crystalline nanomaterials are applicable under such circumstances to provide tremendous results. These are useful in testing materials and in envisioning frequency in waves (Donisi et al. 2010). Optical imaging is also another application of liquid and crystalline materials. This technique is beneficial in diagnosing and treating various diseases (Needleman and Dogic 2017). The novel sensing properties of these materials are utilized in bio-sensing. New avenues are offered by this application to various researchers (Dong and Yang 2013). High sensitivity alignment benefits the sensing application of these materials (Kaushik et al. 2016a, b). In the present chapter, various environmental applications of liquid and crystalline materials have been listed. The liquid and crystalline materials have been introduced with their various properties beneficial for the environment. The environmental industry has been explained with sub-sectioning the various sectors, such as waste water treatment, water remediation, environmental sensing, energy storage and sources.

2. LIQUID AND CRYSTAL NANOMATERIALS

Soft matter namely liquid crystal has become the interest of the world in this century. Liquid crystals are known as the fourth state of matter. These materials have emerged from the condensed matter physics. In our day-to-day life, there are various examples of liquid crystals, such as milk, lipstick, gluten, cell membranes and mineral slurries (Palffy-Muhoray 2007, Mohanty 2003). These are applied in various things fabrication, like calculators, laptops, smartphones and tablets. Liquid crystals are considered as transitional between an isotropic liquid and crystal solids (Yu 2015). The name liquid crystal was given because it possesses the properties of both solids and liquids. Properties such as electrical and optical anisotropy are similar to the solids, while molecular mobility and fluidity are similar to liquids (Demus et al. 1998). Liquid crystals exhibit some of the arrangement of the particles in one particular direction, while others could be in a random state. This striking feature, made of the intermediate phase, is also known as mesophase (Goodby et al. 2008). Classification of liquid crystals depends upon physic-chemical parameters, such as a transition in the phases (thermotropic and lyotropic) (Hyde 2001). Effect of temperature or pressure results in the formation of thermotropic liquid crystals. Synthesis of materials in the liquid phase has several benefits over other matters (Feng et al. 2015, Gutsch et al. 2004). Synthesis in the liquid phase is common for preparing nanoparticles; synthesis is possible in a fraction of minutes with desired particles size, and methods are generally cost-effective, simple and common reagents can work in the process (Charitidis et al. 2014, Cheng et al. 2014). Besides these functionalization of material is possible through *in situ* fabrication and could be applied to the desired field of application. It is easy to control the size and shape of the nanoparticles (Lu and Yin 2012). Synthesis of liquid phased nanomaterials could be achieved through two commonly known approaches, i.e. top down and bottom up. Liquid phased nanomaterials could rest in liquid suspension or collected through simply filtering. Interest in liquid phase material fabrication has increased due to their fast response speed and higher thermal stability (Cui and Zhao 2004, Hayasaka et al. 2008). Modification of liquid materials to crystalline could be achieved through incorporation of other inorganic matrixes (Shibaev 2009). Most of the attention has been paid to liquid materials due to their application in display and photonics. Therefore, methods have been developed for the fabrication of nanostructured materials in liquid crystalline phase. Fabrication in nano-range can result into better thermal stability, improved light response and electric/magnetic field reaction. However, formation of liquid and crystalline nanomaterials is a very difficult task since retention of a phase could be difficult. Hence, only a limited set of nanoparticles are appropriate with different synthesis path (Gerasin et al. 2013, Sanchez and Sobolev 2010). Various nanoparticles of liquid crystalline nature have been fabricated till now and various processes are still going on for its formation.

Out of various nanomaterials, various nanosized metal oxides are applied mostly in the formation of liquid crystalline nanomaterials (Ng et al. 2013, Kango et al. 2013). Upon increasing the optical and thermal properties of selected materials, these could be applied for liquid crystalline materials. Nanomaterials of metal oxides have been applied in various energy storage devices as well as resonance imaging. Metal oxides such as titanium dioxide have been applied in industries such as the paint industry such that its coupling with liquid crystals can enhance its overall activity as thermochromic paints. Coupling of 25-nm sized TiO_2 with nematic liquid crystals was achieved at different concentrations of TiO_2 (Yadav et al. 2015). After coupling, TiO_2 acted for impurities adsorption while the liquid crystal matrix reduced the number of ions at the alignment with a little bit of agglomeration (Aly et al. 2019). Improvement in the arrangement of particles of TiO_2 doped liquid crystal was achieved (Pathak et al. 2018). The coupling of nanoparticles with the liquid crystals also improved their birefringence depending on the chemical behavior of the mixture. Similar trends with the anatase form of TiO_2 were presented (Katiyar et al. 2019). Liquid crystals-based nanocomposites were fabrication recently using zinc oxide, copper ions and carbon-based materials (Manepalli et al. 2018, Tripathi et al. 2013). Nanomaterials formed had improved electro-optical and electrical properties with negative dielectric anisotropy and overall enhanced switching behavior. Such combinations do not perturb the phase formation of liquid crystals. One of the reports showed the effect of ZnO nanoparticles and copper in suspension form. In this process, two (weakly and highly polar) different liquid matrices were utilized. The addition of copper ions cause the relaxation in planar as well as homeotropic arrangement, and a decrease in dielectric permittivity was noted. However, for both crystals disorder in one phase was found owing to the dielectric anisotropy (Martinez-Miranda et al. 2010). Zinc oxide nanoparticles of a very small size of 5 nm were used to discover the upper limit of crystal phase formation. Better homeotropic arrangements of liquid crystal were detected with nanoparticles of 5 nm added at different amounts (0.125 to 0.4 wt%). Other benefiting properties, such as lesser defects, loss in dielectric and higher conductivity were also noted (Supreet et al. 2013). Growth of zinc oxide nanoparticles was controlled in one study by using macromolecular liquid crystals. Hybrid of nanoparticles and liquid crystal was formed using different crystal moieties. It was discovered that with changing the temperature transition temperature of liquid crystal was also altering. However, nuclear magnetic resonance study did not give any clear difference between the parent and hybrid material. Liquid crystals having a nematic phase enhanced the overall growth of zinc oxide toward nanoworms (Saliba et al. 2012). Tao and Tam described the coupling of zinc oxide nanowires with liquid crystal cells. The nanowires of zinc oxide followed the path of reorientation of its liquid crystal medium. However, upon applying magnetic field the alignment of nanoparticles was changed (Tao and Tam 2013).

Liquid crystalline polymer was coupled with zinc, copper oxide and zinc oxide nanoparticles to alter its optical activeness. Low concentrations of nanoparticles were capable of changing the phase and temperature effect of liquid crystals. The properties, such as thermal and mechanical stability of the material, were improved by coupling with nanoparticles (Mishra et al. 2016).

Other than zinc-based nanoparticles, magnetic iron oxide was utilized for enhancing the activity of liquid crystals. The ferromagnetic nanoparticles were capable of changing the transition temperature of liquid crystals by showing wave-like curves. Moreover, the swelling results indicated a stable hydrogel structure of the overall polymer. Thus, the liquid crystal can help in converting the magnetic nanoparticles into cross-linked material (Zhou et al. 2012).

Arantes and co-workers studied the coupled lyotropic liquid crystal and iron oxide nanoparticles of 10 nm size. Phase transitions in the liquid crystal were observed at eight Kelvin temperatures (Arantes et al. 2010). Magnetic iron oxide nanoparticles were fabricated through grafting and coupled with nematic liquid crystals. After coupling, the magnetic nanoparticles were found captured by the liquid crystal phase (Da Cruz et al. 2005). Similar reports were found on nematic liquid crystal and magnetic grafted nanoparticles (Prodanov et al. 2012, Gdovinová et al. 2014).

3. ENVIRONMENTAL INDUSTRY

The application of nanomaterials has been recently applied exhaustively in the environmental industry. The various sector of the environmental industry includes biomedical, food, energy storage, waste water treatment, pharmaceutical and defense (He and Hwang 2016, Pathakoti et al. 2017). Out of which potential use of nanomaterials is in remediation, energy storage and sensors (Wang and Dai 2013). Besides these, the main focus of the environmental industry is also toward the safe fabrication of nanomaterials to promote the sustainable development of technology (Hutchison 2008). In 2014, the global market of nanotechnology was ~23 billion USD in the environmental industry. Speculations were made that the market might rise to ~41 billion USD in 2020 (BCC research, 2015). Various industries involved in the environment have been mentioned below and in Figure 1.

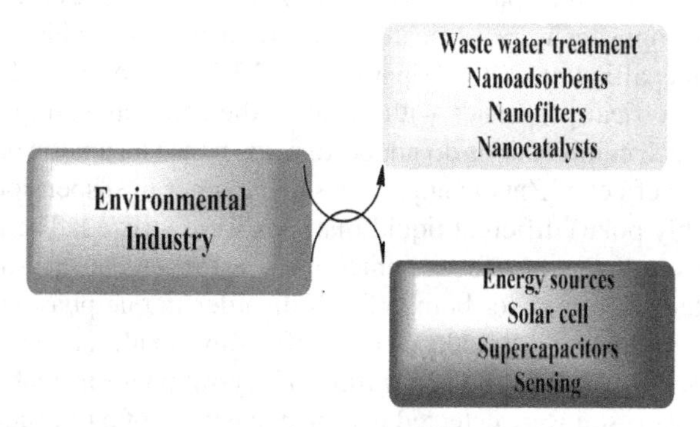

FIGURE 1 Industrial involvement in the environment

3.1 Wastewater Treatment

World health organization in its report cleared that around 844 million people live in scarcity of sufficient water for drinking. Daily required water for household chores is limiting due to energy cost, climate change and population growth (Grey et al. 2013). Most of the contaminants are produced from the effluents or utilization of personal care products (Houtman 2010). Nanotechnology has progressed enough to provide new prospects for the enlargement of the process to treat wastewater. Nanoparticles could be used in the form of adsorbents, catalysts and membranes for pollutants removal. However, the nanoparticles used must be non-toxic and effective at a lower concentration.

Selection of procedures applicable easily and reusable for multiple cycles will make the overall process cost-effective (Theron et al. 2010, Cloete et al. 2010). All in all the activity of nanoparticles is controlled by their size, surface activity, shape, structure, solubility and chemical composition (OECD 2010). Nanoparticles are applicable for the degradation of contaminants owing to their excellent properties compared to their bulk counterparts. Nanoparticles of transition metal oxides are generally utilized for the treatment of contaminants and environmental remediation. These include iron oxide, nickel-zinc ferrite, platinum nickel alloys (Dutta et al. 2014, Kurian and Nair 2015, Ma et al. 2015). Oxides of titanium and zinc have been extensively used based on their semiconducting properties resulting in strong photocatalytic efficiency. These have been in the remediation of organic pollutants from the water as well as sediments (Keller et al. 2005, Pathakoti et al. 2014).

3.2 Remediation

Surface and ground water has been contaminated by the effluents from industrialization and the growth of the population (Chong et al. 2010). The main pollutants found in water are heavy metals, complex organic compounds, inorganic pollutants and other numerous compounds (Fatta-Kassinos et al. 2011). Remediation is the removal of pollutants from water and other natural resources, like soil and sediment. It could be *in situ* or *ex situ* remediation. Treatment of contaminants at their source is called *in situ* remediations, while collecting the samples and treating them in the laboratories are termed as *ex situ* remediation. Various nanomaterials have been applied for remediation processes, such as carbon nanotubes, zeolites, metal oxides and fibers (Karn et al. 2009). In various *in situ* remediation processes, zerovalent iron nanoparticles have been used owing to their cost effectiveness, environmentally benign and easy formation (Fu et al. 2014, Ponder et al. 2000). For more than two decades iron nanoparticles ranging from 10 to 100 nm diameter have been employed for the remediation of contaminated water (Crane and Scott 2012, Guan et al. 2015). Other than water its activity in soil and sediment has also been investigated (Satapanajaru et al. 2008). Out of metal oxides, titanium dioxide is quite cheap, safe and used for remediation purposes many times. Degradation of contaminants, such as polystyrene, polyethylene and dyes has been studied with titanium dioxides (Masciangioli and Zhang 2003, Thomas et al. 2013). Chlorinated compounds were degraded with titanium dioxide based nanotubes (Chen et al. 2005). In some *ex situ* processes titanium dioxide has been involved for effective illumination (Tratnyek and Johnson 2006). Coupled form of titanium dioxide was recently employed for the *in situ* remediation of bisphenol a (He et al. 2016).

3.3 Energy Resources

Growth in population and development of industrialization has resulted in increased demand for energy worldwide. Depletion of sources of fossil fuels and minerals has caused an imbalance in the environment. Speculation about the boost in energy demand up to 44% will be faced worldwide from 2006 to 2030 (Khan and Arsalan 2016). These numbers have caused the shift of energy precursors from conventional to renewable energy sources for sustainable energy formation (Prakash and Bhat 2009).

Various renewable energy resources have been recently focused on, including bioenergy, wind energy, solar energy and hydropower as well as ocean energy. The involvement of nanotechnology in these energy resources has enhanced their overall production as well as processes efficiency. Devices required for these energy resources could be formed of nanomaterials. Nanotechnology can also be employed in the formation of lightweight materials that can simplify the transport of devices (Gullapalli and Wong 2011). These energy resources could be classified based on their application in energy conversion, storage or utilization. Solar cells being quite a famous energy source nowadays generates electrical energy from solar energy using silicon-based semiconducting materials. However, these have limited efficiency and require high manufacturing costs. Even though solar energy is sufficient enough on the planet yet it is capable of sharing only 0.04% fuel requirement of the world (IEA 2007). Conventional solar cells were basically silicon wafers with an efficiency of 15 to 20% (Parida et al. 2011). The advanced technology of the second phase utilized silicon and copper indium gallium diselenide in the form of thin films. However, these were lesser efficient than the earlier process and were not successful in solving the technological problems (Hermann 1998). The next-generation cells are formed of nanocrystals and other nanomaterials and are capable of fulfilling the overall requirements. However, these are still in a growing stage. There are three different types of nanotechnology-based solar cells. Out of these the dye-sensitized solar cells are mostly employed. It involves the conversion of light and confinement in cells by changing the surface of the nanostructure with dye. It is based on colloidal TiO_2 films sensitized with dye (O'Regan and Gratzel 1991). There are three different components of dye-sensitized cells including photo anode sensitized with dye, electrode and redox electrolyte. It acts when light falls on the surface of the cell and *e-h* pairs are formed in the dye with the transfer of electrons to the conduction band of titanium dioxide. The hole left behind in the oxidized form of dye molecules remains as it is. The redox electrolyte, acting as a mediator, helps in the regeneration of the oxidized dye. This cost-effective and simple process has shown 11% performance. However, the stability of the overall process restricts its activity (Chang et al. 2011, Wang et al. 2004). Various coupled versions of nanomaterials have been used to increase the efficiency of these solar cells (Ramiro-Manzano et al. 2007).

3.4 Energy Storage

After the development of sustainable energy production resources, their efficient storage is required. Rechargeable batteries and super-capacitors have been used as promising energy storage devices. Recently, electronic devices, electric automobiles and portable power tools have been generated from highly power-effective rechargeable lithium-ion batteries (Kang et al. 2006). In lithium-ion batteries, the cathode is formed of lithium-based metal oxide, while the anode consists of a layered structure of graphitic carbon. The electrolyte used is generated from the lithium salts soluble in organic carbonates (Dı´az-Gonza´lez et al. 2012). In these batteries, electrodes store the energy in the form of Lithium-intercalation compounds. Even though around 80% of the market has been taken over by lithium-ion batteries, yet challenges related to their cost and environmental impact have categorized these among as unsafe class of devices (Li et al. 2014). Nanotechnology has been utilized in battery formation by advancing the electrode material and solutions of electrolytes. For making the process environmentally friendly anode could be formed of materials, like nanoporous, nanorods or nanowires. Out of various developments made, materials such as silicon, germanium nanowires and carbon-based silicon nanowires have shown ultrahigh capacity (Huang et al. 2009). Composites of silicon and graphene have shown better performance, conductance and reusability with carbon-coated graphene (Liang et al. 2011). In lithium-ion batteries, carbon nanotubes have also been used as an electrode material (Centi et al. 2011, Xiong et al. 2013). Cathode made up of nanomaterials has significantly better energy storage and stability (Bazito and Torresi 2006). Other materials, such as oxides of silicon, titanium, aluminum and tin, have shown better electrolytic activity (Liu et al. 2016).

Super capacitors by polarizing an electrolytic solution can store electrical energy. Its activity can be optimized from two basic parameters, i.e. how long it lasts and how much energy is delivered after a specific time. Super capacitors lie between the old capacitors and batteries. It consists of two electrodes dipped in an electrolyte with a separator. The interface between the electrolyte and electrode represents capacitors. Hence, the overall supercapacitor is a set of capacitors placed in series. Super capacitors are quick, have high power density and extraordinary life cycle (Cao and

Wei 2014). These are capable of forming electronic, power stores in industry and memory backup systems (Yang and Tarascon 2012).

3.5 Environmental Sensing

Another important sector of the environmental industry is sensing. Nanomaterials can be used to form low-cost and highly sensitive detectors for monitoring water and air with higher selectivity and stability. Nanoparticle-based sensors have the capacity to detect toxins, organic contaminants and heavy metals at low concentrations. These can be utilized in sensing pathogens and cyanotoxin with an electrical sensor (Hristozov and Ertel 2009). Pollutants in air and water can be monitored with gas or chemical sensors (Butnar and Llop 2007, Zhang et al. 2016). Other contaminants, such as dyes in textile, leather and food, could be detected with these sensors (Saikia et al. 2015). These applications require highly selective and sensitive sensors. The leakage of gases in various industries can also be ensured with sensors (Asad and Sheikhi 2016). Sensors are composed of polymers, metal oxides semiconducting in nature and silicon-like porous materials (Rittersma 2002). Metal oxide and crystalline nanoparticles have been used to a greater extent in sensors owing to their flexible properties (Comini 2006). Tin, tungsten, zinc, indium and niobium oxides have been used for their application in gas sensing (Kumar et al. 2015, Franke et al. 2006, Han et al. 2016). Metal oxides-based sensors can detect gases via redox reaction amongst the target gas and metal oxide surface (Yamazoe and Shimanoe 2008). These sensors work for the target gases through resistance charge response (Batzill and Diebold 2005).

Zinc oxide-based sensors with high sensitivity, stability and economic nature have been utilized to a greater extent for gas detection (Ozgur et al. 2005, Wang et al. 2014a). Electrochemical sensors for copper detection were developed using cysteine. These include carbon foams coupled with gold and also employed for detecting lead and copper ions with a detection limit of 5 to 0.9 nM (Xiong et al. 2016). Other metal ions such as mercury, copper and cadmium can also be detected using nanostructured sensors. Biological species detection is also possible with carbon nanotubes-based detectors (Gao et al. 2016). Other applications are possible if nanoparticles are self environmentally friendly, reusable and involve simple formation methods.

4. APPLICATION OF LIQUID AND CRYSTAL NANOMATERIALS IN ENVIRONMENTAL INDUSTRY

In recent times concern in the environment, the sector has increased for novel developments and treatment of contaminated areas. There are various toxic pollutants in the environment polluting different matrices, such as soil, water, air and sediments. LCN are capable of converting hazardous chemicals into innocuous forms. LCN are also useful in various other applications, such as photocatalysis, energy storage, sensing development of energy resources. The section-wise discussion of these applications is given below.

4.1 LCN in Wastewater Treatment and Remediation

Limited availability of clean water has been termed as a global issue. In order to resolve this issue various approaches, such as separation, desalination of sea water and treat brackish water. LCN has been employed in processes like reverse osmosis, nanofiltration and photocatalyst removal of pollutants (Werber et al. 2016, Gin and Noble 2011, Park et al. 2017, Wang et al. 2017, Fane et al. 2015). Polymers-based LCN, carbon-based, graphene oxide and peptides based are effective in remediation (Werber et al. 2016, Gin and Noble 2011, Park et al. 2017, Wang et al. 2017, Zhou et al. 2007, Carter et al. 2012, Zhou et al. 2005, Henmi et al. 2012, Marets et al. 2017, Abraham et al. 2017, Bui et al. 2016). Thermotropic and lyotropic liquid crystals based LCN having nanopores were reported for remediation of pollutants (Werber et al. 2016, Gin and Noble 2011). Gin et al. used lyotropic LCN as separation membranes. They showed that these LCN demonstrated elimination selectivity based on the size of the nanomaterials (Zhou et al. 2007, Carter et al. 2012, Zhou et al. 2005). Removal of ibuprofen was tested using Fe/C granules. The composite of granules and sodium persulfate successfully removed 96% of ibuprofen involving the formation of eight intermediates. The intermediates are shown in Figure 2 (Yu and Pei 2021).

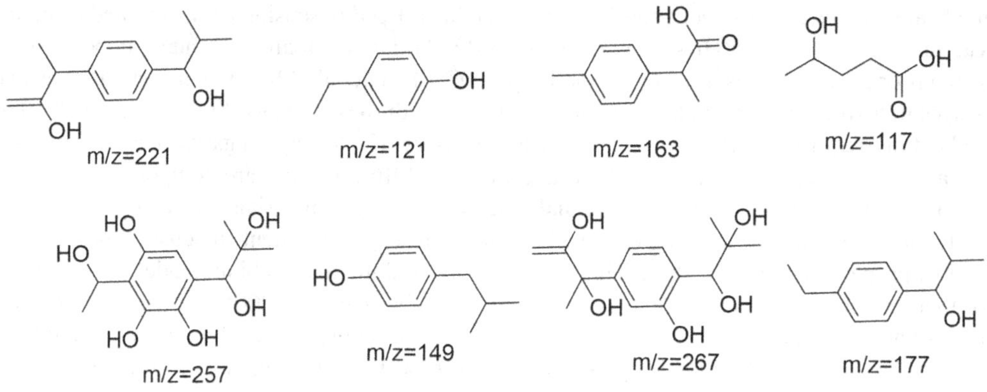

m/z=221 m/z=121 m/z=163 m/z=117

m/z=257 m/z=149 m/z=267 m/z=177

FIGURE 2 Intermediates detected after ibuprofen removal

Meloxin and Naproxen were recently investigated for adsorption using Fe_3O_4@MIL-100(Fe) under different pH conditions. The adsorption followed pseudo first-order kinetics showing that the process was influenced by the chemical reaction. Adsorption was multilayer, spontaneous and endothermic (Liu et al. 2019). Photochemical degradation of four parabens was investigated using crystalline titanium dioxide under UV light. However, at higher concentrations of pollutant degradation efficiency of nanocatalyst was slow (Cuerda-Correa et al. 2016). Diclofenac was degraded using zinc-ferrite nanomaterial using the redox character of nanocatalyst. Within the two minutes of reaction, 90% of diclofenac was degraded through the reduction by electrons and radicals formed on catalysts surface. The degradation pathway of the contaminant has been shown in Figure 3 (Al-Anazi et al. 2020).

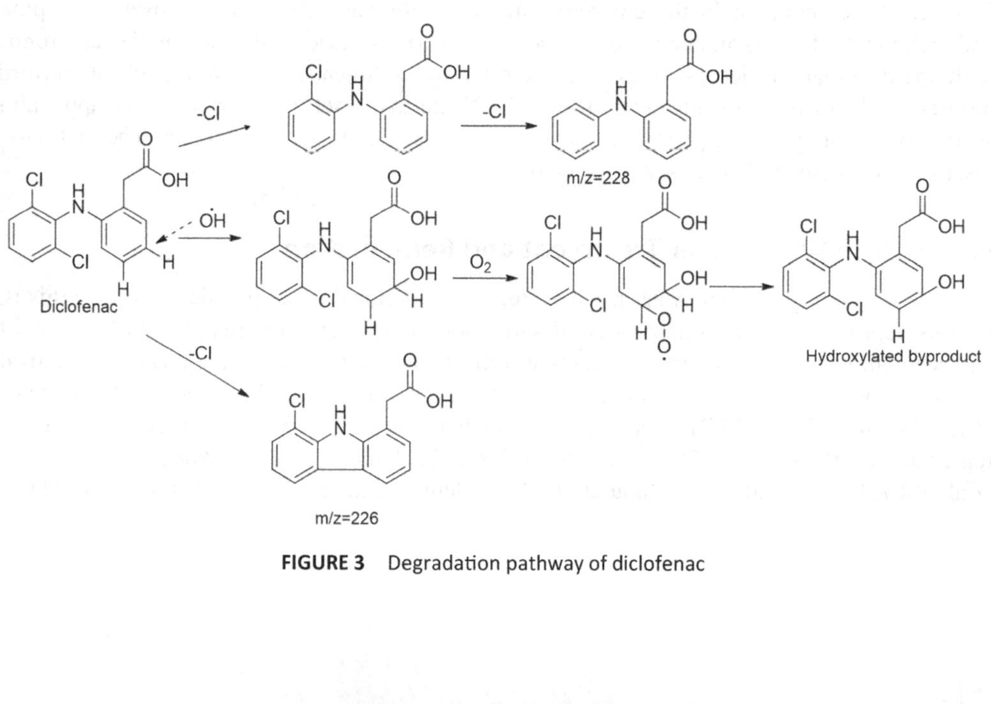

Diclofenac -Cl -Cl $\dot{O}H$ O_2 -Cl m/z=228 Hydroxylated byproduct m/z=226

FIGURE 3 Degradation pathway of diclofenac

$Co/BiFeO_3$ synthesized through the sol-gel method with higher surface area decomposed 81% of tetracycline in water by sulfate radicals formed in presence of persulfate (Zhang et al. 2018). Iron oxide nanoparticles activated with carbon removed 97% of 10 ppm cefriaxone at acidic pH in 90 minutes (Badi et al. 2018). Naproxen affected water was treated with Si/Graphene oxide nanoparticles at pH 5. The results revealed high adsorption capacity (83%) of nanoparticles in the process (Mohammadi Nodeh et al. 2018). Rafati et al. 2019 studied the removal of ibuprofen with functionalized nanoclay composite. The nanocomposite adsorbed 95% of 10 ppm of PCP within two hours. Hematite-based removal of carbamazepine was investigated by Rajendran and Sen (2018). Removal of pollutants was also studied by transition metals doped TiO_2 nanaocatalyst in visible light (Inturi et al. 2014). In one more reports nonmetal doped TiO_2 was utilized in the process (Lin and Yang 2014).

4.2 LCN in Energy Resources

LCN with easy formation process, low cost, superior stability and semiconducting nature have been widely employed in water splitting, photo-voltaic devices and pollutants removal (Hisatomi et al. 2014, Wu et al. 2017, Liu et al. 2015). LCN are used in the electron transporting layer in dye-sensitized solar cells. The involvement of LCN in solar cells is due to their properties, such as wide band gap, electron mobility and band alignment matching with the active materials. Although in the area of photocatalysis and water splitting required features are (a) suitable band gap to confirm the light absorption, (b) appropriate band edge orientation to suit the thermodynamic necessities of water splitting, (c) high mobility for transferring *e-h* to surface and (d) surface structure-activity for confirming the water splitting. LCN having various properties can enhance the efficiency of solar energy conversion of the solar cell (Liu et al. 2014, Lu et al. 2017, Pan et al. 2019, Kanjana et al. 2020). The basic requirement for dye-sensitized solar cells is the generation of electrons and holes through light in active materials. Apart from photovoltaic devices, catalysts and photo-electrochemical, the splitting of water can affect the sunlight to hydrogen and produce hydrogen (Kudo and Miseki 2009, Kitano and Hara 2010). Moreover, sunlight-based degradation of pollutants through catalysis is another tempting way to resolve the issue related to environmental pollution. For these purposes, the basic principle is to excite the light-driven catalyst to produce charge carriers and transport these to encourage water splitting as explained in Figure 4 (Wu et al. 2021).

LCN exhibits superior transportability of charge carriers, high surface area, and enhanced absorbance of incident light and short photo carrier diffusion length. These advanced properties of LCN make them a promising candidate in catalysis and solar cells. Lately, notable advances in applying LCN in such areas have been attained. Hence, it is beneficial to thoroughly understand the effect of exclusive properties of LCN in solar energy conversion processes.

WO_3 has been recently recognized for its long diffusion length and lifetime of charge carrier as well for its stability in acidic aqueous media. Moreover, it is a type of metal oxide having a large band gap lying out of the long spectrum of sunlight (Babu et al. 2015). The valence band maximum of WO_3 can supply excessive potential for photo generation of holes to oxidize water. This LCN is an amorphous hydrated form that cannot be turned into a crystalline form at a temperature above 400°C (Baeck et al. 2002, Monllor-Satoca et al. 2006). The photoanodes of this LCN can form peroxo species and oxygen while, the photo-oxidation process, resulting in a slow loss of its activity. The deposition of Co-Pi on WO_3 can cause notable long-term stability of LCN. Some of the LCN based cells and their efficiency have been listed in the Table 1. The LCN has a large impact upon the catalytic features of photo-electrode instead of its light-harvesting capacity. However, this property cannot be ignored. Moreover, setting up LCN based design is a turning process to create balance in the series of different available processes already in competition.

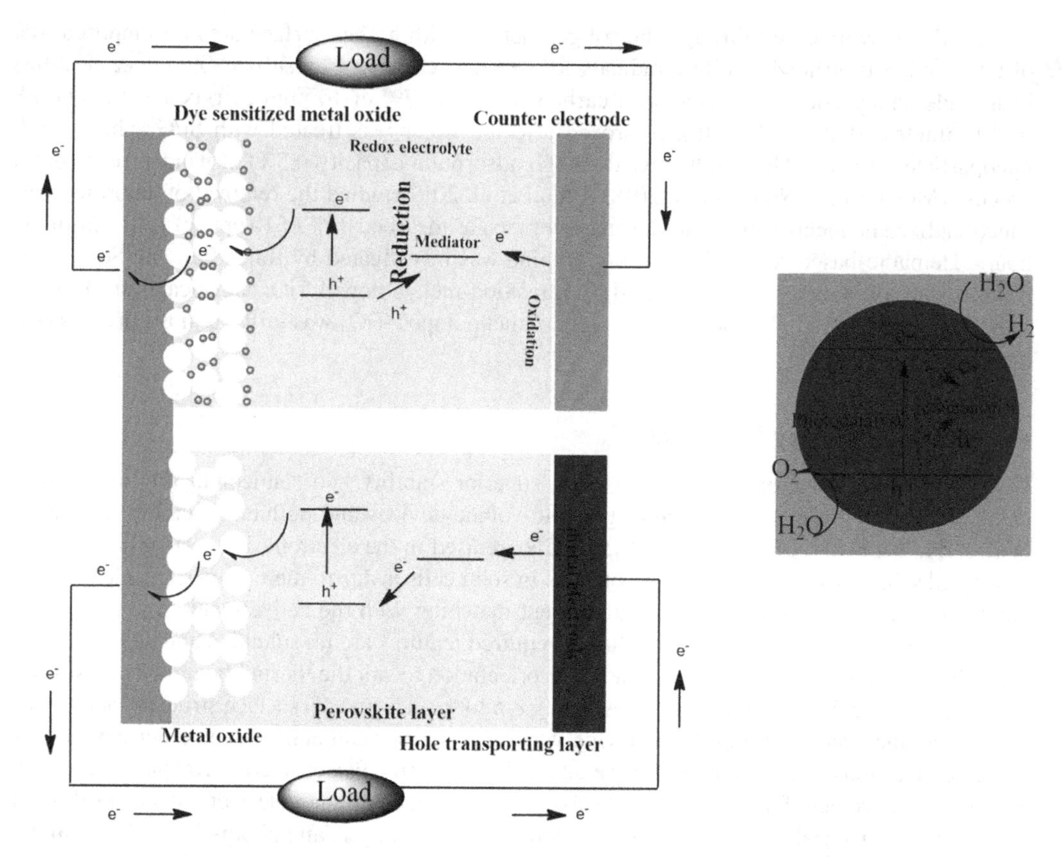

FIGURE 4 Mechanism of water splitting under sunlight

TABLE 1 LCN based cells and their efficiency.

Material	Photocurrent	Reference
TiO_2	3.92	Yun et al. 2016
TiO_2	2.41	Singh et al. 2017
TiO_2	7.7	Chang et al. 2015
TiO_2	46.23	Chang et al. 2015
TiO_2 (N doped)	1.9	Wang et al. 2015b
TiO_2 (S doped)	2.92	Shin et al. 2015
TiO_2 (N-Si doped)	1.77	Zhang et al. 2015a
TiO_2 (F doped)	0.7	Cordova et al. 2015
TiO_2/ Ag/ Ag_2S/Ag_3CuS_2	9.82	Guo et al. 2015
Fe_2O_3@TiO_2	3.39	Li et al. 2015
Fe_2O_3	1	Warwick et al. 2015
Fe_2O_3 (Ti doped)	1.79	Ji et al. 2016
Fe_2O_3(Ti doped)	4.18	Wang et al., 2015a
Fe_2O_3(Ti doped)	2.28	Wang et al., 2015a

Contd.

Fe_2O_3/Fe_2TiO_5	1.4	Bassi et al. 2014
γ-Fe_2O_3/rGO	6.74	Chandrasekaran et al. 2015
In_2O_3	1.9	Liu et al. 2015
Ta_2O_5/Co-Pi	7.5	Su et al. 2014
$BaSnO_3$/CdS	0.4	Zhang et al. 2015b
WO_3	0.45	Zhang et al. 2016
WO_3/SnO_2	0.91	Yun et al. 2016
WO_3/Sb_2S_3	1.79	Zhang et al. 2016
$WO_3/BiVO_4$	3.3	Zhang et al. 2016
$BiVO_4$	0.8	Liu et al. 2015
$BiVO_4$	3	Thalluri et al. 2016
$CuWO_4$	0.75	Tang et al. 2016
ZnO	0.32	Huang et al. 2016
ZnO	0.39	Sohila et al. 2015
ZnO/GO	1.52	Chandrasekaran et al. 2016
ZnO/Au	1.5	Wang et al., 2015b
ZnO/ ZnSe/CdSe/Cu_xS	12	Wang et al., 2015b
CNT/ZnO/Co_3O_4	1.9	Li et al. 2015

4.3 LCN in Energy Storage

The high dielectric constant of LCN makes these applicable for energy storage applications. These storage applications include electronic devices formation, memory, power saver devices as well as energy storage capacitors (Dang et al. 2013, Zhang et al. 2002). Ceramics have been utilized in the energy storage device formation; however, there are certain limitations associated with their synthesis such as the requirement of high temperature. Polymers could be applied as a flexible candidate but their low dielectric constant limits their use for the present purpose (Huang et al. 2005, Dang et al. 2012, Ibrahim et al. 2017). Recently, LCN has been utilized in the formation of super-capacitors and other energy storage devices.

Recently, binders free nickel and cobalt oxide-based LCN was fabricated through hydrothermal and cyclic voltammetry methods. The LCN formed was capable of faster mass transport, stability in cycling and excellent specific capacitance of around 1,400 Fg^{-1}. The supercapacitor-formed LCN has a positive electrode and activated carbon has a negative electrode. The system was capable of delivering around 47 Wh kg^{-1} energy density. These findings were based on the layered morphology of the LCN (Chen et al. 2020). Portable electronics have high impact on societal norms, like the availability of real time information. A low cost fiber shaped LCN was developed with manganese oxide and molybdenum-based LCN. This LCN operating at 1.8 V voltages showed high specific capacitance, cyclic stability of over 3,000 cycles and good rate capability. The process was capable of exhibiting around 80% of capacitance retention. The fabricated device showed excellent stability under different binding conditions (Rafique et al. 2020). Cathode material was recently formed of nickel-rich LCN for the advancement of lithium-ion batteries used in energy storage devices. However, upon continuous cycling and charging at higher voltage, the device showed rigorous capacity fading with structural causing poor cycle stability. Hence, a device was formed by Sn-doping and *in situ* fabricated Li_2SnO_3 modified with $LiNi_{0.5}Co_{0.2}Mn_{0.3}O_2$. The device exhibited excellent cyclic properties with high rate capabilities. The tin doping in LCN formed lithium-ion conductive Li_2SnO_3 on the surface. The tin doping also improved the reversible capacity and rate tendency of the cathode. Overall, LCN formed showed improved electrochemical property of the device and hence better applicability as an energy storage device (Zhu et al. 2020). Carbon-based LCN are considered ideal supercapacitors due to their good conductivity, better pore structure and tremendous connectivity. A green approach was established to form carbon-based LCN through carbonization and activation with potassium hydroxide. The material formed showed high porosity and N content. The resulting capacitance was around 359 F g^{-1}. The specific capacitance obtained in the two electrodes system was also around 254 Fg^{-1}. Clearly, the LCN based device improved the potential for high-performance application (Liu et al. 2020).

A novel vanadium pentaoxide-based LCN was formed coated on polypropylene. This coupling ensured the stable electron and ion path as well as enhance the life span of its cycling performances. The lithium and sulfur battery assembled in this way showed a higher capacity of around 1,100 $mAhg^{-1}$ and reversed capacitance of 1,110 $mAhg^{-1}$. The electrochemical properties obtained were unusual associated with long-term cyclic stability offering a promising design for fine architectures to the separator (Zhang et al. 2020). In another study, a nonbinder-based nickel sulfide LCN was formed with a hydrothermal method via a two-step approach. It was followed by the deposition of nickel-cobalt hydroxide upon the substrate of the nanotube. The product obtained showed high specific capacitance and retention in capacitance (Xin et al. 2020).

Two-dimensional titanium carbide-based LCN has gained attention owing to its excellent chemical and physical properties. The hexamethylenetetramine-based coupled material delivered excellent electrochemical performance as well as the high gravimetric capacitance of value 361 F g^{-1} at 1 A g^{-1}. The cycling stability of around 86% after 10,000 charges/discharge cycles were achieved (Yang et al. 2020).

4.4 LCN as Sensors

The available technique of sensing mainly depends on chromatography, such as gas, mass and liquid chromatography, fluorometry, colorimetry and electrochemical method (Bai et al. 2013, He et al. 2013). These techniques are associated with drawbacks of sample processing, costly equipment, time consumption and toxic chemicals. LCN have gained considerable attention in sensing in this decade due to their properties, such as surface areas, porosity, structural benefits and thermal stability (Wu et al. 2016, Peng et al. 2016). These properties make LCN a promising candidate for various applications, like storing and separating, catalysis and adsorption of materials (Xue et al. 2019, Kazemi and Safarifard 2018, Beheshti et al. 2018, Esrafili et al. 2018). Moreover, LCN have been employed as a leading nominee for bio-sensing in the environment and industries (Moradi et al. 2019, Gharib et al. 2018).

The LCN based sensors will provide well-developed sensing systems with more efficient and advanced structures. These next-generation systems will be cost-effective, adaptable, sensitive and considerable. These demands raised the introduction of nanomaterials-based sensors for selective treatment and sensitive monitoring (Kaushik and Chandra 2016). Nanomaterials in sensors improve various aspects of sensors. Nanomaterials generally used in sensing are ferroelectric in nature, metal-based, semiconductor and carbon-based to enhance the properties of host sensor material (Joshi et al. 2010, Lisetski et al. 2009, Lapanik et al. 2012, Singh et al. 2014). The ferroelectric LCN can significantly increase the order through various interactions between the benefiting parameters (Li et al. 2006). Such materials are one or two-dimensional anisotropic lyotropic LCN (Kato et al. 2006). These sensors were successful in detecting immune-based bio-actives associated with various diseases (Shen and Dierking 2019). LCN based sensors have been found simple and amplified in detecting targets (Zhao et al. 2015, Choudhary et al. 2018, Wei and Jang 2017, Tan et al. 2010, Hartono et al. 2009, Yang et al. 2012, Liao et al. 2012). Wei and co-workers fabricated nickel-based liquid crystalline sensors to investigate the enzymatic activities (Wei and Jang 2017).

Nanomaterials were able to introduce the structure of functionalized materials having size, shape, surface, morphology and optical properties as desired by the method (Kaushik et al. 2014a, b, 2016a, b, 2017, 2018, Tiwari et al. 2019). The small size of LCN can easily merge and align into the liquid crystalline surface to form a new phase of the material. This new class of sensing can exhibit unusual features, which were absent in the conventional sensors. The enhanced order parameters of LCN can help in easy binding to its surface (Choudhary et al. 2018). These properties can again benefit in compatibility, sensing sensibility and selectivity in the overall process of sensors (Vallamkondu et al. 2018). The generally used materials in LCN are carbon nanostructures, tungsten diselenide and molybdenum diselenide owing to their strong pi-stacking interactions (Kim et al. 2011). The LCN is highly dependent on surface effects. Hence, even a small change on the surface can result in the orientation of molecules in different directions. Interaction of light intensity with LCN surface causes a change in the light intensity (Liao et al. 2012). Some of the LCN based biosensors are listed in Table 2.

TABLE 2 LCN-based biosensors.

Pollutant	Electrode	Detection Limit (μM)	Reference
Bisphenol A	Conducting polymer	0.005	(Macholán and Schánl 1977)
2,4-dichlorophenol	Amperometry	0.06	Tungel et al. 1999
Catechol	MWNT/TiO_2	0.087	Yuan et al. 2011
Phenol	MWNT	0.095	Kong et al. 2009
4-chlorophenol	MWNT	0.11	Lee et al. 2007
Nitrate	Pt	10.0	Wang et al. 2010
Nitrites reductases	GC electrode	0.004	Carralero et al. 2006
Phosphate	phosphorylase Au	1.0	Sohail and Adeloju 2009
Chlorpyrifos	SPR	55.0	Joshi et al. 2006
Atrazine	EIS	10.0	Mauriz et al. 2006
Picloram	Amperometric	<1.0	Mauriz et al. 2007
DDT	–	15.0	Zourob et al. 2007
Methyl parathion	MWCT/Au	1.0	Vidal et al. 2008
Testosterone	TIRF	1.7	Hleli et al. 2006
Estradiol	Amperometric	170.0	Chen et al. 2010
Microsystin-LR	TIRF	30.0	Tschmelak et al. 2006

5. CONCLUSIONS AND FUTURE SCOPE

In the present chapter, the main focus was given to the liquid and crystalline materials, their application in the environmental industry. Although events have been made earlier to investigate the application of nanomaterials yet, there is a demand for liquid and crystalline nanomaterials applicability review on the problems related to the natural habitat. Various final points could be concluded drawn from this chapter:

- Liquid and crystalline nanomaterials are capable of increasing the overall available features of nanotechnology and can play a vital role in the remediation of environmental pollutants.
- A lot of liquid and crystalline nanomaterials (coupled with metal oxides, graphene oxide, carbon and other non-metals) were fabricated and applied in various applications of environmental concern.
- Doping and synergism play important roles in better properties of LCN, such as beneficial band gap energy, limit charge recombination, increase the overall surface active sites and charge carriers.
- Mostly oxides of transition metal oxides, like zinc, iron and titanium, were used in environmental applications, such as remediation, energy storage and sensing. More attention is needed to be paid to the other metal oxides and non-conventional materials to form LCN so that bio-saturation of one kind of material could be restricted.
- Liquid and crystalline materials showed applicability in sensing various environmental pollutants through their higher surface activity.
- The fabricated materials were utilized in the formation of capacitors and other energy storage devices.
- Liquid and crystalline nanomaterials were also applicable in the formation of solar cells as well as water splitting reactions.
- A very few reports were available on the utilization of non-conventional metal oxides in the formation of liquid and crystalline nanomaterials.
- Limited reports were available on the degradation of personal care products with the LCN materials.
- Although efforts have been made to explore the various application of LCN in the environmental industry, yet a lot of work is required to have a full profit from the excellent properties provided by the LCN materials.
- Complete knowledge of the LCN materials fabrication is lacking and simple/cost-effective methods development is required.
- To achieve the desired results from the LCN materials better understanding of its features is required. Materials stability under various environmental factors such as temperature, pH and solvent is need to be known.
- Limited data is available on the application of LCN materials in energy storage devices and resources formation. Moreover, most of the results were based on the lab reports, field-based investigation of LCN activity should be investigated to receive the real-life profits of the materials.
- Expansion of more simplistic, one-step and cost-effective processes for industrial application is highly recommended.

Acknowledgments

Authors are grateful to the Ministry of Human Resource of Development for financial assistance.

References

Abraham, J., Vasu, K.S., Williams, C.D., Gopinadhan, K., Su, Y., Cherian, C.T., et al. 2017. Tunable sieving of ions using graphene oxide membranes. Nat. Nanotechnol. 12: 546.

Al-Anazi, A., Abdelraheem, W.H., Scheckel, K., Nadagouda, M.N., Kevin, K. and Dionysiou, D.D. 2020. Novel franklinite-like synthetic zinc-ferrite redox nanomaterial: Synthesis, and evaluation for degradation of diclofenac in water. Appl. Catal. B: Environ. 275: 119098.

Al-Zangana, S., Iliut, M., Turner, M., Vijayaraghavan, A. and Dierking, I. 2016. Properties of a thermotropic nematic liquid crystal doped with graphene oxide. Adv. Opt. Mater. 4: 1541-1548.

Al-Zangana, S., Iliut, M., Turner, M., Vijayaraghavan, A. and Dierking, I. 2017a. Confinement effects on lyotropic nematic liquid crystal phases of graphene oxide dispersions. 2D Mater. 4: 041004.

Al-Zangana, S., Turner, M. and Dierking, I. 2017b. A comparison between size dependent paraelectric and ferroelectric $BaTiO_3$ nanoparticle doped nematic and ferroelectric liquid crystals. J. Appl. Phys. 121: 085105.

Aly, K.I., Al-Muaikel, N.S., Abdel-Rahman, M.A. and Tolba, A.H. 2019. Liquid crystalline polymers XVI*. Thermotropic liquid crystalline copoly(arylidene-ether)/TiO_2 nanocomposites: Synthesis, characterization and applications. Liq. Crys. 46: 1734-1746. https://doi.org/10.1080/02678292.2019.1597192.

Arantes, F.R., Figueiredo Neto, A.M. and Cornejo, D.R. 2010. Study of magnetite nanoparticles embedded in lyotropic crystals. Phys. Proc. 9: 2-5. https://doi.org/10.1016/j.phpro.2010.11.002.

Asad, M. and Sheikhi, M.H. 2016. Highly sensitive wireless H_2S gas sensors at room temperature based on CuO-SWCNT hybrid nanomaterials. Sensor Actuat. B Chem. 231: 474-483.

Babu, S.G., Neppolian, B. and Ashokkumar, M. 2015. Ultrasound-assisted synthesis of nanoparticles for energy and environmental applications. *In*: Ashokkumar, M. [ed.]. Handbook of Ultrasonics and Sonochemistry. Springer Science+Business Media Singapore. doi: 10.1007/978-981-287-470-2_16-1.

Badaire, S., Zakri, C., Maugey, M., Derre, A., Barisci, J.N., Wallace, G., et al. 2005. Liquid crystals of DNA-stabilized carbon nanotubes. Adv. Mater. 13: 1673-1676.

Badi, M.Y., Azari, A., Pasalari, H., Esrafili, A. and Farzadkia, M. 2018. Modification of activated carbon with magnetic Fe_3O_4 nanoparticle composite for removal of ceftriaxone from aquatic solutions. J. Mol. Liquids 261: 146-154.

Baeck, S.H., Jaramillo, T.F., Brändli, C. and McFarland, E.W. 2002. Combinatorial electrochemical synthesis and characterization of tungsten-based mixed-metal oxides. J. Comb. Chem. 4: 563-568.

Bai, J.M., Zhang, L., Liang, R.P. and Qiu, J.D. 2013. Graphene quantum dots combined with europium ions as photoluminescent probes for phosphate sensing. Chemistry–A Euro. J. 19: 3822-3826.

Basu, R. 2014. Soft memory in a ferroelectric nanoparticle-doped liquid crystal. Phys. Rev. E. 89: 022508.

Bassi, P.S., Chiam, S.Y., Gurudyal, Barber, J. and Wong, L.H. 2014. Hydrothermal grown nanoporous iron based titanate, Fe_2TiO_5 for light driven water splitting. ACS. Appl. Mater. Interfaces 6: 22490-22495.

Batzill, M. and Diebold, U. 2005. The surface and materials science of tin oxide. Prog. Surf. Sci. 79: 47-154.

Bazito, F.F. and Torresi, R.M. 2006. Cathodes for lithium ion batteries: The benefits of using nanostructured materials. J. Braz. Chem. Soc. 17: 627-642.

BCC Research, Nanotechnology in environmental applications: The global market, Report Code: NAN039C. BCC Research, Massachusetts, USA, 2015. Source: https://www.bccresearch.com/market-research/nanotechnology/nanotechnology-environmental-applications-market.html.

Beheshti, S., Safarifard, V. and Morsali, A. 2018. Isoreticular interpenetrated pillared-layer microporous metal organic framework as a highly effective catalyst for three-component synthesis of pyrano [2, 3-d] pyrimidines. Inorg. Chem. Comm. 94: 80-84.

Bui, N., Meshot, E.R., Kim, S., PeÇa, J., Gibson, P.W., Wu, K.J., et al. 2016. Ultrabreathable and protective membranes with Sub-5 nm carbon nanotube pores. Adv. Mater. 28: 5871.

Bukowczan, A., Hebda, E. and Pielichowski, K. 2020. The influence of nanoparticles on phase formation and stability of liquid crystals and liquid crystalline polymers. J. Mol. Liq. 2020. https://doi.org/10.1016/j.molliq.2020.114849.

Butnar, I. and Llop, M. 2007. Composition of greenhouse gas emissions in Spain: An input-output analysis. Ecol. Econ. 61: 388-395.

Cao, Z. and Wei, B. 2014. Fragmented carbon nanotube macrofilms as adhesive conductors for lithium-ion batteries. ACS Nano. 8: 3049-3059.

Carralero, V., Mena, M.L., Gonzalez-Cortes, A., Yanez-Sedeno, P. and Pingarron, J.M. 2006. Development of a high analytical performance-tyrosinase biosensor based on composite graphite–Teflon electrode modified with gold nanoparticles. Biosens. Bioelectron. 22: 730-736. https://doi.org/10.1016/j.bios.2006.02.012.

Carter, B.M., Wiesenauer, B.R., Hatakeyama, E.S., Barton, J.L., Noble, R.D. and Gin, D.L. 2012. Glycerol-based bicontinuous cubic lyotropic liquid crystal monomer system for the fabrication of thin-film membranes with uniform nanopores. Chem. Mater. 24: 4005.

Centi, G. and Perathoner, S. 2011. Carbon nanotubes for sustainable energy applications. Chem. Sus. Chem. 4: 913-925.

Chandrasekaran, S., Hur, S.H., Kim, E.J., Rajagopalan, B., Babu, K.F., Senthilkmar, V., et al. 2015. Highly-ordered maghemite/reduced graphene oxide nanocomposites for high-performance photoelectrochemical water splitting. RSC. Adv. 5: 29159-29166.

Chandrasekaran, S., Chung, J.S., Jung, E., Seung, K. and Hur, H. 2016. Exploring complex structural evolution of graphene oxide/ZnO triangles and its impact on photoelectrochemical water splitting. Chem. Eng. J. 290: 465-476.

Chang, C.T., Wang, J., Ouyang, T., Zhang, Q. and Jing, Y.H. 2015. Photocatalytic degradation of acetaminophen in aqueous solutions by TiO_2/ZSM-5 zeolite with low energy irradiation. Mat. Sci. Eng. B. 196: 53-60.

Chang, H., Kao, M.J., Cho, K.C., Chen, S.L., Chu, K.H. and Chen, C.C. 2011. Integration of CuO thin films and dye-sensitized solar cells for thermoelectric generators. Curr. Appl. Phys. 11: S19-S22.

Charitidis, C.A., Georgiou, P., Koklioti, M.A., Trompeta, A.-F. and Markakis, V. 2014. Manufacturing nanomaterials: From research to industry. Manuf. Rev. 1: 11.

Chen, L., Zeng, G., Zhang, Y., Tang, L., Huang, D., Liu, C., et al. 2010. Trace detection of picloram using an electrochemical immune sensor based on three-dimensional gold nanoclusters. Anal. Biochem. 407: 172-179. https://doi.org/10. 1016/j.ab.2010.08.001.

Chen, X., Hui, L., Xu, J., Jaber, F., Musharavati, F., Zalezhad, E., et al. 2020. Synthesis and characterization of a $NiCo_2O_4$@$NiCo_2O_4$ hierarchical mesoporous nanoflake electrode for supercapacitor applications. Nanomaterials 10: 1292. doi:10.3390/nano10071292.

Chen, Y., Crittenden, J.C., Hackney, S., Sutter, L. and Hand, D.W. 2005. Preparation of a novel TiO_2-based *p-n* junction nanotube photocatalyst. Environ. Sci. Technol. 39: 1201-1208.

Cheng, X., Lowe, S.B., Reece, P.J. and Gooding, J.J. 2014. Colloidal silicon quantum dots: From preparation to the modification of self-assembled monolayers (SAMs) for bio-applications. Chem. Soc. Rev. 43: 2680.

Chong, M.N., Jin, B., Chow, C.W.K. and Saint, C. 2010. Recent developments in photocatalytic water treatment technology: A review. Water. Res. 44: 2997-3027.

Choudhary, A., George, T. and Li, G. 2018. Conjugation of nanomaterials and nematic liquid crystals for futuristic applications and biosensors. Biosensors 8: 69.

Cloete, T.E., de Kwaadsteniet, M., Botes, M. and López-Romero, J.M. 2010. Nanotechnology in Water Treatment Applications. Caister Academic Press, Norfolk, UK.

Comini, E. 2006. Metal oxide nano-crystals for gas sensing. Anal. Chim. Acta. 568: 28-40.

Cordova, I.A., Peng, Q., Ferall, I.L., Reith, A.J., Hoertz, P.G. and Glass, J.T. 2015. Enhanced photoelectrochemical water oxidation via atomic layer deposition of TiO_2 on fluorine-doped tin oxide nanoparticle films. Nanoscale 7: 8584-8592.

Crane, R.A. and Scott, T.B. 2012. Nanoscale zero-valent iron: Future prospects for an emerging water treatment technology. J. Hazard Mater. 211: 112-125.

Cuerda-Correa, E.M., Domínguez, J.R., M.J., Muñoz-Peña and González, T. 2016. Degradation of parabens in different aqueous matrices by several O_3-derived advanced oxidation processes. Ind. Eng. Chem. Res. 55: 5161-5172.

Cui, L. and Zhao, Y. 2004. Azopyridine side chain polymers: An efficient way to prepare photoactive liquid crystalline materials through self-assembly. Chem. Mater. 16: 2076-2082. https://doi.org/10.1021/cm0348850.

Da Cruz, C., Sandre, O. and Cabuil, V. 2005. Phase behavior of nanoparticles in a thermotropic liquid crystal. J. Phys. Chem. B 109: 14292-14299. https://doi.org/10.1021/jp0455024.

Dang, Z.M., Yuan, J.K., Zha, J.W., Zhou, T., Li, S.T. and Hu, G.H. 2012. Fundamentals, processes and applications of high-permittivity polymer–matrix composites. Prog. Mater. Sci. 57: 660.

Dang, Z.M., Yuan, J.K., Yao, S.H. and Liao, R.J. 2013. Flexible nanodielectric materials with high permittivity for power energy storage. Adv. Mater. 25: 6334.

Demus, D., Goodby, J., Gray, G.W., Spiess, H.-W. and Vill, V. 1998. Handbook of Liquid Crystals. Weinheim: Wiley-VCH: 1-3.

Dierking, I. 2003. Textures of Liquid Crystals. Wiley-VCH: Weinheim, Germany.

Dierking, I., Scalia, G., Morales, P. and LeClere, D. 2004. Aligning and reorienting carbon nanotubes with nematic liquid crystals. Adv. Mater. 16: 865-869.

Dierking, I., Scalia, G. and Morales, P. 2005. Liquid crystal–carbon nanotube dispersions. J. Appl. Phys. 97: 044309.

Dong, Y. and Yang, Z. 2013. Beyond displays: The recent progress of liquid crystals for bio/chemical detections. Chinese Sci. Bull. 58: 2557-2562.

Donisi, D., Bellini, B., Beccherelli, R., Asquini, R., Gilardi, G., Trotta, M., et al. 2010. Switchable liquid-crystal optical channel wavguide on silicon. IEEE. J. Quant. Electron. 46: 762-768.

Dutta, A., Kumar Maji, S. and Adhikary, B. 2014. γ-Fe_2O_3 nanoparticles: An easily recoverable effective photocatalyst for the degradation of rose Bengal and methylene blue dyes in the waste-water treatment plant. Mat. Res. Bull. 49: 28-34.

Díaz-González, F., Sumper, A., Gomis-Bellmunt, O. and Villafáfila-Robles, R. 2012. A review of energy storage technologies for wind power applications. Renew. Sustain. Energy Rev. 16: 2154-2171.

Esrafili, L., Safarifard, V., Tahmasebi, E., Esrafili, M. and Morsali, A. 2018. Functional group effect of isoreticular metal–organic frameworks on heavy metal ion adsorption. New. J. Chem. 42: 8864-8873.

Esrafili, L., Safarifard, V., Tahmasebi, E., Esrafili, M. and Morsali, A. 2018. Functional group effect of isoreticular metal–organic frameworks on heavy metal ion adsorption. New. J. Chem. 42: 8864-8873.

Fane, A.G., Wang, R. and Hu, M.X. 2015. Synthetic membranes for water purification: Status and future. Angew. Chem. Int. Ed. 54: 3368-3386.

Fatta-Kassinos, D., Kalavrouziotis, I.K., Koukoulakis, P.H. and Vasquez, M.I. 2011. The risks associated with wastewater reuse and xenobiotics in the agroecological environment. Sci. Total Environ. 409: 3555-3563.

Feng, J., Biskos, G. and Schmidt-Ott, A. 2015. Toward industrial scale synthesis of ultrapure singlet nanoparticles with controllable sizes in a continuous gas-phase process. Sci. Rep. 5: 15788.

Franke, M.E., Koplin, T.J. and Simon, U. 2006. Metal and metal oxide nanoparticles in chemiresistors: Does the nanoscale matter? Small 2: 36-50.

Fu, F.L., Dionysiou, D.D. and Liu, H. 2014. The use of zero-valent iron for groundwater remediation and wastewater treatment: A review. J. Hazard Mater. 267: 194-205.

Gao, F., Gao, N.N., Nishitani, A. and Tanaka, H. 2016. Rod-like hydroxyapatite and Nafion nanocomposite as an electrochemical matrix for simultaneous and sensitive detection of Hg^{2+}, Cu^{2+}, Pb^{2+} and Cd^{2+}. J. Electroanal. Chem. 775: 212-218.

Gdovinová, V., Tom šovičová, N., Éber, N., Tóth-Katona, T., Závišová, V., Timko, M., et al. 2014. Influence of the anisometry of magnetic particles on the isotropic–nematic phase transition. Liq. Crys. 41: 1773-1777. https://doi.org/10.1080/02678292.2014.950615.

Gerasin, V.A., Antipov, E.M., Karbushev, V.V., Kulichikhin, V.G., Karpacheva, G.P., Talroze, R.V., et al. 2013. New approaches to the development of hybrid nanocomposites: From structural materials to high-tech applications. Russ. Chem. Rev. 82: 303-332. https://doi.org/10.1070/RC2013v082n04ABEH004322.

Gharib, M., Safarifard, V. and Morsali, A. 2018. Ultrasound assisted synthesis of amide functionalized metalorganic framework for nitroaromatic sensing. Ultrason. Sonochem. 42: 112-118.

Gin, D.L. and Noble, R.D. 2011. Designing the next generation of chemical separation membranes. Science 332: 674.

Goodby, J.W., Saez, I.M., Cowling, S.J., Gortz, V., Draper, M., Hall, A.W., et al. 2008. Transmission and amplification of information and properties in nanostructured liquid crystals. Angew. Chem. Int. Ed. 47: 2754-2787.

Grey, D., Garrick, D., Blackmore, D., Kelman, J., Muller, M. and Sadoff, C. 2013. Water security in one blue planet: Twenty-first century policy challenges for science. Philos Trans. R. Soc., A 371: 20120406.

Guan, X., Sun, Y., Qin, H., Li, J., Lo, I.M.C., He, D., et al. 2015. The limitations of applying zero-valent iron technology in contaminants sequestration and the corresponding countermeasures: The development in zero-valent iron technology in the last two decades (1994-2014). Water Res. 75 (Supplement C): 224-248.

Gullapalli, S. and Wong, M. 2011. Nanotechnology: A guide to nano-objects. Chem. Eng. Prog. 107: 28-32.

Guo, K., Liu, Z., Han, J., Zhang, X., Li, Y., Hong, T., et al. 2015. Higher-efficiency photoelectrochemical electrodes of titanium dioxide-based nanoarrays sensitized simultaneously with plasmonic silver nanoparticles and multiple metal sulfides photosensitizers. J. Power. Sourc. 285: 185-194.

Gutsch, A., Mühlenweg, H. and Krämer, M. 2004. Tailor-made nanoparticles via gas-phase synthesis. Small 1: 30-46.

Han, B.Q., Liu, X., Xing, X.X., Chen, N., Xiao, X.C., Liu, S.Y., et al. 2016. A high response butanol gas sensor based on ZnO hollow spheres. Sensor Actuat. B—Chem. 237: 423-430.

Hartono, D., Qin, W.J., Yang, K.-L. and Yung, L.-Y.L. 2009. Imaging the disruption of phospholipid monolayer by protein-coated nanoparticles using ordering transitions of liquid crystals. Biomaterials 30: 843-849.

Hayasaka, H., Tamura, K. and Akagi, K. 2008. Dynamic switching of linearly polarized emission in liquid-crystallinity-embedded photoresponsive conjugated polymers. Macromolecules 41: 2341-2346. https://doi.org/10.1021/ma800132v.

He, G., Zhao, L., Chen, K., Liu, Y. and Zhu, H. 2013. Highly selective and sensitive gold nanoparticle-based colorimetric assay for PO_4^{3-} in aqueous solution. Talanta. 106: 73-78.

He, X. and Hwang, H.M. 2016. Nanotechnology in food science: Functionality, applicability, and safety assessment. J. Food Drug Anal. 24: 671681.

He, X., Aker, W.G., Pelaez, M., Lin, Y., Dionysiou, D.D. and Hwang, H.-m. 2016. Assessment of nitrogen–fluorine-codoped TiO_2 under visible light for degradation of BPA: Implication for field remediation. J. Photochem. Photobiol. A. 314: 81-92.

Henmi, M., Nakatsuji, K., Ichikawa, T., Tomioka, H., Sakamoto, T., Yoshio, M., et al. 2012. Self-organized liquid-crystalline nanostructured membranes for water treatment: Selective permeation of ions. Adv. Mater. 24: 2238.

Hermann, A.M. 1998. Polycrystalline thin-film solar cells: A review. Sol. Energy Mat. Sol. C. 55: 75-81.

Hisatomi, T., Kubota, J. and Domen, K. 2014. Recent advances in semiconductors for photocatalytic and photoelectrochemical water splitting. Chem. Soc. Rev. 43: 7520-7535.

Hleli, S., Martelet, C., Abdelghani, A., Bessueille, F., Errachid, A., Samitier, J., et al. 2006. Atrazine analysis using an impedimetric immunosensor based on mixed biotinylated self-assembled monolayer. Sensors Actuators B 113: 711-717. https://doi.org/10.1016/j.snb.2005.07.023.

Houtman, C.J. 2010. Emerging contaminants in surface waters and their relevance for the production of drinking water in Europe. J. Integrative Environ. Sci. 7: 271-295.

Hristozov, D. and Ertel, J. 2009. Nanotechnology and sustainability: Benefits and risks of nanotechnology for environmental sustainability. Forum der Forschung 22: 161-168.

Huang, C. and Zhang, Q.M. 2005. Fully functionalized high-dielectric-constant nanophase polymers with high electromechanical response. Adv. Mater. 17: 1153.

Huang, R., Fan, X., Shen, W. and Zhu, J. 2009. Carbon-coated silicon nanowire array films for high-performance lithium-ion battery anodes. Appl. Phys. Lett. 95: 133119.

Huang, Y., Shen, Z., Wu, Y., Wang, X., Zhang, S., Shia, X. and Zeng, H. 2016. Amorphous ZnO based resistive random access memory. RSC Adv. 6: 17867-17872.

Huang, X., Zheng, X., Xu, Z. and Yi, C. 2017. ZnO-based nanocarriers for drug delivery application: From passive to smart strategies. Int. J. Pharm. 534: 190-194.

Hutchison, J.E. 2008. Greener nanoscience: A proactive approach to advancing applications and reducing implications of nanotechnology. ACS Nano. 2: 395-402.

Hyde, S.T. 2001. Identification of lyotropic liquid crystalline mesophases. pp. 299-327. *In*: Holmberg, Krister. [ed.]. Handbook of Applied Surface Chemistry and Colloid Chemistry. John Wiley and Sons. Ltd, Australia.

Ibrahim, S., Labeeb, A., Mabied, A.F., Soliman, O., Ward, A., Abd-El-Messieh, S.L., et al. 2017. Synthesis of super-hydrophobic polymer nanocomposites as a smart self-cleaning coating films. Polym. Compos. 38: E147.

International Energy Agency (IEA) 2007. Renewables in Global Energy Supply: An IEA Facts Sheet. Paris. Head of Publications Service, 9 rue de la Fédération, 75739 Paris Cedex 15, France.

Inturi, S.N.R., Boningari, T., Suidan, M. and Smirniotis, P.G. 2014. Visible-light-induced photodegradation of gas phase acetonitrile using aerosol-made transition metal (V, Cr, Fe, Co, Mn, Mo, Ni, Cu, Y, Ce, and Zr) doped TiO_2. Appl. Catal. B 144: 333-342.

Ji, R., Zhang, Z., Yan, C., Zhu, M. and Li, Z. 2016. Preparation of novel ceramic tiles with high Al_2O_3 content derived from coal fly ash. Constr. Build. Mater. 114: 888-895.

Joshi, K.A., K.M.M. Prouza, Wang, J., Tang, J., Haddon, R., Chen, W., et al. 2006. V-type nerve agent detection using a carbon nanotube-based amperometric enzyme electrode. Anal. Chem. 78: 331-336. https://doi.org/10.1021/ac051052f.

Joshi, T., Kumar, A., Prakash, J. and Biradar, A.M. 2010. Low power operation of ferroelectric liquid crystal system dispersed with zinc oxide nanoparticles. Appl. Phys. Lett. 96: 253109.

Kang, K., Meng, Y.S., Bréger, J., Grey, C.P. and Ceder, G. 2006. Electrodes with high power and high capacity for rechargeable lithium batteries. Science 311: 977-980.

Kango, S., Kalia, S., Celli, A., Njuguna, J., Habibi, Y. and Kumar, R. 2013. Surface modification of inorganic nanoparticles for development of organic–inorganic nanocomposites: A review. Prog. Polym. Sci. 38: 1232-1261. https://doi.org/10.1016/j.progpolymsci.2013.02.003.

Kanjana, N., Maiaugree, W., Poolcharuansin, P. and Laokul, P. 2020. Size controllable synthesis and photocatalytic performance of mesoporous TiO_2 hollow spheres. J. Mater. Sci. Technol. 48: 105-113.

Karn, B., Kuiken, T. and Otto, M. 2009. Nanotechnology and *in situ* remediation: A review of the benefits and potential risks. Environ. Health Persp. 117: 1823-1831.

Katiyar, R., Pathak, G., Agrahari, K., Srivastava, A., Garbat, K. and Manohar, R. 2019. Investigation of dielectric and electro-optical parameters of high birefringent nematic liquid crystal doped with TiO_2 nanoparticles and its applicability toward liquid crystal displays. Mol. Cryst. Liq. Cryst. 691: 50-61. https://doi.org/10.1080/15421406.2019.1702811.

Kato, T., Mizoshita, N. and Kishimoto, K. 2006. Functional liquid-crystalline assemblies: Self-organized soft materials. Angew. Chem. Int. Ed. 45: 38-68.

Kaushik, A., Kumar, R., Huey, E., Bhansali, S., Nair, N. and Nair, M. 2014a. Silica nanowires: Growth, integration, and sensing applications. Micro chimica Acta 181: 1759-1780.

Kaushik, A., Vasudev, A., Arya, S.K., Pasha, S.K. and Bhansali, S. 2014b. Recent advances in cortisol sensing technologies for point-of-care application. Biosens. Bioelectron. 53: 499-512.

Kaushik, A., Jayant, R.D., Tiwari, S., Vashist, A. and Nair, M. 2016a. Nano-biosensors to detect beta-amyloid for Alzheimer's disease management. Biosens. Bioelectron. 80: 273-287.

Kaushik, A., Tiwari, S., Dev Jayant, R., Marty, A. and Nair, M. 2016b. Towards detection and diagnosis of Ebola virus disease at point-of-care. Biosens. Bioelectron. 75: 254-272.

Kaushik, A., Jayant, R.D., Tiwari, S., Vashist, A. and Nair, M. 2016a. Nano-biosensors to detect beta-amyloid for Alzheimer's disease management. Biosens. Bioelectron. 80: 273-287.

Kaushik, A., Tiwari, S., Dev Jayant, R., Marty, A. and Nair, M. 2016b. Towards detection and diagnosis of Ebola virus disease at point-of-care. Biosens. Bioelectron. 75: 254-272.

Kaushik, A., Tiwari, S., Jayant, R.D., Vashist, A., Nikkhah-Moshaie, R., El-Hage, N., et al. 2017. Electrochemical biosensors for early stage Zika diagnostics. Trend. Biotech. 35: 308-317.

Kaushik, A., Yndart, A., Kumar, S., Jayant, R.D., Vashist, A., Brown, A.N., et al. 2018. A sensitive electrochemical immunosensor for label-free detection of Zika-virus protein. Sci. Rep. 8: 9700.

Kaushik, A.K. and Chandra, A.D. 2016. Nanobiotechnology for Sensing Applications: From Lab to Field. First ed. Apple Academic Press, Canada.

Kazemi, S. and Safarifard, V. 2018. Carbon dioxide capture on metal-organic frameworks with amide decorated pores. Nanochem. Res. 3: 62-78.

Keller, V., Keller, N., Ledoux, M.J. and Lett, M.C. 2005. Biological agent inactivation in a flowing air stream by photocatalysis. Chem. Commun. 23: 2918-2920.

Khan, J. and Arsalan, M.H. 2016. Solar power technologies for sustainable electricity generation: A review. Renewable Sustainable Energy Rev. 55: 414-425.

Kim, D.W., Kim, Y.H., Jeong, H.S. and Jung, H.-T. 2011. Direct visualization of large-area graphene domains and boundaries by optical birefringency. Nat. Nanotechnol. 7: 29-34.

Kim, J.E., Han, T.H., Lee, S.H., Kim, J.Y., Ahn, C.W., Yun, J.M., et al. 2011. Graphene oxide liquid crystals. Angew. Chem. Int. Ed. 50: 3043-3047.

Kitano, M. and Hara, M. 2010. Heterogeneous photocatalytic cleavage of water. J. Mater. Chem. 20: 627-641.

Kong, L., Huang, S., Yue, Z., Peng, B. and Zhang, J. 2009. Sensitive mediator-free tyrosinase biosensor for the determination of 2,4-dichlorophenol. Microchim. Acta 165: 203-209. https://doi.org/10.1007/s00604-008-0121-3.

Kudo, A. and Miseki, Y. 2009. Heterogeneous photocatalyst materials for water splitting. Chem. Soc. Rev. 38: 253-278.

Kumar, R., Al-Dossary, O., Kumar, G. and Umar, A. 2015. Zinc oxide nanostructures for NO_2 gas-sensor applications: A review. Nano-Micro Lett. 7: 97-120.

Kumar, S. and Bisoyi, H.K. 2007. Aligned carbon nanotubes in the supramolecular order of discotic liquid crystals. Angew. Chem. Int. 46: 1501-1503.

Kurian, M. and Nair, D.S. 2015. Heterogeneous Fenton behavior of nano nickel zinc ferrite catalysts in the degradation of 4-chlorophenol from water under neutral conditions. J. Water Process Eng. 8: 37-49.

Lagerwall, J., Scalia, G., Haluska, M., Dettlaf-Weglikowska, U., Roth, S. and Giesselmann, F. 2007. Nanotube alignment using lyotropic liquid crystals. Adv. Mater. 19: 359-364.

Lapanik, A., Rudzki, A., Kinkead, B., Qi, H., Hegmann, T. and Haase, W. 2012. Electrooptical and dielectric properties of alkylthiol-capped gold nanoparticle–ferroelectric liquid crystal nanocomposites: Influence of chain length and tethered liquid crystal functional groups. Soft Matter 8: 8722.

Lee, S., Kim, Y.S., Jo, M., Jin, M., Lee, D.-K. and Kim, S. 2007a. Chip-based detection of hepatitis C virus using RNA aptamers that specifically bind to HCV core antigen. Biochem. Biophys. Res. Commun. 358: 47-52. https://doi.org/10.1016/j.bbrc.2007. 04.057.

Li, F., Buchnev, O., Cheon, C.I., Glushchenko, O., Reshetnyak, V., Reznikov, Y., et al. 2006. Orientational coupling amplification in ferroelectric nematic colloids. Phys. Rev. Lett. 97: 147801.

Li, F.H., West, J., Glushchenko, A., Cheon, C.I. and Reznikov, Y. 2006. Ferroelectric nanoparticle/liquid-crystal colloids for display applications. J. Soc. Inf. Disp. 14: 523-527.

Li, L.-S., Walda, J., Manna, L. and Alivisatos, A.P. 2002. Semiconductor nanorod liquid crystals. Nano Lett. 2: 557-560.

Li, M., Chang, K., Wang, T., Liu, L., Zhang, H., Li, P., et al. 2015. Hierarchical nanowire arrays based on carbon nanotubes and CO_3O_4 decorated ZnO for enhanced photoelectrochemical water oxidation. J. Mater. Chem. A. 3: 13731-13737.

Li, R., Jia, Y., Bu, N., Wu, J. and Zhen, Q. 2015. Photocatalytic degradation of methyl blue using Fe_2O_3/TiO_2 composite ceramics. J. Alloy. Comp. 643: 88-93.

Li, Y.W., Zhan, X.J., Xiang, L., Deng, Z.S., Huang, B.H., Wen, H.F., et al. 2014. Analysis of trace microcystins in vegetables using solid-phase extraction followed by high performance liquid chromatography triple-quadrupole mass spectrometry. J. Agric. Food Chem. 62: 11831-11839.

Liang, S., Zhu, X., Lian, P., Yang, W. and Wang, H. 2011. Superior cycle performance of Sn@C/graphene nanocomposite as an anode material for lithium-ion batteries. J. Solid State Chem. 184: 1400-1404.

Liao, S., Qiao, Y., Han, W., Xie, Z., Wu, Z., Shen, G., et al. 2012. Acetylcholinesterase liquid crystal biosensor based on modulated growth of gold nanoparticles for amplified detection of acetylcholine and inhibitor. Anal. Chem. 84: 45-49.

Lin, C.J. and Yang, W.T. 2014. Ordered mesostructured Cu-doped TiO_2 spheres as active visible-light-driven photocatalysts for degradation of paracetamol. Chem. Eng. J. 237: 131-137.

Lisetski, L.N., Minenko, S.S., Zhukov, A.V., Shtifanyuk, P.P. and Lebovka, N.I. 2009. Dispersions of carbon nanotubes in cholesteric liquid crystals. Mol. Cryst. Liq. Cryst. 510: 43-50.

Liu, G., Yang, H.G., Pan, J., Yang, Y.Q., Lu, G.Q. and Cheng, H.M. 2014. Unique electronic structure induced high photoreactivity of sulfur-doped graphitic C_3N_4. Chem. Rev. 114: 9559-9612.

Liu, J., Luo, J., Yang, W.G., Wang, Y.L., Zhu, L.Y., Xu, Y.Y., et al. 2015. Synthesis of single-crystalline anatase TiO_2 nanorods with high-performance dye-sensitized solar cells. J. Mater. Sci. Technol. 31: 106-109.

Liu, S., Zhao, Y., Wang, T., Liang, N. and Hou, X. 2019. Core-shell Fe_3O_4@MIL-100 (Fe) magnetic nanoparticle for effective removal of meloxicam and naproxen in aqueous solution. J. Chem. Eng. Data 64: 2997-3007.

Liu, W., Lin, D., Sun, J., Zhou, G. and Cui, Y. 2016. Improved lithium ionic conductivity in composite polymer electrolytes with oxide-ion conducting nanowires. ACS Nano 10: 11407-11413.

Liu, Y., Qu, X., Huang, G., Xing, B., Zhang, F., Li, B., et al. 2020. 3-dimensional porous carbon with high nitrogen content obtained from longan shell and its excellent performance for aqueous and all-solid-state supercapacitors. Nanomaterials 10: 808.

Lu, T.L., Wang, Y.Q., Wang, Y.L., Zhou, L.P., Yang, X.M. and Su, Y.L. 2017. Synthesis of mesoporous anatase TiO_2 sphere with high surface area and enhanced photocatalytic activity. J. Mater. Sci. Technol. 33: 300-304.

Lu, Z. and Yin, Y. 2012. Colloidal nanoparticle clusters: Functional materials by design. Chem. Soc. Rev. 41: 6874-6887.

Ma, H.Y., Wang, H.T. and Na, C.Z. 2015. Microwave-assisted optimization of platinum-nickel nanoalloys for catalytic water treatment. Appl. Catal. B—Environ. 163: 198-204.

Macholán, L. and Schánl, L. 1977. Enzyme electrode with immobilized polyphenol oxidase for determination of phenolic substrates. Collect. Czechoslov. Chem. Commun. 42: 3667. https://doi.org/10.1135/cccc19773667.

Manepalli, R.K.N.R, Giridhar, G., Pardhasaradhi, P., Jayaprada, P., Tejaswi, M., Sivaram, K., et al. 2018. Influence of ZnO nanoparticles dispersion in liquid crystalline compounds-experimental studies. Mater. Today 5: 2666-2676. https://doi.org/10.1016/j.matpr.2018.01.047.

Marets, N., Kuo, D., Torrey, J., Sakamoto, T., Henmi, M., Katayama, H., et al. 2017. Virus filtration: Highly efficient virus rejection with self-organized membranes based on a crosslinked bicontinuous cubic liquid crystal (Adv. Healthcare Mater. 14/2017). Adv. Healthcare Mater. 6: 1700252.

Martinez-Miranda, L.J., Traister, K.M., Melendez-Rodriguez, I. and Salamanca-Riba, L. 2010. Liquid crystal-ZnO nanoparticles photovoltaics: Role of nanoparticles in ordering the liquid crystal. Appl. Phys. Lett. 97: 223301. https://doi.org/10.1063/1.3511736.

Masciangioli, T. and Zhang, W.X. 2003. Environmental technologies at the nanoscale. Environ. Sci. Technol. 37: 102a-108a.

Mauriz, E., Calle, A., Lechuga, L.M., Quintana, J., Montaya, A. and Manclus, J.J. 2006. Realtime detection of chlorpyrifos at part per trillion levels ground, surface, and drinking water samples by a portable surface plasmon resonance immunosensor. Anal. Chim. Acta 561: 40-47. https://doi.org/10.1016/j.aca.2005.12.069.

Mauriz, E., Calle, A., Manclus, J.J., Montoya, A., Hildebrandt, A., Barcelo, D., et al. 2007. Optical immunosensor for fast and sensitive detection of DDT and related compounds in river water samples. Biosens. Bioelectron. 22: 1410-1418. https://doi. org/10.1016/j.bios.2006.06.016.

Mertelj, A., Lisjak, D., Drofenik, M. and Copic, M. 2013. Ferromagnetism in suspensions of magnetic platelets in liquid crystal. Nature 504: 237.

Mishra, K.G., Dubey, S.K., Mani, S.A. and Pradhan, M.S. 2016. Comparative study of nanoparticles doped in liquid crystal polymer system. J. Mol. Liq. 224: 668-671. https://doi.org/10.1016/j.molliq.2016.10.075.

Mohammadi Nodeh, M.K., Radfard, M., Zardari, L.A. and Rashidi Nodeh, A. 2018. Enhanced removal of naproxen from wastewater using silica magnetic nanoparticles decorated onto graphene oxide: Parametric and equilibrium study. Sep. Sci. Technol. 53: 1-10.

Mohanty, S. 2003. Liquid Crystals – The 'fourth' phase of matter. Resonance 8: 52-70.

Monllor-Satoca, D., Borja, L., Rodes, A., Gómez, R. and Salvador, P. 2006. Photoelectrochemical behavior of nanostructured WO_3 thin-film electrodes: The oxidation of formic acid. https://doi.org/10.1002/cphc.200600379.

Moradi, E., Rahimi, R. and Safarifard, V. 2019. Sonochemically synthesized microporous metal–organic framework representing unique selectivity for detection of Fe^{3+} ions. Polyhedron 159: 251-258.

Narayan, R., Kim, J.E., Kim, J.Y., Lee, K.E. and Kim, S.O. 2016. Liquid crystals: Graphene oxide liquid crystals: Discovery, evolution and applications (Adv. Mater. 16/2016). Adv. Mater. 28: 3044.

Needleman, D. and Dogic, Z. 2017. Active matter at the interface between materials science and cell biology. Nat. Rev. Mat. 2: 17048.

Neto, A.M.F. and Salinas, S.R.A. 2005. The Physics of Lyotropic Liquid Crystals. Oxford University Press: Oxford, UK.

Ng, L.N., Mohammad, A.W., Leo, C.P. and Hilal, N. 2013. Polymeric membranes incorporated with metal/metal oxide nanoparticles: A comprehensive review. Desalination 308: 15-33. https://doi.org/10.1016/j.desal.2010.11.033.

OECD, Testing Programme of Manufactured Nanomaterials, https://www.oecd.org/chemicalsafety/nanosafety/testing-programme-manufactured-nanomaterials.html.

Ozgur, U., Alivov, Y.I., Liu, C., Teke, A., Reshchikov, M.A., Dogan, S., et al. 2005. A comprehensive review of ZnO materials and devices. J. Appl. Phys. 98: 1-103.

O'Regan, B. and Gratzel, M. 1991. A low-cost, high-efficiency solar cell based on dye-sensitized colloidal TiO_2 films. Nature 353: 737-740.

Palffy-Muhoray, P. 2007. The diverse world of liquid crystals. Physics Today 60: 54-60.

Pan, Z., Zhang, G. and Wang, X. 2019. Polymeric carbon nitride/reduced graphene oxide/Fe_2O_3: All-solid-state Z-scheme system for photocatalytic overall water splitting. Angew. Chem. Int. Ed. Engl. 58: 7102-7106.

Parida, B., Iniyan, S. and Goic, R. 2011. A review of solar photovoltaic technologies. Revew. Sustain. Energy. Rev. 15: 1625-1636.

Park, H.B., Kamcev, J., Robeson, L.M., Elimelech, M. and Freeman, B.D. 2017. Maximizing the right stuff: The trade-off between membrane permeability and selectivity. Science 356: 1137.

Pathak, G., Katyiar, R., Agrahari, K., Srivastava, A., Dabrowski, R., Garbat, K., et al. 2018. Analysis of birefringence property of three different nematic liquid crystals dispersed with TiO_2 nanoparticles. Opto-Electr. Rev. 26: 11-18. https://doi.org/10.1016/j.opelre.2017.11.005.

Pathakoti, K., Huang, M.-J., Watts, J.D., He, X. and Hwang, H.-M. 2014. Using experimental data of *Escherichia coli* to develop a QSAR model for predicting the photo-induced cytotoxicity of metal oxide nanoparticles. J. Photochem. Photobiol. B. 130: 234-240.

Pathakoti, K., Manubolu, M. and Hwang, H.-M. 2017. Nanostructures: Current uses and future applications in food science. J. Food Drug Anal. 25: 245-253.

Peng, Z., Jiang, Z., Huang, X. and Li, Y. 2016. A novel electrochemical sensor of tryptophan based on silver nanoparticles/metal–organic framework composite modified glassy carbon electrode. RSC. Adv. 6: 13742-13748.

Podoliak, N., Buchnev, O., Bavykin, D.V., Kulak, A.N., Kaczmarek, M. and Sluckin, T.J. 2012. Magnetite nanorod thermotropic liquid crystal colloids: Synthesis, optics and theory. J. Colloid Interface Sci. 386: 158-166.

Ponder, S.M., Darab, J.G. and Mallouk, T.E. 2000. Remediation of Cr(VI) and Pb(II) aqueous solutions using supported, nanoscale zero-valent iron. Environ. Sci. Technol. 34: 2564-2569.

Prodanov, M.F., Kolosov, M.A., Krivoshey, A.I., Fedoryako, A.P., Yarmolenko, S.N. and Semynozhenko, V.P. 2012. Dispersion of magnetic nanoparticles in a polymorphic liquid crystal. Liq. Crys. 39: 1512-1526. https://doi.org/10.1080/02678292.2012.725867.

Rafati, L., Ehrampoush, M.H., Rafati, A.A., Mokhtari, M. and Mahavi, A.H. 2019. Fixed bed adsorption column studies and models for removal of ibuprofen from aqueous solution by strong adsorbent Nano-clay composite. J. Environ. Health. Sci. Eng. 17: 753-765.

Rafique, A., Zubair, U., Serrapede, M., Fontana, M., Bianco, S., Rivolo, P., et al. 2020. Binder free and flexible asymmetric supercapacitor exploiting Mn_3O_4 and MoS_2 nanoflakes on carbon fibers. Nanomaterials 10: 1084.

Rajendran, K. and Sen, S. 2018. Adsorptive removal of carbamazepine using biosynthesized hematite nanoparticles. Environ. Nanotechnol. Monit. Manage. 9: 122-127.

Ramiro-Manzano, F., Atienzar, P., Rodriguez, I., Meseguer, F., Garcia, H. and Corma, A. 2007. Apollony photonic sponge based photoelectrochemical solar cells. Chem. Commun. 3: 242-244.

Ren, Z., Chen, C., Hu, R., Mai, K., Qian, G. and Wang, Z. 2012. Two-step self-assembly and lyotropic liquid crystal behavior of TiO_2 nanorods. J. Nanomater. 2012: 180989.

Rittersma, Z.M. 2002. Recent achievements in miniaturized humidity sensors: A review of transduction techniques. Sensor Actuat. A—Phys. 96: 196-210.

Saikia, L., Bhuyan, D., Saikia, M., Malakar, B., Dutta, D.K. and Sengupta, P. 2015. Photocatalytic performance of ZnO nanomaterials for self sensitized degradation of malachite green dye under solar light. Appl. Catal. A—Gen. 490: 42-49.

Saliba, S., Coppel, Y., Mignotaud, C., Marty, J.-D. and Kahn, M.L. 2012. ZnO/liquid crystalline nanohybrids: From properties in solution to anisotropic growth. Chem. A Eurp. J. 18: 8084-8091. https://doi.org/10.1002/chem.201200233.

Sanchez, F. and Sobolev, K. 2010. Nanotechnology in concrete: A review. Constr. Build. Mater. 24: 2060-2071. https://doi.org/10.1016/j.conbuildmat.2010.03.014.

Satapanajaru, T., Anurakpongsatorn, P., Pengthamkeerati, P. and Boparai, H. 2008. Remediation of atrazine-contaminated soil and water by nano zerovalent iron. Water Air Soil Pollut. 192: 349-359.

Shen, T.Z., Hong, S.H. and Song, J.K. 2014. Electro-optical switching of graphene oxide liquid crystals with an extremely large Kerr coefficient. Nat. Mater. 13: 394.

Shen, Y. and Dierking, I. 2019. Perspectives in liquid-crystal-aided nanotechnology and nanoscience. Appl. Sci. 9: 2512.

Shibaev, V., Bobrovsky, A. and Boiko, N. 2003. Photoactive liquid crystalline polymer systems with light-controllable structure and optical properties. Prog. Polym. Sci. 28: 729-836. https://doi.org/10.1016/S0079-6700(02)00086-2.

Shibaev, V.P. 2009. Liquid-crystalline polymers: Past, present and future. Polym. Sci. Ser. A. 1: 1131-1193. https://doi.org/10.1134/S0965545X09110029.

Singh, J., Khan, S.A., Shah, J., Kotnala, R.K. and Mohapatra, S. 2017. Nanostructured TiO_2 thin films prepared by RF magnetron sputtering for photocatalytic applications. Appl. Surf. Sci. 422: 953-961.

Shin, K., Kim. H. and Chang, S. 2015. Transition-metal-catalyzed C–N bond forming reactions using organic azides as the nitrogen source: A journey for the mild and versatile C–H amination. Acc. Chem. Res. 48: 1040-1052.

Singh, U.B., Dhar, R., Dabrowski, R. and Pandey, M.B. 2014. Enhanced electro-optical properties of a nematic liquid crystals in presence of $BaTiO_3$ nanoparticles. Liq. Cryst. 41: 953-959.

Sohail, M. and Adeloju, S.B. 2009. Fabrication of redox-mediator supported potentiometric nitrate biosensor with nitrate reductase. Electroanal. 21: 1411-1418. https://doi. org/10.1002/elan.200804542.

Sohila, S., Rajendran, R., Yaakob, Z., Teridi, M.A.M. and Sopian, K. 2015. Photoelectrochemical water splitting performance of flower like ZnO nanostructures synthesized by a novel chemical method. J. Mater. Sci.: Mater. Elect. 27: 2846-2851.

Song, W., Kinloch, I.A. and Windle, A.H. 2003. Nematic liquid crystallinity of multiwall carbon nanotubes. Science 302: 1363.

Sonin, A.A. 2018. Pierre-Gilles de Gennes and physics of liquid crystals. Liq. Cryst. Rev. 109-128.

Supreet, S.K., Raina, K.K. and Pratibha, R. 2013. Enhanced stability of the columnar matrix in a discotic liquid crystal by insertion of ZnO nanoparticles. Liq. Crys. 40: 228-236. https://doi.org/10.1080/02678292.2012.737479.

Su, Z., Wang, L., Grigorescu, S., Lee, K. and Schmuki, P. 2014. Hydrothermal growth of highly oriented single crystalline Ta_2O_5 nanorod arrays and their conversion to Ta_3N_5 for efficient solar driven water splitting. Chem. Commun. 50: 15561-15564.

Tan, H., Yang, S., Shen, G., Yu, R. and Wu, Z. 2010. Signal-enhanced liquid-crystal DNA biosensors based on enzymatic metal deposition. Angew. Chem. Int. Ed. 49: 8608-8611.

Tang, Z., Yin, W., Zhang, L., Wen, B., Zhang, D., Liu, L., et al. 2016. Spatial separation of photo-generated electron-hole pairs in BiOBr/BiOI bilayer to facilitate water splitting. Sci. Rep. 6: 32764.

Tao, Y. and Tam, Y.H. 2013. Dynamics of ZnO nanowires immersed in in-plane switching liquid crystal cells. Appl. Phys. Lett. 103: 203102. https://doi.org/10.1063/1.4829998.

Theron, J., Cloete, T.E. and de Kwaadsteniet, M. 2010. Current molecular and emerging nano biotechnology approaches for the detection of microbial pathogens. Crit. Rev. Microbiol. 36: 318-339.

Thomas, R.T., Nair, V. and Sandhyarani, N. 2013. TiO_2 nanoparticle assisted solid phase photocatalytic degradation of polythene film: A mechanistic investigation. Colloids Surf. A. 422: 1-9.

Thorkelsson, K., Bai, P. and Xu, T. 2015. Self-assembly and applications of anisotropic nanomaterials: A review. Nano Today 10: 48-66.

Thalluri, S.M., Hernandez, S., Bensaid, S. and Saracco, G. 2016. Green-synthesized W- and Mo-doped $BiVO_4$ oriented along the {0 4 0} facet with enhanced activity for the sun-driven water oxidation. Appl. Catal. B: Environ. 180: 630-636.

Tiwari, S., Sharma, V., Mujawar, M., Mishra, Y.K., Kaushik, A. and Ghosal, A. 2019. Biosensors for epilepsy management: State-of-art and future aspects. Sensors 19: 1525.

Tratnyek, P.G. and Johnson, R.L. 2006. Nanotechnologies for environmental cleanup. Nano Today 1: 44-48.

Tripathi, P.Kr., Misra, A.Kr., Manohar, S., Gupta, S.Kr. and Manohar, R. 2013. Improved dielectric and electro-optical parameters of ZnO nano-particle (8% Cu^{2+}) doped nematic liquid crystal. J. Mol. Struct. 1035: 371-377. https://doi.org/10.1016/j.molstruc.2012.10.052.

Tschmelak, J., Kumpf, M., Kappel, N., Proll, G. and Gauglitz, G. 2006. Total internal reflectance fluorescence (TIRF) biosensor for environmental monitoring of testosterone with commercially available immunochemistry: Antibody characterization, assay development and real sample measurements. Talanta. 69: 343-350. https://doi.org/ 10.1016/j.talanta.2005.09.048.

Tungel, R., Rinken, T., Rinken, A. and Tenno, T. 1999. Immobilization and kinetic study of tyrosinase for biosensor construction. Anal. Lett. 32: 235. https://doi.org/10.1080/ 00032719908542818.

Vallamkondu, J., Corgiat, E., Buchaiah, G., Kandimalla, R. and Reddy, P. 2018. Liquid crystals: A novel approach for cancer detection and treatment. Cancers 10: 462.

Vidal, J.C., Bonel, L. and Castillo, J.R.A. 2008. Modulated tyrosinase enzyme-based biosensor for application to the detection of dichlorvos and atrazine pesticides. Electroanal. 20: 865-873. https://doi.org/10.1002/elan.200704115.

Wang, H.L. and Dai, H.J. 2013. Strongly coupled inorganic-nano-carbon hybrid materials for energy storage. Chem. Soc. Rev. 42: 3088-3113.

Wang, L., Ran, Q., Tian, Y., Ye, S., Xu, J., Xian, Y., et al. 2010. Covalent grafting tyrosinase and its application in phenolic compounds detection. Microchim. Acta 171: 217-223. https://doi.org/10.1007/s00604-010-0433-y.

Wang, L., Boutilier, M.S.H., Kidambi, P.R., Jang, D., Hadjiconstantinou, N.G. and Karnik, R. 2017. Fundamental transport mechanisms, fabrication and potential applications of nanoporous atomically thin membranes. Nat. Nanotechnol. 12: 509.

Wang, S.R., Gao, X.L., Yang, J.D., Zhu, Z.Y., Zhang, H.X. and Wang, Y.S. 2014a. Synthesis and gas sensor application of $ZnFe_2O_4$-ZnO composite hollow microspheres. RSC Adv. 4: 57967-57974.

Wang, T., Lv, R., Zhang, P., Li, C. and Gong, J. 2015b. Au nanoparticle sensitized ZnO nanopencil arrays for photoelectrochemical water splitting. Nanoscale 7: 77-81.

Wang, Z., Liu, G., Ding, C., Zhang, F., Shi, J. and Li, C. 2015a. Synergetic effect of conjugated $Ni(OH)_2$/ IrO_2 cocatalyst on titanium-doped hematite photoanode for solar water splitting. J. Phys. Chem. C. 119: 19607: 19612.

Wang, Z.S., Kawauchi, H., Kashima, T. and Arakawa, H. 2004. Significant influence of TiO_2 photoelectrode morphology on the energy conversion efficiency of N719 dye-sensitized solar cell. Coordin. Chem. Rev. 248: 1381-1389.

Warwick, M.E.A., Carraro, G., Gasparotto, A., Maccato, C., Barreca, D., Sada, C., et al. 2015. Interplay of thickness and photoelectrochemical properties in nanostructured α-Fe_2O_3 thin films. Phys. Stat. Solid. A. 212: 1501-1507.

Wei, Y. and Jang, C.-H. 2017. Visualization of cholylglycine hydrolase activities through nickel nanoparticle-assisted liquid crystal cells. Sensor. Actuator. B Chem. 239: 1268-1274.

Werber, J.R., Osuji, C.O. and Elimelech, M. 2016. Materials for next-generation desalination and water purification membranes. Nat. Rev. Mater. 1: 16018.

Wu, L.L., Wang, Z., Zhao, S.N., Meng, X., Song, X.Z., Feng, J., et al. 2016. A metal–organic framework/DNA hybrid system as a novel fluorescent biosensor for mercury(II) ion detection. Chemistry–A Euro. J. 22: 477-480.

Wu, T., Deng, G. and Zhen, C. 2021. Metal oxide mesocrystals and mesoporous single crystals: Synthesis, properties and applications in solar energy conversion. J. Mater. Sci. Technol. 73: 9-22.

Wu, W.Q., Chen, D.H., Caruso, R.A. and Cheng, Y.B. 2017. Recent progress in hybrid perovskite solar cells based on n-type materials. J. Mater. Chem. A 5: 10092-10109.

Xin, C., Ang, L., Musharavati, F., Jaber, F., Hui, L., Zalnezhad, E., et al. 2020. Supercapacitor performance of nickel-cobalt sulfide nanotubes decorated using Ni co-layered double hydroxide nanosheets grown *in situ* on Ni foam. Nanomaterials 10: 584.

Xiong, W., Zhou, L. and Liu, S.T. 2016. Development of gold-doped carbon foams as a sensitive electrochemical sensor for simultaneous determination of Pb(II) and Cu(II). Chem. Eng. J. 284: 650-656.

Xiong, Z., Yun, Y. and Jin, H.-J. 2013. Applications of carbon nanotubes for lithium ion battery anodes. Materials 6: 1138.

Xu, Z. and Gao, C. 2011. Aqueous liquid crystals of graphene oxide. ACS Nano 5: 2908-2915.

Xue, D.-X., Wang, Q. and Bai, J. 2019. Amide-functionalized metal–organic frameworks: Syntheses, structures and improved gas storage and separation properties. Coordin. Chem. Rev. 378: 2-16.

Yadav, S.P., Pande, M., Manohar, R. and Singh, S. 2015. Applicability of TiO_2 nanoparticle towards suppression of screening effect in nematic liquid crystal. J. Mol. Liq. 208: 34-37. https://doi.org/10.1016/j.molliq.2015.04.031.

Yamazoe, N. and Shimanoe, K. 2008. Theory of power laws for semiconductor gas sensors. Sensor Actuat. B—Chem. 128: 566-573.

Yang, B., She, Y., Zhang, C., Kang, S., Zhou, J. and Hu, W. 2020. Nitrogen doped intercalation TiO_2/TiN/ $Ti_3C_2T_x$ nanocomposite electrodes with enhanced pseudocapacitance. Nanomaterials 10: 345.

Yang, P. and Tarascon, J.-M. 2012. Towards systems materials engineering. Nat. Mater. 11: 560.

Yang, S., Liu, Y., Tan, H., Wu, C., Wu, Z., Shen, G., et al. 2012. Gold nanoparticle based signal enhancement liquid crystal biosensors for DNA hybridization assays. Chem. Comm. 48: 2861.

Yu, H. 2015. Dancing with Light: Advances in Photofunctional Liquid-Crystalline Materials. Singapore: Pan Stanford Publishing Co. Pte. Ltd.

Yu, D. and Pei, Y. 2021. Removal of ibuprofen by sodium alginate–coated iron-carbongranules combined with the ultrasound and Fenton technologies: Influencing factors and degradation intermediates. Env. Sci. Poll. Res. 28: 21183-21192.

Yuan, C.J., Wang, C.L., Wu, T.Y., Hwang, K.C. and Chao, W.C. 2011. Fabrication of a carbon fiber paper as the electrode and its application toward developing a sensitive unmediated amperometric biosensor. Biosens. Bioelectron. 26: 2858-2863. https://doi. org/10.1016/j.bios.2010.11.023.

Yun, G., Balamurugan, M., Kim, H., Ahn, K. and Kang, S.H. 2016. Role of WO_3 layers electrodeposited on SnO_2 inverse opal skeletons in photoelectrochemical water splitting. J. Phys. Chem. C. 120: 5906-5915.

Yun, H.J., Paik, T., Diroll, B., Edley, M.E., Baxter, J.B. and Murray, C.B. 2016. Nanocrystal size-dependent efficiency of quantum dot sensitized solar cells in the strongly coupled CdSe nanocrystals/TiO_2 system. ACS. Appl. Mater. Interfaces. 8: 14692-14700.

Zhang, H., Cheng, S., Li, B., Cheng, X. and Cheng, Q. 2018. Fabrication of magnetic Co/$BiFeO_3$ composite and its advanced treatment of pharmaceutical waste water by activation of peroxysulphate. Sep. Purif. Technol. 202: 242-247.

Zhang, J., Liu, X.H., Neri, G. and Pinna, N. 2016. Nanostructured materials for room temperature gas sensors. Adv. Mater. 28: 795-831.

Zhang, Q.M., Li, H.F., Poh, M., Xia, F., Cheng, Z.Y., Xu, H.S., et al. 2002. An all-organic composite actuator material with a high dielectric constant. Nature 419: 284.

Zhang, S., Majewski, P.W., Keskar, G., Pfefferle, L.D. and Osuji, C.O. 2011. Lyotropic self-assembly of high-aspect-ratio semiconductor nanowires of single-crystal ZnO. Langmuir 27: 11616-11621.

Zhang, Z., Wu, G., Ji, H., Chen, D., Xia, D., Gao, K., et al. 2020. 2D/1D V_2O_5 nanoplates anchored carbon nanofibers as efficient separator interlayer for highly stable lithium–sulfur battery. Nanomaterials 10: 705.

Zhao, D., Peng, Y., Xu, L., Zhou, W., Wang, Q. and Guo, L. 2015. Liquid-crystal biosensor based on nickel-nanosphere-induced homeotropic alignment for the amplified detection of thrombin. ACS Appl. Mater. Interfaces 7: 23418-23422.

Zhou, M., Kidd, T.J., Gin, D.L. and Noble, R.D. 2005. Supported lyotropic liquid-crystal polymer membranes: Promising materials for molecular-size-selective aqueous nanofiltration. Adv. Mater. 17: 1850.

Zhou, M., Nemade, P.R., Lu, X., Zeng, X., Hatakeyama, E.S., Noble, R.D., et al. 2007. New type of membrane material for water desalination based on a cross-linked bicontinuous cubic lyotropic liquid crystal assembly. J. Am. Chem. Soc. 129: 9574.

Zhou, Y., Sharma, N., Deshmukh, P., Lakhman, R.K., Jain, M. and Kasi, R.M. 2012. Hierarchically structured free-standing hydrogels with liquid crystalline domains and magnetic nanoparticles as dual physical cross-linkers. J. Am. Chem. Soc. 134: 1630-1641. https://doi.org/10.1021/ja208349x.

Zhu, H., Shen, R., Tang, Y., Yan, X., Liu, J., Song, L., et al. 2020. Sn-doping and Li_2SnO_3 nano-coating layer co-modified $LiNi_{0.5}Co_{0.2}Mn_{0.3}O_2$ with improved cycle stability at 4.6 V cut-off voltage. Nanomaterials 10: 868.

Zourob, M., Simonian, A., Wild, J., Mohr, S., Fan, X., Abdulhalime, I., et al. 2007. Optical leaky waveguide biosensors for the detection of organophosphorus pesticides. Analyst. 13: 114-120. https://doi.org/10.1039/B612871H.

Zhang, X., Zhang, B., Zuo, Z., Wang, M. and Shen, Y. 2015a. N/Si co-doped oriented single crystalline rutile TiO_2 nanorods for photoelectrochemical water splitting. J. Mater. Chem. 3: 10020-10025.

Zhang, Z., Li, X., Gao, C., Tang, F., Wang, Y., and Chen, L. 2015b. Synthesis of cadmium sulfide quantum dot-decorated barium stannate nanowires for photoelectrochemical water splitting. RSC. 10.1039/c0xx00000x, 2015b.

14

CHAPTER

Environmental, Health, and Safety Issues of Liquid and Crystal Nanomaterials

Manviri Rani[1], Keshu[2] and Uma Shanker[2,*]

1. INTRODUCTION

The term 'nanotechnology' used worldwide is a domain of new technology executed with several types of applications in the area of production and is used in biological, chemical, and physical systems at the nanoscale. Particles that are considered nanoscales have at least one dimension smaller than 100 nm (Rothen et al. 2006, Jitendra et al. 2016). For the synthesis of nanoparticles, various techniques, such as hydrothermal, pyrolysis, ion implantation, and chemical precipitation synthesis are used (Pokropivny and Skorokhod 2007). Nanotechnology has contributed to advanced ideas that can boost up various fields of research and applications for the modification of living beings as well as environmental health. In different areas of science, such as chemistry, biology, material science, medicine pharmacy, and engineering, nanotechnology is an emerging multi-disciplinary field connecting (Kania et al. 2018). Moreover, nanoparticles have been swiftly attracted because of their unique morphological and physicochemical characteristics, such as ultra-small size, shape (sheets, rods, tubes, and wires), and size distribution. They also show thermal, optical, and magnetic mechanical significant properties. Presently, nanomaterials are utilized on large scale in the area of material sciences, substance, environmental remediation, security, data advances, gadgets, energy generation, storage, transport, space, indicative, and restorative applications in the drug. Nanomaterials have found applications in the computer parts, medicines, cosmetics, drug delivery, diagnosis methods, treatment methods of various diseases, etc. (Ueli et al. 2011).

Due to its vast application nature, the utilization of nanomaterials in different fields generates worry about these nanomaterials. These nanoscale particles may enter human organs, such as lungs and cerebrums. The major problem arises because of the inhalation of these carcinogenic nanoparticles. A report concluded that there are several adverse effects observed in animals due to engineered nanomaterials, such as fibrosis, carcinogenicity, and inflammation. The major disadvantage of nanomaterials is that the dispersion of nanoscale particles in the earth may damage the ecosystem of our environment (NOISH 2009, Yokel and Macphail 2011, Osman 2019). It has been seen that nanotechnology can have health, safety, and environmental benefits, such as decreasing vitality utilization, discharge of ozone-depleting substances, contamination, remediating ecological harm, pre-empting sicknesses, and providing a new range of safety intensifying substances that are

[1] Department of Chemistry, Malaviya National Institute of Technology, Jaipur, Rajasthan, India-302017.
[2] Department of Chemistry, Dr B R Ambedkar National Institute of Technology, Jalandhar, Punjab, India-144011.
[*] Corresponding author: shankeru@nitj.ac.in, umaorganic29@gmail.com.

self-repairing, stronger, and known to provide safety (Auffan et al. 2009). A new branch termed nanotoxicology, demonstrates the possible health risks of nanomaterials. The key parameters of nanotoxicology are shown in Figure 1.

The unusually tiny measurements of green nanomaterials imply that these are considerably more directly absorbed by individuals rather than large particles. A report concluded that there are several adverse effects observed in animals due to nanomaterials, such as fibrosis, carcinogenicity, and inflammation. Direct and indirect exposure of these toxic nanomaterials to the skin is a major threat to human and animal life. The smaller size nanomaterials easily entered into the human body as compared to the bigger particles. The size, morphology, and reactivity of nanomaterials more often than not affect the tissues of human beings. Nanomaterials are mostly non-degradable and easily enter human organs, and the natural procedures inside the human body are thus directly affected due to their high/low surface area to volume ratio. Due to the contact of nanomaterials with living tissue and liquids, they adsorb on the surface and generate health risks among living beings. Mostly, people working in the laboratories of nanomaterials are adversely affected.

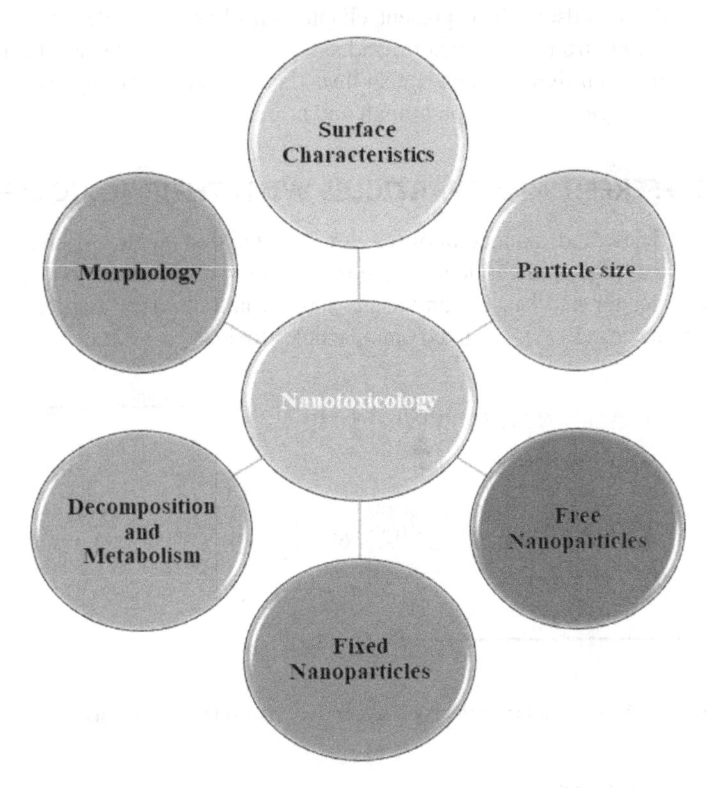

FIGURE 1 Illustration of characteristics of nanoparticles affecting nanotoxicology.

Green nanomaterials can transform nanotechnology and nanoscience into a sustainable platform. The association and better understanding among the researchers, industrialists, government, and society are required for the safer implementation of nanotechnology to manufacture green nanomaterials. This all, in turn, can surely benefit the environment and society. Green nanomaterials can be exploited for applications in the bioremediation of the environment or as transforming components, detectors, and biosensors (Wuana and Okieimen 2011). Nanomaterials of zinc oxide are reported to be toxic toward oyster larvae under seawater conditions (Noventa et al. 2018). There are different methods by which toxic nanomaterials may enter the human body and after some time, it circulates via blood (Sahu and Hayes 2017). If a person inhales gold nanoparticles, it enters the blood in approximately 15 minutes and can remain in the body for nearly three months and thus generate cardiovascular disease (Miller et al. 2017). Nanomaterials of silica directly affect the hepatic, dermal, respiratory, neural, cardiovascular genetic, immune, reproductive, and renal systems. The toxic effects of nanoparticles not only affect human life but cell viability growth, seed germination, and reduce microbial biodiversity of plants are also affected (Chen et al. 2018). The dose of nanomaterials and toxicity both are interconnected to different physical properties, which also affect the toxicity of the nanomaterials. Hence, the size and morphology of nanomaterials are the major parameters that decide the toxicity of nanomaterials. Thus, monitoring nanomaterials is a major factor in the risk assessment of various nanomaterials. To date, the lack of extensive exposure data makes the risk assessment of nanomaterials more difficult, although some hazard data on nanomaterials are available. Due to toxic behavior, it is a problematic issue related to nanomaterials while industries are utilizing nanomaterials exhaustively; they are unwilling to provide information on the properties as well as the quantity of the nanomaterials produced by them (Piccinno et al. 2012). Therefore, many countries that have familiarized the reporting phenomenon or mechanism are primarily voluntary. However, as the voluntary calls are not done precisely, various countries have now made it compulsory. The present chapter familiarizes readers with a comprehensive detailed study of the environmental, health, and safety issues of nanomaterials. The adverse effect of these nanomaterials on living things is also investigated with many suggestions and legal rules, which minimize the negative impact on human and animal healths.

2. DIFFERENT NANOPARTICLES WITH THEIR TOXIC EFFECTS

Various nanoparticles and nanocomposites are synthesized with different methods. These nanoparticles have a large application area due to their unique properties but apart from this, they also have toxic behavior and cause many acute and chronic diseases. Figure 2 clearly shows the Reactive oxygen species (ROS) – Related nanoparticles toxicity.

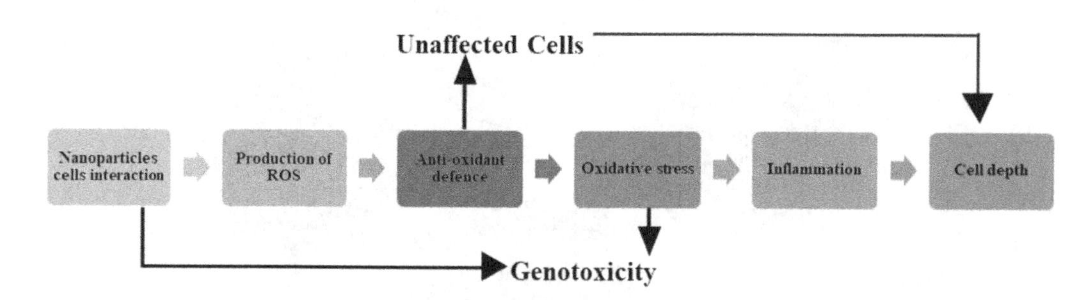

FIGURE 2 Illustration of reactive oxygen species – related nanoparticles toxicity.

2.1 Gold Nanoparticles

In many biological activities, gold nanoparticles can effectively control cellular processes. Gold nanoparticles are widely explored in photothermal therapy and drug delivery applications, as well as their light-reflecting ability, that enhance surface Plasmon resonance phenomenon. It has a great ability to bind up with thiols and amino groups for modification of drugs and other molecules (Goodman et al. 2004). But according to the publication, its toxic effect on immune dendritic cells is extracted from the bone marrow of mice. From the literature, it is revealed that gold nanoparticles with positive charge have a major impact on the toxicity of cells due to their property of easy movement from negatively charged membrane. Among cationic, anionic, and neutral counterparts of ammonium nanoparticles, cationic nanoparticles have maximum cytotoxicity. Acute and chronic administration of gold nanoparticles causes DNA damage in the cerebral cortex of the adult rat.

2.2 Silver Nanoparticles

Antiviral, antimicrobial, and antibacterial properties of silver nanoparticles make it significant for the treatment of trophic sores, acne, chronic ulcers, open wounds, and bums (Beyene et al. 2017). It has been highly used in shampoos, soaps, detergents, toothpaste, and air sanitizer sprays as well as for storage and packaging of foodstuff (Ahmed et al. 2016). Other than this, many researchers have confirmed the noncyto-toxicity of PVPs and ammonia stabilized silver nanoparticles at a minimum concentration. However, silver nanoparticles can cause *in vitro* toxicity, impairment of cell tissues, and DNA damage. Many acute and chronic diseases have been caused by silver nanoparticles that affect endothelial and microvascular cells (Gupta and Xie 2018).

2.3 ZnO Nanoparticles

Zinc oxide nanoparticles are effectively used in biosensors, cosmetic products, and food preservatives (Rani and Shanker 2018). The genotoxicity and cytotoxicity properties of engineered ZnO nanoparticles have been observed in animal and human cell models (*in vivo* and *in vitro*) (Esmaeillou et al. 2013). A high dose of ZnO nanoparticles causes cell viability, hepatotoxicity, an increase in oxidative stress, and the expression of the metallothionein gene. It can also reduce the amount of glutathione peroxidase (GPx) and superoxide dismutase (SOD) enzymes activities (Vandebriel and De Jong 2012). Recently it was observed that zinc oxide nanoparticles are toxic to oyster larvae under seawater conditions (Noventa et al. 2018).

2.4 TiO_2 Nanoparticles

TiO_2 NPs are effectively used as cosmetic products, UV absorbers, pigment, and artificial medical implantation of bone. But some researchers showed skin penetration and toxicity of TiO_2 nanoparticles in hairless mice and porcine skin after subchronic dermal exposure (Wu et al. 2009). TiO_2 nanoparticles also showed genotoxicity and cytotoxicity effect on plants, cell lines, and brains of mice. Acute chronic symptoms are also observed in goldfish (Carassius auratus) and C57BL/6 mice (Husain et al. 2015). The amount of titanium concentration reported across a wastewater treatment plant is shown in Table 1.

TABLE 1 Titanium concentrations across a wastewater treatment plant.

Sampling Location	Titanium Conc. (μg/L)	Biosolids Conc. (μg/g-solids)
Raw sewage	180 ± 51	–
After primary settling	113 ± 63	–
After activated sludge and secondary settling	50	–
After tertiary filtration	39	–
Biosolids from primary settling	–	257
Biosolids from secondary settling	–	8,139

2.5 Silica Nanoparticles

For the last several decades, synthetic silica nanoparticles (amorphous) have been used as a food preservative. Additionally, it is highly used in beverages for controlling foams, thickener in pastes as well as in carrier of flavors (Jeelani et al. 2020). Size and dose suggested the toxicity of SiO_2 nanoparticles on gastrointestinal cells (Kim et al. 2015). The morphology of SiO_2 nanoparticles varies the toxicity, excretion, and misdistribution after oral administration.

3. CYTOTOXICITY MEASUREMENT

The ratio in between absorbance of used cells and unused cells are called viability,

$$\% \text{ Cell Viability} = (\text{Test Absorbance}/\text{Control Absorbance}) \times 100$$

To determine the relative cell viability graph is plotted in between concentration and cell viability (Kamiloglu et al. 2020).

4. ENVIRONMENTAL IMPACT OF NANOMATERIALS

The NIOSH (US National Institute for Occupational Safety and Health) has initiated to conduct examinations on the reaction of nanoparticles with the human body and different organ systems and to know that how the employees are exposed to such very small tiny particles. Nanoparticles may enter the human body by direct injection through the skin, orally through the mouth, or by implantation of any substance. This, in turn, evades the gastrointestinal tract and lungs. There is an increased tendency that nanoparticles can get in direct- or indirect contact with the immune system of the human body and can cause different diseases, like immune system disorders, cancers, etc. Phagocytes are the cells of the immune system that inhale or destroy foreign materials, activating some stress reactions that lead to inflammation and also weaken the body's immune system against other pathogens. One must have complete knowledge about the potential toxicity of nanomaterials. The nature of nanomaterials is a major function of their shape, the reactivity of the surface with the environment as well their size. Nanoparticles possess higher bulk volume and on basis of the bulk material from which they are made; they can have various effects on human health. An increased rate of biological activity may be helpful, harmful, or a combination of both (Oberdörster et al. 2005). However, information regarding the adverse impact of nanoparticles' exposure through either direct or indirect methods is insufficient. Even so, analyzing the safety of nanoproducts and techniques employed for the synthesis, especially nanoparticles is of the highest priority. Currently, there is no standardization on the nanoparticles' toxicology. Furthermore, the contamination of the environment through landfills may also occur.

Different materials, including carbon, boron, and silica, can be used to make nanotubes. Nanotubes on exposure to biological environments may or may not release entrapped molecules. Depending on the nature of the material, material aggregates, capacity to disaggregate into small particles, and efforts required to become airborne; a nanotube may be hazardous. This may also become hazardous by the availability of various small-sized metal particles or catalysts present in the tubes as an outcome of the fabrication method of carbon nanotubes (Maynard et al. 2004).

These particles may enter the human body during the process of respiration. Other than respiration, their entry into the body may occur by means of:

(i) **Supra-Molecular Entanglement:** Muffling up the nanotubes with materials having non-covalent interactions.

(ii) **Covalent Functionalization:** Attaching the polar groups to the surface of the materials.

Carbon nanotubes (CNTs) having water solubilizing peptide groups are known to pass through the membrane of the cell and concentrates in the cell cytoplasm and may enter the cell nucleus without causing any harm to the cell (Davide et al. 2004). If compared, graphite seems to be compatible with cells and *in vivo,* and the plastic valves covered with graphite do not show any blood clotting. The mentioned feature has resulted in the evolution of artificial dental implants and heart valves consisting of carbon fiber strengthened carbon composites (Motohiro et al. 2011). Carbon nanotubes have been asserted to cause harm to life forms (Warheit et al. 2004). They are present in different lengths and an investigation of 220 nm and 825 nm long carbon nanotubes revealed no serious inflammatory responses, though the inflammation rate was more for nanotubes with more length. This provides the idea that macrophages can wrap the short nanotubes more quickly.

This is the reason that long nanotubes can cause more severe health issues than shorter nanotubes. An investigation on mice revealed, for similar quantities of carbon nanotubes and quartz inside the lungs, carbon nanotubes are highly toxic in comparison to quartz, which is reported to reason behind several diseases like silicosis after long term exposure (Lee et al. 2007). Nanotubes show significant pulmonary toxicity. Generally, operationalized nanotubes are non-toxic by themselves, but they have the ability to transport cytotoxic molecules into cells (Shi-Kam et al. 2004). A single-walled carbon nanotube-biotin conjugate (SWCNT-biotin conjugate) causes cell death at a very large scale and at a very fast rate (Hoet et al. 2004). There is very little information available regarding the transportation of compounds on nanoparticulate surfaces, but this all depends upon the nature of the surfaces. Surface-bound substances with airborne nanoparticles can easily penetrate the human body. Moreover, nanoparticles containing toxic and carcinogenic substances, which present persistently in the human lungs for many years, resulting in increased chances of developing cancer (Dalby et al. 2004).

The effortless entry of carbon nanotubes into cells is of great concern (Bianco et al. 2005), and it correlates to the fact that how DNA muffles up the nanotubes. In the literature, nanosilica was tested against murine fibroblasts and it was concluded that cells do not replicate and become rounded in shape. Their activity returns to normal after being returned onto a regular surface, spreading out, and proliferating (Cousins et al. 2004). This result is helpful in demonstrating the surface effects on living cells. The scale of nm, as well as micrometer surface topography of Ti materials, controls cellular bonding and the resulting living functions. Increment in the bonded cell to micron as well as sub-micron patterned surfaces as the substituent of smooth surfaces, which plays the role of positive sites for attachment (Zhu et al. 2004). Chitosan is a biocompatible polysaccharide, allergic reactions do not occur, and breaks harmful products into safer as well as smaller and simpler components that can be removed effortlessly. Drug delivery systems based on chitosan nanoparticulate material have the advantages of improved efficacy, improved patient compliance, and reduced toxicity. The negative impact of nanomaterials is shown in Figure 3.

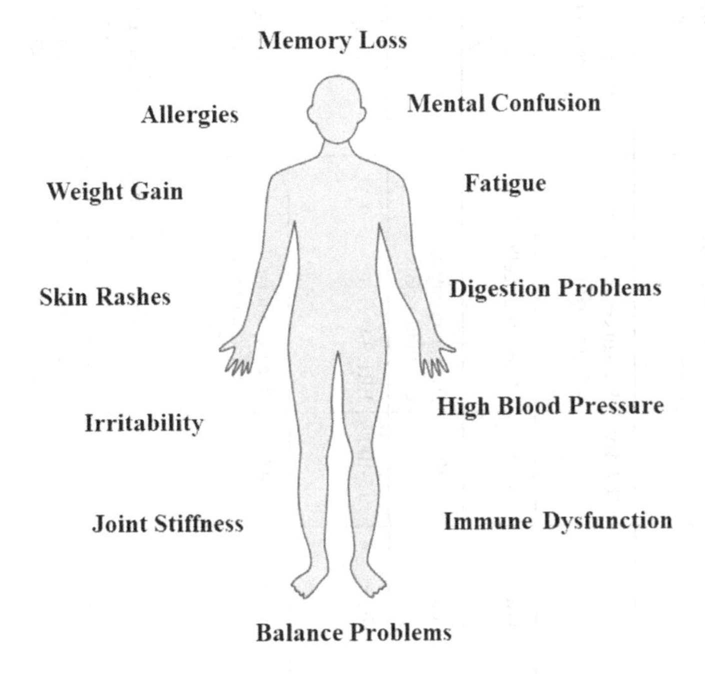

FIGURE 3 Illustration of the negative impact of nanomaterials on human beings.

TABLE 2 List of nanomaterials with the adverse effect of nanomaterials on living beings and analysis method.

S. No.	Nanoparticles	Size (nm)	Cell Culture	Amount and Exposure Medium	Assay Technique	Results	Reference
1	ZnO	50-70	Human colon carcinoma cells	11.5 μg/mL 24 h	ELISA Flow-cytometry	Oxidative stress decreases; Cell viability decreases; Inflammatory biomarkers	Berardis et al. 2010
2	ZnO	307-419	Human cervix carcinoma cell line (HEp-2)	10-100 μg/mL 24-48 h	Comet micronucleus test MTT	DNA damage; cell viability decreases	Osman et al. 2010
3	Al_2O_3	50-80	Mammalian cells	-	EZ4U	No major toxic effect on cell viability	Radziun et al. 2011
4	Fe_2O_3	30	Human hepatocellular carcinoma cells	123.52 μg/mL 12 h	MTT	Cell viability decreases	Ge et al. 2009
5	Silver	15-100	BRL 3A	10-50 μg/mL 24 h	MTT Glutathione DCFH-DA	Cell viability decreases; LDH decreases; ROS decreases	Hussain et al. 2005
6	Silica	43	Hepatocellular carcinoma cells (HepG2)	25-200 μg/mL 3-24 h	DCFH-DA 5,5,6,6-tetraethyl-benzimidazo-lylcarbo-cyanide iodine	ROS decreases; Mitochondrial damage Oxidative stress increases	Sun et al. 2011
7	Silver	30-50	Human alveolar cell line	0-20 μg/mL 24 h	MTT DCFH-DA	Cell viability decreases; ROS increases	Foldbjerg et al. 2011
8	Silica	15-46	Human bronchoalveolar carcinoma cells	10-100 μg/mL 48 h	DCFH-DA Commercial kit	ROS increases; LDH increases; Malondialdehyde increases	Lin et al. 2006

Contd.

9	Copper oxide	50	Human lung epithelial cells	10, 25, 50 μg/mL 24 h	MTT LDH	Cell viability decreases LDH decreases Lipid peroxidation decreases	Ahamed et al. 2010
10	Titanium oxide	160	*in vivo*	–	Comet micronucleus test	DNA damage, genotoxicity	Liu et al. 2009
11	MWCNTs	20	Lung cancer cells	0.002-0.2 μg/mL 4 days	MTT	Cell viability decreases	Magrez et al. 2006
12	SWCNTs	10-30	*in vivo*	40 and 200 μg/mouse, 1 mg/mouse, 90 days	Commercial kits	LDH decreases AST decreases ALT decreases	Yang et al. 2008
13	Fullerenes	178	CHO HELA HEK293	1 ng/mL 80 days	Micronucleus test	DNA strand breakage Chromosomal damage	Dhawan et al. 2006 Niwa et al. 2006
14	Iron oxide	13.8	Human hepatocellular carcinoma cells	123.52 μg/mL 12 h	MTT	Cell viability decreases	Ge et al. 2009
15	Single-wall CNTs	800	HACECs NHBECs	–	MTT	Cell death	Herzog et al. 2007

5. DESTRUCTIVE FACTS OF NANOTECHNOLOGY

Accidental interaction with nanoparticles may occur with the skin, lungs, or intestinal tract. Various nanoparticles can cause some unacceptable genetic changes as negative impact on the human body, e.g. mutagenicity. Nanorobot-related carcinogenicity is based upon the behavior of the surface (Freitas 2003).

Nanoparticles are the major constituents of air pollution that are produced as a result of combustion. If these particles are made up of insoluble and non-toxic materials, they cause higher inflammation rather than fine particles of equivalent material. This is because the ultrafine particles can obstruct the body's cellular-level defense systems. Many reveal that in urban areas, increased airborne particulate matter results in adverse effects on the health of the population, especially in susceptible groups (Zhang et al. 2003).

However, reports show that nanoparticles can enter the skin, but the entry of nanoparticulate material via the skin is not significant as compared to the entry via the lungs or intestinal tract because the skin is an obstacle and does not permit the interchange of materials (Hoet et al. 2004). The toxicity of various nanoparticles reported in the literature is shown in Table 3.

6. MECHANISMS OF TOXICITY

The biological environment of creatures gets affected by hydrophilicity and hydrophobicity properties of advanced nanoparticles by interacting with phagocytosis, plasma protein, cellular uptake, and stimulation of the immune system (Ajdary et al. 2018). This interaction may induce oxidative stress on the surface of nanoparticles and produce some ions that cause toxicity. Cellular toxicity is directly related to the diameter of engineered nanoparticles. The cell membrane is complex and dynamic comprising proteins and extracellular polymeric materials. The penetration of liquid crystals and nanoparticles occurs in the dynamic and complex structured cell membrane through endocytosis, diffusion, and membrane proteins, such as the phospholipid layer (Ajdary et al. 2018). NPs enter the cell through endocytosis and sheltered in Endosomes and nucleus then degraded in lysosomes or further recycled back to the plasma membrane.

7. CONTROLLING THE HAZARD

One of the most important methods of preventing the manpower associated with nanomaterials is to control the exposure of hazards. Exposure to hazard can be achieved by a hierarchical framework so as to lower the risk of suffering (Rout and Sikdar 2017). The different types of hazards caused by nanomaterials as shown in Figure 4.

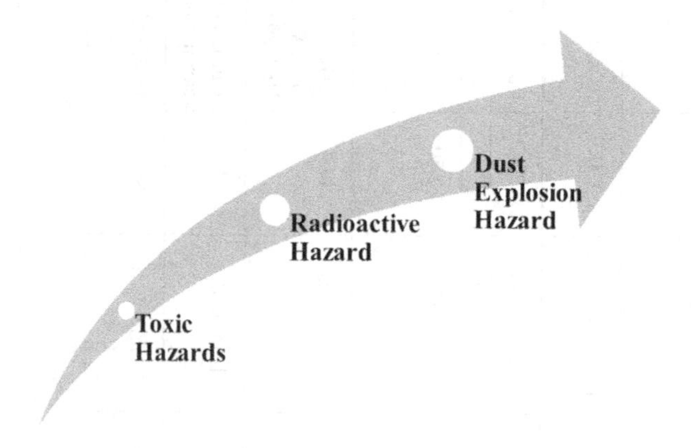

FIGURE 4 Main hazards caused by nanomaterials.

TABLE 3 Toxicity reported by various types of nanomaterials.

S. No.	Type of Nanocomposite	Quantity	Exposure Quantity	Toxicity Observed	References
1	CdSe/ZnS–SSA	0.1-0.4 mg/mL	1×10^6 cells/well; for 0-24 hr	Cytotoxic: 0.1 mg/mL altered cell growth; most cells nonviable at 0.4 mg/mL	Hoshino et al. 2004a
2	CdSe/ZnS	10 pmol QDs/1×10^5 cells (~ 10 nM)	1×10^6 cells; for 10 days (cell culture)	10 nM QD had minimal impact on cell survival	Chen and Gerion 2004
3	CdSe–MAA, TOPO QDs	62.5–1,000 μg/mL	1-8 hr	Cytotoxic: 62.5 μg/mL cytotoxic under oxidative/ photolytic conditions No toxicity on the addition of ZnS cap	Derfus 2004
4	Avidin-conjugated CdSe/ZnS QDs	0.5–1.0 μM	15 min	No effect on cell growth, development	Jaiswal et al. 2003
5	CdSe/ZnS–MUA QDs; QD–SSA complexes	0.24 mg/mL	0.4 mg/mL; 2 hr	0.4 mg/mL MUA/SSA–QD complexes did not affect viability of Vero cells	Hanaki et al. 2003
6	QD micelles: CdSe/ZnS QDs in (PEG–PE) and phosphatydilcholine	1.5–3 nL of 2.3 μM QDs injected, ~ 2.1×10^9 to 4.2×10^9 injected QDs/cell	5×10^9 QDs/cell (~ 0.23 pmol/ cell)	5×10^9 QDs/cell: cell abnormalities, altered viability and motility No toxicity at 2×10^9 QDs/cell	Dubertret et al. 2002

These hierarchical methods control the exposure to hazards as shown in Figure 5 and are discussed below in increasing order of their effectiveness.

i) Personnel Protection Equipments: Personnel protection equipment is a must requirement for the workers. The workers must wear long trousers, full-sleeve shirts, close-toed footwear along with compulsory use of safety gloves and goggles. Personnel protection equipment also includes respirators, which have been known for the effective capture of nanoparticles.
ii) Administrative Controls: This stage of control includes training workers on safer handling, transportation, and storage of nanomaterials. It also includes awareness on proper disposal off of the hazards. Proper labels and warning signs are also be encouraged so as to promote awareness among workers and society. Workers should be told about the importance of good manufacturing and work practices.
iii) Engineered Controls: This includes the isolation of workers from the hazards by isolating them in an enclosed workplace or by the removal of polluted or contaminated air out of working via proper ventilation and filtration techniques.
iv) Substitution and Elimination: This is the most effective approach to control any hazard. However, it is not possible to eliminate or directly substitute any nanomaterial with traditional materials. Rather green nanomaterials with improved properties and desired functionalities must be chosen (Gan 2019).

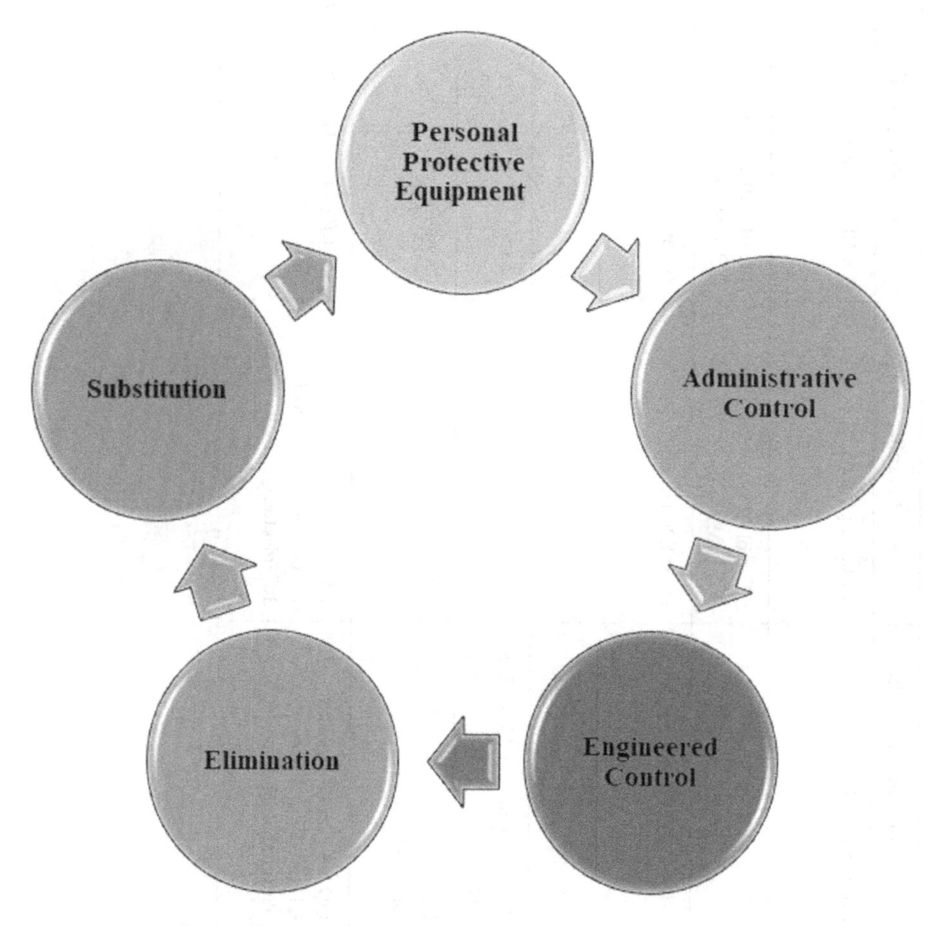

FIGURE 5 Illustration of Methods employed in controlling the hazards.

8. SAFETY ASPECTS

Workers have always been affected by being exposed to chemicals. The history of career and health safety has been interrupted by research analyzing the aspects of chemical vulnerability on workers and employees. Society moves ahead toward valuation, economic essentials, industrial productivity, etc. In this case, we can generate new ideas that existing joint career safety and health risk administrative programs should now be continued to include the green chemistry principles. Therefore, companies have a special opportunity of being proactive by demonstrating the safety of the appearing bio-based nanomaterials for their commercialization, ensuring that their market potentiality is not retarded by unreliability about safety issues (Freitas 2003, Raveendran et al. 2003, Shatkin et al. 2016, Kaur et al. 2018). A study has been conducted by National Institute for Occupational Health and Safety related to the toxic effects and health problems related to nanoparticle exposure at the workplace addresses several health problems as discussed below:

(i) Untreated carbon nanotube's ability to liberate airborne particulate matter while handling.
(ii) Respiratory toxicity of particles from carbon nanotubes.
(iii) Inhaled particles' surface activity.
(iv) Exposure of nanoparticles in the workplace and an evaluation of the risks involved.

9. ASSESSING THE RISK

Risk assessment describes a process of identifying hazards, evaluating and analyzing the risk associated with the hazard (Bos et al. 2015). Risk assessment is an important component of a health and safety plan. Risk assessment is important since it creates awareness about the hazards and risks from those hazards, identifies the manpower at risk, tells about the necessity of a control program for any hazard, monitors the safety measures, prevents injury and illness, and meets legal and ethical necessities (BAuA 2013). The main goal of risk assessment is to evaluate the hazard and then removing that particular hazard or minimizing the risk level by applying proper control and preventive measures. As a result of all this, a safe and healthy workplace can be achieved (Aitken et al. 2011, Giese et al. 2018). Risk assessment should be done before the introduction of new techniques or procedures; it can also be before the introduction of any changes to existing procedures or whenever any hazard has been identified (Gottschalk and Nowack 2013).

10. REGULATION

In the United States, the nanomaterials used in food items and drugs and medicines, and cosmetic items are regulated by 'Food and Drug Administration' under supervision of the 'Federal Food, Drug and Cosmetic Act'. The 'Consumer Product Safety Commission' necessitates that all the products must be tested and certified for the safety of consumers and put cautious labeling of harmful materials under 'Federal hazardous Substances Act'. According to the 'General Duty Clause' of the 'Occupational Safety and Health Act', the workers, as well as other persons, must keep the working area free from dangerous and harmful hazards (Frater et al. 2006, Seaton et al. 2010). In European Union, hazardous materials are regulated by European Commission. Biocidal items and cosmetics, which contain nanomaterials are regulated, under the 'Biocidal Products Regulation'. In the United Kingdom, nanomaterials that can cause a dust explosion come under the 'Chemicals (Hazard Information and Packaging for Supply) Regulations, 2002' and 'Dangerous Substances and Explosive Atmosphere Regulations, 2002'.

11. SAFETY ASPECTS OF STORAGE AND DISPOSAL OF NANOMATERIALS

Nanomaterials are beneficial not only for the scientific community but for the general population as well. For the comprehensive study, the potential impact of nanomaterials on the environment is very important to look at.

The lifecycle point of view can be used to analyze the potential effect of any nanomaterial on the environment. The lifecycle approach includes studying impacts through an items' complete life i.e., production, circulation, usage, recollection, and disposal. As per the US Department of Energy (DOE 2007) and UK Environment Agency Guidelines, the transfer of hazardous material requires an arrangement that must create a storing and disposal of nanomaterial and nanomaterial-contaminated waste. This must include hazardous associated with the nanomaterial and nanomaterial bearing waste streams, which include:

- Pure Nanomaterial
- Items contaminated with nanomaterial e.g. wipes, Dispensable PPE.
- Nanomaterial containing liquid suspensions.

Hence, the storage of nanomaterials is a very important aspect.

12. CONCLUSIONS AND FUTURE RECOMMENDATIONS

Various type of nanomaterials with different size, shape, morphologies, and chemical arrangements that shows an important role for deciding the characteristics. The synthesis method, as well as the modification procedure, directly gives a significant impact on the peculiar properties of nanomaterials. There is a need to highlight the positive as well as negative effects of manufactured nanoparticles, nanotubes, and nanofibers. However, this becomes difficult to determine the potentially toxic effects of nanoparticles by the evaluation of matter in bulk, rather a relationship can be established between the toxicity threshold and exposures at the workplace and environment. It is necessary to address the potential health and environmental aspects of nanotechnology in order to achieve economic and social assistance out of it. Modification in the utilization of the nanoparticles is major demand due to the harmful effect in particular situations, mainly when the nanoparticles are converted into other products. The connectivity between anthropogenic and nanomaterials have close connectivity toward toxicity of nanomaterials, such as particle size, structure, morphology, the surface to volume ratio, and agglomeration state. For the safety purpose of nanomaterials, the scientific community tries to introduce significant models efficient of expecting the release, accumulation, transport, uptake as well as the transformation of nanomaterials in the surrounding. These models connect chemical as well as physical properties of nanoparticles that permit an integrated technique to investigate impacts within susceptible populations. Liquid and crystal nanomaterials are essential dialogues for the scientific community but managing and communicating research on nanotechnology risks outside the scientific world is very difficult. Hence, developing and communicating research activities that allow technical information in a summarized manner for society is important. Finally, nanotechnology is risky if small, as well as large-scale industries, do not follow the safety rules. However, developing economies are not to be ignored and they should know important information on producing safe and eco-friendly nanomaterials. If the scientific community can take benefits from these situations, then in the future we surely look more toward liquid and crystal nanomaterials. Hence, nanoparticles with toxic-free nature are still a challenging area for our scientific community. Other parameters, such as fabrication conditions and catalytic dosage, are also playing a significant role that directly hits degrees of toxicity. As a result, from the above discussions, we can conclude that in-depth studies linked to the physicochemical characteristics with the toxicity of nanomaterials are a major challenge among the scientific community. However, the interaction between nanomaterials and biological systems may vary with particle size and shape.

Acknowledgments

One of the authors Dr. Manviri Rani is grateful for the funding from DST-SERB, New Delhi (Sanction order no. SRG/2019/000114) and TEQIP-III MNIT Jaipur, India. Dr. Uma Shanker thanks TEQIP-III, NIT Jalandhar for partial funding.

Conflict of Interest

The authors declare that there is no conflict of interest regarding the publication of this chapter.

References

Ahmed, S., Ahmad, M., Swami, B.L. and Ikram, S. 2016. A review on plants extract mediated synthesis of silver nanoparticles for antimicrobial applications: A green expertise. J. Adv. Res. 7: 17-28.

Aitken, A., Bassan, A., Friedrichs, S., Hankin, S.M., Hansen, S.F., Holmqvist, J., et al. 2011. Specific advice on exposure assessment and hazard/risk characterisation for nanomaterials under REACH (RIP-oN3). Final Project ReportRNC/RIP-oN3/FPR/1/FINAL.

Ajdary, M., Moosavi, M., Rahmati, M., Falahati, M., Mahboubi, M., Mandegary, A., et al. 2018. Health concerns of various nanoparticles: A review of their *in vitro* and *in vivo* toxicity. Nanomaterials 8(9): 634-668.

Arul Prakash, F., Dushendra Babu, G., Lavanya, M., Shenbaga Vidhya, K. and Devasena, T. 2011. Toxicity studies of aluminium oxide nanoparticles in cell lines. Int. J. Nano. App. 5: 99-107.

Auffan, M., Rose, J., Bottero, J.Y., Lowry, G.V., Jolivet, J.P. and Wiesner, R. 2009. Towards a definition of inorganic nanoparticles from an environmental, health and safety perspective. Nature Nanotechnol. 4: 634-641.

BAuA. 2013. Announcements on Hazardous Substances-Manufactured Nanomaterials Announcement 527. Jt. Ministerial Gaz. 25: 498-511.

Berardis De, B., Civitelli, G., Condello, M., Lista, P., Pozzi, R., Arancia, G., et al. 2010. Exposure to ZnO nanoparticles induces oxidative stress and cytotoxicity in human colon carcinoma cells. Toxicol. Appl. Pharmcol. 246: 116-127.

Beyene, H.D., Werkneh, A.A., Bezabh, H.K. and Ambaye, T.G. 2017. Synthesis paradigm and applications of silver nanoparticles (AgNPs): A review. Sustain. Mater. Technol. 13: 18-23.

Bianco, A., Kostarelos, K., Partidos, C.D. and Prato, M. 2005. Biomedical applications of functionalised carbon nanotubes. Chemi. Communi. 5: 571-577.

Bos, P.M., Gottardo, S., Scott-Fordsmand, J.J., van Tongeren, M., Semenzin, E., Fernandes, T.F., et al. 2015. The MARINA risk assessment strategy: A flexible strategy for efficient information collection and risk assessment of nanomaterials. Int. J. Environ. Res. Public Health. 12: 15007-15021.

Chen, F. and Gerion, D. 2004. Fluorescent CdSe/ZnS nanocrystal–peptide conjugates for long-term, nontoxic imaging and nuclear targeting in living cells. Nano Letters 4: 1827-1832.

Chen, M., Zhou, S., Zhu, Y., Sun, Y., Zeng, G., Yang, C., et al. 2018. Toxicity of carbon nanomaterials to plants, animals and microbes: Recent progress from 2015-present. Chemosphere 206: 255-264.

Cousins, B.G., Doherty, P.J., Williams, R,L., Fink, J. and Garvey, M.J. 2004. The effect of silica nanoparticulate coatings on cellular response. Jour. Mat. Sci. Mat. Medi. 15: 355-359.

Dalby, M.J., Gadegaard, N., Riehle, M.O., Wilkinson, C.D. and Curtis, A.S. 2004. Investigating filopodia sensing using arrays of defined nano-pits down to 35 nm diameter in size. Int. J. Biochem. & Cell Biol. 36: 2005-2015.

Davide, P., Jean-Paul, B., Maurizio, P. and Alberto, B. 2004. Translocation of bioactive peptides across cell membranes by carbon nanotubes. Chem. Comm. 1: 16-17.

Esmaeillou, M., Moharamnejad, M., Hsankhani, R., Tehrani, A.A. and Maadi, H. 2013. Toxicity of ZnO nanoparticles in healthy adult mice. Environmental Toxicology and Pharmacology 35: 67-71.

Frater, L., Stokes, E., Lee, R. and Oriola, T. 2006. An Overview of the Framework of Current Regulation Affecting the Development and Marketing of Nanomaterials. Cardiff, UK: ESRC Centre for Business Relationships Accountability Sustainability and Society (BRASS), Cardiff University.

Robert, A. Freitas Jr. 2003. Nanomedicine, Vol. IIA: Biocom. ISBN: 978-3-8055-7722-9.

Gan, S.L. 2019. Importance of hazard identification in risk management. Ind. Health 57: 281-282.

Giese, B., Klaessig, F., Park, B., Kaegi, R., Steinfeldt, M., Wigger, H., et al. 2018. Risks, release and concentrations of engineered nanomaterial in the environment. Sci. Rep. 8: 1565.

Goodman, C.M., McCusker, C.D., Yilmaz, T. and Rotello, V.M. 2004. Toxicity of gold nanoparticles functionalized with cationic and anionic side chains. Bioconjugate Chemistry 15: 897-900.

Gottschalk, F. and Nowack, B. 2013. A probabilistic method for species sensitivity distributions taking into account the inherent uncertainty and variability of effects to estimate environmental risk. Integr. Environ. Assess. Manag. 9: 79-86.

Gupta, R. and Xie, H. 2018. Nanoparticles in daily life: Applications, toxicity and regulations. J. Environ. Pathol. Toxicol. Oncol. 37(3): 209-230.

Hoet, P., Nemmar, A. and Nemery, B. 2004. Health impact of nanomaterials. Nat. Biotechnol. 22: 19. https://doi.org/10.1038/nbt0104-19.

Husain, M., Wu, D., Saber, A.T., Decan, N., Jacobsen, N.R., Williams, A., et al. 2015. Halappanavar, Intratracheally instilled titanium dioxide nanoparticles translocate to heart and liver and activate complement cascade in the heart of C57BL/6 mice. Nanotoxicology 9: 1013-1022.

Jeelani, P.G., Mulay, P., Venkat, R. and Ramalingam, C. 2020. Multifaceted application of silica nanoparticles: A review. Silicon 12: 1337-1354.

Jitendra, S.T., Roger, A.H., Hanks, C., Trybula, W. and Fazarro, D. 2016. Addressing ethical and safety issues of nanotechnology in health and medicine in undergraduate engineering and technology curriculum. Glob. J. Eng. Educ. 18: 30-34.

Kania, G., Sternak, M., Jasztal, A., Chlopicki, S., Agnieszka, B. and Nasulewicz-goldeman, A. 2018. Uptake and bioreactivity of charged chitosan-coated superparamagnetic nanoparticles as promising contrast agents for magnetic resonance imaging. Nanomed. Nanotechnol. Biol. Med. 14: 131-140.

Kaur, M., Mehta, A. and Gupta, R. 2018. Biomedical applications of synthetic and natural biodegradable polymers. pp. 281-310. *In*: Shakeel Ahmed, et al. [eds]. Green and Sustainable Advanced Materials, vol 2. Scrivener Pub. LLC.

Kamiloglu, S., Sari, G., Ozdal, T. and Capanoglu, E. 2020. Guideline for cell viability assays. Food. Front. 1: 332-349.

Kim, I.Y., Joachim, E., Choi, H. and Kim, K. 2015. Toxicity of silica nanoparticles depends on size, dose, and cell type. Nanomedicine: Nanotechnology, Biology and Medicine 11: 1407-1416.

Lee, M.H., McClellan, W., Candela, J., Andrews, D. and Biswas, P. 2007. Reduction of nanoparticle exposure to welding aerosols by modification of the ventilation system in a workplace. J. Nano. Res. 9: 127-136. 10.1007/s11051-006-9181-7.

Lima, R. de, Seabra, A.B. and Durán, N. 2012. Silver nanoparticles: A brief review of cytotoxicity and genotoxicity of chemically and biogenically synthesized nanoparticles. J. Appl. Toxicol. 32: 867-879.

Liu, R., Yin, L., Pu, Y., Liang, G., Zhang, J., Su, Y., et al. 2009. Pulmonary toxicity induced by three forms of titanium dioxide nanoparticles via intra-tracheal instillation in rats. Prog. Nat. Sci. 19(5): 573-579.

Maynard, A.D., Baron, P.A., Foley, M., Shvedova, A.A., Kisin, E.R. and Castranova, V. 2004. Exposure to carbon nanotube material: Aerosol release during the handling of unrefined single-walled carbon nanotube material. J. Toxic. Environ. Heal. Part A 67: 87-107.

Miller, M.R., Raftis, J.B., Langrish, J.P., Mclean, S.G., Samutrtai, P., Connell, S.P., et al. 2017. Inhaled nanoparticles accumulate at sites of vascular disease. ACS Nano 11(5): 4542-4552.

Motohiro, U.O., Tsukasa, A., Fumio, W., Yoshinori, S. and Kazuyuki, T. 2011. Toxicity evaluations of various carbon nanomaterials. Dent. Mater. J. 30: 245-263.

Noventa, S., Rowe, D. and Galloway, T. Mitigating effect of organic matter on the *in vivo* toxicity of metal oxide nanoparticles in the marine environment. Environ. Sci. Nano 5: 1764-1777.

Oberdörster, G., Oberdörster, E. and Oberdörster, J. 2005. Nanotoxicology: An emerging discipline evolving from studies of ultrafine particles. Environ. Health Perspect. 113: 823-839.

Osman, E.M. 2019. Environmental and health safety considerations of nanotechnology: Nano safety. Biomedical J. Scientific & Technical Research 19: 14501-14515.

Osman, I.F., Baumgartner, A., Cemeli, E., Fletcher, J.N. and Anderson, D. 2010. Genotoxicity and cytotoxicity of zinc oxide and titanium dioxide in HEp-2 cells. Nanomedicine 5: 1193-1203.

Piccinno, F., Gottschalk, F., Seeger, S. and Nowack, B. 2012. Industrial production quantities and uses of ten engineered nanomaterials in Europe and the world. J. Nanopart. Res. 14(9): 1109.

Pokropivny, V.V. and Skorokhod, V.V. 2007. Classification of nanostructures by dimensionality and concept of surface forms engineering in nanomaterial science. Mater. Sci. Eng. C. 27: 990-993.

Radziun, E., Dudkiewicz Wilczyńska, J., Książek, I., Nowak, K., Anuszewska, E.L., Kunicki, A., et al. 2011. Assessment of the cytotoxicity of aluminium oxide nanoparticles on selected mammalian cells. Toxicology *in vitro* 25(8): 1694-1700.

Rani, M. and Shanker, U. 2018. Insight in to the degradation of bisphenol A by doped ZnO@ZnHCF nanocubes: High photocatalytic performance. J. Colloid and Interface Sci. 530: 16-28.

Raveendran, P., Fu, J. and Wallen, S.L. 2003. Completely "green" synthesis and stabilization of metal nanoparticles. J. Am. Chem. Soc. 125: 13940-13941.

Rothen Rutishauser, B.M., Schurch, S. and Haenni, B. 2006. Interaction of fine particles and nanoparticles with red blood cells visualized with advanced microscopic techniques. Environ. Sci. Technol. 40: 4353-4359.

Rout, B.K. and Sikdar, B.K. 2017. Hazard identification, risk assessment, and control measures as an effective tool of occupational health assessment of hazardous process in an iron ore pelletizing industry. Indian J. Occup. Environ. Med. 21: 56-76.

Sahu, S.C. and Hayes, A.W. 2017. Toxicity of nanomaterials found in human environment. Toxicol. Res. App. 1: 1-13.

Seaton, A., Tran, L., Aitken, R. and Donaldson, K. 2010. Nanoparticles, human health hazard and regulation. J. R. Soc. Interface 7: S119-S129.

Shatkin, J.A., Ong, K.J., Ede, J.D., Wegner, T.H. and Goergen, M. 2016. Toward cellulose nanomaterial commercialization: Knowledge gap analysis for safety data sheets according to the globally harmonized system. Tappi. Journal 15: 425-437.

Shi-Kam, N.W., Jessop, T.C., Wender, P.A. and Dai, H. 2004. Nanotube molecular transporters: Internalization of carbon nanotube-protein conjugates into Mammalian cells. J. Am. Chem. Soc. 126: 6850-6851.

Ueli, A., Elke, A., Anders, B., Ken, D., Bengt, F., Robin, F., et al. 2011. Impact of Engineered Nanomaterials on Health: Considerations for Benefit-Risk Assessment, Luxembourg: European Union, EASAC Policy Report, No. 15.

Vandebriel, R.J. and De Jong, W.H. 2012. A review of mammalian toxicity of ZnO nanoparticles. Nanotechnol. Sci. Appl. 5: 61.

Warheit, D.B., Laurence, B.R., Reed, K.L., Roach, D.H., Reynolds, G.A. and Webb, T.R. 2004. Comparative pulmonary toxicity assessment of single-wall carbon nanotubes in rats. Toxico. Sci. 77: 117-125.

Wu, J., Liu, W., Xue, C., Zhou, S., Lan, F., Bi, L., et al. 2009. Toxicity and penetration of TiO_2 nanoparticles in hairless mice and porcine skin after subchronic dermal exposure. Toxicol. Lett. 191: 1-8.

Wuana, R.A. and Okieimen, F.E. 2011. Heavy metals in contaminated soils: A review of sources, chemistry, risks and best available strategies for remediation. ISRN Ecology 2011: 1-20.

Yokel, R.A. and MacPhail, R.C. 2011. Engineered nanomaterials: Exposures, hazards, and risk prevention. Journal of Occupational Medicine and Toxicology 6(1): 7.

Zhang, Q., Kusaka, Y., Zhu, X., Sato, K., Mo, Y., Kluz, T., et al. 2003. Comparative toxicity of standard nickel and ultrafine nickel in lung after intratracheal instillation. J. Occup. Health. 45: 23-30.

Zhu, X., Chen, J., Scheideler, L., Altebaeumer, T., Geis-Gerstorfer, J. and Kern, D. 2004. Cellular reactions of osteoblasts to micron- and submicron-scale porous structures of titanium surfaces. Cells Tissues Organs 178: 13-22.

Index

9780367549909